Texts in Theoretical Computer Science
An EATCS Series

Springer
Berlin
Heidelberg
New York
Barcelona
Hong Kong
London
Milan
Paris
Singapore
Tokyo

Stasys Jukna

Extremal Combinatorics

With Applications in Computer Science

With 36 Figures and 315 Exercises

 Springer

Author

Dr. Stasys Jukna
Informatik – FB 20
Theoretische Informatik
Johann Wolfgang Goethe-Universität
Robert-Mayer-Str. 11–15
60235 Frankfurt, Germany
jukna@thi.informatik.uni-frankfurt.de

and

Institute of Mathematics and Informatics
Akademijos 4
2600 Vilnius, Lethuania

Series Editors

Prof. Dr. Wilfried Brauer
Institut für Informatik
Technische Universität München
Arcisstr. 2, 80333 München
Germany
brauer@informatik.tu-muenchen.de

Prof. Dr. Grzegorz Rozenberg
Leiden Institute
of Advanced Computer Science
University of Leiden
Niels Bohrweg 1, 2333 CA Leiden
The Netherlands
rozenber@liacs.nl

Prof. Dr. Arto Salomaa
Turku Centre for Computer Science
Lemminkäisenkatu 14 A, 20 520 Turku
Finland
asalomaa@utu.fi

Library of Congress Cataloging-in-Publication Data

Jukna, Stasys, 1953–
    Extremal combinatorics: with applications in computer science/Stasys Jukna.
        p. cm. – (Texts in theoretical computer science)
    Includes bibliographical references and indexes.
    ISBN 3540663134 (alk. paper)
        1. Combinatorial analysis. 2. Extremal problems (Mathematics) I. Title. II. Series.
    QA164 .J84 2000
        511'.6–dc21   00-055604

ACM Computing Classification (1998): G.2–3, F.1–2, F.4.1
AMS Mathematics Subject Classification (2000): 05-01,
05Dxx, 06A07, 15A03, 68Rxx, 68Q17

ISBN 3-540-66313-4 Springer-Verlag Berlin Heidelberg New York

Springer-Verlag Berlin Heidelberg New York,
a member of BertelsmannSpringer Science+Business Media GmbH
http://www.springer.de

© Springer-Verlag Berlin Heidelberg 2001
Printed in Germany

Cover Design: design & production, Heidelberg
Typesetting: Camera ready by the author
Printed on acid-free paper   SPIN 10739318   45/3142SR – 5 4 3 2 1 0

To Indrė

Preface

Combinatorial mathematics has been pursued since time immemorial, and at a reasonable scientific level at least since Leonhard Euler (1707–1783). It rendered many services to both pure and applied mathematics. Then along came the prince of computer science with its many mathematical problems and needs – and it was combinatorics that best fitted the glass slipper held out. Moreover, it has been gradually more and more realized that combinatorics has all sorts of deep connections with "mainstream areas" of mathematics, such as algebra, geometry and probability. This is why combinatorics is now a part of the standard mathematics and computer science curriculum.

This book is as an introduction to *extremal combinatorics* – a field of combinatorial mathematics which has undergone a period of spectacular growth in recent decades. The word "extremal" comes from the nature of problems this field deals with: if a collection of finite objects (numbers, graphs, vectors, sets, etc.) satisfies certain restrictions, how large or how small can it be?

For example, how many people can we invite to a party where among each three people there are two who know each other and two who don't know each other? An easy Ramsey-type argument shows that at most five persons can attend such a party. Or, suppose we are given a finite set of nonzero integers, and are asked to mark an as large as possible subset of them under the restriction that the sum of any two marked integers cannot be marked. It appears that (independent of what the given integers actually are!) we can always mark at least one-third of them.

Besides classical tools, like the pigeonhole principle, the inclusion-exclusion principle, the double counting argument, induction, Ramsey argument, etc., some recent weapons – the probabilistic method and the linear algebra method – have shown their surprising power in solving such problems. With a mere knowledge of the concepts of linear independence and discrete probability, completely unexpected connections can be made between algebra, probability, and combinatorics. These techniques have also found striking applications in other areas of discrete mathematics and, in particular, in the theory of computing.

Nowadays we have comprehensive monographs covering different parts of extremal combinatorics. These books provide an invaluable source for students and researchers in combinatorics. Still, I feel that, despite its great potential and surprising applications, this fascinating field is not so well known for students and researchers in computer science. One reason could be that, being comprehensive and in-depth, these monographs are somewhat too difficult to start with for the beginner. I have therefore tried to write a "guide tour" to this field – an introductory text which should

- be self-contained,
- be more or less up-to-date,
- present a wide spectrum of basic ideas of extremal combinatorics,
- show how these ideas work in the theory of computing, and
- be accessible for graduate and motivated undergraduate students in mathematics and computer science.

Even if not all of these goals were achieved, I hope that the book will at least give a first impression about the power of extremal combinatorics, the type of problems this field deals with, and what its methods could be good for. This should help students in computer science to become more familiar with combinatorial reasoning and so be encouraged to open one of these monographs for more advanced study.

Intended for use as an introductory course, the text is, therefore, far from being all-inclusive. Emphasis has been given to theorems with elegant and beautiful proofs: those which may be called the gems of the theory and may be relatively easy to grasp by non-specialists. Some of the selected arguments are possible candidates for *The Book*, in which, according to Paul Erdős, God collects the perfect mathematical proofs.* I hope that the reader will enjoy them despite the imperfections of the presentation.

Extremal combinatorics itself is much broader. To keep the introductory character of the text and to minimize the overlap with existing books, some important and subtle ideas (like the shifting method in extremal set theory, applications of Janson's and Talagrand's inequalities in probabilistic existence proofs, use of tensor product methods, etc.) are not even mentioned here. In particular, only a few results from extremal graph theory are discussed and the presentation of the whole Ramsey theory is reduced to the proof of one of its core results — the Hales–Jewett theorem and some of its consequences. Fortunately, most of these advanced techniques have an excellent treatment in existing monographs by Bollobás (1978) on extremal graph theory, by Babai and Frankl (1992) on the linear algebra method, by Alon and Spencer (1992) on the probabilistic method, and by Graham, Rothschild, and Spencer (1990) on Ramsey theory. We can therefore pay more attention to the recent *applications* of combinatorial techniques in the theory of computing.

A possible feature and main departure from traditional books in combinatorics is the choice of topics and results, influenced by the author's twenty

---

* "You don't have to believe in God but, as a mathematician, you should believe in The Book." (Paul Erdős)

For the first approximation see M. Aigner and G.M. Ziegler, *Proofs from THE BOOK*. Second Edition, Springer, 2000.

years of research experience in the theory of computing. Another departure is the inclusion of combinatorial results that originally appeared in computer science literature. To some extent, this feature may also be interesting for students and researchers in combinatorics. The corresponding chapters and sections are: 2.3, 4.8, 6.2.2, 7.2.2, 7.3, 10.4–10.6, 12.3, 14.2.3, 14.5, 15.2.2, 16, 18.6, 19.2, 20.5–20.9, 22.2, 24, 25, 26.1.3, and 29.3. In particular, some recent applications of combinatorial methods in the theory of computing (a new proof of Haken's exponential lower bound for the resolution refutation proofs, a non-probabilistic proof of the switching lemma, a new lower bounds argument for monotone circuits, a rank argument for boolean formulae, lower and upper bounds for span programs, highest lower bounds on the multi-party communication complexity, a probabilistic construction of surprisingly small boolean formulas, etc.) are discussed in detail.

Teaching. The text is *self-contained.* It assumes a certain mathematical maturity but *no* special knowledge in combinatorics, linear algebra, probability theory, or in the theory of computing — a standard mathematical background at undergraduate level should be enough to enjoy the proofs. All necessary concepts are introduced and, with very few exceptions, all results are proved before they are used, even if they are indeed "well-known." Fortunately, the problems and results of combinatorics are usually quite easy to state and explain, even for the layman. Its accessibility is one of its many appealing aspects.

The book contains much more material than is necessary for getting acquainted with the field. I have split it into 29 relatively short chapters, each devoted to a particular proof technique. I have tried to make the chapters almost *independent*, so that the reader can choose his/her own order to follow the book. The (linear) order, in which the chapters appear, is just an extension of a (partial) order, "core facts first, applications and recent developments later." Combinatorics is broad rather than deep, it appears in different (often unrelated) corners of mathematics and computer science, and it is about techniques rather than results – this is where the independence of chapters comes from.

Each chapter starts with results demonstrating the particular technique in the simplest (or most illustrative) way. The relative importance of the topics discussed in separate chapters is not reflected in their length – only the topics which appear for the first time in the book are dealt with in greater detail. To facilitate the understanding of the material, over 300 exercises of varying difficulty, together with hints to their solution, are included. This is a vital part of the book – many of the examples were chosen to complement the main narrative of the text. I have made an attempt to grade them: problems marked by "−" are particularly easy, while the ones marked by "+" are more difficult than unmarked problems. The mark "(!)" indicates that the exercise may be particularly valuable, instructive, or entertaining. Needless to say, this grading is subjective. Some of the hints are quite detailed so that they actually sketch the entire solution; in these cases the reader should try to fill out all missing details.

Feedback to the author. I have tried to eliminate errors, but surely some remain. I hope to receive mail offering suggestions, praise, and criticism, comments on attributions of results, suggestions for exercises, or notes on typographical errors. I am going to maintain a website that will contain a (short, I hope) list of errata, solutions to exercises, feedback from the readers, and any other material to assist instructors and students. The link to this site as well as my email address can be obtained from the Springer website

    http://www.springer.de/comp/

Please send your comments either to my email address or to my permanent address: Institute of Mathematics, Akademijos 4, 2600 Vilnius, Lithuania.

Acknowledgments. I would like to thank everybody who was directly or indirectly involved in the process of writing this book. First of all, I am grateful to Alessandra Capretti, Anna Gál, Thomas Hofmeister, Daniel Kral, G. Murali Krishnan, Martin Mundhenk, Gurumurthi V. Ramanan, Martin Sauerhoff and P.R. Subramania for comments and corrections.

Although not always directly reflected in the text, numerous earlier discussions with Anna Gál, Pavel Pudlák, and Sasha Razborov on various combinatorial problems in computational complexity, as well as short communications with Noga Alon, Aart Blokhuis, Armin Haken, Johan Håstad, Zoltan Füredi, Hanno Lefmann, Ran Raz, Mike Sipser, Mario Szegedy, and Avi Wigderson, have broadened my understanding of things. I especially benefited from the comments of Aleksandar Pekec and Jaikumar Radhakrishnan after they tested parts of the draft version in their courses in the BRICS International Ph.D. school (University of Aarhus, Denmark) and Tata Institute (Bombay, India), and from valuable comments of László Babai on the part devoted to the linear algebra method.

I would like to thank the Alexander von Humboldt Foundation and the German Research Foundation (Deutsche Forschungsgemeinschaft) for supporting my research in Germany since 1992. Last but not least, I would like to acknowledge the hospitality of the University of Dortmund, the University of Trier and the University of Frankfurt; many thanks, in particular, to Ingo Wegener, Christoph Meinel and Georg Schnitger, respectively, for their help during my stay in Germany. This was the time when the idea of this book was born and realized. I am indebted to Hans Wössner and Ingeborg Mayer of Springer-Verlag for their editorial help, comments and suggestions which essentially contributed to the quality of the presentation in the book.

My deepest thanks to my wife, Daiva, and my daughter, Indrė, for being there.

Frankfurt/Vilnius, March 2001   Stasys Jukna

Contents

Part II. Extremal Set Theory

Razborov

Wigderson

Part III. The Linear Algebra Method

Prolog: What This Book Is About

Many combinatorial problems have the following "extremal" formulation. Given a finite n-element set of points, the goal is to find the maximum (or minimum) possible cardinality of a system of its subsets satisfying certain assumptions. To get a feeling about what kind of problems this book deals with, we list several typical examples. (Although long, the list is far from being exhaustive.) The number(s) in brackets indicate the section(s), where the corresponding problem is discussed.

Graphs: acquaintances and strangers

○ In a town with n inhabitants, how many acquaintances can there be if we know that among any k inhabitants at least two of them are strangers? For $k = 3$ the answer "at most $n^2/4$ acquaintances" was found by Mantel in 1907. Note that this is only about a half of all $n(n - 1)/2$ possible acquaintances. For an arbitrary k the answer was found by Turán in 1941, and this fundamental result initiated the field, currently known as the *extremal graph theory*. [4.3, 4.4]

○ We want to avoid the situation that some k of inhabitants are either mutually acquainted or are mutual strangers. Ramsey's theorem says that in any town with at least 4^k inhabitants this bad situation will definitely occur. On the other hand, using the probabilistic argument, Erdős has proved that in every town with up to $2^{k/2}$ inhabitants, there *exists* an arrangement of mutual acquaintances and strangers such that this bad situation will not appear. Using the linear algebra method, Frankl and Wilson were able even to *construct* such an arrangement if the town has up to about $k^{\log k}$ inhabitants. [27.2, 18.1, 14.3.3]

Set systems: clubs

○ A town has n inhabitants and some number of clubs; each inhabitant may be a member of several (or none) of them. If no club contains all the members of another club, then we can have at most $\binom{n}{\lfloor n/2 \rfloor}$ clubs in the town. This is the classical *Sperner's theorem*. [9.2.1]

○ We have m clubs $A_1, \ldots, A_m$ with s members in each, and want to know their number m. We can try to form m new "fictive" clubs $B_1, \ldots, B_m$, each with r members such that A_i and B_j will share a member if and only if $i \neq j$. If we succeed in doing so, then we know the answer: $m \leqslant \binom{s+r}{s}$. This result, due to Bollobás, generalizes Sperner's theorem and is one of the corner-stones in *extremal set theory*. [9.2.2]

○ A collection of clubs forms a "sunflower" if each inhabitant, participating in at least one of them, is either a member of all or of precisely one of these clubs. A classical result of Erdős and Rado says that if each club has s members and we have more than $s!(k-1)^s$ clubs in a town, then some k of them will form a sunflower. [7.1]

○ We want to form as few clubs as possible with the property that if we take any set of k inhabitants and arbitrarily split them in two groups, then there will be a club which contains among its members all the inhabitants from the first group and none from the other. It is clear that 2^n clubs are enough and that we need at least 2^k clubs (or more, if $k < n$). Using the *probabilistic method* it can be shown that, somewhat surprisingly, it is possible to achieve this situation with only about $k2^k \log n$ clubs. Such collections of clubs are important in many applications, such as testing logical circuits, construction of k-wise independent random variables, etc. [11.3, 14.4.1, 18.5]

○ Each of n inhabitants participates in the lottery, where he/she can win with equal probability some amount x of points, $0 \leqslant x \leqslant N$. After that, each club calculates the total sum of points gained by its members. What is the probability that precisely one club will have the smallest (or the largest) total yield? The *isolation lemma*, due to K. Mulmuley, U. Vazirani, and V. Vazirani, says that (independent of how the clubs are formed) this will happen with probability at least $1 - n/N$. [12.3]

○ The city council selects some s numbers and passes a rule that if a pair of clubs share ℓ members, then this ℓ must be among the given s numbers. How many clubs can be formed under this rule? Using the *linear algebra method* it can be proved that (no matter what the selected numbers are) the inhabitants can form at most $\sum_{i=0}^{s} \binom{n}{i}$ clubs. This far reaching extension of Fisher's inequality is the celebrated Ray-Chaudhuri–Frankl–Wilson theorem. [14.3.2]

Numbers

○ A set of integers is sum-free if the sum of every two (not necessarily distinct) of its elements does not belong to it. In 1965 Erdős, using a *probabilistic argument*, proved that *every* set of N nonzero integers always contains a sum-free subset of at least size $N/3$. [20.2]

○ Given an integer k, how long must a sequence of integers $a_1, \ldots, a_n$ be in order to be sure that it contains a subsequence of (not necessarily consecutive) elements whose sum is divisible by k? The sequence $0\ldots01\ldots1$ of $k-1$ subsequent 0's and 1's shows that the sequence must have at least $2k-1$ numbers. Using an *algebraic argument*, it can be shown that every sequence of $2k-1$ numbers will already have the desired subsequence. [28.2]

○ If somebody gives us a sequence of more than sr integers, then we (without looking at that sequence) can be sure that it contains either a subsequence of s (not necessarily consecutive) increasing numbers or a subsequence of r decreasing numbers. This result was first proved by Erdős and Szekeres in 1935. In 1959 Seidenberg found a very short proof using the *pigeonhole principle*. [4.2]

Geometry

○ What is the maximal set of points in the n-dimensional Euclidean space $\mathbb{R}^n$, such that all angles determined by three points from the set are strictly less than $\pi/2$? It was an old conjecture of Danzer and Grünbaum that any such set can have at most $2n-1$ points. Using the *probabilistic method*, Erdős and Füredi disproved this conjecture: there is such a set with about 1.15^n points. [21.4]

○ In 1944 Hadwiger proposed the following question: how many colors do we need in order to color the points of the n-dimensional Euclidean space $\mathbb{R}^n$ so that each monochromatic set of points misses some distance? A set is said to "miss distance" d if no two of its points are at distance d apart from each other. This turns out to be quite a hard problem; the exact answer is not known even for the plane (where $n = 2$). In 1972 Larman and Rogers proved that about 2.8^n colors are enough. Using the *linear algebra method*, in 1981 Frankl and Wilson were able to prove that this exponential bound is not far from the truth: at least 1.2^n colors are necessary. [14]

Complexity theory

○ Let f be a boolean function and a be an input vector. A certificate of a is a set of its bits such that looking at only these bits of a we can determine the value $f(a)$. A decision tree for f is a deterministic algorithm which, given an input a, tests its bits one-by-one in a prescribed order and outputs the value $f(a)$. Suppose we know that all inputs have certificates of size at most k. How many tests must a decision tree make on the worst case input? It turns out that k^2 tests are always enough. [10.4]

○ Given a set of m 0-1 vectors, how many of their bits must be exposed in order to distinguish every single vector from the remaining vectors in the set? It turns our that, on average, it is enough to expose at most $\sqrt{m}$ bits, and there are sets for which this bound cannot be improved. [12.2] o With every boolean function f on $2n$ variables we can associate a graph G_f whose vertices are 0-1 vectors of length n, and two vertices a, b are joined by an edge precisely when $f(a, b) = 1$. If the graph G_f has a "complicated" structure, then (intuitively) the function f should be hard to compute, that is, should require a large formula or circuit. Using the *probabilistic argument*, Razborov has proved that this intuition may be false! [24.3]

o Given a boolean function, how many And, Or and Not operations do we need to represent it as a formula? The difficulty in proving that a given boolean function has high complexity (i.e., requires large formulas, or large circuits, etc.) seems to lie in the nature of the adversary: the algorithm. Fast algorithms may work in a counterintuitive fashion, using deep, devious, and fiendishly clever ideas. How could one prove that there is no clever way to quickly solve a given problem? This has been the main issue confronting the complexity theorists since the early 1950's. We will show how, using non-trivial combinatorial arguments, this task can be solved for different models of computation – like DeMorgan formulas, combinational circuits, and span programs – under additional restrictions on the use of Not gates. [10.6, 15.2.2, 16.4, 16.5]

In this book we will learn some of the most powerful combinatorial tools which have proved useful in attacking such and similar problems:

1. basic methods: the double counting argument, the pigeonhole principle, the inclusion-exclusion formula, the averaging argument, etc.
2. the linear algebra method
3. the probabilistic method
4. Ramsey arguments.

These tools are presented in a form acceptable also to a reader from other fields of mathematics and computer science. (However, the reader should not be immediately disappointed if some of the seemingly "simple" proofs would require a half an hour of thinking – bright brains have spent maybe months to produce them!) The emphasis is made on learning methods rather than the results themselves – these are chosen to illustrate the way of reasoning in elegant and simple form.

Most of the results and techniques presented in this book are motivated by applications in the theory of computing. A fundamental problem of this theory – known as the *lower bounds problem* – is to prove that a given function cannot be computed within a given amount of resourses (time, space, chip-area, etc.). This is an extremal problem per se and we will demonstrate the role of combinatorial reasoning in its solution for different models of computation: resolution refutation proofs, boolean formulas, circuits, span programs and multi-party communication protocols (Sects. 10.4, 10.5, 10.6, 15.2.2, 16, 24, 29.3).

Notation

In this section we give the notation that shall be standard throughout the book.

Sets

We deal exclusively with finite objects. We use the standard set-theoretical notation:

$|X|$ denotes the *size* (the *cardinality*) of a set X.

A *k-set* or *k-element set* is a set of k elements.

$[n] = \{1, 2, \ldots, n\}$ is often used as a "standard" n-element set.

$A - B = \{x : x \in A \text{ and } x \notin B\}$.

$\overline{A} = X \backslash A$ is the complement of A.

$A \oplus B = (A - B) \cup (B - A)$ (symmetric difference).

$A \times B = \{(a, b) : a \in A, b \in B\}$ (Cartesian product).

$A \subseteq B$ if B contains all the elements of A.

$A \subset B$ if $A \subseteq B$ and $A \neq B$.

2^X is the set of all subsets of the set X. If $|X| = n$ then $\left|2^X\right| = 2^n$.

A *permutation* of X is a one-to-one mapping (a bijection) $f : X \to X$.

$\{0, 1\}^n = \{(v_1, \ldots, v_n) : v_i \in \{0, 1\}\}$ is the (binary) n-cube.

0-1 vector (matrix) is a vector (matrix) with entries 0 and 1.

An $m \times n$ matrix is a matrix with m rows and n columns.

The *incidence vector* of a set $A \subseteq \{x_1, \ldots, x_n\}$ is a 0-1 vector $v = (v_1, \ldots, v_n)$, where $v_i = 1$ if $x_i \in A$, and $v_i = 0$ if $x_i \notin A$.

The *characteristic function* of a subset $A \subseteq X$ is the function $f : X \to \{0, 1\}$ such that $f(x) = 1$ if and only if $x \in A$.

Arithmetic

Some of the results are asymptotic, and we use the standard asymptotic notation: for two functions f and g, we write $f = O(g)$ if $f \leqslant c_1 g + c_2$ for all possible values of the two functions, where c_1, c_2 are absolute constants. We write $f = \Omega(g)$ if $g = O(f)$, and $f = \Theta(g)$ if $f = O(g)$ and $g = O(f)$. If the limit of the ratio f/g tends to 0 as the variables of the functions tend to infinity, we write $f = o(g)$. Finally, $f \lesssim g$ means that $f \leqslant (1 + o(1))g$, and $f \sim g$ denotes that $f = (1 + o(1))g$, i.e., that f/g tends to 1 when the variables tend to infinity. If x is a real number, then $\lceil x \rceil$ denotes the smallest integer not less than x, and $\lfloor x \rfloor$ denotes the greatest integer not exceeding x. As customary, $\mathbb{Z}$ denotes the set of integers, $\mathbb{R}$ the set of reals, $\mathbb{Z}_n$ an additive group of integers modulo n, and $\mathrm{GF}(q)$ (or $\mathbb{F}_q$) a finite Galois field with q elements. Such a field exists as long as q is a prime power. If $q = p$ is a prime then $\mathbb{F}_p$ can be viewed as the set $\{0, 1, \ldots, p-1\}$ with addition and multiplication performed modulo p. The sum in $\mathbb{F}_2$ is often denoted by $\oplus$, that is, $x \oplus y$ stands for $x + y \,(\mathrm{mod}\,2)$. We will often use the so-called *Cauchy–Schwarz inequality* (see Proposition 14.4 for a proof): if $a_1, \ldots, a_n$ and $b_1, \ldots, b_n$ are real numbers then

$$\left(\sum_{i=1}^{n} a_i b_i \right)^2 \leqslant \left(\sum_{i=1}^{n} a_i^2 \right) \left(\sum_{i=1}^{n} b_i^2 \right).$$

If not stated otherwise, $\mathrm{e} = 2.718...$ will always denote the base of the natural logarithm.

Graphs

A *graph* is a pair $G = (V, E)$ consisting of a set V, whose members are called *vertices* (or *nodes*), and a family E of 2-element subsets of V, whose members are called *edges*. A vertex v is *incident* with an edge e if $v \in e$. The two vertices incident with an edge are its *endvertices* or *endpoints*, and the edge *joins* its ends. Two vertices u, v of G are *adjacent*, or *neighbors*, if $\{u, v\}$ is an edge of G. The number $d(u)$ of neighbors of a vertex u is its *degree*. A *walk* of length k in G is a sequence $v_0, e_1, v_1 \ldots, e_k, v_k$ of vertices and edges such that $e_i = \{v_{i-1}, v_i\}$. A walk without repeated vertices is a *path*. A walk without repeated edges is a *trail*. A *cycle* of length k is a path $v_0, \ldots, v_k$ with $v_0 = v_k$. A (connected) *component* in a graph is a set of its vertices such that there is a path between any two of them. A graph is *connected* if it consists of one component. A *tree* is a connected graph without cycles. A *subgraph* is obtained by deleting edges and vertices. A *spanning subgraph* is obtained by deleting edges only. An *induced subgraph* is obtained by deleting vertices (together with all the edges incident to them).

A *complete graph* or *clique* is a graph in which every pair is adjacent. An *independent set* in a graph is a set of vertices with no edges between them. The greatest integer r such that G contains an independent set of size r is the *independence number* of G, and is denoted by $\alpha(G)$. A graph is *bipartite* if its vertex set can be partitioned into two independent sets.

A *legal coloring* of $G = (V, E)$ is an assignment of colors to each vertex so that adjacent vertices receive different colors. In other words, this is a partition of the vertex set V into independent sets. The minimum number of colors required for that is the *chromatic number* $\chi(G)$ of G.

Set systems

A *set system* or *family of sets* $\mathcal{F}$ is a collection of sets. Because of their intimate conceptual relation to graphs, a set system is often called a *hypergraph*. A family is *k-uniform* if all its members are k-element sets. Thus, graphs are k-uniform families with $k = 2$. The *rank* of a family is the maximum cardinality of its member. A *blocking set* (or transversal) of $\mathcal{F}$ is a set which intersects every member of $\mathcal{F}$. The minimum number of elements in a blocking set of a family $\mathcal{F}$ is its *blocking number*, and is denoted by $\tau(\mathcal{F})$.

The notions of independent set and chromatic number extend to set systems. For a set system $\mathcal{F}$ over the universe X, the subset $S \subseteq X$ is called *independent* if S does not contain any member of $\mathcal{F}$. An *r-coloring* of $\mathcal{F}$ is a map $h: X \to \{1, 2, \ldots\}$ which assigns to each point $x \in X$ its "color" $h(x)$. Such a coloring is *legal* if none of the members of $\mathcal{F}$ is monochromatic, i.e., if for all $A \in \mathcal{F}$ there exist $x, y \in A$ such that $h(x) \neq h(y)$. The independence number $\alpha(\mathcal{F})$ and the chromatic number $\chi(\mathcal{F})$ are defined as for graphs.

Three representations

In order to prove something about families of sets (as well as to interpret the results) it is often useful to keep in mind that any family can be looked at either as a 0-1 matrix or as a bipartite graph.

Let $\mathcal{F} = \{A_1, \ldots, A_m\}$ be a family of subsets of a set $X = \{x_1, \ldots, x_n\}$. The *incidence matrix* of $\mathcal{F}$ is an $n \times m$ 0-1 matrix $M = (m_{i,j})$ such that $m_{i,j} = 1$ if and only if $x_i \in A_j$. Hence, the jth column of M is the incidence vector of the set A_j. The *incidence graph* of $\mathcal{F}$ is a bipartite graph with parts X and $\mathcal{F}$, where x_i and A_j are joined by an edge if and only if $x_i \in A_j$.

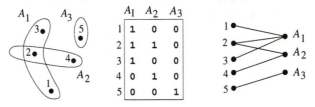

Fig. 0.1. Three representations of the family $\mathcal{F} = \{A_1, A_2, A_3\}$ over the set of points $X = \{1, 2, 3, 4, 5\}$ with $A_1 = \{1, 2, 3\}$, $A_2 = \{2, 4\}$ and $A_3 = \{5\}$

Projective planes

To justify the optimality of some results, we will often refer to the following regular families. Let $n = q^2 + q + 1$. A *projective plane* of order q is a family of n subsets (called *lines*) of an n-element set X (of *points*) satisfying the following four conditions:

- every line has exactly $q + 1$ points;
- every point belongs to exactly $q + 1$ lines;
- every two lines meet in exactly one point;
- any two points lie on a unique line.

Such a family exists for any prime power q; see Chap. 13 for more details.

Boolean functions

A *boolean function* $f = f(x_1, \ldots, x_n)$ on the n variables $x_1, \ldots, x_n$ is simply a function $f: \{0, 1\}^n \to \{0, 1\}$. In particular,

$$0, \quad 1, \quad x_1 \wedge \cdots \wedge x_n, \quad x_1 \vee \cdots \vee x_n, \quad x_1 \oplus \cdots \oplus x_n$$

denote, as usual, the two constant functions, the *And* function (whose value is 1 iff $x_i = 1$ for all i), the *Or* function (whose value is 0 iff $x_i = 0$ for all i), and the *Parity* function (whose value is 0 iff $x_i = 1$ for an even number of variables x_i). For a function f, we let $\overline{f} = f \oplus 1$ denote its complement, *Not* f. The functions x_i and $\overline{x}_i$ are called *literals* (or *atoms*).

A *monomial* is an And of literals, and a *clause* is an Or of literals. The number of literals in a clause or monomial is its *length* (or *size*). The Or of an arbitrary number of monomials is a *disjunctive normal form (DNF)*. Dually, the And of an arbitrary number of clauses is a *conjunctive normal form (CNF)*. A boolean function f is a *t-And-Or* function if it can be written as an And of an arbitrary number of clauses, each being an Or of at most t literals. That is, a function is t-And-Or function if it can be represented by a CNF, all whose clauses have length at most t. Dually, a boolean function f an *s-Or-And* if it can be written as an Or of an arbitrary number of monomials, each being an And of at most s literals.

Part I

The Classics

1. Counting

We start with the oldest combinatorial tool — *counting*.

1.1 The binomial theorem

Given a set of n elements, how many of its subsets have exactly k elements? This number (of k-element subsets of an n-element set) is usually denoted by $\binom{n}{k}$ and is called the *binomial coefficient*. The following identity was proved by Sir Isaac Newton in about 1666, and is known as the *Binomial theorem*.

Binomial Theorem. *Let n be a positive integer. Then for all x and y,*

$$(x+y)^n = \sum_{k=0}^{n} \binom{n}{k} x^k y^{n-k}.$$

Proof. By definition, $\binom{n}{k}$ is precisely the number of ways to get the term $x^k y^{n-k}$ when multiplying $(x+y)(x+y)\cdots(x+y)$. □

The *factorial* of n is the product $n! \rightleftharpoons n(n-1)\cdots 2\cdot 1$. This is extended to all non-negative integers by letting $0! = 1$. The *k-th factorial* of n is the product of the first k terms:

$$(n)_k \rightleftharpoons \frac{n!}{(n-k)!} = n(n-1)\cdots(n-k+1).$$

Note that $\binom{n}{0} = 1$ (the empty set) and $\binom{n}{n} = 1$ (the whole set). In general, binomial coefficients can be written as quotients of factorials:

Proposition 1.1. $\qquad \binom{n}{k} = \frac{(n)_k}{k!} = \frac{n!}{k!(n-k)!}.$

Proof. Observe that $(n)_k$ is the number of (ordered!) strings $(x_1, x_2, \ldots, x_k)$ consisting of k different elements of a fixed n-element set: there are n possibilities to choose the first element x_1; after that there are still $n-1$ possibilities to choose the next element x_2, etc. Another way to produce such strings is to choose a k-element set and then arrange its elements in an arbitrary order. Since each of $\binom{n}{k}$ k-element subsets produces exactly $(k)_k = k!$ such strings, we conclude that $(n)_k = \binom{n}{k}k!$. □

There are a lot of useful equalities concerning binomial coefficients. In most situations, using their *combinatorial* nature (instead of algebraic, as given by the previous proposition) we obtain the desired result fairly easily. For example, if we observe that each subset is uniquely determined by its complement, then we immediately obtain the equality

$$\binom{n}{n-k} = \binom{n}{k}.$$ (1.1)

In a similar way other useful identities can be established (see Exercises for more examples).

Proposition 1.2. $\binom{n}{k} = \binom{n-1}{k-1} + \binom{n-1}{k}.$

Proof. The first term $\binom{n-1}{k-1}$ is the number of k-sets containing a fixed element, and the second term $\binom{n-1}{k}$ is the number of k-sets avoiding this element; their sum is the whole number $\binom{n}{k}$ of k-sets.   □

For growing n and k, exact values of binomial coefficients $\binom{n}{k}$ are hard to compute. In applications, however, we are often interested only in their rate of growth, so that (even rough) estimates suffice. Such estimates can be obtained, using the Taylor series of the exponential function:

$$e^t = 1 + t + \frac{t^2}{2!} + \frac{t^3}{3!} + \cdots$$ (1.2)

This, in particular, implies one useful estimate, which we will use later quite often:

$$1 + t < e^t \quad \text{for all } t \in \mathbb{R}, \, t \neq 0.$$ (1.3)

Proposition 1.3. $\left(\frac{n}{k}\right)^k \leqslant \binom{n}{k} < \left(\frac{en}{k}\right)^k.$

Proof. Lower bound:

$$\left(\frac{n}{k}\right)^k = \frac{n}{k} \cdot \frac{n}{k} \cdots \frac{n}{k} \leqslant \frac{n}{k} \cdot \frac{n-1}{k-1} \cdots \frac{n-k+1}{1} = \binom{n}{k}.$$

Upper bound. By (1.3) and binomial theorem,

$$e^{nt} > (1+t)^n = \sum_{i=1}^{n} \binom{n}{i} t^i > \binom{n}{k} t^k.$$

Substituting $t = k/n$ we obtain

$$e^k > \binom{n}{k} \left(\frac{k}{n}\right)^k,$$

as desired.   □

Tighter (asymptotic) estimates can be obtained using the famous *Stirling formula* for the factorial:

$$n! = \left(\frac{n}{e}\right)^n \sqrt{2\pi n}\, e^{\alpha_n}, \tag{1.4}$$

where $1/(12n+1) < \alpha_n < 1/12n$. This leads, for example, to the following elementary but very useful asymptotic formula for the kth factorial:

$$(n)_k = n^k e^{-\frac{k^2}{2n} - \frac{k^3}{6n^2} + o(1)} \quad \text{valid for } k = o(n^{3/4}), \tag{1.5}$$

and hence, for binomial coefficients:

$$\binom{n}{k} = \frac{n^k e^{-\frac{k^2}{2n} - \frac{k^3}{6n^2}}}{k!}(1 + o(1)). \tag{1.6}$$

1.2 Selection with repetitions

In the previous section we considered the number of ways to choose r *distinct* elements from an n-element set. It is natural to ask what happens if we can choose the same element repeatedly? In other words, we may ask how many integer solutions does the equation $x_1 + \cdots + x_n = r$ have under the condition that $x_i \geqslant 0$ for all $i = 1, \ldots, n$? (Look at x_i as the number of times the ith element was chosen.) The following more entertaining formulation of this problem was suggested by Lovász, Pelikán, and Vesztergombi (1977).

Suppose we have r sweets (of the same sort), which we want to distribute to n children. In how many ways we can do this? Letting x_i denote the number of sweets we give to the ith child, this question is equivalent to that stated above.

The answer depends on how many sweets we have and how fair we are. If we are fair but have only $r \leqslant n$ sweets, then it is natural to allow no repetitions and give each child no more than one sweet (each x_i is 0 or 1). In this case the answer is easy: we just choose those r (out of n) children who will get a sweet, and we already know that this can be done in $\binom{n}{r}$ ways.

Suppose now that we have enough sweets, i.e., that $r \geqslant n$. Let us first be fair, that is, we want every child gets at least one sweet. We lay out the sweets in a single row of length r (it does not matter in which order, they all are alike), and let the first child pick them up from the left to right. After a while we stop him/her and let the second child pick up sweets, etc. The distribution of sweets is determined by specifying the place (between consecutive sweets) of where to start with a new child. There are $r-1$ such places, and we have to select $n-1$ of them (the first child always starts at the beginning, so we have no choice here). For example, if we have $r = 9$ sweets and $n = 6$ children, a typical situation looks like this:

$$\square\, \lambda\, \square\,\square\,\square\, \lambda\, \square\, \lambda\, \square\,\square\, \lambda\, \square\, \lambda\, \square$$
$$\quad 2 \qquad\qquad 3 \quad 4 \qquad 5 \quad 6$$

Thus, we have to select an $(n-1)$-element subset from an $(r-1)$-element set. The number of possibilities to do so is $\binom{r-1}{n-1}$. If we are unfair, we have more possibilities:

Proposition 1.4. *The number of integer solutions to the equation*

$$x_1 + \cdots + x_n = r$$

under the condition that $x_i \geqslant 0$ for all $i = 1, \ldots, n$, is $\binom{n+r-1}{r}$.

Proof. In this situation we are unfair and allow that some of the children may be left without sweet. With the following trick we can reduce the problem of counting the number of such distributions to the problem we just solved: we borrow one sweet from each child, and then distribute the whole amount of $n + r$ sweets to the children so that each child gets at least one sweet. This way every child gets back the sweet we borrowed from him/her, and the lucky ones get some more. This "more" is exactly r sweets distributed to n children. We already know that the number of ways to distribute $n + r$ sweets to n children in a fair way is $\binom{n+r-1}{n-1}$, which by (1.1) equals $\binom{n+r-1}{r}$.   □

1.3 Partitions

A *partition* of n objects is a collection of its mutually disjoint subsets, called *blocks*, whose union gives the whole set. Let $S(n; k_1, k_2, \ldots, k_n)$ denote the number of all partitions of n objects with k_i i-element blocks ($i = 1, \ldots, n$; $k_1 + 2k_2 + \ldots + nk_n = n$).

Proposition 1.5. $S(n; k_1, k_2, \ldots, k_n) = \dfrac{n!}{k_1! \cdots k_n! (1!)^{k_1} \cdots (n!)^{k_n}}.$

Proof. If we consider any arrangement (i.e., a permutation) of the n objects we can get such a partition by taking the first k_1 elements as 1-element blocks, the next $2k_2$ elements as 2-element blocks, etc. Since we have $n!$ possible arrangements, it remains to show that we get any given partition exactly

$$k_1! \cdots k_n! (1!)^{k_1} \cdots (n!)^{k_n}$$

times. Indeed, we can construct an arrangement of the objects by putting the 1-element blocks first, then the 2-element blocks, etc. However, there are $k_i!$ possible ways to order the i-element blocks and $(i!)^{k_i}$ possible ways to order the elements in the i-element blocks.   □

1.4 Double counting

The *double counting* principle states the following "obvious" fact: if the elements of a set are counted in two different ways, the answers are the same.

In terms of matrices the principle is as follows. Let M be an $n \times m$ matrix with entries 0 and 1. Let r_i be the number of 1's in the ith row, and c_j be the number of 1's in the jth column. Then

$$\sum_{i=1}^{n} r_i = \sum_{j=1}^{m} c_j = \text{the total number of 1's in } M.$$

The next example is a standard demonstration of double counting. Suppose a finite number of people meet at a party and some shake hands. Assume that no person shakes his or her own hand and furthermore no two people shake hands more than once.

Handshaking Lemma. *At a party, the number of guests who shake hands an odd number of times is even.*

Proof. Let $P_1, \ldots, P_n$ be the persons. We apply double counting to the set of ordered pairs (P_i, P_j) for which P_i and P_j shake hands with each other at the party. Let x_i be the number of times that P_i shakes hands, and y the total number of handshakes that occur. On one hand, the number of pairs is $\sum_{i=1}^{n} x_i$, since for each P_i the number of choices of P_j is equal to x_i. On the other hand, each handshake gives rise to two pairs (P_i, P_j) and (P_j, P_i); so the total is $2y$. Thus $\sum_{i=1}^{n} x_i = 2y$. But, if the sum of n numbers is even, then evenly many of the numbers are odd. (Because if we add an odd number of odd numbers and any number of even numbers, the sum will be always odd). □

This lemma is also a direct consequence of the following general identity, whose special version for graphs was already proved by Euler. For a point x, its *degree* or *replication number* $d(x)$ in a family $\mathcal{F}$ is the number of members of $\mathcal{F}$ containing x.

Proposition 1.6. *Let $\mathcal{F}$ be a family of subsets of some set X. Then*

$$\sum_{x \in X} d(x) = \sum_{A \in \mathcal{F}} |A|. \tag{1.7}$$

Proof. Consider the *incidence matrix* $M = (m_{x,A})$ of $\mathcal{F}$. That is, M is a 0-1 matrix with $|X|$ rows labeled by points $x \in X$ and with $|\mathcal{F}|$ columns labeled by sets $A \in \mathcal{F}$ such that $m_{x,A} = 1$ if and only if $x \in A$. Observe that $d(x)$ is exactly the number of 1's in the x-th row, and $|A|$ is the number of 1's in the A-th column. □

Graphs are families of 2-element sets, and the degree of a vertex x is the number of edges incident to x, i.e., the number of vertices in its neighborhood. Proposition 1.6 immediately implies

Theorem 1.7 (Euler 1736). *In every graph the sum of degrees of its vertices is two times the number of its edges, and hence, is even.*

The following identities can be proved in a similar manner (we leave their proofs as exercises):

$$\sum_{x \in Y} d(x) = \sum_{A \in \mathcal{F}} |Y \cap A| \quad \text{for any } Y \subseteq X. \tag{1.8}$$

$$\sum_{x \in X} d(x)^2 = \sum_{A \in \mathcal{F}} \sum_{x \in A} d(x) = \sum_{A \in \mathcal{F}} \sum_{B \in \mathcal{F}} |A \cap B|. \tag{1.9}$$

Turán's number $T(n, k, l)$ $(l \leqslant k \leqslant n)$ is the smallest number of l-element subsets of an n-element set X such that every k-element subset of X contains at least one of these sets.

Proposition 1.8. *For all positive integers* $l \leqslant k \leqslant n$,

$$T(n, k, l) \geqslant \binom{n}{l} \Big/ \binom{k}{l}.$$

Proof. Let $\mathcal{F}$ be a smallest l-uniform family over X such that every k-subset of X contains at least one member of $\mathcal{F}$. Take a 0-1 matrix $M = (m_{A,B})$ whose rows are labeled by sets A in $\mathcal{F}$, columns by k-element subsets B of X, and $m_{A,B} = 1$ if and only if $A \subseteq B$. Let r_A be the number of 1's in the A-th row and c_B be the number of 1's in the B-th column. Then, $c_B \geqslant 1$ for every B, since B must contain at least one member of $\mathcal{F}$. On the other hand, r_A is precisely the number of k-element subsets B containing a fixed l-element set A; so $r_A = \binom{n-l}{k-l}$ for every $A \in \mathcal{F}$. By the double counting principle,

$$|\mathcal{F}| \cdot \binom{n-l}{k-l} = \sum_{A \in \mathcal{F}} r_A = \sum_B c_B \geqslant \binom{n}{k},$$

which yields

$$T(n, k, l) = |\mathcal{F}| \geqslant \binom{n}{k} \Big/ \binom{n-l}{k-l} = \binom{n}{l} \Big/ \binom{k}{l},$$

where the last equality is another property of binomial coefficients (see Exercise 1.10). $\qquad\square$

1.5 The averaging principle

Suppose we have a set of m objects, the ith of which has "size" l_i, and we would like to know if at least one of the objects is large, i.e., has size $l_i \geqslant t$ for some given t. In this situation we can try to consider the average size $\bar{l} = \sum l_i / m$ and try to prove that $\bar{l} \geqslant t$. This would immediately yield the result, because we have the following

Averaging Principle. *Every set of numbers must contain a number at least as large* $(\geqslant)$ *as the average and a number at least as small* $(\leqslant)$ *as the average.*

This principle is a prototype of a very powerful technique – the probabilistic method – which we will study in Part 4. The concept is very simple, but the applications can be surprisingly subtle. We will use this principle quite often.

To demonstrate the principle, let us prove the following sufficient condition that a graph is disconnected.

A (connected) *component* in a graph is a set of its vertices such that there is a path between any two of them. A graph is *connected* if it consists of one component; otherwise it is *disconnected*.

Proposition 1.9. *Every graph on n vertices with fewer than $n - 1$ edges is disconnected.*

Proof. Induction by n. When $n = 1$, the claim is vacuously satisfied, since no graph has a negative number of edges.

When $n = 2$, a graph with less than 1 edge is evidently disconnected.

Suppose now that the result has been established for graphs on n vertices, and take a graph $G = (V, E)$ on $|V| = n + 1$ vertices such that $|E| \leqslant n - 1$. By Euler's theorem (Theorem 1.7), the average degree of its vertices is

$$\frac{1}{|V|} \sum_{x \in V} d(x) = \frac{2|E|}{|V|} \leqslant \frac{2(n-1)}{n+1} < 2.$$

By the averaging principle, some vertex x has degree 0 or 1. If $d(x) = 0$, x is a component disjoint from the rest of G, so G is disconnected. If $d(x) = 1$, suppose the unique neighbor of x is y. Then, the graph H obtained from G by deleting x and its incident edge has $|V| - 1 = n$ vertices and $|E| - 1 \leqslant (n - 1) - 1 = n - 2$ edges; by the induction hypothesis, H is disconnected. The restoration of an edge joining a vertex y in one component to a vertex x which is outside of a second component cannot reconnect the graph. Hence, G is also disconnected.   □

We mention one important inequality, which is especially useful when dealing with averages.

A real-valued function $f(x)$ is *convex* if

$$f(\lambda x + (1 - \lambda)y) \leqslant \lambda f(x) + (1 - \lambda)f(y),$$

for any $0 \leqslant \lambda \leqslant 1$. From a geometrical point of view, the convexity of f means that if we draw a line l through points $(x, f(x))$ and $(y, f(y))$, then the graph of the curve $f(z)$ must lie below that of $l(z)$ for $z \in [x, y]$. Thus, for a function f to be convex it is sufficient that its second derivative is nonnegative.

Proposition 1.10 (Jensen's Inequality). *If $0 \leqslant \lambda_i \leqslant 1$, $\sum_{i=1}^{r} \lambda_i = 1$ and f is convex, then*

$$f\left(\sum_{i=1}^{r} \lambda_i x_i\right) \leqslant \sum_{i=1}^{r} \lambda_i f(x_i).$$

Proof. Easy induction on the number of summands r. For $r = 2$ this is true, so assume the inequality holds for the number of summands up to r, and prove it for $r + 1$. For this it is enough to replace the sum of the first two terms in $\lambda_1 x_1 + \lambda_2 x_2 + \ldots + \lambda_{r+1} x_{r+1}$ by the term

$$(\lambda_1 + \lambda_2)\left(\frac{\lambda_1}{\lambda_1 + \lambda_2}x_1 + \frac{\lambda_2}{\lambda_1 + \lambda_2}x_2\right),$$

and apply the induction hypothesis.   □

Jensen's inequality immediately yields the following useful inequality between the arithmetic and geometric means.

Proposition 1.11. *Let $a_1, \ldots, a_n$ be non-negative numbers. Then*

$$\frac{1}{n}\sum_{i=1}^{n} a_i \geqslant \left(\prod_{i=1}^{n} a_i\right)^{1/n}. \tag{1.10}$$

Proof. Let $f(x) = 2^x$, $\lambda_1 = \ldots = \lambda_n = 1/n$ and $x_i = \log_2 a_i$, for all $i = 1, \ldots, n$. By Jensen's inequality

$$\frac{1}{n}\sum_{i=1}^{n} a_i = \sum_{i=1}^{n} \lambda_i f(x_i) \geqslant f\left(\sum_{i=1}^{n} \lambda_i x_i\right) = 2^{(\sum_{i=1}^{n} x_i)/n} = \left(\prod_{i=1}^{n} a_i\right)^{1/n}.$$

□

To give a less direct illustration, consider the following question: given a binary tree with m leaves, what can be said about the average length of its paths from a root to a leaf? This question may be restated in terms of so-called "prefix-free codes."

A string a is a *prefix* of a string b if $b = ac$ for some string c. A *prefix-free code* is a set $C = \{c_1, \ldots, c_m\}$ of 0-1 strings none of which is a prefix of another. Let l_i be the length of c_i. The famous Kraft inequality from Information Theory states that

$$\sum_{i=1}^{m} 2^{-l_i} \leqslant 1 \tag{1.11}$$

(see Exercise 1.24 for the proof). Together with Jensen's inequality, this implies that the average length $\bar{l} = \sum l_i/m$ of strings in C is at least $\log m$ (see Exercise 1.25). This fact may be also easily derived using the direct convexity argument.

Proposition 1.12. *If C is a prefix-free code then the average length of its strings is at least $\log_2 |C|$.*

Note that prefix-free codes are in one-to-one correspondence with binary trees: given a binary tree, the set of paths from its root to leaves forms a prefix-free code (just because no such path can be prolonged). Thus, the proposition says that the average depth of a binary tree with m leaves is at least $\log_2 m$.

Proof. Let $C = \{c_1, \ldots, c_m\}$ be a prefix-free code and let l_i be the length of c_i. Consider the total length $\ell(C) = l_1 + l_2 + \cdots + l_m$. Our goal is to show that $\ell(C) \geqslant |C| \cdot \log_2 |C|$.

We argue by induction on m. The basic cases $m = 1$ and $m = 2$ are trivial. Now take a prefix-free code C with more than two strings. Depending on what the first bit is, split C into two subsets C_0 and C_1. If we attach a fixed string as a prefix to all the strings in C, then the total length can only increase, whereas the number of strings remains the same. Therefore, we can assume w.l.o.g. that both C_0 and C_1 are non-empty.

Since the first bit contributes exactly $|C_0| + |C_1|$ to the total length $\ell(C)$, we have, by the induction hypothesis, that

$$\ell(C) = \big(\ell(C_0) + |C_0|\big) + \big(\ell(C_1) + |C_1|\big)$$

$$\geqslant |C_0| \cdot \log_2 |C_0| + |C_1| \cdot \log_2 |C_1| + |C|.$$

For $x > 0$ the function $f(x) = x \log_2 x$ is convex (its second derivative is $f''(x) = (\log_2 e)/x > 0$). Applying Jensen's inequality to the previous estimate (with $x_1 = |C_0|$, $x_2 = |C_1|$ and $\lambda_1 = \lambda_2 = 1/2$), we conclude that

$$\ell(C) \geqslant \big(|C_0| + |C_1|\big) \cdot \log_2 \frac{|C_0| + |C_1|}{2} + |C|$$

$$= |C| \cdot \log_2 \frac{|C|}{2} + |C| = |C| \cdot \log_2 |C|,$$

as desired. $\qquad\qquad\qquad\qquad\qquad\qquad\qquad\qquad\qquad\qquad\qquad\qquad\square$

Exercises

1.1.$^-$ In how many ways can we distribute k balls to n boxes so that each box has at most one ball?

1.2.$^-$ Show that for every k the product of any k subsequent natural numbers is divisible by $k!$. *Hint:* Consider $\binom{n+k}{k}$.

1.3.$^-$ Show that the number of pairs (A, B) of distinct subsets of $\{1, \ldots, n\}$ with $A \subset B$ is $3^n - 2^n$.

Hint: Use the binomial theorem to evaluate $\sum_{k=0}^{n} \binom{n}{k}(2^k - 1)$.

1.4.$^-$ Prove that

$$\binom{n}{k} = \frac{n}{k}\binom{n-1}{k-1}.$$

Hint: Count in two ways the number of pairs (x, M), where M is a k-element subset of $\{1, \ldots, n\}$ and $x \in M$.

1.5.⁻ Prove that $\sum_{k=1}^{n} k\binom{n}{k} = n2^{n-1}$.

Hint: Count in two ways the number of pairs (x, M) with $x \in M \subseteq \{1, \ldots, n\}$.

1.6.⁻ Show that $\binom{n}{0}^2 + \binom{n}{1}^2 + \binom{n}{2}^2 + \cdots + \binom{n}{n}^2 = \binom{2n}{n}$.

Hint: Apply the binomial theorem for $(1 + x)^{2n}$.

1.7.⁻ There is a set of $2n$ people: n male and n female. A good party is a set with the same number of male and female. How many possibilities are there to build such a good party?

1.8.⁻ Prove the Cauchy–Vandermonde identity: $\binom{m+n}{k} = \sum_{i=0}^{k} \binom{m}{i}\binom{n}{k-i}$.

Hint: Take a set of $m + n$ people (m male and n female) and make a set of k people (with i male and $k - i$ female).

1.9. Prove the following analogy of the binomial theorem for factorials:

$$(x + y)_n = \sum_{k=0}^{n} \binom{n}{k} (x)_k (y)_{n-k}.$$

Hint: Divide both sides by $n!$, and use the Cauchy–Vandermonde identity.

1.10.⁽¹⁾ Let $0 \leqslant l \leqslant k \leqslant n$. Show that

$$\binom{n}{k}\binom{k}{l} = \binom{n}{l}\binom{n-l}{k-l}.$$

Hint: Count in two ways the number of all pairs (L, K) of subsets of $\{1, \ldots, n\}$ such that $L \subseteq K$, $|L| = l$ and $|K| = k$.

1.11. Use combinatorics (not algebra) to prove that, for $0 \leqslant k \leqslant n$, $\binom{n}{2} = \binom{k}{2} + k(n - k) + \binom{n-k}{2}$.

Hint: $\binom{n}{2}$ is the number of edges in a complete graph on n vertices.

1.12. Prove Fermat's Little theorem: if p is a prime and if a is a natural number, then $a^p \equiv a \pmod{p}$. In particular, if p does not divide a, then $a^{p-1} \equiv 1 \pmod{p}$.

Hint: Apply the induction on a. For the induction step, use the binomial theorem to show that $(a + 1)^p \equiv a^p + 1 \pmod{p}$.

1.13. Let $0 < \alpha < 1$ be a real number, and αn be an integer. Using Stirling's formula show that

$$\binom{n}{\alpha n} = \frac{1 + o(1)}{\sqrt{2\pi\alpha(1 - \alpha)n}} \cdot 2^{n \cdot H(\alpha)},$$

where $H(\alpha) = -\alpha \log_2 \alpha - (1 - \alpha) \log_2(1 - \alpha)$ is the binary entropy function.
Hint: $H(\alpha) = \log_2 h(\alpha)$, where $h(\alpha) = \alpha^{-\alpha}(1 - \alpha)^{-(1-\alpha)}$.

1.14. Prove that, for $s \leqslant n/2$,

(1) $\displaystyle \sum_{k=0}^{s} \binom{n}{k} \leqslant \binom{n}{s} \left(1 + \frac{s}{n - 2s + 1} \right)$;

(2) $\displaystyle \sum_{k=0}^{s} \binom{n}{k} \leqslant 2^{n \cdot H(s/n)}$.

Hint: To (1): observe that $\binom{n}{k-1}/\binom{n}{k} = k/(n - k + 1)$ does not exceed $\alpha \rightleftharpoons s/(n - s + 1)$, and use the identity $\sum_{i=0}^{\infty} \alpha^i = 1/(1 - \alpha)$.
To (2): set $p = s/n$ and apply the binomial theorem to show that

$$p^s (1 - p)^{n-s} \sum_{k=0}^{s} \binom{n}{k} \leqslant 1.$$

See also Corollary 23.6 for another proof.

1.15.⁺ Prove the following estimates: If $k \leqslant k + x < n$ and $y < k \leqslant n$, then

$$\left(\frac{n - k - x}{n - x} \right)^x \leqslant \binom{n - x}{k} \binom{n}{k}^{-1} \leqslant \left(\frac{n - k}{n} \right)^x \leqslant e^{-(k/n)x} \qquad (1.12)$$

and

$$\left(\frac{k - y}{n - y} \right)^y \leqslant \binom{n - y}{k - y} \binom{n}{k}^{-1} \leqslant \left(\frac{k}{n} \right)^y \leqslant e^{-(1 - k/n)y}.$$

1.16.⁺ Prove that if $1 \leqslant k \leqslant n/2$, then

$$\binom{n}{k} \geqslant \gamma \cdot \left(\frac{ne}{k} \right)^k, \quad \text{where } \gamma = \frac{1}{\sqrt{2\pi k}} e^{-k^2/n - 1/(6k)}. \qquad (1.13)$$

Hint: Use Striling's formula to show that

$$\binom{n}{k} \geqslant \frac{1}{\sqrt{2\pi} e^{1/(6k)}} \left(\frac{n}{k} \right)^k \left(\frac{n}{n - k} \right)^{n - k} \left(\frac{n}{k(n - k)} \right)^{1/2},$$

and apply the estimate $\ln(1 + t) \geqslant t - t^2/2$ valid for all $t \geqslant 0$.

1.17. In how many ways can we choose a subset $S \subseteq \{1, 2, \ldots, n\}$ such that $|S| = k$ and no two elements of S precede each other, i.e., $x \neq y + 1$ for all $x \neq y \in S$?

Hint: If $S = \{a_1, \ldots, a_k\}$ is such a subset with $a_1 < a_2 < \ldots < a_k$, then $a_1 < a_2 - 1 < \ldots < a_k - (k - 1)$.

1.18. Let $k \geqslant 2n$. In how many ways can we distribute k sweets to n children, if each child is supposed to get at least 2 of them?

1.19. *Bell's number* B_n is the number of all possible partitions of an n-element set X (we assume that $B_0 = 1$). Prove that $B_{n+1} = \sum_{i=1}^{n} \binom{n}{i} B_i$.

Hint: For every subset $A \subseteq X$ there are precisely $B_{|X - A|}$ partitions of X containing A as one of its blocks.

1.20. Let $|N| = n$ and $|X| = x$. Show that there are x^n mappings from N to X, and that $S(n,k)x(x-1)\cdots(x-k+1)$ of these mappings have a range of cardinality k; here $S(n,k)$ is the Stirling number (the number of partitions of an n-element set into exactly k blocks).

Hint: We have $x(x-1)\cdots(x-k+1)$ possibilities to choose a sequence of k elements in X, and we can specify $S(n,k)$ ways in which elements of N are mapped onto these chosen elements.

1.21.⁻ Let $\mathcal{F}$ be a family of subsets of an n-element set X with the property that any two members of $\mathcal{F}$ meet, i.e., $A \cap B \neq \emptyset$ for all $A, B \in \mathcal{F}$. Suppose also that no other subset of X meets all of the members of $\mathcal{F}$. Prove that $|\mathcal{F}| = 2^{n-1}$. *Hint*: Consider sets and their complements.

1.22. Let $\mathcal{F}$ be a family of k-elements subsets of an n-element set X such that every l-element subset of X is contained in at least one member of $\mathcal{F}$. Show that $|\mathcal{F}| \geqslant \binom{n}{l}/\binom{k}{l}$. *Hint*: Argue as in the proof of Proposition 1.8.

1.23 (Sperner 1928). Let $\mathcal{F}$ be a family of k-element subsets of $\{1,\ldots,n\}$. Its *shadow* is the family of all those $(k-1)$-element subsets which lie entirely in at least one member of $\mathcal{F}$. Show that the shadow contains at least $k|\mathcal{F}|/(n-k+1)$ sets. *Hint*: Argue as in the proof of Proposition 1.8.

1.24. Prove the Kraft inequality (1.11).

Hint: Let $l = \max l_i$, and let A_i be the set of vectors in the cube $\{0,1\}^l$ for which c_i is a prefix. These sets are disjoint and each A_i is a 2^{-l_i}-fraction of the cube.

1.25.⁻ Use Jensen's inequality to derive Proposition 1.12 from the Kraft inequality (1.11). *Hint*: Apply Jensen's inequality with $\lambda_i = 1/m$ and $f(x) = 2^{-x}$.

1.26 (Quine 1988). The famous Fermat's Last Theorem states that if $n > 2$, then $x^n + y^n = z^n$ has no solutions in nonzero integers x, y and z. This theorem can be stated in terms of sorting objects into a row of bins, some of which are red, some blue, and the rest unpainted. The theorem amounts to saying that when there are more than two objects, the following statement is newer true: *The number of ways of sorting them that shun both colors is equal to the number of ways that shun neither.* Show that this statement is equivalent to Fermat's equation $x^n + y^n = z^n$.

Hint: Let n be the number of objects, z the number of bins, x the number of bins that are not red and y the number of bins that are not blue.

2. Advanced Counting

When properly applied, the (double) counting argument can lead to more subtle results than those discussed in the previous chapter.

2.1 Bounds on intersection size

How many r-element subsets of an n-element set can we choose under the restriction that no two of them share more than k elements? Intuitively, the smaller k is, the fewer sets we can choose. This intuition can be made precise as follows. (We address the optimality of this bound in Exercise 2.5.)

Lemma 2.1 (Corrádi 1969). *Let $A_1, \ldots, A_N$ be r-element sets and X be their union. If $|A_i \cap A_j| \leqslant k$ for all $i \neq j$, then*

$$|X| \geqslant \frac{r^2 N}{r + (N-1)k}. \tag{2.1}$$

Proof. Just count. By (1.8), we have for each $i = 1, \ldots, N$,

$$\sum_{x \in A_i} d(x) = \sum_{j=1}^{N} |A_i \cap A_j| = |A_i| + \sum_{j \neq i} |A_i \cap A_j| \leqslant r + (N-1)k.$$

Summing over all sets A_i we get

$$\sum_{i=1}^{N} \sum_{x \in A_i} d(x) = \sum_{x \in X} d(x)^2 = |X| \left(\sum_{x \in X} d(x)^2 \right) \left(\sum_{x \in X} \frac{1}{|X|^2} \right)$$

$$\geqslant |X| \left(\frac{\sum_{x \in X} d(x)}{|X|} \right)^2 = \frac{\left(\sum_i |A_i| \right)^2}{|X|} = \frac{(Nr)^2}{|X|}.$$

Here the first equality is (1.9) and the inequality follows from the Cauchy–Schwarz inequality: $\left(\sum_{i=1}^{n} a_i^2 \right) \left(\sum_{i=1}^{n} b_i^2 \right) \geqslant \left(\sum_{i=1}^{n} a_i b_i \right)^2$. Using the previous upper bound on $\sum_{x \in A_i} d(x)$, we obtain

$$(Nr)^2 \leqslant N \cdot |X| \left(r + (N-1)k \right),$$

which gives the desired lower bound on $|X|$. $\qquad\square$

Given a family of sets $A_1, \ldots, A_N$, their *average size* is

$$\frac{1}{N} \sum_{i=1}^{N} |A_i|.$$

The following lemma says that, if the average size of sets is large, then some two of them must share many elements.

Lemma 2.2. *Let X be a set of n elements, and let $A_1, \ldots, A_N$ be subsets of X of average size at least n/w. If $N \geqslant 2w^2$, then there exist $i \neq j$ such that*

$$|A_i \cap A_j| \geqslant \frac{n}{2w^2}. \tag{2.2}$$

Proof. Again, let us just count. By (1.7),

$$\sum_{x \in X} d(x) = \sum_{i=1}^{N} |A_i| \geqslant \frac{nN}{w}.$$

Hence, the sum $\sum_{x \in X} d(x)^2$ is minimized when $d(x) = N/w$ for all x, which means that

$$\sum_{x \in X} d(x)^2 \geqslant \frac{nN^2}{w^2}. \tag{2.3}$$

On the other hand, assuming that (2.2) is false and using (1.8) and (1.9) we would obtain

$$\sum_{x \in X} d(x)^2 = \sum_{i=1}^{N} \sum_{j=1}^{N} |A_i \cap A_j| = \sum_{i} |A_i| + \sum_{i \neq j} |A_i \cap A_j|$$

$$< nN + \frac{nN(N-1)}{2w^2} = \frac{nN^2}{2w^2}\left(1 + \frac{2w^2}{N} - \frac{1}{N}\right) \leqslant \frac{nN^2}{w^2},$$

a contradiction with (2.3). $\square$

Lemma 2.2 is a very special (but still illustrative) case of the following more general result.

Lemma 2.3 (Erdős 1964b). *Let X be a set of n elements $x_1, \ldots, x_n$, and let $A_1, \ldots, A_N$ be N subsets of X of average size at least n/w. If $N \geqslant 2kw^k$, then there exist $A_{i_1}, \ldots, A_{i_k}$ such that $|A_{i_1} \cap \cdots \cap A_{i_k}| \geqslant n/(2w^k)$.*

The proof is a generalization of the one above and we leave it as an exercise (see Exercises 2.8 and 2.9).

2.2 Zarankiewicz's problem

At most how many 1's can an $n \times n$ 0-1 matrix contain if it has no $a \times b$ submatrix whose entries are all 1's? Zarankiewicz (1951) raised the problem of the estimation of this number for $a = b = 3$ and $n = 4, 5, 6$ and the general problem became known as *Zarankiewicz's problem.*

It is worth reformulating this problem in terms of bipartite graphs. A bipartite graph with parts of size n is a triple $G = (V_1, V_2, E)$, where V_1 and V_2 are disjoint n-element sets of *vertices* (or *nodes*), and $E \subseteq V_1 \times V_2$ is the set of *edges*. We say that the graph contains an $a \times b$ *clique* if there exist an a-element subset $A \subseteq V_1$ and a b-element subset $B \subseteq V_2$ such that $A \times B \subseteq E$.

Let $k_a(n)$ be the minimal integer k such that *any* bipartite graph with parts of size n and more than k edges contains at least one $a \times a$ clique. Using the probabilistic argument, it can be shown (see Exercise 21.5) that

$$k_a(n) \geqslant c \cdot n^{2-2/a},$$

where $c > 0$ is a constant, depending only on a. It appears that this bound is not very far from the best possible, and this can be proved using the double counting argument. The result is essentially due to Kővári, Sós and Turán (1954).

Theorem 2.4. *For all natural numbers n and a we have*

$$k_a(n) \leqslant (a - 1)^{1/a} n^{2-1/a} + (a - 1)n.$$

Proof. Our goal is to prove the following: let $G = (V_1, V_2, E)$ be a bipartite graph with parts of size n, and suppose that G *does not* contain an $a \times a$ clique; then $|E| \leqslant (a - 1)^{1/a} n^{2-1/a} + (a - 1)n$.

By a *star* in the graph G we will mean a set of any of its a edges incident with one vertex $x \in V_1$, i.e., a set of the form

$$S(x, B) = \{(x, y) \in E : y \in B\},$$

where $B \subseteq V_2$, $|B| = a$. Let Δ be the total number of such stars in G. We may count the stars $S(x, B)$ in two ways, by fixing either the vertex x or the subset B.

For a fixed subset $B \subseteq V_2$, with $|B| = a$, we can have at most $a - 1$ stars of the form $S(x, B)$, because otherwise we would have an $a \times a$ clique in G. Thus,

$$\Delta \leqslant (a - 1) \cdot \binom{n}{a}. \tag{2.4}$$

On the other hand, for a fixed vertex $x \in V_1$, we can form $\binom{d(x)}{a}$ stars $S(x, B)$, where $d(x)$ is the degree of vertex x in G (i.e., the number of vertices adjacent to x). Therefore,

$$\Delta = \sum_{x \in V_1} \binom{d(x)}{a} \geqslant n \cdot \binom{(\sum d(x))/n}{a} = n \cdot \binom{|E|/n}{a} \tag{2.5}$$

where the inequality follows from Jensen's inequality by taking $f(x) = \binom{d(x)}{a}$ and $\lambda_x = 1/n$ for all $x \in U$. Comparing (2.4) with (2.5) we obtain

$$n \cdot \binom{|E|/n}{a} \leqslant (a-1) \cdot \binom{n}{a}.$$

Expressing the binomial coefficients as quotients of factorials, this inequality implies

$$n \left(|E|/n - (a-1)\right)^a \leqslant (a-1)(n - (a-1))^a,$$

and therefore $|E|/n \leqslant (a-1)^{1/a} n^{1-1/a} + a - 1$, from which the desired upper bound on $|E|$ follows.   □

The theorem above says that any bipartite graph with many edges has large cliques. In order to destroy such cliques we can try to remove some of their vertices. We would like to remove as few vertices as possible. Just how few says the following result.

Theorem 2.5 (Ossowski 1993). *Let $G = (V_1, V_2, E)$ be a bipartite graph with no isolated vertices, $|E| < (k+1)r$ edges and $d(y) \leqslant r$ for all $y \in V_2$. Then we can delete at most k vertices from V_1 so that the resulting graph has no $(r-a+1) \times a$ clique for $a = 1, 2, \ldots, r$.*

For a vertex x, let $N(x)$ denote the set of its neighbors in G, that is, the set of all vertices adjacent to x; hence, $|N(x)|$ is the degree $d(x)$ of x. We will use the following lemma relating the degree with the total number of vertices.

Lemma 2.6. *Let (X, Y, E) be a bipartite graph with no isolated vertices, and $f : Y \to [0, \infty)$ be a function. If the inequality $d(y) \leqslant d(x) \cdot f(y)$ holds for each edge $(x, y) \in E$, then $|X| \leqslant \sum_{y \in Y} f(y)$.*

Proof. By double counting,

$$|X| = \sum_{x \in X} \sum_{y \in N(x)} \frac{1}{d(x)} \leqslant \sum_{x \in X} \sum_{y \in N(x)} \frac{f(y)}{d(y)}$$

$$= \sum_{y \in Y} \sum_{x \in N(y)} \frac{f(y)}{d(y)} = \sum_{y \in Y} \frac{f(y)}{d(y)} \cdot |N(y)| = \sum_{y \in Y} f(y).$$

□

Proof of Theorem 2.5 (due to F. Galvin 1997). For a set of vertices $Y \subseteq V_2$, let $\Gamma(Y) = \bigcap_{y \in Y} N(y)$ denote the set of all its *common neighbors* in G, that is, the set of all those vertices in V_1 which are joined to each vertex of Y; hence $|\Gamma(Y)| \leqslant r$ for all $Y \subseteq V_2$. Let $X \subseteq V_1$ be a minimal set with the property that $|\Gamma(Y) - X| \leqslant r - |Y|$ whenever $Y \subseteq V_2$ and $1 \leqslant |Y| \leqslant r$. Put otherwise, X is a minimal set of vertices in V_1, the removal of which leads to a graph without $(r-a+1) \times a$ cliques, for all $a = 1, \ldots, r$.

Our goal is to show that $|X| \leqslant k$.

Note that, for each $x \in X$ we can choose $Y_x \subseteq V_2$ so that $1 \leqslant |Y_x| \leqslant r$, $x \in \Gamma(Y_x)$ and

$$|\Gamma(Y_x) - X| = r - |Y_x|;$$

otherwise X could be replaced by $X - \{x\}$, contradicting the minimality of X. We will apply Lemma 2.6 to the bipartite graph $G' = (X, V_2, F)$, where

$$F = \{(x, y) : y \in Y_x\}.$$

All we have to do is to show that the hypothesis of the lemma is satisfied by the function (here $N(y)$ is the set of neighbors of y in the original graph G):

$$f(y) = \frac{|N(y)|}{r},$$

because then

$$|X| \leqslant \sum_{y \in V_2} f(y) = \frac{1}{r} \sum_{y \in V_2} |N(y)| = \frac{|E|}{r} < k + 1.$$

Consider an edge $(x, y) \in F$; we have to show that $d(y) \leqslant d(x) \cdot f(y)$, where

$$d(x) = |Y_x| \quad \text{and} \quad d(y) = |\{x \in X : y \in Y_x\}|$$

are the degrees of x and y in the graph $G' = (X, V_2, F)$. Now, $y \in Y_x$ implies $\Gamma(Y_x) \subseteq N(y)$, which in its turn implies

$$|N(y) - X| \geqslant |\Gamma(Y_x) - X| = r - |Y_x|;$$

hence

$$
\begin{aligned}
d(y) &\leqslant |N(y) \cap X| = |N(y)| - |N(y) - X| \\
&\leqslant |N(y)| - r + |Y_x| = r \cdot f(y) - r + d(x),
\end{aligned}
$$

and so

$$
\begin{aligned}
d(x) \cdot f(y) - d(y) &\geqslant d(x) \cdot f(y) - r \cdot f(y) + r - d(x) \\
&= (r - d(x)) \cdot (1 - f(y)) \geqslant 0.
\end{aligned}
$$

$\square$

2.3 Density of 0-1 matrices

Let H be an $m \times n$ 0-1 matrix. We say that H is α-*dense* if at least an α-fraction of all its mn entries are 1's. Similarly, a row (or column) is α-*dense* if at least an α-fraction of all its entries are 1's.

The next result says that any dense 0-1 matrix must either have one "very dense" row or there must be many rows which are still "dense enough."

Lemma 2.7 (Grigni–Sipser 1995). *If H is 2α-dense then either*
(a) *there exists a row which is $\sqrt{\alpha}$-dense, or*
(b) *at least $\sqrt{\alpha} \cdot m$ of the rows are α-dense.*

Note that $\sqrt{\alpha}$ is larger than α when $\alpha < 1$.

Proof. Suppose that the two cases do not hold. We calculate the density of the entire matrix. Since (b) does not hold, less than $\sqrt{\alpha} \cdot m$ of the rows are α-dense. Since (a) does not hold, each of these rows has less than $\sqrt{\alpha} \cdot n$ 1's; hence, the fraction of 1's in α-dense rows is strictly less than $(\sqrt{\alpha})(\sqrt{\alpha}) = \alpha$. We have at most m rows which are not α-dense, and each of them has less than αn 1's. Hence, the fraction of 1's in these rows is also less than α. Thus, the total fraction of 1's in the matrix is less than 2α, a contradiction with the 2α-density of H. □

Now consider a slightly different question: if H is α-dense, how many of its rows *or* columns are "dense enough"? The answer is given by the following general estimate due to Johan Håstad. This result appeared in the paper of Karchmer and Wigderson (1990) and was used to prove that the graph connectivity problem cannot be solved by monotone circuits of logarithmic depth.

For a subset H of a universe Ω, its *density* is the fraction $\mu(H) \coloneqq \frac{|H|}{|\Omega|}$.

Suppose that our universe is a Cartesian product $\Omega = A_1 \times \cdots \times A_k$ of some finite sets $A_1, \ldots, A_k$. Hence, elements of Ω are strings $a = (a_1, \ldots, a_k)$ with $a_i \in A_i$. Given a set $H \subseteq \Omega$ of such strings and a point $b \in A_i$, we denote by $H_{i \to b}$ the set of all strings $(a_1, \ldots, a_{i-1}, a_{i+1}, \ldots, a_k)$ for which the extended string $(a_1, \ldots, a_{i-1}, b, a_{i+1}, \ldots, a_k)$ belongs to H.

We stress that the density of a set depends on the size of the underlying universe Ω. Since the universe for $H_{i \to b}$ is $|A_i|$ times smaller than that for H, we have $\mu(H_{i \to b}) = |A_i| \cdot \mu\{a \in H : a_i = b\}$.

Lemma 2.8 (J. Håstad). *Let $B = B_1 \times \cdots \times B_k$, where*

$$B_i \coloneqq \left\{ b \in A_i : \mu(H_{i \to b}) \geq \frac{\mu(H)}{2k} \right\}.$$

Then $\mu(B) \geq \frac{1}{2}\mu(H)$.

Proof. We have $\mu(B) \geq \mu(H \cap B) = \mu(H) - \mu(H - B)$, where

$$\mu(H - B) \leq \sum_{i=1}^{k} \sum_{b \notin B_i} \mu\{a \in H : a_i = b\} = \sum_{i=1}^{k} \sum_{b \notin B_i} \frac{\mu(H_{i \to b})}{|A_i|}$$

$$< \sum_{i=1}^{k} \sum_{b \notin B_i} \frac{\mu(H)}{2k \cdot |A_i|} \leq \sum_{i=1}^{k} \frac{\mu(H)}{2k} = \frac{\mu(H)}{2}.$$

□

Corollary 2.9. *In any 2α-dense matrix either a $\sqrt{\alpha}$-fraction of its rows or a $\sqrt{\alpha}$-fraction of its columns (or both) are $(\alpha/2)$-dense.*

Proof. The case $k = 2$ of Lemma 2.8 corresponds to 0-1 matrices, and in this case the lemma says that $\mu(B_i) \geqslant \left(\frac{\mu(H)}{2}\right)^{1/2}$ for some $i \in \{1, 2\}$.   □

Exercises

2.1. Let $A_1, \ldots, A_m$ be subsets of an n-element set such that $|A_i \cap A_j| \leqslant t$ for all $i \neq j$. Prove that $\sum_{i=1}^{m} |A_i| \leqslant n + t \cdot \binom{m}{2}$.

2.2.[(!)] Let $A = (a_{ij})$ be an $n \times n$ matrix ($n \geqslant 4$). The matrix is filled with integers and each integer appears exactly twice. Show that there exists a permutation π of $\{1, \ldots, n\}$ such that all the numbers $a_{i,\pi(i)}$, $i = 1, \ldots, n$ are distinct. (Such a permutation π is also called a *Latin transversal* of A.)

> *Hint*: Look at how many pairs of entries are "bad," i.e., contain the same number, and show that strictly less than $n!$ of all permutations can go through such pairs.

2.3.[−] Let $\mathcal{F}$ be a family of m subsets of a finite set X. For $x \in X$, let $p(x)$ be the number of pairs (A, B) of sets $A \neq B \in \mathcal{F}$ such that either $x \in A \cap B$ or $x \notin A \cup B$. Prove that $p(x) \geqslant m^2/2$ for every $x \in X$.

> *Hint*: Let $d(x)$ be the degree of x in $\mathcal{F}$, and observe that $p(x) = d(x)^2 + (m - d(x))^2$.

2.4.[+] Let $\mathcal{F}$ be a family of nonempty subsets of a finite set X that is closed under union (i.e., $A, B \in \mathcal{F}$ implies $A \cup B \in \mathcal{F}$). Prove or give a counterexample: there exists $x \in X$ such that $d(x) \geqslant |\mathcal{F}|/2$. (Open conjecture, due to Peter Frankl.)

2.5.[−] A *projective plane* of order $r - 1$ is a family of $n = r^2 - r + 1$ r-element subsets (called *lines*) of an n-element set of points such that each two lines intersect in precisely one point and each point belongs to precisely r lines (cf. Sect. 13.4). Use this family to show that the bound given by Corrádi's lemma (Lemma 2.1) is optimal.

2.6. Theorem 2.4 gives a sufficient condition for a bipartite graph with parts of the same size n to contain an $a \times a$ clique. Extend this result to not necessarily balanced graphs. Let $k_{a,b}(m, n)$ be the minimal integer k such that any bipartite graph with parts of size m and n and more than k edges contains at least one $a \times b$ clique. Prove that for any $0 \leqslant a \leqslant m$ and $0 \leqslant b \leqslant n$,

$$k_{a,b}(m, n) \leqslant (a - 1)^{1/b}(n - b + 1)m^{1 - 1/b} + (b - 1)m.$$

2.7 (Paturi–Zane 1998). Extend Theorem 2.4 to r-partite graphs as follows. An r-partite m-clique is a Cartesian product $V_1 \times V_2 \times \cdots \times V_r$ of m-element sets $V_1, \ldots, V_r$. An r-partite graph with parts of size m is a subset E of an r-partite m-clique. Let $\mathrm{ex}(m, r, 2)$ denote the maximum size $|E|$ of such a graph E which does not contain an r-partite 2-clique. Erdős (1959, 1964b) proved that

$$cm^{r-r/2^{r-1}} \leqslant \mathrm{ex}(m, r, 2) \leqslant m^{r-1/2^{r-1}},$$

where $c = c(r) > 0$ is a constant depending only on r. A slightly weaker upper bound $\mathrm{ex}(m, r, 2) < 2m^{r-1/2^{r-1}}$ can be derived from Lemma 2.2. Show how to do this.

Hint: Argue by induction on r. For the induction step take $X = V_1 \times \cdots \times V_{r-1}$ and consider m subsets $A_v = \{x \in X : (x, v) \in E\}$ with $v \in V_k$. Apply Lemma 2.2 with $n = m^{r-1}$, $N = m$ and $w = \frac{1}{2}m^{1/2^{r-1}}$, to obtain a pair of points $u \neq v \in V_k$ for which the graph $E' = A_u \cap A_v$ is large enough, and use the induction hypothesis.

2.8.⁻ Let $\mathcal{F} = \{A_1, \ldots, A_N\}$ be a family of subsets of some set X. Use (1.8) to prove that for every $1 \leqslant s \leqslant N$,

$$\sum_{x \in X} d(x)^s = \sum_{(i_1, i_2, \ldots, i_s)} |A_{i_1} \cap A_{i_2} \cap \cdots \cap A_{i_s}|,$$

where the last sum is over all s-tuples $(i_1, i_2, \ldots, i_k)$ of (not necessarily distinct) indices.

2.9. Use the previous exercise and the argument of Lemma 2.2 to prove Lemma 2.3.

2.10.⁺ Let $A_1, \ldots, A_N$ be subsets of some n-element set X, and suppose that these sets have average size at least αn. Show that for every every $s \leqslant (1 - \epsilon)\alpha N$ with $0 < \epsilon < 1$, there are indices $i_1, i_2, \ldots, i_s$ such that

$$|A_{i_1} \cap A_{i_2} \cap \cdots \cap A_{i_s}| \geqslant (\epsilon\alpha)^s n.$$

Hint: Consider the bipartite graph $G = (X, V, E)$ where $V = \{1, \ldots, N\}$, and $(x, i) \in E$ if and only if $x \in A_i$. Observe that $|E| \geqslant \alpha n N$ and argue as in the proof of Theorem 2.4.

2.11.⁽¹⁾ Prove the following very useful averaging principle for partitions. Let $X = A_1 \cup A_2 \cup \cdots \cup A_m$ be a partition of a finite set X into m mutually disjoint sets (blocks), and $a = \sum_{i=1}^{m} |A_i|/m$ be the average size of a block in this partition. Show that for every $1 \leqslant b \leqslant a$, at least $(1 - 1/b)|X|$ elements of X belong to blocks of size at least a/b. How many elements of X belong to blocks of size at most ab? *Hint*: $m \cdot (a/b) \leqslant |X|/b$.

2.12.[+] Let $A_1, \ldots, A_r$ be a sequence of (not necessarily distinct) subsets of an n-element set X such that each set has size n/s and each element $x \in X$ belongs to least one and to at most k of them; hence $r \leqslant ks$. Let $K = \sum_{i=0}^{k} \binom{r}{i}$ and assume that $s > 2k$. Prove that there exist two disjoint subsets X_1 and X_2 of X such that $|X_i| \geqslant n/(2K)$ for both $i = 1, 2$, and none of the sets $A_1, \ldots, A_r$ contains points from both sets X_1 and X_2.

Hint: Associate with each $x \in X$ its *trace* $T(x) = \{i : x \in A_i\}$ and partite the elements of X according to their traces. Use the previous exercise to show that at least $n/2$ elements belong to blocks of size at least $n/(2K)$. Show that some two of these elements x and y must have *disjoint* traces, $T(x) \cap T(y) = \emptyset$.

2.13. Let $C_1, \ldots, C_r$ be sets of 0-1 strings, each of which forms a prefix-free code (we can think of each C_i as the set of all paths from a root to a leaf in a particular binary tree). Define the *average length* of strings in these sets by $\bar{l} = \left(\sum_{i=1}^{r} \ell(C_i) \right)/N$, where $N = \sum_{i=1}^{r} |C_i|$ and $\ell(C_i)$ is the total length of strings in C_i (i.e., the sum of the lengths of all strings in C_i). Use Jensen's inequality to prove that the average length of strings is at least the logarithm of the average size of sets C_i: $\bar{l} \geqslant \log_2(N/r)$.

Hint: Apply Jensen's inequality with $\lambda_i = 1/r$, $x_i = |C_i|$ and $f(x) = x \log_2 x$.

2.14.[−] Let $X = A_1 \cup A_2 \cup \cdots \cup A_m$ be a partition of a finite set X into mutually disjoint blocks. Given a subset $Y \subseteq X$, we obtain its partition $Y = B_1 \cup B_2 \cup \cdots \cup B_m$ into blocks $B_i = A_i \cap Y$. Say that a block B_i is λ-*large* if $|B_i|/|A_i| \geqslant \lambda \cdot |Y|/|X|$. Show that, for every $\lambda > 0$, at least $(1-\lambda) \cdot |Y|$ elements of Y belong to λ-large blocks.

2.15. Given a family $S_1, \ldots, S_n$ of subsets of $V = \{1, \ldots, n\}$, its *intersection graph* $G = (V, E)$ is defined by: $\{i, j\} \in E$ if and only if $S_i \cap S_j \neq \emptyset$. Suppose that: (i) the sets have average size at least r, and (ii) the average size of their pairwise intersections does not exceed k. Show that $|E| \geqslant \frac{n}{k} \cdot \binom{r}{2}$.

Hint: Consider the sum $\sum_{i<j} |S_i \cap S_j|$.

2.16. Let H be a 2α-dense 0-1 matrix. Prove that at least an $\alpha/(1-\alpha)$ fraction of its rows must be α-dense.

2.17 (Alon 1986). Let S be a set of strings of length n over some alphabet. Suppose that every two strings of S differ in at least d coordinates. Let k be such that $d > n(1 - 1/\binom{k}{2})$. Show that any k distinct strings $v_1, \ldots, v_k$ of S attain k distinct values in at least one coordinate.

Hint: Assume the opposite and count the sum of distances between the $\binom{k}{2}$ pairs of v_i's.

3. The Principle of Inclusion and Exclusion

The *principle of inclusion and exclusion* (sieve of Eratosthenes) is a powerful tool in the theory of enumeration as well as in number theory. This principle relates the cardinality of the union of certain sets to the cardinalities of intersections of some of them, these latter cardinalities often being easier to handle.

3.1 The principle

For any two sets A and B we have

$$|A \cup B| = |A| + |B| - |A \cap B|.$$

In general, given n subsets $A_1, \ldots, A_n$ of a set X, we want to calculate the number $|A_1 \cup \cdots \cup A_n|$ of points in their union. As the first approximation of this number we can take the sum

$$|A_1| + \cdots + |A_n|. \tag{3.1}$$

However, in general, this number is too large since if, say, $A_i \cap A_j \neq \emptyset$ then each point of $A_i \cap A_j$ is counted two times in (3.1): once in $|A_i|$ and once in $|A_j|$. We can try to correct the situation by subtracting from (3.1) the sum

$$\sum_{1 \leqslant i < j \leqslant n} |A_i \cap A_j|. \tag{3.2}$$

But then we get a number which is too small since each of the points in $A_i \cap A_j \cap A_l \neq \emptyset$ is counted three times in (3.2): once in $|A_i \cap A_j|$, once in $|A_j \cap A_k|$, and once in $|A_i \cap A_k|$. We can therefore try to correct the situation by adding the sum

$$\sum_{1 \leqslant i < j < k \leqslant n} |A_i \cap A_j \cap A_k|,$$

but again we will get a too large number, etc. Nevertheless, it appears that after n steps we will get the correct result. This result is known as the *inclusion-exclusion principle*. The following notation will be handy: if I is a subset of the index set $\{1, \ldots, n\}$, we set

$$A_I = \bigcap_{i \in I} A_i,$$

with the convention that $A_\emptyset = X$.

Proposition 3.1 (Inclusion-Exclusion Principle). *Let $A_1, \ldots, A_n$ be subsets of X. Then the number of elements of X which lie in none of the subsets A_i is*

$$\sum_{I \subseteq \{1,\ldots,n\}} (-1)^{|I|} |A_I|. \tag{3.3}$$

Proof. The sum is a linear combination of cardinalities of sets A_I with coefficients $+1$ and -1. We can re-write this sum as

$$\sum_I (-1)^{|I|} |A_I| = \sum_I \sum_{x \in A_I} (-1)^{|I|} = \sum_x \sum_{I : x \in A_I} (-1)^{|I|}.$$

We calculate, for each point of X, its contribution to the sum, that is, the sum of the coefficients of the sets A_I which contain it.

First suppose that $x \in X$ lies in none of the sets A_i. Then the only term in the sum to which x contributes is that with $I = \emptyset$; and this contribution is 1.

Otherwise, the set $J = \{i : x \in A_i\}$ is non-empty; and $x \in A_I$ precisely when $I \subseteq J$. Thus, the contribution of x is

$$\sum_{I \subseteq J} (-1)^{|I|} = \sum_{i=0}^{|J|} \binom{|J|}{i} (-1)^i = (1-1)^{|J|} = 0$$

by the binomial theorem.

Thus, points lying in no set A_i contribute 1 to the sum, while points in some A_i contribute 0; so the overall sum is the number of points lying in none of the sets, as claimed. □

For some applications the following form of the inclusion-exclusion principle is more convenient.

Proposition 3.2. *Let $A_1, \ldots, A_n$ be a sequence of (not necessarily distinct) sets. Then*

$$|A_1 \cup \cdots \cup A_n| = \sum_{\emptyset \neq I \subseteq \{1,\ldots,n\}} (-1)^{|I|+1} |A_I|. \tag{3.4}$$

Proof. The left-hand of (3.4) is $|A_\emptyset|$ minus the number of elements of $X = A_\emptyset$ which lie in none of the subsets A_i. By Proposition 3.1 this number is

$$|A_\emptyset| - \sum_{I \subseteq \{1,\ldots,n\}} (-1)^{|I|} |A_I| = \sum_{\emptyset \neq I \subseteq \{1,\ldots,n\}} (-1)^{|I|+1} |A_I|,$$

as desired. □

Suppose we would like to know, given a set of indices I, how many elements belong to all the sets A_i with $i \in I$ and do not belong to any of the remaining sets. Proposition 3.1 (which corresponds to the case when $I = \emptyset$) can be generalized for this situation.

Proposition 3.3. *Let $A_1, \ldots, A_n$ be sets, and I a subset of the index set $\{1, \ldots, n\}$. Then the number of elements which belong to A_i for all $i \in I$ and for no other values is*

$$\sum_{J \supseteq I} (-1)^{|J-I|} |A_J|. \tag{3.5}$$

Proof. Consider the set $X = \bigcap_{i \in I} A_i$ and its subsets $B_k = X \cap A_k$, for all $k \in N - I$, where $N = \{1, \ldots, n\}$. The proposition asks us to calculate the number of elements of X lying in none of B_k. By Proposition 3.1, this number is

$$\sum_{K \subseteq N-I} (-1)^{|K|} \left| \bigcap_{k \in K} B_k \right| = \sum_{K \subseteq N-I} (-1)^{|K|} \left| \bigcap_{i \in K \cup I} A_i \right|$$

$$= \sum_{J \supseteq I} (-1)^{|J-I|} |A_J|.$$

□

Next, we turn this result into a more abstract form, referring to arbitrary set functions rather than cardinalities of sets.

Proposition 3.4. *Let f and g be two functions defined on subsets of a finite set such that $f(A) = \sum_{B \subseteq A} g(B)$. Then $g(A) = \sum_{B \subseteq A} (-1)^{|A-B|} f(B)$.*

Proof.

$$\sum_{B \subseteq A} (-1)^{|A-B|} f(B) = \sum_{B \subseteq A} \sum_{C \subseteq B} (-1)^{|A-B|} g(C)$$

$$= \sum_{C \subseteq A} g(C) \sum_{C \subseteq B \subseteq A} (-1)^{|A-B|} = g(A),$$

because, for fixed sets $C \subseteq A$, the sum $\sum_{C \subseteq B \subseteq A} (-1)^{|A-B|}$ equals 1 if $C = A$, and is 0 otherwise (Exercise 3.5).   □

3.2 The number of derangements

What is the probability that if n people randomly search a dark closet to retrieve their hats, no person will pick his own hat? Using the principle of inclusion and exclusion it can be shown that this probability is very close to $e^{-1} = 0.3678\ldots$

This question can be formalized as follows. A *permutation* is a bijective mapping f of the set $\{1,\ldots,n\}$ into itself. We say that f *fixes* a point i if $f(i) = i$. A *derangement* is a permutation which fixes none of the points. We have exactly $n!$ permutations. How many of them are derangements?

Proposition 3.5. *The number of derangements of $\{1,\ldots,n\}$ is equal to*

$$\sum_{i=0}^{n}(-1)^i\binom{n}{i}(n-i)! = n!\sum_{i=0}^{n}\frac{(-1)^i}{i!}. \tag{3.6}$$

The sum $\sum_{i=0}^{n}\frac{(-1)^i}{i!}$ is the initial part of the Taylor expansion of e^{-1}; so about an e^{-1} fraction of all permutations are derangements.

Proof. We are going to apply the inclusion-exclusion formula (3.3). Let X be the set of all permutations, and A_i the set of permutations fixing the point i; so $|A_i| = (n-1)!$, and more generally, $|A_I| = (n-|I|)!$, since permutations in A_I fix every point in I and permute the remaining points arbitrarily. A permutation is a derangement if and only if it lies in none of the sets A_i; so by (3.3), the number of derangements is

$$\sum_{I\subseteq\{1,\ldots,n\}}(-1)^{|I|}(n-|I|)! = \sum_{i=0}^{n}(-1)^i\binom{n}{i}(n-i)!$$

of putting $i = |I|$.   □

Exercises

3.1. Use the principle of inclusion and exclusion to determine the number of ways in which three women and their three spouses may be seated around a round table, so that:

(i) no woman sits beside her spouse (on either side);
(ii) no two women may sit opposite one another at the table (i.e., with two people between them on either side).

Hint: To (i): two seatings are equivalent if one can be rotated into the other; so the underlying set consists of all circular permutations, 5! in number. Let A_i ($i = 1,2,3$) be the subset of permutations in which the members of the ith couple sit side by side. Show that $|A_i| = 2\cdot4!$, $|A_i \cap A_j| = 2^2\cdot s!$, $|A_1 \cap A_2 \cap A_3| = 2^3\cdot2!$ and apply the inclusion-exclusion formula. To (ii): distinguish two cases, according to whether there exist two women sitting side by side or not.

3.2. Let $m \geqslant n$. A function $f\colon [m] \to [n]$ is a *surjection* (or a mapping of $[m]$ *onto* $[n]$) if f maps at least one element of $[m]$ to each element of $[n]$. Prove that the number of such functions is $\sum_{k=0}^{n-1}(-1)^k\binom{n}{k}(n-k)^m$.

Hint: Let $A_i = \{f : f(j) \neq i \text{ for all } j\}$ and apply the inclusion-exclusion formula.

3.3.+ Let n and $k \geqslant l$ be positive integers. How many different integer solutions are there to the equation $x_1 + x_2 + \cdots + x_n = k$, with all $0 \leqslant x_i < l$?

Hint: Consider the universum $X = X_{n,k}$ of all solutions with all $x_i \geqslant 0$, let A_i be the set of all solutions with $x_i \geqslant l$, and apply the inclusion-exclusion formula (3.3). Observe that $|A_i| = |X_{n,k-l}|$, where the size of $X_{n,k}$ is given by Proposition 1.4.

3.4.⁻ Let $r \geqslant 5$. How many ways are there to color the vertices with r colors in the following graphs such that adjacent vertices get different colors?

Hint: For the first graph, the universe X is the set of all r^4 ways to color the vertices. Associate with each edge e the set A_e of all colorings, which assign the same color to its ends, and apply the inclusion-exclusion formula (3.3).

3.5. Prove that for any two sets $I \subseteq J$,

$$\sum_{I \subseteq K \subseteq J} (-1)^{|K-I|} = \begin{cases} 1, & \text{if } I = J \\ 0, & \text{if } I \neq J. \end{cases}$$

3.6. Let $\mathcal{F}$ be a family of subsets of some set X, and $f: \mathcal{F} \to \mathbb{R}$ be a mapping which associates with each set $A \in \mathcal{F}$ a real number $f(A) \in \mathbb{R}$. Define

$$\delta(S) = \sum_{A \in \mathcal{F}, A \supseteq S} f(A).$$

Prove that for every subset Y of X and for any $B \subseteq Y$,

$$\sum_{B \subseteq S \subseteq Y} (-1)^{|S-B|} \delta(S) = \sum_{A \in \mathcal{F}, A \cap Y = B} f(A).$$

Hint: Argue as in the proof of Proposition 3.4.

3.7.+ Prove the following *Bonferroni inequalities* for each even $k \geqslant 2$:

$$\sum_{\nu=1}^{k} (-1)^{\nu+1} \sum_{|I|=\nu} |A_I| \leqslant \left| \bigcup_{i=1}^{n} A_i \right| \leqslant \sum_{\nu=1}^{k+1} (-1)^{\nu+1} \sum_{|I|=\nu} |A_I|$$

where $A_I = \bigcap_{i \in I} A_i$. What about an odd k?

3.8.⁽ˡ⁾ The *determinant* $\det(A)$ of an $n \times n$ matrix $A = (a_{ij})$ is a sum of $n!$ signed products $\pm a_{1i_1} a_{2i_2} \cdots a_{ni_n}$, where $(i_1, i_2, \ldots, i_n)$ is a permutation of $(1, 2, \ldots, n)$, the sign being $+1$ or -1, according to whether the number of *inversions* of $(i_1, i_2, \ldots, i_n)$ is even or odd. An inversion occurs when $i_r > i_s$ but $r < s$. Prove the following: let A be a matrix of even order n with 0's on the diagonal and arbitrary entries from $\{+1, -1\}$ elsewhere. Then $\det(A) \neq 0$.

Hint: Observe that for such matrices, $\det(A)$ is congruent modulo 2 to the number of derangements on n points, and show that for even n, the sum (3.6) is odd.

4. The Pigeonhole Principle

The *pigeonhole principle* (also known as *Dirichlet's principle*) states the "obvious" fact that $n + 1$ pigeons cannot sit in n holes so that every pigeon is alone in its hole. More generally, the pigeonhole principle states the following:

> *If a set consisting of more than kn objects is partitioned into n classes, then some class receives more than k objects.*

Its truth is easy to verify: if every class receives at most k objects, then a total of at most kn objects have been distributed.

This is one of the oldest "non-constructive" principles: it states only the *existence* of a pigeonhole with more than k items and says nothing about how to *find* such a pigeonhole. Today we have powerful and far reaching generalizations of this principle (Ramsey-like theorems, the probabilistic method, etc.). We will talk about them later.

As trivial as the pigeonhole principle itself may sound, it has numerous nontrivial applications. The hard part in applying this principle is deciding what are the pigeons and what are the pigeonholes. Let us illustrate this by several examples.

4.1 Some quickies

To "warm-up," let us start with the simplest applications. The *degree* of a vertex x in a graph G is the number $d(x)$ of edges of G adjacent to x.

Proposition 4.1. *In any graph there exist two vertices of the same degree.*

Proof. Given a graph G on n vertices, make n pigeonholes labeled from 0 up to $n - 1$ and put a vertex x into the kth pigeonhole iff $d(x) = k$. If some pigeonhole contains more than one vertex, we are done. So, assume that no pigeonhole has more than one vertex. There are n vertices going into the n pigeonholes; hence each pigeonhole has *exactly one* vertex. Let x and y be the vertices lying in the pigeonholes labeled 0 and $n - 1$, respectively. The vertex x has degree 0 and so has no connection with other vertices, including y. But y has degree $n - 1$ and hence, is connected with all the remaining vertices, including x, a contradiction. $\square$

If G is a finite graph, the *independence number* $\alpha(G)$ is the maximum number of pairwise nonadjacent vertices of G. The *chromatic number* $\chi(G)$ of G is the minimum number of colors in a coloring of the vertices of G with the property that no two adjacent vertices have the same color.

Proposition 4.2. *In any graph G with n vertices, $n \leqslant \alpha(G) \cdot \chi(G)$.*

Proof. Consider the vertices of G partitioned into $\chi(G)$ color classes (sets of vertices with the same color). By the pigeonhole principle, one of the classes must contain at least $n/\chi(G)$ vertices, and these vertices are pairwise nonadjacent. Thus $\alpha(G) \geqslant n/\chi(G)$, as desired.   □

A graph is *connected* if there is a path between any two of its vertices.

Proposition 4.3. *Let G be an n-vertex graph. If every vertex has a degree of at least $(n-1)/2$ then G is connected.*

Proof. Take any two vertices x and y. If these vertices are not adjacent, then at least $n-1$ edges join them to the remaining vertices, because both x and y have a degree of at least $(n-1)/2$.

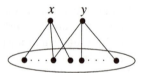

Fig. 4.1. There are only $n-2$ vertices and at least $n-1$ edges going to them.

Since there are only $n-2$ other vertices, the pigeonhole principle implies that one of them must be adjacent to both x and y (see Fig. 4.1). We have proved that every pair of vertices is adjacent or has a common neighbor, so G is connected.   □

Remark. A result is *best possible* if the conclusion no longer holds when we weaken one of the conditions. Such is, for example, the result above: let n be even and G be a union of two vertex disjoint complete graphs on $n/2$ vertices; then every vertex has degree $(n-2)/2$, but the graph is disconnected.

4.2 The Erdős–Szekeres theorem

Let $A = (a_1, a_2, \ldots, a_n)$ be a sequence of n different numbers. A *subsequence of k terms* of A is a sequence B of k distinct terms of A appearing in the same order in which they appear in A. In symbols, we have $B = (a_{i_1}, a_{i_2}, \ldots, a_{i_k})$, where $i_1 < i_2 < \cdots < i_k$. A subsequence B is said to be *increasing* if $a_{i_1} < a_{i_2} < \cdots < a_{i_k}$, and *decreasing* if $a_{i_1} > a_{i_2} > \cdots > a_{i_k}$.

We will be interested in the length of the *longest* increasing and decreasing subsequences of A. It is intuitively plausible that there should be some kind of tradeoff between these lengths. If the longest increasing subsequence is short, say has length s, then *any* subsequence of A of length $s + 1$ must contain a pair of decreasing elements, so there are lots of pairs of decreasing elements. Hence, we would expect the longest decreasing sequence to be large. An extreme case occurs when $s = 1$. Then the whole sequence A is decreasing.

How can we quantify the feeling that the length of *both*, longest increasing and longest decreasing subsequences, cannot be small? A famous result of Erdős and Szekeres (1935) gives an answer to this question and was one of the first results in extremal combinatorics.

Theorem 4.4 (Erdős–Szekeres 1935). *Let $A = (a_1, \ldots, a_n)$ be a sequence of n different real numbers. If $n \geqslant sr + 1$ then either A has an increasing subsequence of $s + 1$ terms or a decreasing subsequence of $r + 1$ terms (or both).*

Proof (due to Seidenberg 1959). Associate to each term a_i of A a pair of "scores" (x_i, y_i) where x_i is the number of terms in the longest *increasing* subsequence *ending* at a_i, and y_i is the number of terms in the longest *decreasing* subsequence *starting* at a_i. Observe that no two terms have the same score, i.e., that $(x_i, y_i) \neq (x_j, y_j)$ whenever $i \neq j$. Indeed, if we have $\cdots a_i \cdots a_j \cdots$, then either $a_i < a_j$ and the longest increasing subsequence ending at a_i can be extended by adding on a_j (so that $x_i < x_j$), or $a_i > a_j$ and the longest decreasing subsequence starting at a_j can be preceded by a_i (so that $y_i > y_j$).

Now make a grid of n^2 pigeonholes:

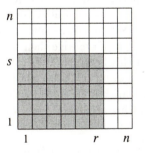

Place each term a_i in the pigeonhole with coordinates (x_i, y_i). Each term of A can be placed in some pigeonhole, since $1 \leqslant x_i, y_i \leqslant n$ for all $i = 1, \ldots, n$. Moreover, no pigeonhole can have more than one term because $(x_i, y_i) \neq (x_j, y_j)$ whenever $i \neq j$. Since $|A| = n \geqslant sr + 1$, we have more items than the pigeonholes shaded in the above picture. So some term a_i will lie outside this shaded region. But this means that either $x_i \geqslant s + 1$ or $y_i \geqslant r + 1$ (or both), exactly what we need.   □

The set of real numbers is *totally ordered*. That is, for any two distinct numbers x and y, either $x < y$ or $y < x$. The following lemma, due to Dilworth, generalizes the Erdős–Szekeres theorem to sets in which two elements may or may not be comparable.

A (weak) *partial order* on a set P is a binary relation $<$ between its elements. We say that elements x and y are *comparable* if either $x < y$ or $y < x$ (or both) hold. A *chain* is a set $Y \subseteq P$ in which every two elements are comparable. If no two distinct elements of Y are comparable, then Y is an *antichain* .

Lemma 4.5 (Dilworth 1950). *In any partial order on a set P of $n \geqslant sr + 1$ elements, there exists a chain of length $s + 1$ or an antichain of size $r + 1$.*

Proof. Suppose there is no chain of length $s+1$. Then we may define a function $f : P \to \{1, \ldots, s\}$ where $f(x)$ equals the maximal number of elements in a chain with greatest element x. By the pigeonhole principle, some $r + 1$ elements of P have the same image under f. By the definition of f, these elements are incomparable; that is, they form an antichain of size $r + 1$. $\square$

This lemma implies the Erdős–Szekeres theorem (we address this question in Exercise 4.9).

4.3 Mantel's theorem

Here we discuss one typical *extremal* property of graphs. How many edges are possible in a triangle-free graph G on $2n$ vertices? A triangle is a set $\{x, y, z\}$ of three vertices, each two of which are connected by an edge. Certainly, G can have n^2 edges without containing a triangle: just let G be the bipartite complete graph consisting of two sets of n vertices each and all the edges between the two sets. Indeed, n^2 turns out to be the maximum possible number of edges: if we take one more edge then the graph will have a triangle.

We give four proofs of this beautiful result: the first uses the pigeonhole principle, the second (original proof) is based on double counting, the third uses the inequality of the arithmetic and geometric mean, and the fourth employs the so-called "shifting argument" (we will give this last proof in the Sect. 4.7 devoted to this argument).

Theorem 4.6 (Mantel 1907). *If a graph G on $2n$ vertices contains $n^2 + 1$ edges, then G contains a triangle.*

First proof. We argue by induction on n. If $n = 1$, then G cannot have $n^2 + 1$ edges; hence the statement is true. Assuming the result for n, we now consider a graph G on $2(n+1)$ vertices with $(n+1)^2 + 1$ edges. Let x and y be adjacent vertices in G, and let H be the induced subgraph on the remaining $2n$ vertices. If H contains at least $n^2 + 1$ edges then we are done by the induction hypothesis. Suppose that H has at most n^2 edges, and therefore at least $2n + 1$ edges of G emanate from x and y to vertices in H:

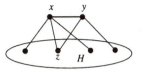

By the pigeonhole principle, among these $2n + 1$ edges there must be an edge from x and an edge from y to the same vertex z in H. Hence G contains the triangle $\{x, y, z\}$.   □

Second proof. Let G be a graph on a set V of $2n$ vertices containing $m \geqslant n^2 + 1$ edges. Assume that G has no triangles. Then adjacent vertices have no common neighbors, so $d(x) + d(y) \leqslant 2n$ for each edge $\{x, y\} \in E$. Summing over all edges of G, we have (cf. equation (1.9))

$$\sum_{x \in V} d(x)^2 = \sum_{\{x,y\} \in E} (d(x) + d(y)) \leqslant 2mn.$$

On the other hand, using Cauchy–Schwarz inequality (see Notation or Proposition 14.4) and Euler's equality $\sum_{x \in V} d(x) = 2m$ (see Theorem 1.7), we obtain

$$\sum_{x \in V} d(x)^2 \geqslant \frac{\left(\sum_{x \in V} d(x)\right)^2}{|V|} = \frac{2m^2}{n}.$$

These two inequalities imply that $m \leqslant n^2$, contradicting the hypothesis.   □

Third proof. Let $G = (V, E)$ be a graph on a set V of $2n$ vertices and assume that G has no triangles. Let $A \subseteq V$ be the largest independent set, i.e., a maximal set of vertices, no two of which are adjacent in G. Since G is triangle-free, the neighbors of a vertex $x \in V$ form an independent set, and we infer $d(x) \leqslant |A|$ for all x.

The set $B = V - A$ meets every edge of G. Counting the edges of G according to their end-vertices in B, we obtain $|E| \leqslant \sum_{x \in B} d(x)$. The inequality of the arithmetic and geometric mean (Proposition 1.11) yields

$$|E| \leqslant \sum_{x \in B} d(x) \leqslant |A| \cdot |B| \leqslant \left(\frac{|A| + |B|}{2}\right)^2 = n^2.$$

     □

4.4 Turán's theorem

A *k-clique* is a graph on k vertices, every two of which are connected by an edge. For example, triangles are 3-cliques. Mantel's theorem says that, if a graph on n vertices has no 3-clique then it has at most $n^2/4$ edges. What about $k > 3$?

The answer is given by a fundamental result of Paul Turán, which initiated extremal graph theory.

Theorem 4.7 (Turán 1941). *If a graph $G = (V, E)$ on n vertices has no $(k + 1)$-clique, $k \geqslant 2$, then*

$$|E| \leqslant \left(1 - \frac{1}{k}\right) \frac{n^2}{2}. \tag{4.1}$$

Like Mantel's theorem, this result was rediscovered many times with various different proofs. Here we present the original one due to Turán. The proof based on so-called "weight shifting" argument is addressed in Exercise 4.8. In Sect. 20.4 we will give a proof which employs ideas of a totally different nature – the probabilistic argument.

Proof. We use induction on n. Inequality (4.1) is trivially true for $n = 1$. The case $k = 2$ is Mantel's theorem. Suppose now that the inequality is true for all graphs on at most $n - 1$ vertices, and let $G = (V, E)$ be a graph on n vertices without $(k + 1)$-cliques and with a maximal number of edges. This graph certainly contains k-cliques, since otherwise we could add edges. Let A be a k-clique, and set $B = V - A$.

Since each two vertices of A are joined by an edge, A contains $e_A = \binom{k}{2}$ edges. Let e_B be the number of edges joining the vertices of B and $e_{A,B}$ the number of edges between A and B. By induction, we have

$$e_B \leqslant \left(1 - \frac{1}{k}\right) \frac{(n - k)^2}{2}.$$

Since G has no $(k+1)$-clique, every $x \in B$ is adjacent to at most $k-1$ vertices in A, and we obtain

$$e_{A,B} \leqslant (k - 1)(n - k).$$

Summing up and using the identity

$$\left(1 - \frac{1}{k}\right) \frac{n^2}{2} = \binom{k}{2} \left(\frac{n}{k}\right)^2$$

we conclude that

$$|E| \leqslant e_A + e_B + e_{A,B} \leqslant \binom{k}{2} + \binom{k}{2} \left(\frac{n - k}{k}\right)^2 + (k - 1)(n - k)$$

$$= \binom{k}{2} \left(1 - \frac{n - k}{k}\right)^2 = \left(1 - \frac{1}{k}\right) \frac{n^2}{2}.$$

$\square$

4.5 Dirichlet's theorem

Here is the application of the pigeonhole principle which Dirichlet made, resulting in his name being attached to the principle. It concerns the existence of good rational approximations to irrational numbers. The result belongs to number theory, but the argument is combinatorial.

Theorem 4.8 (Dirichlet 1879). *Let x be a real number. For any natural number n, there is a rational number p/q such that $1 \leqslant q \leqslant n$ and*

$$\left| x - \frac{p}{q} \right| < \frac{1}{nq} \leqslant \frac{1}{q^2}.$$

Proof. For this proof, we let $\{x\}$ denote the *fractional part* of the real number x, that is, $\{x\} = x - \lfloor x \rfloor$.

If x is rational, there is nothing to prove. So, suppose that x is irrational and consider the $n + 1$ numbers $\{ax\}$, $a = 1, 2, \ldots, n + 1$. We put these numbers into the n pigeonholes

$$\left(0, \frac{1}{n} \right), \left(\frac{1}{n}, \frac{2}{n} \right), \ldots, \left(\frac{n-1}{n}, 1 \right).$$

(None of the numbers coincides with an end-point of the intervals, since x is irrational.) By the pigeonhole principle, some interval contains more than one of the numbers, say $\{ax\}$ and $\{bx\}$ with $a > b$, which therefore differ by less than $1/n$. Letting $q = a - b$, we see that there exists an integer p such that $|qx - p| < 1/n$, from which the result follows on division by q. Moreover, q is the difference between two integers in the range $1, \ldots, n + 1$, so $q \leqslant n$. $\square$

4.6 Swell-colored graphs

Let us color the edges of the complete graph K_n on n vertices. We say that the graph is *swell-colored* if each triangle contains exactly 1 or 3 colors, but never 2 colors and if the graph contains more than one color. That is, we must use at least two colors, and for every triangle, either all its three edges have the same color or each of them has a different color.

It can be shown (do this!) that K_n can never be swell-colored with exactly two colors. A simple investigation shows that K_3 and K_4 are the only K_n swell-colorable with 3 colors; the other K_n require more colors since they are more highly connected.

Using the pigeonhole principle we can prove the following lower bound.

Theorem 4.9 (Ward–Szabó 1994). *The complete graph on n vertices cannot be swell-colored with fewer than $\sqrt{n} + 1$ colors.*

Proof. Let K_n be swell-colored with r distinct colors. Let $N(x, c)$ denote the number of edges incident to vertex x which have color c. Fix x_0 and c_0 for which $N(x_0, c_0)$ is maximal, and denote this maximum by N.

The $n - 1$ edges incident to x_0 can be sorted into $\leqslant r$ color classes, each of which with N or fewer members. By the pigeonhole principle,

$$N \cdot r \geqslant n - 1.$$

Let $x_1, x_2, \ldots, x_N$ be the vertices connected to x_0 by the N edges of color c_0. Let G denote the (complete) subgraph of K_n induced by the vertex set $\{x_0, x_1, \ldots, x_N\}$. The swell-coloredness of K_n is inherited by G and so all edges of G have color c_0. Since K_n is assumed to have at least two colors, there must be some vertex y of K_n not in subgraph G and such that at least one edge joining y to G has a color different from c_0.

Claim 4.10. *The $N + 1$ edges connecting y to G all are distinctly colored with colors other than c_0.*

The claim implies that $r \geqslant N + 2$, which together with $N \cdot r \geqslant n - 1$ yields $r(r - 2) \geqslant n - 1$, and hence, $r \geqslant \sqrt{n} + 1$, as desired. So, it remains to prove the claim.

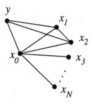

If an edge connecting y to G, say $\{y, x_1\}$ (see the figure above), has color c_0 then by the swell-coloredness of G, edge $\{y, x_0\}$ would have color c_0, contrary to the definition of y (recall that $x_1, x_2, \ldots, x_N$ are *all* the edges incident to x_0 and colored by c_0). Furthermore, if any two edges connecting y to G, say $\{y, x_1\}$ and $\{y, x_2\}$, have the same color, then the swell-coloredness of K_n implies that the edge $\{x_1, x_2\}$ shares the same color. But $\{x_1, x_2\}$ belongs to G, and hence has color c_0 and so $\{y, x_1\}$ would have color c_0 which we have seen is impossible. This completes the proof of the claim, and thus, of the theorem. □

The optimality of the lower bound given by Theorem 4.9, can be shown using a configuration known as "affine plane." We will investigate these configurations in Chap. 13. For our current purposes it is enough to know that an *affine plane* $AG(2, q)$ of order q contains exactly q^2 points and exactly $q + 1$ classes (also called "pencils") of parallel lines, each containing q lines (two lines are parallel if they share no point). Moreover, each two points lie on a unique line.

Having such a plane, we can construct a swell-coloring of K_{q^2} with $q + 1$ colors as follows. Identify the vertices of K_{q^2} with the points in $AG(2, q)$ and associate some unique color with each of the $q + 1$ pencils of parallel lines. In order to define a swell-coloring, consider two distinct vertices x and y of K_{q^2}. These points lie on a unique line which, in its turn, belongs to exactly one of the pencils. Color the edge $\{x, y\}$ with the color of this pencil. Since any two points lie on a unique line and parallel lines do not meet in a point, all three edges of a triangle will receive different colors, and hence, the coloring is swell, as desired.

In fact, Ward and Szabó (1994) have proved that the converse also holds: if the graph K_{q^2} ($q \geqslant 2$) can be swell-colored using $q + 1$ colors then this coloring can be used to construct an affine plane of order q.

4.7 The weight shifting argument

A version of the pigeonhole principle is the *averaging principle* which we formulated in Sect. 1.5: *every set of numbers contains a number at least as large as the average (and one at least as small).*

Trying to show that some "good" object exists, we can try to assign objects their "weights" so that objects with a large enough (or small enough) weight are good, and try to show that the average weight is large (or small). The averaging principle then guarantees that at least one of the objects is good. The main difficulty is to *define* the weights relevant for the desired application. After this we face the problem of how to *compute* the weights and accumulate their sum. At this step the so-called "shifting argument" can help. Let us illustrate this by three examples (the first is trivial, whereas the next two are not).

Proposition 4.11. *Let $n \leqslant m < 2n$. Then for any distribution of m pigeons among n pigeonholes so that no hole is left empty, at most $2(m - n)$ of the pigeons will be happy, i.e., will sit not alone in their holes.*

Proof. If some hole contains more than two pigeons then, by removing a pigeon from this hole and placing it in a hole which had contained exactly one pigeon, we arrive to a new distribution with one more happy pigeon. Thus, the maximum number of happy pigeons is achieved when each hole has at most two pigeons, and in this case this number is $\leqslant 2(m - n)$, as desired.   □

A *trail* in a graph is a walk without repeated edges.

Theorem 4.12 (Graham–Kleitman 1973). *If the edges of a complete graph on n vertices are labeled arbitrarily with the integers $1, 2, \ldots, \binom{n}{2}$, then there is a trail of length at least $n - 1$ with an increasing sequence of edge-labels.*

Proof. To each vertex x, assign its weight w_x equal to the length of the longest increasing trail ending at x. If we can show that $\sum_x w_x \geqslant n(n-1)$, then the averaging principle guarantees a vertex with a large enough weight.

We accumulate the weights and their sum iteratively, growing the graph from the trivial graph; at each step we add a new edge whose label is *minimal* among the remaining ones. Initially, the graph has no edges, and the weights are all 0. At the ith step we take a new edge $e = \{x, y\}$ labeled by i. Let w_x and w_y be the weights of x and y accumulated so far.

If $w_x = w_y$ then increase both weights by 1. If $w_x < w_y$ then the edge e prolongs the longest increasing trail ending at y by 1; so the new weights are $w'_x = w_y + 1$ and $w'_y = w_y$. In either case, when an edge is added, the sum of the weights of the vertices increases by at least 2. Therefore, when all the $\binom{n}{2}$ steps are finished, the sum of the vertex weights is at least $n(n-1)$, as desired.   □

Finally, we illustrate the shifting argument by the fourth proof of Mantel's theorem: *If a graph G on $2n$ vertices contains $n^2 + 1$ edges, then G contains a triangle.*

Fourth proof of Mantel's theorem (Motzkin–Straus 1965). Let G be a graph on $2n$ vertices. Assume that G has no triangles. Our goal is to prove that then $m \leqslant n^2$. We assign a nonnegative w_x to each vertex x such that $\sum_x w_x = 1$. We seek to maximize

$$S = \sum w_x w_y,$$

where the sum is taken over all edges $\{x, y\}$ of G. One way of assigning the weights is to let $w_x = 1/(2n)$ for each x. This gives

$$S \geqslant \frac{m}{(2n)^2}. \tag{4.2}$$

We are going to show that, on the other hand, S never exceeds $1/4$, which together with the previous lower bound will imply that $m \leqslant n^2$, as desired.

And now comes the "shifting argument." Suppose that x and y are two nonadjacent vertices and W_x and W_y are the total weights of vertices connected to x and y, respectively. Suppose also that $W_x \geqslant W_y$. Then for any $\epsilon \geqslant 0$,

$$(w_x + \epsilon)W_x + (w_y - \epsilon)W_y \geqslant w_x W_x + w_y W_y.$$

This, in particular, means that we do not decrease the value of S if we shift all of the weight of vertex y to the vertex x. It follows that S is maximized when all of the weight is concentrated on a complete subgraph of G. But we have assumed that G has no triangles; so G cannot have complete subgraphs other than single edges. Hence, S is maximized when all of the weight is concentrated on two adjacent vertices, say x and y. Therefore

$$S \leqslant \max \left\{ w_x \cdot w_y \; : \; w_x + w_y = 1 \right\} = 1/4$$

which, together with (4.2), yield the desired upper bound $m \leqslant n^2$.   □

4.8 Pigeonhole and resolution

In this section we will present a result of a different nature – we will use the pigeonhole principle to show that the principle *itself* is hard to deduce in a classical proof system, known as Resolution.

4.8.1 Resolution refutation proofs

The resolution proof system was introduced by Blake (1937) and has been made popular as a theorem-proving technique by Davis and Putnam (1960) and Robinson (1965). This system operates with clauses, i.e., with Or's of literals, where each literal is either a variable x_i or its negation $\overline{x}_i$. A *truth-assignment* is an assignment of constants 0 and 1 to all the variables. Such an assignment *satisfies* (*falsifies*) a clause if it evaluates at least one (respectively, none) of its literals to 1. A set of clauses is satisfiable if there is an assignment which satisfies all its clauses.

Let F be a set of clauses and suppose that F is not satisfiable. A *resolution refutation proof* for F is a sequence of clauses $\mathcal{R} = (C_1, \ldots, C_t)$ where C_t is the empty clause (which, by definition, is satisfied by no assignment) and each intermediate clause C_i either belongs to F or is derived from some previous two clauses using the following *resolution rule*:

> The clause $C \vee C'$ can be inferred from two clauses $C \vee x_i$ and $C' \vee \overline{x}_i$.

In this case one also says that the variable x_i was *resolved* to derive the clause $C \vee C'$. The *length* of such a proof is equal to the number t of clauses in the derivation.

Observe that the resolution rule is *sound* in the following sense: if some assignment (of constants to all the variables) falsifies the derived clause $C \vee C'$, then it must falsify at least one of the clauses $C \vee x_i$ and $C' \vee \overline{x}_i$ from which it was derived. It is also known (and easy to show) that Resolution is *complete*: every unsatisfiable set of clauses has a resolution refutation proof.

What about the length of such derivations? Due to its practical importance, this question bothered complexity theoreticians and logicians for a long time.

The first exponential lower bound for the length of regular resolution was proved by Tseitin (1968) already 30 years ago. (These are resolution proofs with the additional restriction that along every path every particular variable x_i can be resolved at most once; a path in a derivation is just a sequence of clauses, each of which is one of the two hypotheses from which the next clause is derived.) However, despite its apparent simplicity, the first lower bounds for non-regular resolution were only proved in 1985 by Haken. These bounds were achieved for the set of clauses PHP_n^{n+1} formalizing the pigeonhole principle. Subsequently, Haken's argument was refined and extended to other principles as well as to proof systems generalizing Resolution.

One may also consider the generalized pigeonhole principle PHP_n^m saying that m pigeons $(m \geqslant n+1)$ cannot sit in n holes so that every pigeon is alone in its hole. The larger the difference $m - n$, the "more true" is the principle itself, and its proof might be shorter. Buss and Pitassi (1998) have proved that, for $m \geqslant 2^{\sqrt{n \log n}}$, PHP_n^m has a resolution proof of length polynomial in m. But for a long time, no non-trivial lower bound was known for $m > n^2$. Overcomming this "n square" barrier was one of the most challenging open problems about the power of Resolution. This problem was recently resolved by Ran Raz (2001) who proved that for any $m \geqslant n+1$, any Resolution proof of PHP_n^m requires length 2^{n^ϵ}, where $\epsilon > 0$ is an absolute constant.

4.8.2 Haken's lower bound

Recall that the pigeonhole principle states that n pigeons cannot sit in $n - 1$ holes so that every pigeon is alone in its hole. To formalize the principle, let us introduce boolean variables $x_{i,j}$ interpreted as:

$x_{i,j} = 1$ if and only if the ith pigeon sits in the jth hole.

Let PHP_{n-1}^n denote the set of clauses:

(i) $x_{i,1} \vee x_{i,2} \vee \cdots \vee x_{i,n-1}$ for each $i = 1, \ldots, n$;
(ii) $\overline{x}_{i,k} \vee \overline{x}_{j,k}$ for each $1 \leqslant i \neq j \leqslant n$ and $1 \leqslant k \leqslant n - 1$.

Note that the And of all clauses of the first sort is satisfiable if and only if every pigeon sits in at least one hole, whereas the And of the clauses of the second sort can be satisfied if and only if no two pigeons sit in the same hole. Thus, by the pigeonhole principle(!), the And of all clauses in PHP_{n-1}^n is not satisfiable.

Theorem 4.13 (Haken 1985). *For a sufficiently large n, any Resolution proof of PHP_{n-1}^n requires length $2^{\Omega(n)}$.*

Originally, Haken's proof used the so-called "bottleneck counting" argument and was quite involved. Here we present a new and simple proof of his result found by Beame and Pitassi (1996).

Proof. We will concentrate on a particular subset of truth assignments. Look at the set of underlying variables $X = \{x_{i,j} : 1 \leqslant i \leqslant n, 1 \leqslant j \leqslant n - 1\}$ as an $n \times (n - 1)$ matrix. Say that a truth assignment $\alpha \colon X \to \{0,1\}$ to the underlying variables $x_{i,j}$ is *critical* if it defines a one-to-one map from $n - 1$ pigeons to $n - 1$ holes, with the remaining pigeon not mapped to any hole. A critical assignment, where i is the left-out pigeon, is called *i-critical* (see Fig. 4.2). In what follows we will be interested only in these critical truth assignments. (How many such assignments do we have?)

Take an arbitrary resolution refutation proof $\mathcal{R} = (C_1, \ldots, C_t)$ for PHP_{n-1}^n. As a first step, we get rid of negations: we replace each clause C in $\mathcal{R}$ by a *positive* clause C^+, i.e., by a clause without negated variables.

$$
i
\begin{array}{|ccccc|}
\hline
1 & 0 & 0 & 0 & 0 \\
0 & 1 & 0 & 0 & 0 \\
0 & 0 & 1 & 0 & 0 \\
0 & 0 & 0 & 0 & 0 \\
0 & 0 & 0 & 1 & 0 \\
0 & 0 & 0 & 0 & 1 \\
\hline
\end{array}
\qquad
i
\begin{array}{|ccccc|}
\hline
0 & 1 & 0 & 0 & 0 \\
0 & 0 & 0 & 1 & 0 \\
1 & 0 & 0 & 0 & 0 \\
0 & 0 & 0 & 0 & 0 \\
0 & 0 & 0 & 0 & 1 \\
0 & 0 & 1 & 0 & 0 \\
\hline
\end{array}
$$

Fig. 4.2. Two i-critical truth assignments for $i = 4$ in the case of 6 pigeons and 5 holes

The idea of this transformation is due to Buss (1987) and works as follows: replace each occurrence of $\overline{x}_{i,j}$ in the clause C by the Or

$$
C_{i,j} \rightleftharpoons x_{1,j} \vee \cdots \vee x_{i-1,j} \vee x_{i+1,j} \vee \cdots \vee x_{n,j}
$$

of all the variables, corresponding to the jth hole, except $x_{i,j}$. The resulting sequence of positive clauses $\mathcal{R}^+ = (C_1^+, \ldots, C_t^+)$ is no longer a valid resolution refutation proof – it is just a *sequence* which we will call a *positive pseudo-proof* of PHP_{n-1}^n. For the rest of the proof it will only be important that this sequence has the property that, with respect to critical assignments, the rules in it are still sound. That is, if C is derived from C_1 and C_2 in the original proof $\mathcal{R}$ then, for every critical α,

$$
C_1^+(\alpha) \cdot C_2^+(\alpha) \leqslant C^+(\alpha).
$$

This is an immediate consequence of the following claim.

Claim 4.14. *For every critical truth assignment α, $C^+(\alpha) = C(\alpha)$.*

Proof. Suppose there is a critical assignment α such that $C^+(\alpha) \neq C(\alpha)$. This can only happen if C contains a literal $\overline{x}_{i,j}$ such that $\overline{x}_{i,j}(\alpha) \neq C_{i,j}(\alpha)$. But this is impossible, since α has precisely one 1 in the jth column. $\square$

We will use this property (the soundness with respect to critical assignments) to show that the pseudo-proof $\mathcal{R}^+$ (and, hence, also the original proof $\mathcal{R}$) must be long, namely – that $t \geqslant 2^{n/32}$. For the sake of contradiction, assume that we have fewer than $2^{n/32}$ clauses in $\mathcal{R}^+$. Say that a clause is *long* if it has at least $n^2/8$ variables, i.e., if it includes more that $1/8$ fraction of all $n(n-1)$ possible variables. Let ℓ be the number of long clauses in $\mathcal{R}$; hence

$$
\ell < 2^{n/32}.
$$

Since each long clause has at least a $1/8$ fraction of all the variables, there must be (by the pigeonhole principle!) a variable $x_{i,j}$ which occurs in at least $\ell/8$ of the long clauses. Set this variable to 1, and at the same time set to 0 all the variables $x_{i,j'}$ and $x_{i',j}$ for all $j' \neq j, i' \neq i$ (see Fig. 4.3). After this setting, all the clauses containing $x_{i,j}$ will disappear from the proof (they all get the value 1) and the variables which are set to 0 will disappear from the remaining clauses.

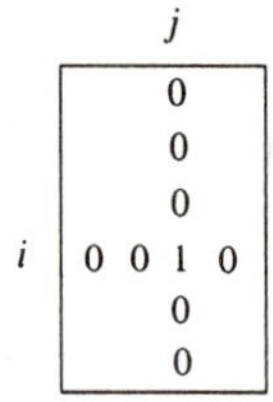

Fig. 4.3. Setting of constants to eliminate long clauses containing $x_{i,j}$

Applying this restriction to the entire proof $\mathcal{R}^+$ leaves us with a new positive pseudo-proof of PHP^{n-1}_{n-2}, where the number of long clauses is at most $\ell(1 - 1/8)$. Continue in this fashion until we have set all long clauses to 1. Applying this argument iteratively $d = 8 \ln \ell$ many times, we are guaranteed to have knocked out all long clauses, because

$$\ell(1 - 1/8)^d < e^{\ln \ell - d/8} = 1.$$

Thus, we are left with a positive pseudo-proof $\mathcal{R}'$ of PHP^m_{m-1}, where $m = n - 8 \ln \ell$, and where *no* clause is long, i.e., has length at least $n^2/8$. But this contradicts the following claim which states that such a pseudo-proof must have a clause of size

$$2m^2/9 = 2(n - 8 \ln \ell)^2/9 > 2(n - n/4)^2/9 = n^2/8.$$

So, it remains to prove the claim.

Claim 4.15. *Any positive pseudo-proof of* PHP^m_{m-1} *must have a clause with at least* $2m^2/9$ *variables.*

Proof. Let $\mathcal{R}'$ be a positive pseudo-proof of PHP^m_{m-1}. Recall that $\mathcal{R}'$ contains no negated literals and that the rules in $\mathcal{R}'$ are sound with respect to critical assignments. This implies that for every clause C in $\mathcal{R}'$ there is a set of clauses $\mathcal{W}$ from PHP^m_{m-1} whose conjunction implies C on all critical truth assignments. That is, every critical assignment satisfying all the clauses in $\mathcal{W}$ must also satisfy the clause C. We call such a set of clauses $\mathcal{W}$ a *witness* of C. One clause C may have several witnesses. We define the *weight* of C as the minimal number of clauses in its witness.

Let us make several observations about this measure. Since we are considering only critical truth assignments, only the "pigeon" clauses of type (i), saying that some pigeon must be mapped to a hole, will be included in a minimal witness, just because all other clauses are satisfied by every critical assignment (no column has two 1's). The weight of these initial "pigeon" clauses is 1, and the weight of the final clause is m (since this clause outputs 0 for *all* critical assignments). Since (by soundness) the weight of a clause is at most the sum of weights of the two clauses from which it is derived, there must exist a clause C in the proof whose weight s is between $m/3$ and $2m/3$.

(This is a standard and useful trick, and we address it in Exercise 4.14.) We will prove that this clause C must contain at least $2m^2/9$ variables.

To show this, let $W = \{C_i : i \in S\}$, where $|S| = s$, be a minimal set of pigeon clauses

$$C_i = x_{i,1} \vee x_{i,2} \vee \cdots \vee x_{i,m-1}$$

in PHP^m_{m-1} whose conjunction implies C. We will show that C has at least

$$(m-s)s \geqslant 2m^2/9$$

distinct literals.

Take an $i \in S$, and let α be an i-critical truth assignment falsifying C. (Such an assignment exists by the minimality of W; check this!) For each $j \notin S$, consider the j-critical assignment α' obtained from α by replacing i by j. This assignment differs from α only in two places: if α mapped the pigeon j to the hole k, then α' maps the pigeon i to this hole k (see Fig. 4.4).

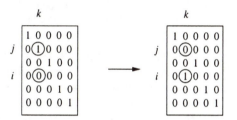

Fig. 4.4. Assignment α' is obtained from α by interchanging the ith and jth rows.

Since $j \notin S$, the assignment α' satisfies all the clauses of the witness W of C, and hence, must satisfy the clause C. Since $C(\alpha) = 0$ and the assignments α, α' differ only in the variables $x_{i,k}$ and $x_{j,k}$, this can only happen when C contains the variable $x_{i,k}$ (remember that the clause C has no negated literals). Running the same argument over all $m - s$ pigeons $j \notin S$ (using the same α), it follows that C must contain at least $m - s$ distinct variables $x_{i,k_1}, x_{i,k_2}, \ldots, x_{i,k_{m-s}}$ corresponding to the ith pigeon. Repeating the argument for all pigeons $i \in S$ shows that C contains at least $(m - s)s$ variables, as claimed.

This completes the proof of the claim, and thus, the proof of the theorem.
□

Exercises

4.1.$^-$ Suppose five points are chosen inside an equilateral triangle with side-length 1. Show that there is at least one pair of points whose distance apart is at most $1/2$. *Hint*: Divide the triangle into four suitable boxes.

4.2 (D.R. Karger). Jellybeans of 8 different colors are in 6 jars. There are 20 jellybeans of each color. Use the pigeonhole principle to prove that there must be a jar containing two pairs of jellybeans from two different colors of jellybeans.

Hint: For each color there is a jar containing a pair of jellybeans of that color, and we have more colors than jars.

4.3.⁻ Prove that every set of $n+1$ distinct integers chosen from $\{1, 2, \ldots, 2n\}$ contains a pair of subsequent numbers and a pair whose sum is $2n + 1$. For each n, exhibit two sets of size n to show that these results are the best possible.

Hint: Use pigeonholes $(i, i+1)$ and $(i, 2n-i+1)$, $i = 1, \ldots, n-1$.

4.4. Prove that every set of $n+1$ distinct integers chosen from $\{1, 2, \ldots, 2n\}$ contains two numbers such that one divides the other.

Sketch (due to Lajos Pósa): Write every number x in the form $x = k_x 2^a$, where k_x is an odd number between 1 and $2n - 1$. Take odd pigeonholes $1, 3, 5, \ldots, 2n - 1$ and put x into the pigeonhole number k_x. Some hole must have two numbers $x < y$.

4.5. Coin-weighing problem (Erdős–Spencer 1974). Let n coins of weights 0 and 1 be given. We are also given a scale with which we may weigh any subset of the coins. The information from previous weighings may be used. The object is to determine the weights of the coins with the minimal number of weighings. Formally, the problem may be stated as follows. A collection $S_1, \ldots, S_m$ of subsets of $[n]$ is called *determining* if an arbitrary subset T of $[n]$ can be uniquely determined by the cardinalities $|S_i \cap T|$, $1 \leqslant i \leqslant m$. Let $D(n)$ be the minimum m for which such a determining collection exists. By weighting each coin separately ($S_i = \{i\}$) we see that $D(n) \leqslant n$. Show that $D(n) \geqslant n/(\log_2(n+1))$.

Hint: Take a determining collection $S_1, \ldots, S_m$, observe that for each i there are only $n + 1$ possible $|S_i \cap T|$, and apply the pigeonhole principle.

4.6.⁻ Suppose that n is a multiple of k. Construct a graph without $(k + 1)$-cliques, in which the number of edges achieves the upper bound (4.1) given by Turán's theorem.

Hint: Split the n vertices into k equal size parts and join all pairs of vertices from different parts (this is a complete k-partite graph).

4.7. Recall that the *independence number* $\alpha(G)$ of a graph G is the maximum number of pairwise nonadjacent vertices of G. Prove the following dual version of Turán's theorem: if G is a graph with n vertices and $nk/2$ edges, $k \geqslant 1$, then $\alpha(G) \geqslant n/(k+1)$.

4.8[!] (Motzkin–Straus 1965). Prove Turán's theorem using the shifting argument described in the fourth proof of Mantel's theorem.

Sketch: Let G be a graph with n vertices and m edges, and suppose that G has no $(k+1)$-clique. Assign weights w_x to the vertices as before. Setting $w_x = 1/n$ for all vertices, we obtain $S \geqslant m/n^2$. On the other hand, the same shifting argument yields that the weight is concentrated on some clique U with $|U| = t \leqslant k$ vertices. Setting $w_x = 1/t$ for $x \in U$, and $w_x = 0$ otherwise, the total weight becomes $\binom{t}{2}/t^2 = (1 - 1/t)/2$. Since this expression is increasing in t, the best we can do is to set $t = k$.

4.9.$^{(!)}$ Derive the Erdős–Szekeres theorem from Lemma 4.5.

Hint: Given a sequence $A = (a_1, \ldots, a_n)$ of $n \geqslant rs + 1$ real numbers, define a partial order $\preccurlyeq$ on A by $a_i \preccurlyeq a_j$ if $a_i \leqslant a_j$ and $i \leqslant j$, and apply Dilworth's lemma.

4.10. Let $n^2 + 1$ points be given in $\mathbb{R}^2$. Prove that there is a sequence of $n+1$ points $(x_1, y_1), \ldots, (x_{n+1}, y_{n+1})$ for which $x_1 \leqslant x_2 \leqslant \cdots \leqslant x_{n+1}$ and $y_1 \geqslant y_2 \geqslant \cdots \geqslant y_{n+1}$, or a sequence of $n+1$ points for which $x_1 \leqslant x_2 \leqslant \cdots \leqslant x_{n+1}$ and $y_1 \leqslant y_2 \leqslant \cdots \leqslant y_{n+1}$.

4.11. Show that, if $n > srp$, then any sequence of n real numbers must contain either a strictly increasing subsequence of length greater than s, a strictly decreasing subsequence of length greater than r, or a constant sequence of length greater than p.

Hint: By the pigeonhole principle, if only sr or fewer distinct values occur, then some value must be taken by more than p numbers in the sequence. Otherwise, we can argue as in the Erdős–Szekeres theorem.

4.12. Let $0 < a_1 < a_2 < \cdots < a_{sr+1}$ be $sr + 1$ integers. Prove that we can select either $s + 1$ of them, no one of which divides any other, or $r + 1$ of them, each dividing the following one.

Hint: Apply Dilworth's lemma.

4.13.$^-$ Show that the bound in the Erdős–Szekeres' theorem is best possible.

Hint: Consider the sequence $A = (B_{s-1}, B_{s-2}, \ldots, B_0)$, where

$$B_i = (ir + 1, ir + 2, \ldots, ir + r).$$

4.14. Let G be a directed acyclic graph of outdegree 2. That is, for every vertex, except for the leaves (whose outdegree is 0), there are two edges leaving it. Define the *weight* of a vertex to be the number of leaves which are reachable from it. Assume that there is a vertex (a root) from which all the leaves are reachable. Prove that then there exists a vertex whose weight lies between $m/3$ and $2m/3$, where m is the total number of leaves.

Hint: Start from the root, and each time see whether some successor has weight at most $2m/3$. If not, then take any one of them and continue the walk. Observe that the weight of a vertex is at most the sum of the weights of its two successors.

4.15. Use the pigeonhole principle to prove the following fact, known as *Chinese remainder theorem*. Let $a_1, \ldots, a_k, b$ be integers, and $m = m_1 \cdots m_k$ where m_i and m_j are relatively prime, for all $i \neq j$. Then there exists exactly one integer a, $b \leqslant a < b + m$, such that $a \equiv a_i \bmod m_i$ for all $i = 1, \ldots, k$.

Hint: The integers $x \in \{b, b+1, \ldots b+m-1\}$ are different modulo m; hence their residues $(x \bmod m_1, \ldots, x \bmod m_k)$ run through all m possible values.

4.16$^+$ (Moon–Moser 1962). Let $G = (V, E)$ be a graph on n vertices and $t(G)$ the number of triangles in it. Show that

$$t(G) \geqslant \frac{|E|}{3n} \left(4 \cdot |E| - n^2 \right).$$

Sketch: For an edge $e = \{x, y\}$, let $t(e)$ be the number of triangles containing e. Let $B = V - \{x, y\}$. Among the vertices in B there are precisely $t(e)$ vertices which are adjacent to both x and y. Every other vertex in B is adjacent to at most one of these two vertices. We thus obtain $d(x) + d(y) - t(e) \leqslant n$. Summing over all edges $e = \{x, y\}$ we obtain

$$\sum_{e \in E} (d(x) + d(y)) - \sum_{e \in E} t(e) \leqslant n \cdot |E|.$$

Apply the Cauchy–Schwarz inequality to estimate the first sum.

Comment: This implies that a graph G on an even number n of vertices with $|E| = n^2/4 + 1$ edges not only contains one triangle (as it must be by Mantel's theorem), but more than $n/3$.

4.17. A set $S \subseteq V$ of vertices in a graph $G = (V, E)$ *spans* an edge $e \in E$ if both endpoints of e belong to S. Say that a graph is (k, r)-*sparse* if every subset of k vertices spans at most r of its edges. Turán's theorem (Theorem 4.7) gives an upper bound on the maximal possible number of edges in a (k, r)-sparse graph for $r = \binom{k}{2} - 1$. Show that every (k, r)-sparse graph on n vertices has at most $\alpha \cdot \binom{n}{2}$ edges, where $\alpha = r \cdot \binom{k}{2}^{-1}$.

Hint: Observe that every edge is spanned by precisely $\binom{n-2}{k-2}$ of k-element subsets and use Exercise 1.10.

5. Systems of Distinct Representatives

A *system of distinct representatives* for a sequence of (not necessarily distinct) sets $S_1, S_2, \ldots, S_m$ is a sequence of distinct elements $x_1, x_2, \ldots, x_m$ such that $x_i \in S_i$ for all $i = 1, 2, \ldots, m$.

When does such a system exist? This problem is called the "marriage problem" because an easy reformulation of it asks whether we can marry each of m girls to a boy she knows; boys are the elements and S_i is the set of boys known to the ith girl.

Clearly, if the sets $S_1, S_2, \ldots, S_m$ have a system of distinct representatives then the following *Hall's Condition* is fulfilled:

$(*)$ for every $k = 1, 2, \ldots, m$ the union of any k sets has at least k elements:

$$\left| \bigcup_{i \in I} S_i \right| \geqslant |I| \quad \text{for all } I \subseteq \{1, \ldots, m\}.$$

Surprisingly, this obvious necessary condition is also sufficient.

5.1 The marriage theorem

The following fundamental result is known as *Hall's marriage theorem* (Hall 1935), though an equivalent form of it was discovered earlier by König (1931) and Egerváry (1931), and the result is also a special case of Menger's theorem (1927). The case when we have the same number of girls as boys was proved by Frobenius (1917).

Theorem 5.1 (Hall's Theorem). *The sets $S_1, S_2, \ldots, S_m$ have a system of distinct representatives if and only if $(*)$ holds.*

Proof. We prove the sufficiency of Hall's condition $(*)$ by induction on m. The case $m = 1$ is clear. Assume that the claim holds for any collection with less than m sets.

Case 1: For each k, $1 \leqslant k < m$, the union of any k sets contains more than k elements.

Take any of the sets, and choose any of its elements x as its representative, and remove x from all the other sets. The union of any $s \leqslant m - 1$ of the remaining $m - 1$ sets has at least s elements, and therefore the remaining sets have a system of distinct representatives, which together with x give a system of distinct representatives for the original family.

Case 2: The union of some k, $1 \leqslant k < m$, sets contains exactly k elements.

By the induction hypothesis, these k sets have a system of distinct representatives. Remove these k elements from the remaining $m - k$ sets. Take any s of these sets. Their union contains at least s elements, since otherwise the union of these s sets and the k sets would have less than $s + k$ elements. Consequently, the remaining $m - k$ sets also have a system of distinct representatives by the induction hypothesis. Together these two systems of distinct representatives give a system of distinct representatives for the original family. □

In general, Hall's condition $(*)$ is hard to verify: we must check if the union of *any* k, $1 \leqslant k \leqslant m$, of the sets $S_1, \ldots, S_m$ contains at least k elements. But if we know more about these sets, then (sometimes) the situation is much better. Here is an example.

Corollary 5.2. *Let $S_1, \ldots, S_m$ be r-element subsets of an n-element set such that each element belongs to the same number d of these sets. If $m \leqslant n$, then the sets $S_1, \ldots, S_m$ have a system of distinct representatives.*

Proof. By the double counting argument (1.7), $mr = nd$, and hence, $m \leqslant n$ implies that $d \leqslant r$. Now suppose that $S_1, \ldots, S_m$ does not have a system of distinct representatives. By Hall's theorem, the union $Y = S_{i_1} \cup \cdots \cup S_{i_k}$ of some k ($1 \leqslant k \leqslant m$) sets contains strictly less than k elements. For $x \in Y$, let d_x be the number of these sets containing x. Then, again, using (1.7), we obtain

$$rk = \sum_{j=1}^{k} |S_{i_j}| = \sum_{x \in Y} d_x \leqslant d |Y| < dk,$$

a contradiction with $d \leqslant r$. □

Hall's theorem was generalized in different ways. Suppose, for example, that each of the elements of the underlying set is colored either in red or in blue. Interpret red points as "bad" points. Given a system of subsets of this (colored) set, we would like to come up with a system of distinct representatives which has as few bad elements as possible.

Theorem 5.3 (Chvátal–Szemerédi 1988). *The sets $S_1, \ldots, S_m$ have a system of distinct representatives with at most t red elements if and only if they have a system of distinct representatives and for every $k = 1, 2, \ldots, m$ the union of any k sets has at least $k - t$ blue elements.*

Proof. The "only if" part is obvious. To prove the "if" part, let R be the set of red elements. We may assume that $|R| > t$ (otherwise the conclusion is trivial). Now enlarge $S_1, \ldots, S_m$ to $S_1, \ldots, S_m, S_{m+1}, \ldots, S_{m+r}$ by adding

$r = |R| - t$ copies of the set R. Observe that the sequence $S_1, \ldots, S_m$ has a system of distinct representatives with at most t red elements if and only if the extended sequence has a system of distinct representatives (without any restriction). Hence, Hall's theorem reduces our task to proving that the extended sequence fulfills Hall's condition $(*)$, i.e., that for any set of indices $I \subseteq \{1, \ldots, m+r\}$, the union $Y = \bigcup_{i \in I} S_i$ contains at least $|I|$ elements. Let $J = I \cap \{1, \ldots, m\}$. If $J = I$ then, by the first assumption, the sets S_i $(i \in I)$ have a system of distinct representatives, and hence, $|Y| \geqslant |I|$. Otherwise, by the second assumption,

$$|Y| = \left| \bigcup_{i \in J} (S_i - R) \right| + |R| \geqslant (|J| - t) + |R|$$
$$= |J| + (|R| - t) \geqslant |J| + |I - J| = |I| \, ;$$

hence $(*)$ holds again.   □

5.2 Two applications

In this section we present two applications of Hall's theorem to prove results whose statement does not seem to be related at all to set systems and their representatives.

5.2.1 Latin rectangles

An $r \times n$ *Latin rectangle* is an $r \times n$ matrix with entries in $\{1, \ldots, n\}$ such that each of the numbers $1, 2, \ldots, n$ occurs once in each row and at most once in each column. A *Latin square* is a Latin $r \times n$-rectangle with $r = n$. This is one of the oldest combinatorial objects, whose study goes back to ancient times.

Suppose somebody gives us an $n \times n$ matrix, *some* of whose entries are filled with the numbers from $\{1, \ldots, n\}$ so that no number occurs more than once in a row or column. Our goal is to fill the remaining entries so that to get a Latin square. When is this possible? Of course, the fewer entries are filled, the more chances we have to complete the matrix. Fig. 5.1 shows that, in general, it is possible to fill n entries so that the resulting partial matrix cannot be completed.

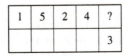

Fig. 5.1. A partial 2×5 Latin square that cannot be completed

In 1960, Trevor Evans raised the following question: if fewer than n entries in an $n \times n$ matrix are filled, can one then always complete it to obtain a Latin square? The assertion that a completion is always possible became known as the Evans conjecture, and was proved by Smetaniuk (1981) using a quite subtle induction argument.

On the other hand, it was long known that if a partial Latin square has no partially filled rows (that is, each row is either completely filled or completely free) then it can always be completed. That is, we can build Latin squares by adding rows one-by-one. And this can be easily derived from Hall's theorem.

Theorem 5.4 (Ryser 1951). *If $r < n$, then any given $r \times n$ Latin rectangle can be extended to an $(r + 1) \times n$ Latin rectangle.*

Proof. Let R be an $r \times n$ Latin rectangle. For $j = 1, \ldots, n$, define S_j to be the set of those of integers $1, 2, \ldots, r$ which do *not* occur in the jth column of R. It is sufficient to prove that the sets $S_1, \ldots, S_n$ have a system of distinct representatives. But this follows immediately from Corollary 5.2, because: every set S_j has precisely $n-r$ elements, and each element belongs to precisely $n - r$ sets S_j (since it appears in precisely r columns of the rectangle R).   □

5.2.2 Decomposition of doubly stochastic matrices

Using Hall's theorem we can obtain a basic result of polyhedral combinatorics, due to Birkhoff (1949) and von Neumann (1953).

An $n \times n$ matrix $A = \{a_{ij}\}$ with real non-negative entries $a_{ij} \geqslant 0$ is *doubly stochastic* if the sum of entries along any row and any column equals 1. A *permutation matrix* is a doubly stochastic matrix with entries 0 and 1; such a matrix has exactly one 1 in each row and in each column. Doubly stochastic matrices arise in the theory of Markov chains: a_{ij} is the transition probability from the state i to the state j. A matrix A is a convex combination of matrices $A_1, \ldots, A_s$ if there exist non-negative reals $\lambda_1, \ldots, \lambda_s$ such that $A = \sum_{i=1}^{s} \lambda_i A_i$ and $\sum_{i=1}^{s} \lambda_i = 1$.

Birkhoff–Von Neumann Theorem. *Every doubly stochastic matrix is a convex combination of permutation matrices.*

Proof. We will prove a more general result that every $n \times n$ non-negative matrix $A = (a_{ij})$ having all row and column sums equal to some positive value $\gamma > 0$ can be expressed as a linear combination $A = \sum_{i=1}^{s} \lambda_i P_i$ of permutation matrices $P_1, \ldots, P_s$, where $\lambda_1, \ldots, \lambda_s$ are non-negative reals such that $\sum_{i=1}^{s} \lambda_i = \gamma$.

To prove this, we apply induction on the number of non-zero entries in A. Since $\gamma > 0$, we have at least n such entries. If there are exactly n non-zero entries then $A = \gamma P$ for some permutation matrix P, and we are done. Now suppose that A has more than n non-zero entries and that the result holds for matrices with a smaller number of such entries. Define

$$S_i = \{j : a_{ij} > 0\}, \ i = 1, 2, \ldots, n,$$

and observe that the sets $S_1, \ldots, S_n$ fulfill Hall's condition. Indeed, if the union of some k $(1 \leqslant k \leqslant n)$ of these sets contained less than k elements, then all the non-zero entries of the corresponding k rows of A would occupy no more than $k - 1$ columns; hence, the sum of these entries by columns would be at most $(k - 1)\gamma$, whereas the sum by rows is $k\gamma$, a contradiction.

By Hall's theorem, there is a system of distinct representatives

$$j_1 \in S_1, \ldots, j_n \in S_n.$$

Take the permutation matrix $P_1 = \{p_{ij}\}$ with entries $p_{ij} = 1$ if and only if $j = j_i$. Let $\lambda_1 = \min\{a_{1j_1}, \ldots, a_{nj_n}\}$, and consider the matrix $A_1 = A - \lambda_1 P_1$. By the definition of the sets S_i, $\lambda_1 > 0$. So, this new matrix A_1 has less non-zero entries than A. Moreover, the matrix A_1 satisfies the condition of the theorem with $\gamma_1 = \gamma - \lambda_1$. We can therefore apply the induction hypothesis to A_1, which yields a decomposition $A_1 = \lambda_2 P_2 + \cdots + \lambda_s P_s$, and hence, $A = \lambda_1 P_1 + A_1 = \lambda_1 P_1 + \lambda_2 P_2 + \cdots + \lambda_s P_s$, as desired. $\square$

5.3 Min–max theorems

The early results of Frobenius and König have given rise to a large number of *min-max theorems* in combinatorics, in which the minimum of one quantity equals the maximum of another. Celebrated among these are:

- *Menger's theorem* (Menger 1927), which states that the minimum number of vertices separating two given vertices in a graph is equal to the maximum number of vertex-disjoint paths between them;
- *König–Egerváry's min-max theorem* (König 1931, Egerváry 1931), that the size of a largest matching in a bipartite graph is equal to the smallest set of vertices which together touch every edge;
- *Dilworth's theorem* for partially ordered sets (Dilworth 1950), that the minimum number of *chains* (totally ordered sets) which cover a partially ordered set is equal to the maximum size of an antichain (set of incomparable elements).

Here we present the proof of König–Egerváry's theorem (stated not for bipartite graphs but for their incidence matrices); the proof of Dilworth's theorem is given in Sect. 9.1.

By Hall's theorem, we know whether each of the girls can be married to a boy she knows. If so, all are happy (except for the boys not chosen ...). But what if not? In this sad situation it would be nice to make as many happy marriages as possible. So, given a sequence of sets $S_1, S_2, \ldots, S_m$, we try to find a system of distinct representatives for as many of these sets as possible. In terms of 0-1 matrices this problem is solved by the following result.

Let A be an $m \times n$ matrix, all whose entries have value 0 or 1. Two 1's are *dependent* if they are on the same row or on the same column; otherwise, they are *independent*. The size of the largest set of independent 1's is also known as the *term rank* of A.

Theorem 5.5 (König 1931, Egerváry 1931). *Let A be an $m \times n$ 0-1 matrix. The maximum number of independent 1's is equal to the minimum number of rows and columns required to cover all the 1's in A.*

Proof. Let r denote the maximum number of independent 1's and R the minimum number of rows and columns required to cover all the 1's. Clearly, $R \geqslant r$, because we can find r independent 1's in A, and any row or column covers at most one of them.

We need to prove that $r \geqslant R$. Assume that some a rows and b columns cover all the 1's and $a + b = R$. Because permuting the rows and columns changes neither r nor R, we may assume that the first a rows and the first b columns cover the 1's. Write A in the form

$$A = \begin{pmatrix} B_{a \times b} & C_{a \times (n-b)} \\ D_{(m-a) \times b} & E_{(m-a) \times (n-b)} \end{pmatrix}.$$

We know that there are no 1's in E. We will show that there are a independent 1's in C. The same argument shows – by symmetry – that there are b independent 1's in D. Since altogether these $a + b$ 1's are independent, this shows that $r \geqslant a + b = R$, as desired.

We use Hall's theorem. Define

$$S_i = \{j : c_{ij} = 1\} \subseteq \{1, 2, \ldots, n - b\},$$

as the set of locations of the 1's in the ith row of $C = (c_{ij})$. We claim that the sequence $S_1, S_2, \ldots, S_a$ has a system of distinct representatives, i.e., we can choose a 1 from each row, no two in the same column. Otherwise, Hall's theorem tells us that the 1's in some k ($1 \leqslant k \leqslant a$) of these rows can all be covered by less than k columns. But then we obtain a covering of all the 1's in A with fewer than $a + b$ rows and columns, a contradiction. □

5.4 Matchings in bipartite graphs

Let G be a bipartite graph with bipartition A, B. Two edges are *disjoint* if they have no vertex in common. A *matching* in G is a set of pairwise disjoint edges. The vertices belonging to the edges of a matching are *matched*, others are *free*. We may ask whether G has a matching which matches all the vertices from A; we call this a matching of A *into* B. A *perfect matching* is a matching of A into B in the case when $|A| = |B|$.

The answer is given by Hall's theorem. A vertex $x \in A$ is a *neighbor* of a vertex $y \in B$ in the graph G if $(x, y) \in E$. Let S_x be the set of all neighbors of x in G. Observing that there is a matching of A into B if and only if the sets S_x with $x \in A$ have a system of distinct representatives, Hall's theorem immediately yields the following:

Theorem 5.6. *If G is a bipartite graph with bipartition A, B, then G has a matching of A into B if and only if, for every $k = 1, 2, \ldots, |A|$, every subset of k vertices from A has at least k neighbors.*

To illustrate this form of Hall's theorem, we prove the following (simple but non-trivial!) fact.

Proposition 5.7. *Let X be an n-element set. For any $k \leqslant (n-1)/2$ it is possible to extend every k-element subset of X to a $(k+1)$-element subset (by adding some element to that set) so that the extensions of no two sets coincide.*

Proof. Consider the bipartite graph $G = (A, B, E)$, where A consists of all k-element subsets, B consists of all $(k+1)$-element subsets of X and $(x, y) \in E$ if and only if $x \subset y$. What we need is to prove that this graph has a matching of A into B. Is the condition of Theorem 5.6 satisfied? Certainly, since for $I \subseteq A$, every vertex of I is joined to $n - k$ vertices in B and every vertex of B is joined to at most $k + 1$ vertices in I. So, if $S(I)$ is the union of all neighbors of the vertices from I, and $E' = E \cap (I \times B)$ is the set of edges in the corresponding subgraph, then

$$|I| \, (n - k) = |E'| \leqslant |S(I)| \, (k + 1).$$

Thus,

$$|S(I)| \geqslant |I| \, (n - k)/(k + 1) \geqslant |I|$$

for every $I \subseteq A$, and Theorem 5.6 gives the desired matching of A into B. $\square$

In terms of (bipartite) graphs, the König–Egerváry theorem is as follows. A *vertex cover* in a bipartite graph G with bipartition A, B is a set of vertices $S \subseteq A \cup B$ such that every edge is incident to at least one vertex from S. A *maximum matching* is a matching of maximum size.

Theorem 5.8. *The maximum size of a matching in a bipartite graph equals the minimum size of a vertex cover.*

How can we find such a matching of maximal size? To obtain a large matching, we could iteratively select an edge disjoint from those previously selected. This yields a matching which is "maximal" in a sense that no more edges can be added to it. But this matching does not need to be a maximum matching: some other matching may have more edges. A better idea is to jump between different matchings so that the new matching will always have one edge more, until we exhaust the "quota" of possible edges, i.e., until we reach the maximal possible number of edges in a matching. This idea employs the notion of "augmenting paths."

Assume that M is a (not necessarily maximum) matching in a given graph G. The edges of M are called *matched* and other edges are called *free*. Similarly, vertices which are endpoints of edges in M are called *matched* (in M); all other vertices are called *free* (in M). An *augmenting path* with respect to M (or *M-augmenting path*) is a path in G such that its edges are alternatively matched and free, and the endpoints of the path are free.

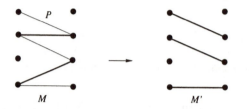

Fig. 5.2. Enlarging the matching M by the M-augmenting path P

If P is an M-augmenting path, then M is certainly not a maximum size matching: the set M' of all *free* edges along this path form a matching with one more edge (see Fig. 5.2). Thus, the presence of an augmenting path implies that a matching is not a maximum matching. Interestingly (and it is a key for the matching algorithm), the converse is also valid: the *absence* of an augmenting path implies that the matching is, in fact, a maximum matching. This result was proved by Berge (1957), and holds for arbitrary graphs.

Theorem 5.9 (Berge 1957). *A matching M in a graph G is a maximum matching if and only if G has no M-augmenting path.*

Proof. We have noted that an M-augmenting path produces a larger matching. For the converse, suppose that G has a matching M' larger than M; we want to construct an M-augmenting path. Consider the graph $H = M \oplus M'$, where $\oplus$ is the symmetric difference of sets. That is, H consists of precisely those edges which appear in *exactly one* of the matchings M and M'.

Since M and M' are matchings, every vertex has at most one incident edge in each of them. This means that in H, every vertex has at most degree 2, and hence, the graph H consists of disjoint paths and cycles. Furthermore, every path or cycle in H alternates between edges of M and edges of M'. This implies that each cycle in H has even length. As $|M'| > |M|$, the graph H must have a component with more edges of M' than of M. Such a component can only be a path that starts and ends with an edge of M'; it remains to observe that every such path in H is an M-augmenting path in G.   □

This theorem suggests the following algorithm to find a maximum matching in a graph G: start with the empty matching $M = \emptyset$, and at each step search for an M-augmenting path in G. In one step the matching is enlarged by one, and we can have at most ℓ such steps, where ℓ is the size of a maximum matching. In general, the computation of augmenting paths is not a trivial task, but for bipartite graphs it is quite easy.

Given a bipartite graph $G = (A, B, E)$ and a matching M in it, construct a *directed* graph G_M by directing all matched edges from A to B and other edges from B to A. Let A_0, B_0 denote the sets of free vertices in A and B, respectively.

Proposition 5.10. *A bipartite graph G has an M-augmenting path if and only if there is a directed path in G_M from a vertex in B_0 to a vertex in A_0.*

We leave the proof of this fact as an exercise.

Using this fact, one may easily design an augmenting path algorithm running in time $O(n^2)$, where n is the total number of vertices. (One can apply, for example, the "depth-first search" algorithm to find a path from A_0 to B_0.) We need to find an augmenting path at most $n/2$ times, hence, the complexity of this matching algorithm is $O(n^3)$. Using a trickier augmenting path algorithm, Hopcroft and Karp (1973) have found a faster algorithm using time $O(n^{5/2})$.

Exercises

5.1.⁻ Let $S_1, \ldots, S_m$ be a sequence of sets such that: (i) each set contains at least r elements (where $r > 0$) and (ii) no element is in more than r of the sets. Show that these sets have a system of distinct representatives.

Hint: See the proof of Corollary 5.2.

5.2.⁻ Show that in a group of m girls and n boys there exist some t girls for whom husbands can be found if and only if any subset of the girls (k of them, say) between them know at least $k + t - m$ of the boys.

Hint: Invite additional $m - t$ "very popular" boys who are known to all the girls. Show that at least t girls can find husbands in the original situation if and only if *all* the girls can find husbands in the new situation. Then apply Hall's theorem to the new situation.

5.3.⁻ Let $S_1, \ldots, S_m$ be a sequence of sets satisfying Hall's condition $(*)$. Suppose that for some $1 \leqslant k < m$, the union $S_1 \cup \cdots \cup S_k$ of the first k sets has precisely k elements. Show that none of the remaining sets $S_{k+1}, \ldots, S_m$ can lie entirely in this union.

5.4.⁺ In Theorem 5.4 we have shown that (as long as $r < n$) we can a new row add to every $r \times n$ Latin rectangle such that the resulting $(r + 1) \times n$ matrix is still Latin. Prove that this can be done in at least $(n - r)!$ ways.

5.5.[+] Let $S_1, \ldots, S_m$ be a sequence of sets each of cardinality at least r. Prove that it has at least

$$f(r, m) = \prod_{i=1}^{\min\{r,m\}} (r + 1 - i)$$

systems of distinctive representatives.

Hint: Follow the proof of Hall's theorem. Case 1 gives at least $r \cdot f(r-1, m-1) = f(r, m)$ and Case 2 at least $f(r, k) \cdot f(\max\{r - k, 1\}, m - k) = f(r, m)$ systems of distinctive representatives.

5.6.[−] Prove that every bipartite graph G with ℓ edges has a matching of size at least $\ell/\Delta(G)$, where $\Delta(G)$ is the maximum degree of a vertex in G.

Hint: Use Theorem 5.8.

5.7. Suppose that M, M' are matchings in a bipartite graph G with bipartition A, B. Suppose that all the vertices of $S \subseteq A$ are matched by M and that all the vertices of $T \subseteq B$ are matched by M'. Prove that G contains a matching that matches all the vertices of $S \cup T$.

5.8 (Lovász et al. 1995). Let $\mathcal{F}$ be a family of sets, each of size at least 2. Let A, B be two sets such that $|A| = |B|$, both A and B intersect all the members of $\mathcal{F}$, and no set of fewer than $|A|$ elements does this. Consider a bipartite graph G with parts A and B, where $a \in A$ is connected to $b \in B$ if there is an $F \in \mathcal{F}$ containing both a and b. Show that this graph has a perfect matching.

Hint: For $I \subseteq A$, let $S(I) \subseteq B$ be the set of neighbors of I in G; show that the set $A' = (A - I) \cup S(I)$ intersects all the members of $\mathcal{F}$.

5.9 (Sperner 1928). Let $t < n/2$ and let $\mathcal{F}$ be a family of subsets of an n-element set X. Suppose that: (i) each member of $\mathcal{F}$ has size at most t, and (ii) $\mathcal{F}$ is an *antichain*, i.e., no member of $\mathcal{F}$ is a subset of another one. Let $\mathcal{F}_t$ be the family of all those t-element subsets of X, which contain at least one member of $\mathcal{F}$. Prove that then $|\mathcal{F}| \leqslant |\mathcal{F}_t|$.

Hint: Use Proposition 5.7 to extend each member of $\mathcal{F}$ to a unique member in the family $\mathcal{F}_t$.

5.10.[−] Let A be a 0-1 matrix with m 1's. Let s be the maximal number of 1's in a row or column of A, and suppose that A has no square $r \times r$ all-1 sub-matrix. Use the König–Egerváry theorem to show that we then need at least $m/(sr)$ all-1 (not necessarily square) sub-matrices to cover all 1's in A.

Hint: There are at least m/s independent 1's, and at most r of them can be covered by one all-1 sub-matrix.

6. Colorings

A town has several clubs, with at least two members in each. One day, two famous lecturers visit the town and make two lectures. The problem is that the lectures are hold in parallel, in two different places and at the same time. Every club would like each of the lectures be visited by at least one of its members. Is it possible to arrange the attendance of inhabitants so that every club will know the contents of both lectures? This is a typical coloring problem: we want to distribute red and blue cards among the inhabitants so that each club will have cards of both colors.

In general, we have a set of points X and a set of $r \geqslant 2$ colors $1, 2, \ldots, r$. An *r-coloring* of X is a mapping $c: X \to \{1, \ldots, r\}$. A subset of X is *monochromatic* (under this coloring) if all its elements get the same color. Given a family $\mathcal{F}$ of subsets of X, a coloring is *legal* (with respect to this family) if no member is monochromatic. If a family admits a legal r-coloring then it is called *r-colorable*. The minimal number of colors r, for which such a legal coloring exists, is the *chromatic number* $\chi(\mathcal{F})$ of $\mathcal{F}$. A family $\mathcal{F}$ is also said to have Property B (in honor of Felix Bernstein) if it is 2-colorable, that is, if $\chi(\mathcal{F}) = 2$. Thus, the question we asked at the beginning is whether the family of clubs in the town has this property.

Graphs are families of 2-element sets (called edges). The coloring problem for graphs has a long history, and has been most intensively studied (see any monograph in graph theory). In this chapter we will mainly consider arbitrary families, but will pay more attention to graphs in Exercises.

6.1 Property B

Which families are 2-colorable? Below we will give several partial answers to this question.

Theorem 6.1. *If $|A \cap B| \neq 1$ for any two sets $A \neq B \in \mathcal{F}$ then $\mathcal{F}$ is 2-colorable.*

Proof (Lovász 1979). Let $X = \{x_1, \ldots, x_n\}$. We will color the points $x_1, \ldots, x_n$ one-by-one so that we do not color all points of any set in $\mathcal{F}$ with the same color. Color the first point x_1 arbitrarily. Suppose that $x_1, \ldots, x_i$

are already colored. If we cannot color the next element x_{i+1} in red then this means that there is a set $A \in \mathcal{F}$ such that $A \subseteq \{x_1, \ldots, x_{i+1}\}$, $x_{i+1} \in A$ and all the points in $A - \{x_{i+1}\}$ are red. Similarly, if we cannot color the next element x_{i+1} in blue, then there is a set $B \in \mathcal{F}$ such that $B \subseteq \{x_1, \ldots, x_{i+1}\}$, $x_{i+1} \in B$ and all the points in $B - \{x_{i+1}\}$ are blue. But then $A \cap B = \{x_{i+1}\}$, a contradiction. Thus, we *can* color the point x_{i+1} either red or blue. Proceeding in this way we will finally color all the points and no set of $\mathcal{F}$ becomes monochromatic. □

Theorem 6.2 (Erdős 1963b). *Let $\mathcal{F}$ be a family of sets, each of size at least k. If $|\mathcal{F}| < 2^{k-1}$ then $\mathcal{F}$ is 2-colorable.*

Proof. Let n be the number of underlying points, and consider all possible colorings of these points in red and blue. For a set $A \in \mathcal{F}$, let $C(A)$ denote the number of colorings after which the set A is either entirely red or entirely blue. Then $C(A) = 2 \cdot 2^{n-|A|} \leqslant 2^{n-k+1}$. Therefore, the number of 2-colorings, which are illegal for $\mathcal{F}$, does not exceed

$$\sum_{A \in \mathcal{F}} C(A) \leqslant \sum_{A \in \mathcal{F}} 2^{n-k+1} = |\mathcal{F}| \cdot 2^{n-k+1}.$$

Since $|\mathcal{F}| < 2^{k-1}$, this number is strictly smaller than 2^n, the total number of colorings. Hence, at least one coloring must be legal. □

Recall that a family if *k-uniform* if each member has exactly k elements. By the previous theorem, small uniform families are colorable. We will now prove that large uniform families are also 2-colorable if they are *intersecting*, i.e., if $A \cap B \neq \emptyset$ for any two sets $A, B \in \mathcal{F}$.

Theorem 6.3 (Erdős–Lovász 1974). *Let $\mathcal{F}$ be k-uniform and intersecting. If $|\mathcal{F}| > k^k$ then $\mathcal{F}$ is 2-colorable.*

Proof. Suppose that $\mathcal{F}$ is *not* 2-colorable. We will prove that then $|\mathcal{F}| \leqslant k^k$. For a set B, define its *degree* in $\mathcal{F}$ as the number $d(B)$ of members of $\mathcal{F}$ containing B. We will inductively construct a sequence of sets $B_i = \{x_1, \ldots, x_i\}$ $(i = 1, \ldots, k)$ such that $d(B_i) \geqslant |\mathcal{F}| \cdot k^{-i}$. For $i = k$ this will immediately give us the desired upper bound $|\mathcal{F}| \leqslant k^k$ since

$$1 \geqslant d(B_k) \geqslant |\mathcal{F}| \cdot k^{-k}.$$

For $i = 1$, let x_1 be a point of maximal degree in $\mathcal{F}$. Then $d(x_1) \geqslant |\mathcal{F}|/k$, and we can take $B_1 \rightleftharpoons \{x_1\}$.

Now suppose that the set B_i is already chosen. This set must avoid at least one set $E \in \mathcal{F}$, since otherwise we could get a legal 2-coloring of $\mathcal{F}$ by coloring all points of B_i in one color, and the rest in the other. Recall that $d(B_i)$ is the number of members of $\mathcal{F}$, all of which contain the set B_i. Since $\mathcal{F}$ is intersecting, the set E must intersect all these members; hence, there must be a point $x \in E$ which belongs to at least $d(B_i)/|E|$ of these members. Since $E \cap B_i = \emptyset$, this point x does not belong to B_i, and we can take $B_{i+1} \rightleftharpoons B_i \cup \{x\}$. Then

$$d(B_{i+1}) \geqslant d(B_i)/|E| \geqslant |\mathcal{F}| \cdot k^{-(i+1)},$$

as desired.   □

6.2 The averaging argument

The next three theorems use the argument known as the *averaging method* (see Sect. 1.5). This method seems to be trivial, but its direct application leads to conclusions which are not so trivial.

6.2.1 Almost good colorings

Suppose we have a family which does not admit a legal coloring. In this situation we would like to legally color as many of its members as possible.

Theorem 6.4. *For every k-uniform family $\mathcal{F}$ there exists a 2-coloring of its points which colors at most $|\mathcal{F}| \cdot 2^{1-k}$ of the sets of $\mathcal{F}$ monochromatically.*

Proof. Let n be the number of underlying points. For a set of points S, let $M(S)$ be the number of monochromatic sets in $\mathcal{F}$ under the coloring which colors all the points in S with one of the colors and the rest with the other. Summing over all subsets S we obtain

$$\sum_{S \subseteq X} M(S) = \sum_{A \in \mathcal{F}} 2 \cdot \big| \{S : S \supseteq A\} \big| = |\mathcal{F}| \cdot 2 \cdot 2^{n-k} = |\mathcal{F}| \cdot 2^{n-k+1}.$$

Thus, there must be at least one set S for which $M(S) \leqslant |\mathcal{F}| \cdot 2^{1-k}$.   □

Let us now consider the situation when we have more than two colors. We say that a subset $A \subseteq X$ is *differently colored* by an r-coloring c if $|c(A)| = |A|$, i.e., if $c(x) \neq c(y)$ for any $x \neq y \in A$. The question about such colorings is related to partitions of families: if some r-coloring colors *all* the sets of an r-uniform family $\mathcal{F}$ differently, then this means that we can partite the points into r mutually disjoint blocks so that every set of $\mathcal{F}$ has exactly one point in each block.

The following result was proved by Erdős (1967) for $r = 2$, and by Erdős and Kleitman (1968) for all r.

Theorem 6.5. *Let $\mathcal{F}$ be an r-uniform family. Then there exists an r-coloring of its points which colors at least $(r!/r^r)|\mathcal{F}| > |\mathcal{F}|/e^r$ of the sets in $\mathcal{F}$ differently.*

Proof. Let n be the number of underlying points. For an r-coloring c of points, let $D(c)$ denote the number of sets in $\mathcal{F}$ which are differently colored by c. Then summing over all r-colorings we obtain

$$\sum_c D(c) = \sum_{A \in \mathcal{F}} |\{c : |c(A)| = r\}| = \sum_{A \in \mathcal{F}} r! \cdot r^{n-r} = |\mathcal{F}| \cdot r! \cdot r^{n-r}.$$

Since there are exactly r^n r-colorings of underlying points, there must be a coloring c with $D(c) \geqslant |\mathcal{F}| \cdot r!/r^r$, as desired. $\qquad\square$

6.2.2 The number of mixed triangles

Let K_n be a complete graph on n vertices and color its edges with two colors, red and blue. It can be shown (see Exercises 27.3 and 27.4) that (independent of the coloring) at least one quarter of all possible $\binom{n}{3}$ triangles will be monochromatic (all three edges of the same color). Here we ask the opposite question: how many rectangles will be *mixed*, i.e., will have edges of different colors? To make this question non-trivial, we have to somehow restrict the set of all possible colorings. A natural restriction is to require that a coloring is *balanced*, i.e., that the number of red edges is equal to the number of blue edges ± 1.

To motivate this question, let us mention that it arises, for example, when dealing with the (nondeterministic) communication complexity of the triangle-freeness property of graphs. In this game we have three players, Alice, Bob and Carole. Before the game starts, the players choose some balanced coloring of the edges of K_n. After that the first two players have only partial information about the subgraphs of K_n: Alice can see only the subgraph corresponding to the red edges, and Bob can see only the subgraph corresponding to the blue edges of K_n.

A graph is *triangle-free* if it has no triangles. Given a triangle-free subgraph G of K_n, Carole's goal is to convince Alice and Bob that G indeed has no triangles. For this purpose, she announces some binary string, a *certificate*, to both players. Having this certificate, Alice and Bob verify it *independently* and respond with either Yes or No. Alice and Bob accept the graph (as triangle-free) if and only if they both replied with Yes. The communication complexity of this game is the length of the certificate in the worst case.

Of course, Carole can always convince the players that an input graph has *any* particular property using a certificate of length $\binom{n}{2}$: she just announces the binary code of the entire graph to both players. However, for some graph properties much fewer bits are enough.

For example, Carole can easily convince Alice and Bob that a graph G *has* a triangle: using only $3\lceil \log_2 n \rceil$ bits she announces the binary code of a triangle in G; Alice and Bob can locally check whether the edges of this triangle she/he should see are indeed present. But to convince the players that a graph has *no* triangles, Carole must announce almost entire graph!

Theorem 6.6. (Papadimitriou–Sipser 1984) *The communication complexity of the triangle-freeness property of graphs on n vertices is $\Omega(n^2)$.*

The crucial step in the proof of this theorem is to show that if we color the edges of a complete n-vertex graph K_n in red and blue, and if the coloring is balanced, then at least $\Omega(n^2)$ of triangles will be mixed and, moreover, these triangles do not "collide."

We specify a triangle by exhibiting one of its edges (the *fixed* edge of the triangle) and the vertex (the *top* vertex of the triangle) adjacent with both endpoints of that edge; the remaining two edges are *free* edges of the triangle. Thus, given any three vertices, there are three possibilities to specify a triangle on them. Such a triangle is *mixed* if its free edges have different colors. A set Δ of triangles is *collision-free* if no two triangles share a free edge and no new triangle can be formed by taking edges from different triangles in Δ.

Lemma 6.7. *There is a constant $\epsilon > 0$ such that every balanced coloring of the edges of K_n produces a collision-free set of at least ϵn^2 triangles.*

Proof. Call a vertex *blue* (resp., *red*) if more than $0.8n$ of its incident edges are blue (resp., red). If a vertex is neither blue nor red then it is *mixed*. Let B, R and M denote, respectively, the set of blue, red and mixed vertices.

Claim 6.8. *There are at least $0.1n$ mixed vertices.*

Proof of Claim 6.8. First, observe that at most $0.5n$ vertices are blue and at most $0.5n$ vertices are red, for otherwise more than half of all $\binom{n}{2} \leqslant 0.5n^2$ edges would have the same color. To see this, let $v_1, \ldots, v_m$ be blue vertices, and let d_i be the number of blue edges incident to v_i and to none of the vertices $v_1, \ldots, v_{i-1}$. As $d_i \geqslant 0.8n - i + 1$, we have at least

$$\sum_{i=1}^{m} d_i \geqslant 0.8nm - \binom{m}{2}$$

blue edges, which is more than $0.25n^2$ if, say, $m \geqslant 0.5n$. (As we are interested only in the order, we are quite generous in the choice of concrete constants.)

Now assume that there are *fewer* than $0.1n$ mixed vertices. Then there must be at least $n - (|R| + |M|) \geqslant 0.4n$ blue vertices and at least $0.4n$ red vertices. Hence, $|B|, |R| \geqslant 0.4n$. But for each blue vertex $v \in B$, more than $|R| - 0.2n \geqslant \frac{1}{2}|R|$ of the edges going to the vertices in R must be blue, for otherwise at least $0.2n$ of its incident edges would be red. Hence, for the blue vertices to meet their quota of blue edges, *more* than half of all edges between B and R must be blue. Symmetrically, *more* than half of edges between B and R must be red. The obtained contradiction finishes the proof of the claim. $\square$

Select now $\lfloor 0.1n \rfloor$ of mixed vertices and call them *top* vertices. Each such vertex has at least $0.2n$ blue edges and at least $0.2n$ red edges, implying that each top vertex has at least $0.1n$ blue edges and at least $0.1n$ red edges to *bottom* (non-top) vertices. Let E be the set of all edges between bottom vertices. The *weight* $w(e)$ of an edge $e \in E$ is the number of top vertices that are connected to this edge by edges of two different colors. Take a sufficiently small constant $c > 0$ ($c = 0.0005$ is sufficient for our purposes), and let $P \subseteq E$ be the set of edges of weight at least cn; we call such edges *popular*.

The sum of all weights is

$$\sum_{e \in E} w(e) \geqslant 2cn^3,$$

since each of the $0.1n$ top vertices contributes at least $(0.1n)^2$ to this sum. Non-popular edges contribute at most

$$\sum_{e \in E-P} w(e) \leqslant |E - P| \cdot (cn) \leqslant cn^3.$$

As there are only $0.1n$ top vertices, each popular edge $e \in P$ can also contribute at most $w(e) \leqslant 0.1n$; hence,

$$\sum_{e \in P} w(e) \leqslant |P| \cdot (0.1n).$$

This implies that many of edges in E must be popular:

$$|P| \geqslant \frac{2cn^3 - cn^3}{0.1n} \geqslant 10cn^2.$$

Let $M \subseteq P$ be a maximal set of popular edges where the pairs are mutually disjoint, i.e., form a matching. Since the set M is maximal, each of $10cn^2$ popular edges must share a vertex with an edge in M. But one edge can be incident with at most $2(n-2) < 2n$ other edges, implying that $|M| \geqslant |P|/(2n) \geqslant 5cn > cn$. Each edge $e \in M$, together with at least cn of top vertices, induces at least cn mixed triangles; e is the fixed edge of these triangles. Hence, the total number of mixed triangles induced by the edges in M is at least $|M| \cdot (cn) \geqslant c^2 n^2$. Since M is a matching and none of top vertices is incident with an edge in M, the set of induced triangles is collision-free.

$\square$

Proof of Theorem 6.6. Let Δ be a collision-free set of triangles, guaranteed by Lemma 6.7. We can form a set $\mathcal{G}$ of graphs by picking from each of the triangles in Δ its fixed edge and *precisely one* of its free edges. Collision-freeness of Δ ensures that none of the obtained graphs contains a triangle, and we have $|\mathcal{G}| = 2^{|\Delta|}$ such graphs in total. But by the construction, the union of any two graphs in $\mathcal{G}$ *contains* a triangle!

We claim that for every graph from $\mathcal{G}$, Carole must use a different certificate, implying that the binary length of a certificate must be at least $\log |\Delta| = \Omega(n^2)$.

To show this, assume that Carole uses the same certificate for two different graphs G_1 and G_2 in $\mathcal{G}$. The union $G = G_1 \cup G_2$ of these graphs contains at least one triangle $t \in \Delta$. Each of the players, Alice and Bob, can see only one of its two free edges, and each of them replied with Yes on both G_1 and G_2. Since the players have to verify the certificate *independently* and every free edge of t is present in only one of the graphs G_1 or G_2, the players are forced to reply with Yes also on G, thus (wrongly) accepting a graph with a triangle, a contradiction. $\qquad\square$

6.3 Coloring the cube: the algorithmic aspect

Untill now we have mainly been interested in the mere *existence* of a desired coloring. The next natural question is how to *construct* it. Since the number of possible 2-colorings of n points is finite, we can just exhaustively search all 2^n colorings one-by-one until the first legal coloring is found. In the worst case, this "dummy" strategy will require a huge (exponential in n) number of steps, and it would be interesting to know if any "more clever" strategy could find the coloring quicker. Of course, the answer depends on what we can do in one step. In this section we will show one situation where an exponential number of steps is necessary, independent on how clever the coloring procedure is.

We consider the n-cube $\{0,1\}^n$. A *subcube* $C \subseteq \{0,1\}^n$ is defined by fixing some k $(0 \leqslant k \leqslant n)$ components to constants 0 and 1:

$$C = \{(x_1,\ldots,x_n) : x_{i_1} = \epsilon_{i_1},\ldots,x_{i_k} = \epsilon_{i_k}\}.$$

If we use $*$ in a particular component to indicate that the component is not fixed (i.e., it can take both 0 and 1 as values), then each subcube can be represented as a string $C \in \{0,1,*\}^n$. For example, $C = (*1*0)$ denotes the subcube of $\{0,1\}^4$, where the second component is fixed to 1 and the last is fixed to 0. We say that a 0-1 vector is even (odd) if it contains an even (odd) number of 1's.

The n-cube $\{0,1\}^n$ is said to be *properly colored* if each even vector is colored white and each odd vector is colored black. How many steps do we need in order to produce such a coloring if each primitive step of an algorithm consists in specifying a single subcube and coloring it white or black? (This question aroses in the context of proving lower bounds on the length of width-2 branching programs; see Borodin et al. (1986) for details.) That is, we consider the following model of coloring algorithms:

1. Initially, no vectors are colored.
2. At each step we choose a subcube and either color all its vectors white or color all of them black.
3. Once a vector has been given a color, it keeps that color even if this vector is an element of a subcube colored by a later step.

For example, it is easy to design such an algorithm which properly colors the n-cube in $2^{n-1} + 1$ steps: in the first 2^{n-1} steps, color every singular even vector white, and then color the entire n-cube blue. Can we do better? The question is not that simple as it might seem at first glance.

Consider, for example, the case when $n = km$ for some integers k and m. Then each subcube $C \in \{0, 1, *\}^n$ can be split into k subcubes $C = (C_1, \ldots, C_k)$ with each $C_i \in \{0, 1, *\}^m$. Let us run the following k-parts algorithm:

> For each $j = 0, 1, \ldots, k$, run through all subcubes $C = (C_1, \ldots, C_k)$ such that j of the C_i's are $\{*\}^m$ and the remaining $k - j$ of the C_i's are even vectors in $\{0, 1\}^m$; color C white if j is even, and color C black if j is odd.

For example, if we run the algorithm on the 4-cube, with $k = m = 2$, it will perform the following steps:

> $j = 0$: Color (0000), (0011), (1100), and (1111) white.
> $j = 1$: Color $(00 * *)$, $(11 * *)$, $(* * 00)$, and $(* * 11)$ black.
> $j = 2$: Color $(* * * *)$ white.

Proposition 6.9. *If $n = km$ where both k and m are integers, the k-parts algorithm colors the n-cube in $(2^{m-1} + 1)^k$ steps.*

Proof. It is not difficult to show by induction that, if some vector $y = y_1 y_2 \cdots y_k$, with all $y_i \in \{0, 1\}^m$, is colored by the algorithm for the first time at step j, then j of the y_i's are odd and $k - j$ of the y_i's are even. Now, if j is even, then there are an even number of odd subvectors, so the vector y is even and the algorithm correctly colors it white. If j is odd, the number of odd subvectors is odd, making the vector y odd, and the algorithm correctly colors the vector black. Thus, the algorithm is correct.

To estimate the number of steps, observe that, for each j, there are $\binom{k}{j}$ possibilities for the all-* subcubes C_i, and $(2^{m-1})^{k-j}$ possible even subvectors for the remaining C_i's. Thus, at the jth stage the algorithm colors $\binom{k}{j}(2^{m-1})^{k-j}$ subcubes, and the total number of steps used is

$$\sum_{j=0}^{k} \binom{k}{j} (2^{m-1})^{k-j} = (2^{m-1} + 1)^k.$$

$\square$

When $m = 3$, the k-part algorithm requires only $5^{n/3} \approx 1.7^n$ steps. This algorithm is due to B. Plumstead and J. Plumstead (1985). On the other hand, they have proved that we cannot do much better: any algorithm will require more than 1.5^n steps.

Theorem 6.10. *Every algorithm which colors the n-cube requires at least $2 \left(\frac{3}{2}\right)^{n-1}$ steps.*

Proof. Let S_n be the minimum possible number of steps of a coloring algorithm for the n-cube. We have to prove that $S_n \geqslant 2 \left(\frac{3}{2}\right)^{n-1}$. For the initial values $n = 1, 2$, we have $S_1 = 2$ and $S_2 = 3$. So, the desired lower bound follows directly from the recursion

$$S_n \geqslant \tfrac{3}{2} S_{n-1},$$

which we are going to prove now.

Take a shortest algorithm P for coloring the n-cube. This algorithm can be written as a sequence $(C_1, a_1), (C_2, a_2), \ldots, (C_s, a_s)$, where C_i is the subcube colored at the ith step and a_i is the color it is given; the *length* $|P|$ of this procedure is the number s of steps.

Look at the first component of the subcube C_i. It is either $0, 1$ or $*$. So C_i can be written as $0C'_i, 1C'_i$ or $*C'_i$, where C'_i is a subcube of the $(n-1)$-cube. Let P_0 and P_1 be the algorithms defined by the subsequences of P which include exactly those steps i for which the first component of C_i is 0 or 1, respectively. Similarly, let P_0^* (P_1^*) be the algorithm defined by the subsequence of P which includes exactly those steps i for which the first component of C_i is 0 or $*$ (respectively, 1 or $*$).

It is clear that $|P| = |P_0^*| + |P_1| = |P_1^*| + |P_0|$. Moreover, P_0^* and P_1^* can both be easily modified to color the $(n-1)$-cube. For P_0^* we can simply strip off the first component of every subcube used, and for P_1^* we can strip off the first component and switch all the colors. Thus,

$$|P| \geqslant S_{n-1} + \max(|P_0|, |P_1|).$$

Since $|P| = S_n$, it remains to show that $|P_0| + |P_1| \geqslant S_{n-1}$. For this, it is again enough to show that P can be modified to color the $(n-1)$-cube. Consider a vector y in the $(n-1)$-cube, and look for the first time when $0y$ or $1y$ was colored in the algorithm P. Say this occurred in step i. The subcube C_i can be written as $0C'_i, 1C'_i$ or $*C'_i$, where C'_i is a subcube of the $(n-1)$-cube. Then $y \in C'_i$ and $y \notin C'_j$ for any $j < i$. Moreover, the situation $C_i = *C'_i$ is impossible, because then P would color both vectors $0y$ and $1y$ with the same color; but we assumed that P is correct, so this cannot happen. Thus, for every vector y from the $(n-1)$-cube, during the coloring procedure P, the extended vector $0y$ or $1y$ can appear for the first time only in the cube C_i where the first coordinate is specified. Now, omit from the sequence P all those steps (C_i, a_i) where the first coordinate of C_i was not specified, and then strip off the first component of all the remaining subcubes C_i, and switch all the colors iff $C_i = 1C'_i$. By the previous discussion, the obtained algorithm correctly colors the $(n-1)$-cube and has length $|P_0| + |P_1|$, as desired. $\square$

Exercises

6.1.$^-$ Let G_1, G_2 be two graphs. Prove that $\chi(G_1 \cup G_2) \leqslant \chi(G_1) \cdot \chi(G_2)$.
Hint: Use pairs of colors to color $G_1 \cup G_2$.

6.2. $^-$ Let G be a graph on n vertices. A complement $\overline{G}$ of a graph G is a graph on the same set of vertices in which two vertices are adjacent if and only if they are non-adjacent in G. Prove that $\chi(G) \cdot \chi(\overline{G}) \geqslant n$ and $\chi(G) + \chi(\overline{G}) \geqslant 2\sqrt{n}$. *Hint*: $(\chi(G) - \chi(\overline{G}))^2 \geqslant 0$.

6.3. Prove that $\chi(G) \leqslant \Delta(G) + 1$, where $\Delta(G)$ is the maximum degree of a vertex in G.

> *Hint*: Order the vertices $v_1, \dots, v_n$ and use greedy coloring: assign to v_i the smallest-indexed color not already used on its lower-indexed neighbors.

6.4 (Welsch–Powell 1967). Let G be a graph on n vertices, whose degrees are $d_1 \geqslant d_2 \geqslant \dots \geqslant d_n$. Prove that $\chi(G) \leqslant 1 + \max_i \min\{d_i, i - 1\}$.

> *Hint*: Apply the greedy algorithm from the previous exercise. When we color the ith vertex, at most $\min\{d_i, i - 1\}$ of its neighbors have already been colored, so its color is at most $1 + \min\{d_i, i - 1\}$.

6.5 $^+$ (Füredi–Kahn 1986). Let $\mathcal{F}$ be a family of *rank a*, i.e., each member has at most a points, and suppose that no point belongs to more than b members of $\mathcal{F}$. Prove that then it is possible to color the points in $r = (a - 1)b + 1$ colors so that every member of $\mathcal{F}$ is differently colored, i.e., no member of $\mathcal{F}$ has two points of the same color.

> *Sketch*: By induction on a. The case $a = 2$ is Exercise 6.3. For the induction step, select a sequence of points $V = \{x_1, x_2, \dots, x_m\}$ by the following rule: at the ith step take a set $F \in \mathcal{F}$ disjoint from $\{x_1, \dots, x_{i-1}\}$, and let x_i be an arbitrary point in this set. If we delete the points V from all members of $\mathcal{F}$, we obtain a family $\mathcal{F}'$ of rank at most $a - 1$. By the induction hypothesis, $\mathcal{F}'$ can be differently colored using only $(a - 2)b + 1$ colors. So, it remains to color the deleted points. For this, consider the graph $G = (V, E)$ where two points are joined by an edge iff both these points belong to some member of $\mathcal{F}$. Show that $\Delta(G) \leqslant b - 1$ and apply Exercise 6.3.

6.6. $^-$ Let $G = (V, E)$ be a graph and $S \subseteq V$ a subset of its vertices. The induced subgraph of G is the graph $G[S]$ on vertices S, in which two vertices are adjacent if and only if they are such in the original graph G. Prove that for any graph G we can find a partition $V = S \cup T$ of its vertices into two disjoint non-empty subsets S and T such that $\chi(G[S]) + \chi(G[T]) = \chi(G)$.

6.7. $^-$ A graph G is *k-critical* if $\chi(G) = k$ but $\chi(H) < k$ for every proper subgraph H of G. Let $\delta(G)$ denote the minimum degree of a vertex in G. Prove the following: if G is a k-critical graph, then $\delta(G) \geqslant k - 1$.

> *Sketch*: Assume there is a vertex $x \in V$ of degree at most $k - 2$, and consider the induced subgraph $H = G[V - \{x\}]$. Graph H must have a legal $(k - 1)$-coloring, and at least one of these $k - 1$ colors is not used to color the neighbors of x; we can use it for x.

6.8 (Szekeres–Wilf 1968). Prove that $\chi(G) \leqslant 1 + \max_{H \subseteq G} \delta(H)$ holds for any graph G. *Hint*: Let $k = \chi(G)$, take a k-critical subgraph H of G and use the previous estimate.

6.9.⁻ Let G be a directed graph without cycles and suppose that G has no path of length k. Prove that then $\chi(G) \leqslant k$.

Hint: Let $c(x)$ denote the maximum length of a path starting from x. Then c is a coloration with colors $0, 1, \ldots, k-1$. Show that it is legal.

6.10.⁻ Let G be a graph on n vertices, and $\alpha(G)$ be its *independence number*, i.e., the maximal number of vertices, no two of which are joined by an edge. Show that $n/\alpha(G) \leqslant \chi(G) \leqslant n - \alpha(G) + 1$.

6.11. It is clear that $\chi(G) \geqslant \omega(G)$, where $\omega(G)$ is the *clique number* of G, i.e., the maximum size of a clique in G. Erdős (1947) has proved that, for every large enough n, there exists an n-vertex graph G such that $\omega(G) \leqslant 2\log_2 n$ and $\omega(\overline{G}) \leqslant 2\log_2 n$ (see Theorem 18.1 for a proof). Use this result to show that the gap between $\chi(G)$ and $\omega(G)$ can be quite large: the maximum of $\chi(G)/\omega(G)$ over all n-vertex graphs G is $\Omega\left(n/(\log_2 n)^2\right)$. *Hint*: $\chi(G) \geqslant n/\omega(\overline{G})$.

6.12⁺ (Lovász 1979; problem 13.35). Let $\mathcal{F}$ be a family, each member of which has $\geqslant 3$ points and any two members share exactly one point in common. Suppose also that $\mathcal{F}$ is not 2-colorable, i.e., that $\chi(\mathcal{F}) > 2$. Prove that: (i) every point x belongs to at least two members of $\mathcal{F}$, and (ii) any two points x, y belong to at least one member of $\mathcal{F}$.

Hint: (i) Take $x \in A \in \mathcal{F}$, color $A - \{x\}$ red and the rest blue. (ii) Select sets A, B such that $x \in A - B$ and $y \in B - A$; color $(A \cup B) - \{x, y\}$ red and everything else blue.

6.13 (Lovász 1973). Let $\mathcal{F}$ be 3-uniform family on $n \geqslant 5$ points, in which each pair of points occurs in the same number of sets. Prove that $\mathcal{F}$ is not 2-colorable.

Sketch: Suppose there is a 2-coloring, count the members of $\mathcal{F}$ in two ways: by the monochromatic pairs contained in them and also by the bichromatic pairs contained in them. Let n_1 and n_2 denote the number of red and blue points, respectively, and let a be the number of members of $\mathcal{F}$ containing a given pair of points. We have $a\binom{n_1}{2}$ sets in $\mathcal{F}$ containing a pair $\{x, y\}$ of red points, and $a\binom{n_2}{2}$ sets containing a blue pair of points. Hence, $|\mathcal{F}|$ is the sum of these two numbers. On the other hand, each set of $\mathcal{F}$ contains exactly two pairs $\{x, y\}$ where x is blue and y is red; so $2|\mathcal{F}| = an_1n_2$. Compare these numbers, and use the arithmetic-geometric mean inequality (see Proposition 1.11) to show that the equality can hold only if $n \leqslant 4$.

6.14.⁻ Let $G = (V, E)$ be a graph, and $(C_v)_{v \in V}$ be a sequence of (not necessarily disjoint) sets. We can look at each set C_v as a color set (or a "palette") for the vertex v. Given such a list of color sets, we consider only colorings c such that $c(v) \in C_v$ for all $v \in V$, and call them *list colorings* of G. As before, a coloring is *legal* if no two adjacent vertices receive the same color. The *list chromatic number* $\chi_\ell(G)$ is the smallest number k such that for *any* list of color sets C_v with $|C_v| = k$ for all $v \in V$, there always exists a legal list coloring of G. Of course, $\chi_\ell(G) \leqslant |V|$. Show that $\chi(G) \leqslant \chi_\ell(G) \leqslant \Delta(G) + 1$.

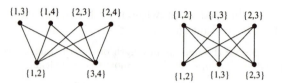

{1,3} {1,4} {2,3} {2,4}   {1,2}   {1,3}   {2,3}

{1,2}   {3,4}   {1,2} {1,3} {2,3}

Fig. 6.1. The graphs $K_{2,4}$ and $K_{3,3}$ with a particular lists of color sets

6.15.⁻ Let $K_{2,4}$ be a complete bipartite graph with parts of size 2 and 4 (see Fig. 6.1). Show that $\chi(K_{2,4}) = 2$ but $\chi_\ell(K_{2,4}) = 3$. What is $\chi_\ell(K_{3,3})$?

Hint: Use the list of color sets given in Fig 6.1.

6.16. Generalize the above construction for $K_{3,3}$ to find graphs G where $\chi(G) = 2$, but $\chi_\ell(G)$ is arbitrarily large. For this, consider the complete bipartite graph $G = V_1 \times V_2$ whose parts V_1 and V_2 consist of all k-subsets v of $\{1, \ldots, 2k - 1\}$. Define the pallete C_v of a vertex (k-subset) v to be the subset v itself. Show that $\chi_\ell(G) > k$.

Hint: Observe that we need at least k colors to color V_1 and at least k colors to color V_2.

6.17. Let S_n be a graph which has vertex set the n^2 entries of an $n \times n$ matrix with two entries adjacent if and only if they are in the same row or in the same column. Show that $\chi_\ell(S_n) \geqslant n$. *Hint*: Any legal coloring of S_n corresponds to Latin square.

Comment: The problem, whether $\chi_\ell(S_n) = n$, was raised by Jeff Dinitz in 1978. Janssen (1992) has proved that $\chi_\ell(S_n) \leqslant n+1$, and the final solution $\chi_\ell(S_n) = n$ was found by Galvin (1995).

6.18.⁺ In Theorem 6.7 we require that the coloring is balanced. Extend this theorem to the case of *almost balanced* colorings where we only require that, say, at least a quarter of edges get blue color and at least a quarter of edges get red color.

6.19. A *maximal triangle-free* graph is one that does not contain a triangle, but the addition of any new edge would create a triangle. Barefoot et al. (1995) have proved that $f(n) \leqslant 2^{n^2/4}$. Show that for n divisible by 4, $f(n) \geqslant 2^{n^2/8}$.

Hint: Fix a partition of the vertex set into two parts of equal size, take a maximal matching M in the first part, and consider the family of graphs, each of which is obtained by joining every vertex in the second part with *precisely one* endpoint of each of the edges in M.

Part II

Extremal Set Theory

7. Sunflowers

One of most beautiful results in extremal set theory is the so-called *Sunflower Lemma* discovered by Erdős and Rado (1960) asserting that in a sufficiently large uniform family, some highly regular configurations, called "sunflowers," must occur, regardless of the size of the universe. In this chapter we will consider this result as well as some of its modifications and applications.

7.1 The sunflower lemma

A *sunflower* (or *Δ-system*) with k petals and a *core Y* is a collection of sets $S_1, \ldots, S_k$ such that $S_i \cap S_j = Y$ for all $i \neq j$; the sets $S_i - Y$ are petals, and we require that none of them is empty. Note that a family of pairwise disjoint sets is a sunflower (with an empty core).

Fig. 7.1. A sunflower with 8 petals

Sunflower Lemma. *Let $\mathcal{F}$ be family of sets each of cardinality s. If $|\mathcal{F}| > s!(k-1)^s$ then $\mathcal{F}$ contains a sunflower with k petals.*

Proof. We proceed by induction on s. For $s = 1$, we have more than $k - 1$ points (disjoint 1-element sets), so any k of them form a sunflower with k petals (and an empty core). Now let $s \geqslant 2$, and take a maximal family $\mathcal{A} = \{A_1, \ldots, A_t\}$ of pairwise disjoint members of $\mathcal{F}$.

If $t \geqslant k$, these sets form a sunflower with $t \geqslant k$ petals (and empty core), and we are done.

Assume that $t \leqslant k - 1$, and let $B = A_1 \cup \cdots \cup A_t$. Then $|B| \leqslant s(k - 1)$. By the maximality of $\mathcal{A}$, the set B intersects every member of $\mathcal{F}$. By the pigeonhole principle, some point $x \in B$ must be contained in at least

$$\frac{|\mathcal{F}|}{|B|} > \frac{s!(k - 1)^s}{s(k - 1)} = (s - 1)!(k - 1)^{s-1}$$

members of $\mathcal{F}$. Let us delete x from these sets and consider the family

$$\mathcal{F}_x = \{S - \{x\} \; : \; S \in \mathcal{F}, x \in S\}.$$

By the induction hypothesis, this family contains a sunflower with k petals. Adding x to the members of this sunflower, we get the desired sunflower in the original family $\mathcal{F}$. $\Box$

It is not known if the bound $s!(k - 1)^s$ is the best possible. Let $f(s, k)$ denote the least integer so that any s-uniform family of $f(s, k)$ sets contains a sunflower with k petals. Then

$$(k - 1)^s < f(s, k) \leqslant s!(k - 1)^s + 1. \tag{7.1}$$

The upper bound is the sunflower lemma, the lower bound is Exercise 7.2. The gap between the upper and lower bound for $f(s, k)$ is still huge (by a factor of $s!$).

Conjecture 1 (Erdős and Rado). *For every fixed k there is a constant $C = C(k)$ such that $f(s, k) < C^s$.*

The conjecture remains open even for $k = 3$ (note that in this case the sunflower lemma requires at least $s! 2^s \approx s^s$ sets). Several authors have slightly improved the bounds in (7.1). In particular, J. Spencer has proved

$$f(s, 3) \leqslant e^{c\sqrt{s}} s!.$$

For s fixed and k sufficiently large, Kostochka et al. (1999) have proved

$$f(s, k) \leqslant k^s \left(1 + ck^{-2^{-s}}\right),$$

where c is a constant depending only on s.

But the proof or disproof of the conjecture is nowhere in sight.

A family $\mathcal{F} = \{S_1, \ldots, S_m\}$ is called a *weak Δ-system* if there is some λ such that $|S_i \cap S_j| = \lambda$ whenever $i \neq j$. Of course, not every such system is a sunflower: in a weak Δ-system it is enough that all the cardinalities of mutual intersections coincide whereas in a sunflower we require that these intersections all have the same elements. However, the following interesting result due to M. Deza states that if a weak Δ-system has many members then it is, in fact, "strong," i.e., forms a sunflower. We state this result without proof.

Theorem 7.1 (Deza 1973). *Let $\mathcal{F}$ be an s-uniform weak Δ-system. If $|\mathcal{F}| \geqslant s^2 - s + 2$ then $\mathcal{F}$ is a sunflower.*

The family of lines in a projective plane of order $s - 1$ shows that this bound is optimal (see Exercise 7.1).

A related problem is to estimate the maximal possible number $F(n, k)$ of members in a family $\mathcal{F}$ of subsets of an n-element set such that $\mathcal{F}$ does not contain a weak Δ-system with k members. It is known that

$$2^{0.01(n \ln n)^{1/3}} \leqslant F(n, 3) \leqslant 1.99^n.$$

The upper bound was proved by Frankl and Rödl (1987), and the lower bound by Kostochka and Rödl (1998).

7.2 Modifications

Due to its importance, the sunflower lemma was modified in various directions. If $S_1, \ldots, S_k$ form a sunflower with a core Y, then we have two nice properties:

(a) the core Y lies *entirely* in all the ~~petals~~ $S_1, \ldots, S_k$;
(b) the sets $S_1 - Y, \ldots, S_k - Y$ are mutually disjoint.

It is therefore natural to look at what happens if we relax any of these two conditions.

7.2.1 Relaxed core

We can relax property (a) and require that only the differences $S_i - Y$ be non-empty and mutually disjoint for *some* set Y.

Given a family $\mathcal{F} = \{S_1, \ldots, S_k\}$, its *common part* is the set

$$Y(\mathcal{F}) = \bigcup_{i \neq j} (S_i \cap S_j).$$

Note that, if $\mathcal{F}$ is s-uniform and if its common part Y has fewer than s elements, then all the sets $S_1 - Y, \ldots, S_k - Y$ are mutually disjoint.

Lemma 7.2 (Füredi 1980). *Let $\mathcal{F}$ be a family of sets each of cardinality at most s. If $|\mathcal{F}| > k^s$ then the common part of some $k + 1$ of its members has fewer than s elements.*

Proof. The cases $k = 1$ and $s = 1$ are trivial. Apply induction on k. Once k is fixed, apply induction on s. We may assume that $\mathcal{F}$ contains at least one set with exactly s elements. Fix such a set $B_0 \in \mathcal{F}$, and define

$$\mathcal{F}(B) = \{S - B \ : \ S \in \mathcal{F}, \ S \cap B_0 = B\}, \quad \text{for all } B \subseteq B_0.$$

We claim that

$$|\mathcal{F}(B)| > (k - 1)^{s - |B|} \tag{7.2}$$

for at least one $B \subseteq B_0$. Indeed, otherwise we would have

$$|\mathcal{F}| = \sum_{B \subseteq B_0} |\mathcal{F}(B)| = \sum_{i=0}^{s} \sum_{B \subseteq B_0, |B|=i} |\mathcal{F}(B)| \leqslant \sum_{i=0}^{s} \binom{s}{i}(k-1)^{s-i} = k^s,$$

a contradiction with $|\mathcal{F}| > k^s$.

Fix the set B, guaranteed by (7.2) and apply the induction hypothesis to $\mathcal{F}(B)$. This gives us a family $S_1 - B, \ldots, S_k - B$ with all $S_i \in \mathcal{F}$, whose common part Y has less than $s - |B|$ elements. Add the set B to all these k sets, and consider the family $S_1, \ldots, S_k, B_0$. These sets belong to $\mathcal{F}$, and their common part has $|Y| + |B| < (s - |B|) + |B| = s$ elements, as desired. □

7.2.2 Relaxed disjointness

What if we relax the disjointness property (b) of sunflowers, and require only that the differences $S_1 - Y, \ldots, S_k - Y$ cannot be intersected (blocked) by a set of size smaller than k? In this case we say that sets $S_1, \ldots, S_k$ form a "flower."

A *blocking set* of a family $\mathcal{F}$ is a set which intersects all the members of $\mathcal{F}$; the minimum number of elements in a blocking set is the *blocking number* of $\mathcal{F}$ and is denoted by $\tau(\mathcal{F})$; if $\emptyset \in \mathcal{F}$ then we set $\tau(\mathcal{F}) = 0$.

A *restriction* of a family $\mathcal{F}$ onto a set Y is the family

$$\mathcal{F}_Y \rightleftharpoons \{S - Y : S \in \mathcal{F}, S \supseteq Y\}.$$

A *flower* with k petals and a *core* Y is a family $\mathcal{F}$ such that $\tau(\mathcal{F}_Y) \geqslant k$.

Not every flower is a sunflower (give an example). Håstad et al. (1995) observed that the proof of the sunflower lemma can be easily modified to yield a similar result for flowers.

Lemma 7.3. *Let $\mathcal{F}$ be a family of sets each of cardinality s. If $|\mathcal{F}| > (k-1)^s$ then $\mathcal{F}$ contains a flower with k petals.*

Proof. Induction on s. The basis $s = 1$ is trivial. Now suppose that the lemma is true for $s-1$ and prove it for s. Take a family $\mathcal{F}$ of sets each of cardinality ~~at most~~ s, and assume that $|\mathcal{F}| > (k-1)^s$. If $\tau(\mathcal{F}) \geqslant k$ then the family $\mathcal{F}$ itself is a flower with at least $(k-1)^s + 1 \geqslant k$ petals (and an empty core). Otherwise, some set of size $k-1$ intersects all the members of $\mathcal{F}$, and hence, at least $|\mathcal{F}|/(k-1)$ of the members must contain some point x. The family $\mathcal{F}_x \rightleftharpoons \{S - \{x\} : S \in \mathcal{F}, x \in S\}$ has

$$|\mathcal{F}_x| \geqslant \frac{|\mathcal{F}|}{k-1} > (k-1)^{s-1}$$

members, each of cardinality $s - 1$. By the induction hypothesis, the family $\mathcal{F}_x$ contains a flower with k petals and some core Y, $x \notin Y$. Adding the element x back to the sets in this flower, we obtain a flower in $\mathcal{F}$ with the same number of petals and the core $Y \cup \{x\}$. □

7.3 Applications

The sunflower lemma and its modifications have many applications in complexity theory. In particular, the combinatorial part of the celebrated lower bounds argument for monotone circuits, found by Razborov (1985), is based on this lemma and on its modification due to Füredi (Lemma 7.2). Andreev (1987) has also used his modification (Exercise 7.5) to prove exponential lower bounds for such circuits. In this section we will show how the last modification (Lemma 7.3) can be used to obtain some information about the number of minterms and to prove lower bounds for small depth non-monotone circuits.

7.3.1 The number of minterms

Let $x_1, \ldots, x_n$ be boolean variables taking their values in $\{0, 1\}$. A *monomial* is an And of literals, and a *clause* is an Or of literals, where a *literal* is either a variable x_i or its negation $\bar{x}_i = x_i \oplus 1$. Thus, we have $2^s \binom{n}{s}$ monomials and that many clauses of size s.

A 1-*term* (0-*term*) of a boolean function $f : \{0,1\}^n \to \{0,1\}$ is a monomial M (a clause C) such that $M(a) \leqslant f(a)$ (resp., $C(a) \geqslant f(a)$) for all inputs $a \in \{0,1\}^n$. A *minterm* of f is a monomial M such that $M \leqslant f$ and which is minimal in the sense that deleting every single literal from M already violates this property.   *minterm = minimal - 1 term*

A boolean function f is a t-*And-Or* if it can be written as an And of an arbitrary number of clauses, each of size at most t.   *maxterm = minimal 0-term*

Lemma 7.4. *Let f be a t-And-Or function on n variables. Then for every $s = 1, \ldots, n$ the function f has at most t^s minterms of size s.*

Proof. Let $f = C_1 \wedge \cdots \wedge C_m$, where each clause C_i has size at most t. We interpret the clauses as sets of their literals, and let $\mathcal{C} = \{C_1, \ldots, C_m\}$ be the corresponding family of these sets. Let $\mathcal{F}$ be the family of all minterms of f that have size s (we again look at minterms as sets of their literals). Then every set in $\mathcal{C}$ intersects each set in $\mathcal{F}$ (see Exercise 7.9).

Suppose that $|\mathcal{F}| > t^s$. Then, by Lemma 7.3, $\mathcal{F}$ has a flower with $t + 1$ petals. That is, there exists a set of literals Y such that no set of at most t literals can intersect all the members of the family

$$\mathcal{F}_Y = \{M - Y \; : \; M \in \mathcal{F}, \, M \supseteq Y\}.$$

The set Y is a proper part of at least one minterm of f, meaning that Y cannot intersect all the clauses in $\mathcal{C}$. Take a clause $C \in \mathcal{C}$ such that $C \cap Y = \emptyset$. Since this clause intersects all the sets in $\mathcal{F}$, this means that it must intersect all the sets in $\mathcal{F}_Y$. But this is impossible because C has size at most t. $\square$

7.3.2 Small depth formulas

An *s-threshold function* is a monotone boolean function T_s^n which accepts a 0-1 vector if and only if it has at least s 1's. That is,

$$T_s^n(x_1, \ldots, x_n) = 1 \text{ if and only if } x_1 + \cdots + x_n \geqslant s.$$

This function can be computed by the following formula:

$$T_s^n(x_1, \ldots, x_n) = \bigvee_{I\,:\,|I|=s} \bigwedge_{i \in I} x_i.$$

This formula is monotone (has no negated literals) and has depth 2 (there are only two alternations between And and Or operations). But the size of this formula (the number of literals in it) is $s\binom{n}{s}$. Can T_s^n be computed by a substantially smaller formula if we allow negated literals and/or a larger depth?

Håstad (1986) has proved that, for $s = \lfloor n/2 \rfloor$, each such formula computing T_s^n must have size exponential in n, even if we allow any *constant* depth, i.e., any constant number of alternations of And's and Or's. Razborov (1987) has proved that the same holds even if we allow sum modulo 2 as an additional operation. Both these proofs employ non-trivial machinery: the switching lemma and approximations of boolean functions by low-degree polynomials (we will discuss these ideas in Sects. 10.5 and 20.5).

On the other hand, Håstad et al. (1995) have shown that, at least for depth-3, one can deduce the same lower bound in an elementary way using the flower lemma (Lemma 7.3). In fact, their proof holds for depth-3 *circuits* but, to demonstrate the idea, it is enough to show how it works for special depth-3 *formulas*.

An *Or-And-Or formula* is a formula of the form

$$F = F_1 \vee F_2 \vee \cdots \vee F_t, \tag{7.3}$$

where each F_i is an And-Or formula, that is, each F_i is an And of an arbitrary number of clauses, each clause being an Or of literals (variables or their negations). We say that such a formula has *bottom fan-in k* if each of its clauses has at most k positive literals (the number of negated variables may be arbitrary). The *size* of a formula is the total number of literals in it.

At this point, let us note that the condition on bottom fan-in is not crucial: if the size of F is not too large then it is possible to set some small number of variables to constant 1 so that the resulting formula will already satisfy this condition (see Exercise 7.10).

The idea of Håstad et al. (1995) is accumulated in the following lemma.

Lemma 7.5. *Let $F = F_1 \vee F_2 \vee \cdots \vee F_t$ be an Or-And-Or formula of bottom fan-in k. Suppose that F rejects all vectors with fewer than s 1's. Then F cannot accept more than tk^s vectors with precisely s 1's.*

Note that this lemma immediately implies that every Or-And-Or formula of bottom fan-in k computing the threshold function T_s^n has size at least

$$\binom{n}{s} k^{-s} > \left(\frac{n}{ks}\right)^s.$$

Proof. Suppose that F accepts more than tk^s vectors with precisely s 1's. Then some of its And-Or subformulas F_i accepts more than k^s of such vectors. Let A be this set of vectors with s 1's accepted by F_i; hence

$$|A| > k^s.$$

The formula F_i has the form

$$F_i = C_1 \wedge C_2 \wedge \cdots \wedge C_r,$$

where $C_1, \ldots, C_r$ are clauses with at most k positive literals in each of them. Let B be the set of all vectors with at most $s - 1$ 1's. All these vectors must be rejected by F_i, since they are rejected by the whole formula F. Our goal is to show that the set B contains a vector v on which each of the clauses $C_1, \ldots, C_r$ outputs the same value as on some vector from A; this will mean that the formula F_i makes an error on this input – it is forced to accept v.

Say that a vector v is a *k-limit* for A if, for every subset S of k coordinates, there exists a vector $u \in A$ such that $v \leqslant u$ and v coincides with u in all the coordinates from S.

Claim 7.6. *There exists a vector $v \in B$ which is a k-limit for A.*

Proof of Claim 7.6. For a vector $u \in \{0,1\}^n$, let E_u be the corresponding subset of $\{1, \ldots, n\}$, whose incidence vector is u, that is, $E_u = \{i \; : \; u_i = 1\}$. Consider the family $\mathcal{F} = \{E_u : u \in A\}$. This family is s-uniform and has more than k^s members. By Lemma 7.3, $\mathcal{F}$ has a flower with $k + 1$ petals. That is, there exists a set Y such that no set of size at most k can intersect all the members of the family $\mathcal{F}_Y = \{E - Y \; : \; E \in \mathcal{F}, E \supseteq Y\}$. Let v be the incidence vector of Y. We claim that v is a k-limit for A.

To show this, take an arbitrary subset S of $\{1, \ldots, n\}$ of size at most k. Then

$$S \cap (E_u - Y) = \emptyset \tag{7.4}$$

for at least one set $E_u \in \mathcal{F}$ such that $Y \subseteq E_u$. The last condition implies that $v \leqslant u$, and hence, v coincides with u on all coordinates from $S - E_u$ and from $S \cap Y$. But, by (7.4), there are no other coordinates in S, and hence, v coincides with u on all coordinates from S, as desired.   □

Fix a vector v guaranteed by the claim. To get the desired contradiction we will show that the formula will be forced to (wrongly) accept this vector. Suppose the opposite that v is rejected by F. Then $C(v) = 0$ for some clause C of F. This clause has a form

$$C = \left(\bigvee_{i \in S} x_i \right) \vee \left(\bigvee_{j \in T} \overline{x}_j \right)$$

for some two disjoint sets of S, T such that $|S| \leqslant k$. By Claim 7.6, there is a vector u in A such that $v \leqslant u$ and v coincides with u on all the coordinates from S. The vector u must be accepted by the formula F, and hence, by the clause C. This can happen only if this vector has a 1 in some coordinate $i \in S$ or has a 0 in some coordinate $j \in T$ (or both). In the first case $C(v) = 1$ because v coincides with u on S, and in the second case $C(v) = 1$ because, due to the condition $v \leqslant u$, vector v has 0's in all coordinates where vector u has them. Thus, in both cases, $C(v) = 1$, a contradiction.   □

Exercises

7.1.⁻ A projective plane of order $s - 1$ is a family of $n = s^2 - s + 1$ s-element subsets (called *lines*) of an n-element set of points such that each two lines intersect in precisely one point and each point belongs to precisely s lines (cf. Sect. 13.4). Show that the equality in Deza's theorem (Theorem 7.1) is attained when a projective plane of order $s - 1$ exists.

7.2. Take s pairwise disjoint $(k-1)$-element sets $V_1, \ldots, V_s$ and consider the family

$$\mathcal{F} = \{S \ : \ |S| = s \text{ and } |S \cap V_i| = 1 \text{ for all } = 1, \ldots, s\}.$$

This family has $(k-1)^s$ sets. Show that it has no sunflower with k petals.

7.3.⁻ Show that the bounds in Lemmas 7.2, and 7.3 are optimal.

Hint: Consider the family defined in the previous exercise.

7.4.⁻ A *matching* of size k in a graph is a set of its k pairwise disjoint edges (two edges are disjoint if they have no vertex in common). A *star* of size k is a set of k edges incident to one vertex. Argue as in the proof of the sunflower lemma to show that any set of more than $2(k-1)^2$ edges either contains a matching of size k or a star of size k.

7.5 (Andreev 1987). Let $\mathcal{F}$ be a family of sets each of cardinality at most s, and suppose that $|\mathcal{F}| > (k-1)^s$. Use the argument of Lemma 7.2 to prove that then there exist k sets $S_1, \ldots, S_k$ in $\mathcal{F}$ such that all the sets $S_i - (S_1 \cap S_2)$, $i = 1, \ldots, k$ are pairwise disjoint.

Hint: Apply the argument of Lemma 7.2.

7.6. Let $n - k + 1 < s \leqslant n$ and consider the family $\mathcal{F}$ of all s-element subsets of a n-element set. Prove that $\mathcal{F}$ has no sunflower with k petals.

Hint: Suppose the opposite and count the number of elements used in such a sunflower.

7.7$^+$ (Håstad et al. 1995). The sunflower lemma says nothing about the core. Using the same induction on s, prove the following "modular" version of this lemma. Let $\mathcal{F}$ be a family of s-element subsets of an n-element set, and suppose that $|\mathcal{F}| > F(n, k, s)$, where

$$
F(n, k, s) = \begin{cases} n^{\frac{s}{2}} k^{\frac{s}{2}} \cdot \frac{1 \cdot 3 \cdots (s-1)}{2 \cdot 4 \cdots s} & \text{if } s \text{ is even,} \\ n^{\frac{s-1}{2}} k^{\frac{s+1}{2}} \cdot \frac{1 \cdot 3 \cdots s}{2 \cdot 4 \cdots (s-1)} & \text{if } s \text{ is odd.} \end{cases}
$$

Then $\mathcal{F}$ contains a sunflower with $k + 1$ petals and a core Y such that $|Y| \equiv s + 1 \pmod 2$.

Hint: In the induction step consider separately the case when s is even and when s is odd.

7.8. Given a graph $G = (V, E)$ and a number $2 \leqslant s \leqslant |V|$, let G^s denote the graph whose vertices are all s-element subsets of V, and two such subsets A and B are connected by an edge if and only if there is an edge $(u, v) \in E$ such that $u \in A - B$ and $v \in B - A$. Suppose that the graph G is "sparse" in the following sense: every subset of at most $s \cdot k$ vertices spans fewer that $\binom{k}{2}$ edges. Use the sunflower lemma to show that then G^s has no clique of size larger than $s!(k - 1)^s$.

7.9.$^-$ Show that every 0-term C and every 1-term K of a boolean function f must share at least one literal in common.

Hint: Take a restriction which evaluates all the literals of K to 1. If C has no literal of K, then this restriction can be extended to an input a such that $f(a) = 0$.

7.10. Let F be a set of clauses on n variables. Say that a clause is *long* if it has at least $k + 1$ positive literals. Let ℓ be the number of long clauses in F, and suppose that

$$
\ell < \left(\frac{n+1}{m+1} \right)^k.
$$

Prove that then it is possible to assign some $n - m$ variables to constant 1 so that the resulting set F' will have no long clauses.

Hint: Construct the desired set assignment via the following "greedy" procedure: Take the variable x_{i_1} which occurs in the largest number of long clauses and set it to 1; then take the variable x_{i_2} which occurs in the largest number of remaining long clauses and set it to 1, and so on, until all long clauses dissapear (get value 1). In computations use the estimate $\sum_{i=1}^{n} i^{-1} \sim \ln n$.

7.11. Consider the following function on $n = sr$ variables:

$$
f = \bigwedge_{i=1}^{s} \bigvee_{j=1}^{r} x_{ij}.
$$

Let F be an Or-And-Or formula of bottom fan-in k ($k \leqslant r$) computing this function. Show that then F has size at least $(r/k)^s$.

Hint: Observe that f rejects all vectors with fewer than s ones and accepts r^s vectors with precisely s ones; apply Lemma 7.5.

7.12⁺ (Håstad et al. 1995). Consider the function on $n = m^2$ variables defined by the formula

$$f = \bigwedge_{i=1}^{m} \bigvee_{j=1}^{m} x_{ij} \wedge y_{ij}.$$

This formula is a depth-3 And-Or-And formula of size only $2n$. Prove that any depth-3 Or-And-Or formula for this function has size at least $2^{\Omega(\sqrt{n})}$.

> *Sketch*: Assume that f has such a formula F of size at most $2^{m/3}$. Reduce the bottom fan-in of F to $k = \lceil m/2 \rceil$ by setting one half of the variables to constants at random as follows: for each pair of variables x_{ij}, y_{ij}, pick one of them at random (with probability $1/2$) and set it to 1. If some clause has more than k positive literals, then none of these literals is set to 1 with probability at most 2^{-k-1}. The probability, that some of the clauses with more than k positive literals is not evaluated to 1, does not exceed $2^{m/3} \cdot 2^{-(k+1)} \leqslant 2^{-m/6} < 1$, and in particular such a setting exists. The resulting function has the same form as that considered in the previous exercise.

7.13. Consider the following more general notion of "finite limits:" a vector v is a k-*limit* of a subset $A \subseteq \{0,1\}^n$ if on every subset of k coordinates, v coincides with at least one vector $u \in A$, $u \neq v$. A set A is k-*closed* if none of its k-limits lies outside A. A boolean function f is k-*local* if it can be written as an And of an arbitrary number of boolean functions, each depending on at most k variables. Prove the following:

(i)   A set A is k-closed if and only if $A = f^{-1}(1)$ for some k-local f.

(ii) At most $2^{2^k \binom{n}{k}}$ of all possible 2^{2^n} sets can be k-closed.

(iii) If $k \geqslant 2$ and if each two vectors from A differ in more than $n - k + 1$ coordinates, then A has no k-limits (and hence, is k-closed).

8. Intersecting Families

A basic interrelation between sets is their intersection. The size (or other characteristics) of mutual intersections between the members of a given family reflects some kind of "dependence" between them. In this chapter we will study the weakest kind of this dependence – the members are required to be non-disjoint. A family is *intersecting* if any two of its sets have a non-empty intersection.

8.1 The Erdős–Ko–Rado theorem

Let $\mathcal{F}$ be an intersecting family of k-element subsets of $\{1, \ldots, n\}$. The basic question is: how large can such a family be? To avoid trivialities, we assume $n \geqslant 2k$ since otherwise any two k-element sets intersect, and there is nothing to prove.

We can obtain an intersecting family by taking all $\binom{n-1}{k-1}$ k-element subsets containing a fixed element. Can we find larger intersecting families? The whole number of k-element subsets is $\binom{n}{k} = \frac{n}{k}\binom{n-1}{k-1}$, so the question is not trivial.

The following result, found by Erdős, Ko, and Rado in 1938 (but published only 23 years later), answers the question.

Theorem 8.1 (Erdős–Ko–Rado 1961). *If $2k \leqslant n$ then every intersecting family of k-element subsets of an n-element set has at most $\binom{n-1}{k-1}$ members.*

Proof (due to G.O.H. Katona 1972). Let X be the underlying n-element set, and assume for definiteness that $X = \{0, 1, \ldots, n-1\}$. The idea is to study all permutations of the elements of X, estimating how often the consecutive elements of these permutations can constitute one of the sets in our family.

For $s \in X$, define $B_s \rightleftharpoons \{s, s+1, \ldots, s+k-1\}$, where the addition is modulo n.

Claim 8.2. *At most k of the sets B_s can belong to $\mathcal{F}$.*

We can suppose that $B_0 \in \mathcal{F}$. The only sets B_s that intersect B_0 other than B_0 itself are the $2k-2$ sets B_s with $-(k-1) \leqslant s \leqslant k-1$, $s \neq 0$ (where the indices are taken modulo n). These sets can be partitioned into $k-1$ pairs of disjoint sets, B_i, B_{i+k}, where $-(k-1) \leqslant i \leqslant -1$.

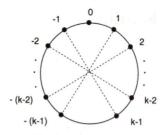

Since $\mathcal{F}$ can contain at most one set of each such pair the assertion of the claim follows.

We now count in two ways the number L of pairs (f, s), where f is a permutation of X and s is a point in X, such that the set

$$f(B_s) \rightleftharpoons \{f(s), f(s+1), \ldots, f(s+k-1)\}$$

belongs to $\mathcal{F}$. By the claim, for each *fixed* permutation f, the family $\mathcal{F}$ can contain at most k of the sets $f(B_s)$. Hence, $L \leqslant kn!$. On the other hand, exactly $nk!(n-k)!$ of the pairs (f, s) yield the same set $f(B_s)$: there are n possibilities for s, and for each fixed s, there are $k!(n-k)!$ possibilities to choose the permutation f. Hence, $L = |\mathcal{F}| \cdot nk!(n-k)!$. Combining this with the previous estimate, we obtain

$$|\mathcal{F}| \leqslant \frac{kn!}{nk!(n-k)!} = \frac{k}{n}\binom{n}{k} = \binom{n-1}{k-1}.$$

$\square$

8.2 Finite ultrafilters

An *ultrafilter* over a set X is a collection $\mathcal{F}$ of its subsets such that:

(i) $\mathcal{F}$ is upwards-closed, that is, if $A \in \mathcal{F}$ then all supersets of A also belong to $\mathcal{F}$;

(ii) for every subset A of X, exactly one of A or its complement $\overline{A} = X - A$ belongs to $\mathcal{F}$.

If $\mathcal{F}$ is an ultrafilter then $\mathcal{F}$ is intersecting. Indeed, if some two members A, B of $\mathcal{F}$ were disjoint, then the complement of B would contain the set A, and hence, would belong to $\mathcal{F}$ (by (i)). But this is impossible since, by (ii), $\mathcal{F}$ cannot contain the set B together with its complement.

The dual property is more interesting.

Proposition 8.3. *Every intersecting family can be extended to an ultrafilter (by adding new members).*

Proof. Take an arbitrary intersecting family and extend it as follows. If there are some sets not in the family and such that their addition does not destruct the intersection property, add all them. After that, add all supersets of the sets we already have. We claim that the resulting family $\mathcal{F}$ is an ultrafilter. Indeed, if it is not, there must be a set A such that neither A nor its complement $\overline{A}$ belongs to $\mathcal{F}$. By the construction, A must be disjoint from at least one member B of our initial family (for otherwise A would be added during the first phase), and hence, B is contained in the complement $\overline{A}$. But $B \in \mathcal{F}$ and $\mathcal{F}$ is upwards-closed, a contradiction. $\square$

8.3 Maximal intersecting families

Let $\mathcal{F}$ be a k-uniform family of sets of some n-element set. Say that $\mathcal{F}$ is *maximal intersecting* if

(i) $\mathcal{F}$ is intersecting;
(ii) the addition of any new k-element set to $\mathcal{F}$ destroys this property, that is, for every k-element subset $E \notin \mathcal{F}$, the family $\mathcal{F} \cup \{E\}$ is no longer intersecting.

The case when $n \leqslant 2k - 1$ is not interesting, because then the only maximal intersecting family is the family of all k-element subsets. But what if $n \geqslant 2k$? Intuitively, any maximal intersecting family must be large enough, because *every* k-element set not in the family must be avoided by at least one of its members. It is therefore interesting to investigate the minimal possible number $f(k)$ of members which such a family can have.

To give an upper bound on $f(k)$, consider the family $\mathcal{F}$ of lines in a projective plane of order $k - 1$ (see Sect. 13.4). For our current purposes it is enough to know that $\mathcal{F}$ is a family of $|\mathcal{F}| = n = k^2 - k + 1$ k-element subsets (called *lines*) of an n-element set of points such that each two lines intersect in precisely one point and each point belongs to precisely k lines. It is easy to show that this family is maximal intersecting (see Exercise 8.6). Hence, $f(k) \leqslant k^2 - k + 1$ for all those values of k for which a projective plane of order $k - 1$ exists.

In the case of projective planes, we have a k-uniform family with about k^2 sets. But the number of points in this case is also the same. What if we take a lot fewer than k^2 points? Can we then still find a k-uniform and maximal intersecting family of size at most k^2? Using double-counting we can answer this question negatively.

Theorem 8.4 (Füredi 1980). *Let $\mathcal{F}$ be a maximal intersecting family of k-element sets of an n-element set. If $n \leqslant k^2/2 \log k$, then $\mathcal{F}$ must have more than k^2 members.*

Proof. To simplify computations, we only prove the theorem under a slightly stronger assumption that $n \leqslant k^2/(1 + 2 \log k)$. The idea is to count in two ways the number N of pairs (F, E) where $F \in \mathcal{F}$ and E is a k-element subset disjoint from F (and hence, $E \notin \mathcal{F}$). Since every such set E must be avoided by at least one member of $\mathcal{F}$,

$$N \geqslant \binom{n}{k} - |\mathcal{F}|.$$

On the other hand, each member of $\mathcal{F}$ can avoid at most $\binom{n-k}{k}$ of the sets E; hence

$$N \leqslant |\mathcal{F}| \cdot \binom{n-k}{k}.$$

These two inequalities, together with the estimate (1.12), imply

$$|\mathcal{F}| \geqslant \frac{\binom{n}{k}}{1 + \binom{n-k}{k}} \geqslant \frac{1}{2} \cdot \left(\frac{n}{n-k}\right)^k > e^{k^2/n-1} \geqslant e^{2\log k} \geqslant k^2.$$

$\square$

Now suppose that $\mathcal{F}_1, \ldots, \mathcal{F}_m$ are intersecting (not necessarily uniform) families of an n-element set $\{1, \ldots, n\}$. How many sets can we have in their union?

Taking each $\mathcal{F}_i$ to be the family of all 2^{n-1} subsets containing the element i, we see that the union will have $2^n - 2^{n-m}$ sets.

A beautiful result, due to Kleitman, says that this bound is best possible.

Theorem 8.5 (Kleitman 1966). *The union of m intersecting families contains at most $2^n - 2^{n-m}$ sets.*

Proof. We apply induction on m. The case $m = 1$ being trivial, we turn to the induction step. We say that a family $\mathcal{A}$ is *monotone increasing* (*monotone decreasing*) if $A \in \mathcal{A}$ and $B \supseteq A$ (respectively, $B \subseteq A$) implies $B \in \mathcal{A}$. A famous result, also due to Kleitman (we will prove it in Sect. 11.2; see Theorem 11.6 and Exercise 11.8) says that, if $\mathcal{A}$ is a monotone decreasing and $\mathcal{B}$ is a monotone increasing family of subsets of an n-element set, then

$$|\mathcal{A} \cap \mathcal{B}| \leqslant 2^{-n} |\mathcal{A}| \cdot |\mathcal{B}|. \tag{8.1}$$

Now let $\mathcal{F} = \bigcup_{i=1}^m \mathcal{F}_i$, with each $\mathcal{F}_i$ being an intersecting family. Since our aim is to bound $|\mathcal{F}|$ from the above, we may assume that each $\mathcal{F}_i$ is maximal intersecting family; in particular, $|\mathcal{F}_m| = 2^{n-1}$. Let $\mathcal{A}$ be the complement of $\mathcal{F}_m$, i.e., the family of all $|\mathcal{A}| = 2^{n-1}$ subsets not in $\mathcal{F}_m$, and $\mathcal{B} = \bigcup_{i=1}^{m-1} \mathcal{F}_i$. Since $\mathcal{F}_1, \ldots, \mathcal{F}_m$ are maximal intersecting families, $\mathcal{A}$ is monotone decreasing and $\mathcal{B}$ is monotone increasing. By the induction hypothesis, $|\mathcal{B}| \leqslant 2^n - 2^{n-m+1}$, and, by (8.1),

$$|\mathcal{A} \cap \mathcal{B}| \leqslant 2^{-n} 2^{n-1} (2^n - 2^{n-m+1}) = 2^{n-1} - 2^{n-m}.$$

Therefore,

$$|\mathcal{B} \cap \mathcal{F}_m| = |\mathcal{B}| - |\mathcal{A} \cap \mathcal{B}| \geqslant |\mathcal{B}| - 2^{n-1} + 2^{n-m}$$

and

$$|\mathcal{F}| = \left| \bigcup_{i=1}^m \mathcal{F}_i \right| = |\mathcal{B}| + |\mathcal{F}_m| - |\mathcal{B} \cap \mathcal{F}_m| \leqslant 2^n - 2^{n-m}.$$

$\square$

8.4 A Helly-type result

In 1923, E. Helly proved the following result: if $n \geqslant k+1$ convex sets in $\mathbb{R}^k$ have the property that any $k+1$ of them have a nonempty intersection, then there is a point common to all of them.

It is natural to ask if objects other than convex sets obey Helly-type laws. For arbitrary families of sets we have the following Helly-type result.

Theorem 8.6. *Let $\mathcal{F}$ be a family and k be the minimum size of its member. If any $k+1$ members of $\mathcal{F}$ intersect (i.e., share a common point) then all of them do.*

Proof. Suppose the opposite that the intersection of all sets in $\mathcal{F}$ is empty, and take a set $A = \{x_1, \ldots, x_k\} \in \mathcal{F}$. For every $i = 1, \ldots, k$ there must be a set $B_i \in \mathcal{F}$ such that $x_i \notin B_i$. Hence, $A \cap B_1 \cap \cdots \cap B_k = \emptyset$, a contradiction. $\square$

8.5 Intersecting systems

The intersection property has a natural extension to systems of families. Such extensions were considered by Ahlswede, Cai, and Zhang (1994) and Ahlswede et al. (1997). In this section we present some of their results illustrating what type of arguments work in this generalized case.

Let $\mathfrak{F}$ be a collection of pairwise disjoint families of k-element subsets of some n-element set X. We call $\mathfrak{F}$ an *intersecting system* if it satisfies the following condition:

> For any ordered pair of distinct families $\mathcal{F}$ and $\mathcal{F}'$ in $\mathfrak{F}$, there is an $F \in \mathcal{F}$ such that $F \cap F' \neq \emptyset$ for all $F' \in \mathcal{F}'$.

Let us stress that *one* set F of $\mathcal{F}$ must intersect *all* the sets in $\mathcal{F}'$. In this case we also say that F is a set of $\mathcal{F}$ *responsible* for the family $\mathcal{F}'$.

It is easy to see that there are intersecting systems $\mathfrak{F}$ containing

$$|\mathfrak{F}| \geqslant \binom{n-1}{k-1}$$

families. Indeed, the intersecting system, which contains as a one-element family each of the $\binom{n-1}{k-1}$ k-element subsets of X that contain one fixed point $i \in X$, shows this. From the other side the following trivial upper bound holds.

Proposition 8.7. *If $\mathfrak{F}$ is an intersecting system then*

$$|\mathfrak{F}| \leqslant k\binom{n-1}{k-1}.$$

Proof. Let $\mathfrak{F}$ be an intersecting system and let $\mathcal{F}$ be the smallest family in $\mathfrak{F}$. Every member of all other families must intersect at least one member of $\mathcal{F}$, implying that $|\bigcup \mathfrak{F}| \leqslant |\mathcal{F}| k\binom{n-1}{k-1}$, and due to the minimality of $\mathcal{F}$, $|\mathfrak{F}| \leqslant \frac{1}{|\mathcal{F}|} |\mathcal{F}| k\binom{n-1}{k-1}$. $\hfill\square$

Ahlswede, Cai, and Zhang (1992) made the conjecture that, for a large enough $n = n(k)$, the maximal intersecting system has $\binom{n-1}{k-1}$ families. Ahlswede et al. (1997) have proved that this conjecture holds for $k \leqslant 3$ but is false for $k \geqslant 8$. On the other hand, they have found several interesting properties of $\mathfrak{F}$ that ensure the inequality

$$|\mathfrak{F}| < \binom{n-1}{k-1};$$

in this case we will also say that $\mathfrak{F}$ is *small*. Let $\ker(\mathcal{F})$ denote the *kernel* of $\mathcal{F}$, i.e., the set of elements in the total intersection

$$\ker(\mathcal{F}) = \bigcap_{F \in \mathcal{F}} F.$$

Proposition 8.8. *Let $\mathfrak{F}$ be an intersecting system on an n-element underlying set with $n \geqslant k^5$. If $\ker(\mathcal{F}_1) = \emptyset$ for at least one family $\mathcal{F}_1 \in \mathfrak{F}$ then $\mathfrak{F}$ is small, i.e., $|\mathfrak{F}| < \binom{n-1}{k-1}$.*

Proof. First, observe that, by Theorem 8.6, the emptiness of total intersection implies that there exists $r \leqslant k+1$ members $F_1, \ldots, F_r$ of $\mathcal{F}_1$ such that $F_1 \cap \cdots \cap F_r = \emptyset$. Let $H = F_1 \cup \cdots \cup F_r$. Again, by the emptiness of total intersection of sets in $\mathcal{F}_1$, for any family $\mathcal{F}_2 \neq \mathcal{F}_1$ the set $F \in \mathcal{F}_2$, which is responsible for the family $\mathcal{F}_1$, must intersect the set H in at least two points (for if not, then the sets $F_1, \ldots, F_r$ would share a point in common). Since each family in $\mathfrak{F} - \{\mathcal{F}_1\}$ must contain a set responsible for $\mathcal{F}_1$, we conclude that

$$|\mathfrak{F}| \leqslant 1 + |\{F \subseteq X : |F| = k \text{ and } F \text{ is responsible for } \mathcal{F}_1\}|$$

$$\leqslant 1 + \binom{|H|}{2}\binom{n-2}{k-2} \leqslant 1 + \binom{k(k+1)}{2}\binom{n-2}{k-2}$$

$$\leqslant 1 + k^4 \binom{n-2}{k-2} = 1 + \frac{k^4(k-1)}{n-1}\binom{n-1}{k-1} < \binom{n-1}{k-1},$$

where in the last step we applied the fact that $\binom{n}{k} = \frac{n}{k}\binom{n-1}{k-1}$ and $k^5 \leqslant n$. $\hfill\square$

By the above proposition, if an intersecting system is *large* (not *small*) then each of its families has a non-empty kernel. Now we will show that each family in a big intersecting system must contain many members.

Call an intersecting system $\mathfrak{F}$ *minimal* if there is no "superfluous" set in any family. More precisely, for every set $F \in \mathcal{F} \in \mathfrak{F}$ there exists a family $\mathcal{F}' \in \mathfrak{F}$ such that only F is responsible in $\mathcal{F}$ for the family $\mathcal{F}'$. Note that if a given intersecting system $\mathfrak{F}$ is not minimal, one can get a minimal one with the same cardinality by repeatedly deleting a superfluous member as long as there is one.

Proposition 8.9. *Let $\mathfrak{F}$ be a minimal intersecting system and assume that* $\ker(\mathcal{F}) \neq \emptyset$ *for all $\mathcal{F} \in \mathfrak{F}$. If $\mathfrak{F}$ contains a family $\mathcal{F}$ satisfying $1 < |\mathcal{F}| \leqslant n/k^3$ and $n \geqslant 2k^4$, then $\mathfrak{F}$ is small, i.e., $|\mathfrak{F}| < \binom{n-1}{k-1}$.*

Proof. Since our system is minimal, for every $F \in \mathcal{F}$ there exist $\mathcal{F}' \in \mathfrak{F} - \{\mathcal{F}\}$ and $F' \in \mathcal{F}'$ such that $F \cap F' = \emptyset$. Indeed, otherwise F would be responsible for every other family in $\mathfrak{F}$, and therefore the system would not be minimal (since $|\mathcal{F}| \geqslant 2$). Let the subsystem $\mathfrak{F}(F)$ contain all the families for which F is responsible. Any family in $\mathfrak{F}(F)$ must contain a member which is responsible for $\mathcal{F}'$, and this set intersects F as well as F'. Hence, the number of k-element subsets of X intersecting F and F' is an upper bound for the cardinality of this subsystem. Thus, $|\mathfrak{F}(F)| \leqslant k^2 \binom{n-2}{k-2}$, and therefore

$$|\mathfrak{F}| \leqslant 1 + |\mathcal{F}| \, k^2 \binom{n-2}{k-2} \leqslant 1 + \frac{n}{k} \binom{n-2}{k-2} < \frac{n-1}{k-1} \binom{n-2}{k-2} = \binom{n-1}{k-1},$$

where we made use of the assumption on the size of the family $\mathcal{F}$. $\square$

Proposition 8.10. *Let $k \geqslant 1$ be a fixed integer, let $\mathfrak{F}$ be an intersecting system, and suppose that it contains a family $\mathcal{F}_1$ whose kernel consists of a single point. Then $|\mathfrak{F}| < (1 + \epsilon)\binom{n-1}{k-1}$ where $\epsilon \to 0$ as $n \to \infty$.*

Proof. Let $\{x\}$ be the kernel of $\mathcal{F}_1$. If we remove the point x from all the members of $\mathcal{F}_1$ then, by Theorem 8.6, some $l \leqslant k$ of the resulting sets must have empty total intersection. So, there must be $l \leqslant k$ sets $F_1, \ldots, F_l \in \mathcal{F}_1$ whose intersection is $\{x\}$. Define $H \rightleftharpoons F_1 \cup \cdots \cup F_l$, and observe that $|H| \leqslant k^2$. Since every family $\mathcal{F}$ in $\mathfrak{F}$, none of whose members contains x, must have a member that intersects H in at least two elements (for otherwise the sets $F_1, \ldots, F_l$ would share a point different from x), we conclude that

$$|\mathfrak{F}| \leqslant \binom{n-1}{k-1} + \binom{k^2}{2}\binom{n-2}{k-2} \leqslant (1 + \epsilon)\binom{n-1}{k-1}.$$

$\square$

Exercises

8.1.$^-$ Let $\mathcal{F}$ be a family of subsets of an n-element set. Prove that if $\mathcal{F}$ is intersecting then $|\mathcal{F}| \leqslant 2^{n-1}$. Is this the best bound? If so, then exhibit an intersecting family with $|\mathcal{F}| = 2^{n-1}$.

Hint: A set and its complement cannot both be the members of $\mathcal{F}$.

8.2. $^-$ Let $n \leqslant 2k$ and let $A_1, \ldots, A_m$ be a family of k-element subsets of $[n]$ such that $A_i \cup A_j \neq [n]$ for all i, j. Show that $m \leqslant \left(1 - \frac{k}{n}\right)\binom{n}{k}$.

Hint: Apply the Erdős–Ko–Rado theorem to the complements $\overline{A_i} = [n] - A_i$.

8.3. $^{(!)}$ The upper bound $\binom{n-1}{k-1}$ given by Erdős–Ko–Rado theorem is achieved by the families of sets containing a fixed element. Show that for $n = 2k$ there are other families achieving this bound.

Hint: Include one set out of every pair of sets formed by a k-element set and its complement.

8.4. One can generalize the intersection property and require that $|A \cap B| \geqslant t$ for all $A \neq B \in \mathcal{F}$. Such families are called *t-intersecting*. The first example of a t-intersecting family which comes to mind, is the family of all subsets of $[n]$ containing some fixed set of t elements. This family has 2^{n-t} sets. Are there larger t-intersecting families?

Hint: Let $n + t$ be even and take $\mathcal{F} = \left\{A \subseteq [n] : |A| = \frac{n+t}{2}\right\}$.

8.5. $^+$ Let $m(n, k, t)$ be the maximal cardinality of a family $\mathcal{F}$ of k-element subsets of $\{1, \ldots, n\}$ such that $|A \cap B| \leqslant t$ for any $A \neq B \in \mathcal{F}$. Prove that

$$m(n, k, t) \geqslant \binom{n}{k} \bigg/ \sum_{i=t+1}^{k} \binom{k}{i}\binom{n-k}{k-i}.$$

Hint: Consider a graph G with the vertex set consisting of all k-element subsets of $[n]$ and in which two subsets are adjacent when they intersect in more than t elements. Use the fact (see Exercise 6.3) that G can be properly colored with at most $\Delta + 1$ colors, where Δ is the maximum degree of a vertex in G.

8.6. Consider the k-uniform family of all $n = k^2 - k + 1$ lines in the set of points of a projective plane of order $k - 1$. Clearly, this family is intersecting. Show that it is also maximal intersecting, i.e., that every k-element set E, which intersects all the lines, must be a line.

Hint: Assume that E is not a line, draw a line L through some two points $x \neq y$ of E, and take a point $z \in L - \{x, y\}$. This point belongs to k lines, and each of them intersect E.

9. Chains and Antichains

Partial ordered sets provide a common frame for many combinatorial configurations. Formally, a *partially ordered set* (or *poset*, for short) is a set P together with a binary relation $<$ between its elements which is transitive and irreflexive: if $x < y$ and $y < z$ then $x < z$, but $x < y$ and $y < x$ cannot both hold. We write $x \leqslant y$ if $x < y$ or $x = y$. Elements x and y are *comparable* if either $x \leqslant y$ or $y \leqslant x$ (or both) hold.

A *chain* in a poset P is a subset $C \subseteq P$ such that any two of its points are comparable. Dually, an *antichain* is a subset $A \subseteq P$ such that no two of its points are comparable. Observe that $|C \cap A| \leqslant 1$, i.e., every chain C and every antichain A can have at most one element in common (for two points in their intersection would be both comparable and incomparable).

Here are some frequently encountered examples of posets: a family of sets is partially ordered by set inclusion; a set of positive integers is partially ordered by division; a set of vectors in $\mathbb{R}^n$ is partially ordered by $(a_1, \ldots, a_n) < (b_1, \ldots, b_n)$ iff $a_i \leqslant b_i$ for all i, and $a_i < b_i$ for at least one i.

Small posets may be visualized by drawings, known as *Hasse diagrams*: x is lower in the plane than y whenever $x < y$ and there is no other point $z \in P$ for which both $x < z$ and $z < y$. For example:

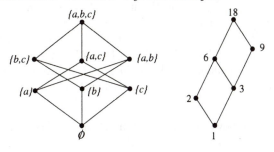

9.1 Decomposition of posets

A decomposition of a poset is its partition into mutually disjoint chains or antichains. Given a poset P, our goal is to decompose it into as few chains (or antichains) as possible. One direction is easy: if a poset P has a chain (antichain) of size r then it cannot be partitioned into fewer than r antichains

(chains). The reason here is simple: any two points of the same chain must lie in different members of a partition into antichains.

Is this optimal? If P has no chain (or antichain) of size greater than r, is it then possible to partition P into r antichains (or chains, respectively)? One direction is straightforward (see Exercise 9.8 for an alternative proof):

Theorem 9.1. *Suppose that the largest chain in the poset P has size r. Then P can be partitioned into r antichains.*

Proof. Let A_i be the set of points $x \in P$ such that the longest chain, whose greatest element is x, has i points (including x). Then, by the hypothesis, $A_i = \emptyset$ for $i \geqslant r + 1$, and hence, $P = A_1 \cup A_2 \cup \cdots \cup A_r$ is a partition of P into r mutually disjoint subsets (some of them may be also empty). Moreover, each A_i is an antichain, since if $x, y \in A_i$ and $x < y$, then the longest chain $x_1 < \ldots < x_i = x$ ending in x could be prolonged to a longer chain $x_1 < \ldots < x_i < y$, meaning that $y \notin A_i$. $\square$

The dual result looks similar, but its proof is more involved. This result, uniformly known as Dilworth's Decomposition Theorem (Dilworth 1950) has played an important role in motivating research into posets. There are several elegant proofs; the one we present is due to F. Galvin.

Theorem 9.2 (Dilworth's theorem). *Suppose that the largest antichain in the poset P has size r. Then P can be partitioned into r chains.*

Proof (due to Galvin 1994). We use induction on the cardinality of P. Let a be a maximal element of P, and let n be the size of a largest antichain in $P' = P - \{a\}$. Then P' is the union of n disjoint chains $C_1, \ldots, C_n$. We have to show that P either contains an $(n+1)$-element antichain or else is the union of n chains. Now, every n-element antichain in P' consists of one element from each C_i. Let a_i be the maximal element in C_i which belongs to some n-element antichain in P'. It is easy to see that $A = \{a_1, \ldots, a_n\}$ is an antichain. If $A \cup \{a\}$ is an antichain in P, we are done. Otherwise, we have $a > a_i$ for some i. Then $K = \{a\} \cup \{x \in C_i : x \leqslant a_i\}$ is a chain in P, and there are no n-element antichains in $P - K$ (since a_i was the maximal element of C_i participating in such an antichain), whence $P - K$ is the union of $n - 1$ chains. $\square$

To recognize the power of this theorem, let us show that it contains Hall's Marriage Theorem 5.1 as a special case!

Suppose that $S_1, \ldots, S_m$ are sets satisfying Hall's condition, i.e., $|S(I)| \geqslant |I|$ for all $I \subseteq \{1, \ldots, m\}$, where $S(I) = \bigcup_{i \in I} S_i$. We construct a poset P as follows. The points of P are the elements of $X = S_1 \cup \cdots \cup S_m$ and symbols $y_1, \ldots, y_m$, with $x < y_i$ if $x \in S_i$, and no other comparabilities. It is clear that X is an antichain in P. We claim that there is no larger antichain. To show this, let A be an antichain, and set $I = \{i : y_i \in A\}$. Then A contains no point of $S(I)$, for if $x \in S_i$ then x is comparable with y_i, and hence,

A cannot contain both of these points. Hence, Hall's condition implies that $|A| \leqslant |I| + |X| - |S(I)| \leqslant |X|$, as claimed.

Now, Dilworth's theorem implies that P can be partitioned into $|X|$ chains. Since the antichain X is maximal, each of the chains in the partition must contain a point of X. Let the chain through y_i be $\{x_i, y_i\}$. Then $(x_1, \ldots, x_m)$ is a desired system of distinct representatives: for $x_i \in S_i$ (since $x_i < y_i$) and $x_i \neq x_j$ (since the chains are disjoint).

9.1.1 Symmetric chains

In general, Dilworth's theorem says nothing more about the chains, forming the partition, except that they are mutually disjoint. However, if we consider special posets then we can extract more information about the partition. To illustrate this, let us consider now the poset 2^X whose points are all subsets of an n-element set X partially ordered by set inclusion. De Bruijn, Tengbergen, and Kruyswijk (1952) have shown that 2^X can be partitioned into disjoint chains that are also "symmetric."

Let $\mathcal{C} = \{A_1, \ldots, A_k\}$ be a chain in 2^X, i.e., $A_1 \subset A_2 \subset \ldots \subset A_k$. This chain is *symmetric* if $|A_1| + |A_k| = n$ and $|A_{i+1}| = |A_i| + 1$ for all $i = 1, \ldots, k - 1$. "Symmetric" here means symmetric positioned about the middle level $\frac{n}{2}$. Symmetric chains with $k = n$ are *maximal*. Maximal chains are in one-to-one correspondence with the permutations of the underlying set: every permutation $(x_1, \ldots, x_n)$ gives the maximal chain

$$\{x_1\} \subset \{x_1, x_2\} \subset \ldots \subset \{x_1, \ldots, x_n\}.$$

Theorem 9.3. *The family of all subsets of an n-element set can be partitioned into at most $\binom{n}{\lfloor n/2 \rfloor}$ mutually disjoint symmetric chains.*

Proof. Take an n-element set X, and assume for a moment that we already have some partition of 2^X into symmetric chains. Every such chain contains exactly one set from the middle level; hence there can be at most $\binom{n}{\lfloor n/2 \rfloor}$ chains in that partition.

Let us now prove that such a partition is possible at all. We argue by the induction on $n = |X|$. Clearly the result holds for the one point set X. So, suppose that it is true for all sets with fewer points then n. Pick a point $x \in X$, and let $Y = X - \{x\}$. By induction, we can partition 2^Y into symmetric chains $\mathcal{C}_1, \ldots, \mathcal{C}_r$. Each of these chains over Y

$$\mathcal{C}_i = A_1 \subset A_2 \subset \ldots \subset A_k$$

produce the following two chains over the whole set X:

$$\mathcal{C}'_i = A_1 \subset A_2 \subset \ldots \subset A_{k-1} \subset A_k \subset A_k \cup \{x\}$$

$$\mathcal{C}''_i = A_1 \cup \{x\} \subset A_2 \cup \{x\} \subset \ldots \subset A_{k-1} \cup \{x\}.$$

These chains are symmetric since

$$|A_1| + |A_k \cup \{x\}| = (|A_1| + |A_k|) + 1 = (n-1) + 1 = n$$

and

$$|A_1 \cup \{x\}| + |A_{k-1} \cup \{x\}| = (|A_1| + |A_{k-1}|) + 2 = (n-2) + 2 = n.$$

Is this a partition? It is indeed. If $A \subseteq Y$ then only C_i' contains A where C_i is the chain in 2^Y containing A. If $A = B \cup \{x\}$ where $B \subseteq Y$ then $B \in C_i$ for some i. If B is the maximal element of C_i then C_i' is the only chain containing A, otherwise A is contained only in C_i''. □

9.1.2 Application: the memory allocation problem

The following problem arises in information storage and retrieval. Suppose we have some list (a sequence) $L = (a_1, a_2, \ldots, a_m)$ of not necessarily distinct elements of some set X. We say that this list *contains* a subset A if it contains A as a subsequence of consecutive terms, that is, if

$$A = \{a_i, a_{i+1}, \ldots, a_{i+|A|-1}\}$$

for some i. A sequence is *universal* for X if it contains all the subsets of X. For example, if $X = \{1, 2, 3, 4, 5\}$ then the list

$$L = (1\,2\,3\,4\,5\,1\,2\,4\,1\,3\,5\,2\,4)$$

of length $m = 13$ is universal for X.

What is the length of a shortest universal sequence for an n-element set? Since any two sets of equal cardinality must start from different places of this string, the trivial lower bound for the length of universal sequence is $\binom{n}{\lfloor n/2 \rfloor}$, which is about $\sqrt{\frac{2}{\pi n}} 2^n$, according to Stirling's formula (1.4). A trivial upper bound for the length of the shortest universal sequence is obtained by considering the sequence obtained simply by writing down each subset one after the other. Since there are 2^n subsets of average size $n/2$, the length of the resulting universal sequence is at most $n2^{n-1}$. Using Dilworth's theorem, we can obtain a universal sequence, which is n times (!) shorter than this trivial one.

Theorem 9.4 (Lipski 1978). *There is a universal sequence for $\{1, \ldots, n\}$ of length at most $\frac{2}{\pi} 2^n$.*

Proof. We consider the case when n is even, say $n = 2k$ (the case of odd n is similar). Let $S = \{1, \ldots, k\}$ be the set of the first k elements and $T = \{k+1, \ldots, 2k\}$ the set of the last k elements. By Theorem 9.3, both S and T have symmetric chain decompositions of their posets of subsets into $m = \binom{k}{k/2}$ symmetric chains: $2^S = C_1 \cup \cdots \cup C_m$ and $2^T = \mathcal{D}_1 \cup \cdots \cup \mathcal{D}_m$. Corresponding to the chain

$$C_i = \{x_1, \ldots, x_j\} \subset \{x_1, \ldots, x_j, x_{j+1}\} \subset \ldots \subset \{x_1, \ldots, x_h\} \quad (j+h = k)$$

we associate the sequence (not the set!) $C_i = (x_1, x_2, \ldots x_h)$. Then every subset of S occurs as an *initial* part of one of the sequences $C_1, \ldots, C_m$. Similarly let $D_1, \ldots, D_m$ be sequences corresponding to the chains $\mathcal{D}_1, \ldots, \mathcal{D}_m$. If we let $\overline{D}_i$ denote the sequence obtained by writing D_i in reverse order, then every subset of T occurs as a *final* part of one of the $\overline{D}_i$. Next, consider the sequence

$$L = \overline{D}_1 C_1 \overline{D}_1 C_2 \ldots \overline{D}_1 C_m \ldots \overline{D}_m C_1 \overline{D}_m C_2 \ldots \overline{D}_m C_m.$$

We claim that L is a universal sequence for the set $\{1, \ldots, n\}$. Indeed, each of its subsets A can be written as $A = E \cup F$ where $E \subseteq S$ and $F \subseteq T$. Now F occurs as the final part of some $\overline{D}_f$ and E occurs as the initial part of some C_e; hence, the whole set A occurs in the sequence L as the part of $\overline{D}_f C_e$. Thus, the sequence L contains every subset of $\{1, \ldots, n\}$. The length of the sequence L is at most $km^2 = k\binom{k}{k/2}^2$. Since, by Stirling's formula, $\binom{k}{k/2} \sim 2^k \sqrt{\frac{2}{k\pi}}$, the length of the sequence is $km^2 \sim k\frac{2}{k\pi} \cdot 2^{2k} = \frac{2}{\pi} 2^n$. $\square$

9.2 Antichains

A set system $\mathcal{F}$ is an *antichain* (or *Sperner system*) if no set in it contains another: if $A, B \in \mathcal{F}$ and $A \neq B$ then $A \not\subseteq B$. It is an antichain in the sense that this property is the other extreme from that of the chain in which every pair of sets is comparable.

9.2.1 Sperner's theorem

Simplest examples of antichains over $\{1, \ldots, n\}$ are the families of all sets of fixed cardinality k, $k = 0, 1, \ldots, n$. Each of these antichains has $\binom{n}{k}$ members. Recognizing that the maximum of $\binom{n}{k}$ is achieved for $k = \lfloor n/2 \rfloor$, we conclude that there are antichains of size $\binom{n}{\lfloor n/2 \rfloor}$. Are these antichains the largest ones?

The positive answer to this question was found by Emanuel Sperner in 1928, and this result is known as Sperner's Theorem.

Theorem 9.5 (Sperner 1928). *Let $\mathcal{F}$ be a family of subsets of an n element set. If $\mathcal{F}$ is an antichain then $|\mathcal{F}| \leqslant \binom{n}{\lfloor n/2 \rfloor}$.*

A considerably sharper result, Theorem 9.6 below, is due to Lubell (1966). The same result was discovered by Meshalkin (1963) and (not so explicitly) by Yamamoto (1954). Although Lubell's result is also a rather special case of an earlier result of Bollobás (see Theorem 9.8 below), inequality (9.1) has become known as the *LYM inequality*.

Theorem 9.6 (LYM Inequality). *Let $\mathcal{F}$ be an antichain over a set X of n elements. Then*

$$\sum_{A \in \mathcal{F}} \binom{n}{|A|}^{-1} \leqslant 1. \tag{9.1}$$

Note that Sperner's theorem follows from this bound: recognizing that $\binom{n}{k}$ is maximized when $k = \lfloor n/2 \rfloor$, we obtain

$$|\mathcal{F}| \cdot \binom{n}{\lfloor n/2 \rfloor}^{-1} \leqslant \sum_{A \in \mathcal{F}} \binom{n}{|A|}^{-1} \leqslant 1.$$

We will give an elegant proof of Theorem 9.6 due to Lubell (1966) together with one its reformulations which is pregnant with further extensions.

First proof. For each subset A, exactly $|A|!(n - |A|)!$ maximal chains over X contain A. Since none of the $n!$ maximal chains meet $\mathcal{F}$ more than once, we have $\sum_{A \in \mathcal{F}} |A|!(n - |A|)! \leqslant n!$. Dividing this inequality by $n!$ we get the desired result. □

Second proof. The idea is to associate with each subset $A \subseteq X$, a permutation on X, and count their number. For an a-element set A let us say that a permutation $(x_1, x_2, \ldots, x_n)$ of X *contains* A if $\{x_1, \ldots, x_a\} = A$. Note that A is contained in precisely $a!(n - a)!$ permutations. Now if $\mathcal{F}$ is an antichain, then each of $n!$ permutations contains at most one $A \in \mathcal{F}$. Consequently, $\sum_{A \in \mathcal{F}} a!(n - a)! \leqslant n!$, and the result follows. To recover the first proof, simply identify a permutation $(x_1, x_2, \ldots, x_n)$ with the maximal chain $\{x_1\} \subset \{x_1, x_2\} \subset \ldots \subset \{x_1, x_2 \ldots, x_n\} = X$. □

9.2.2 Bollobás's theorem

The following theorem due to B. Bollobás is one of the cornerstones in extremal set theory. Its importance is reflected, among other things, by the list of different proofs published as well as the list of different generalizations. In particular, this theorem implies both Sperner's theorem and the LYM inequality.

Theorem 9.7 (Bollobás's theorem). *Let* $A_1, \ldots, A_m$ *be* a*-element sets and* $B_1, \ldots, B_m$ *be* b*-element sets such that* $A_i \cap B_j = \emptyset$ *if and only if* $i = j$. *Then* $m \leqslant \binom{a+b}{a}$.

This is a special case of the following result.

Theorem 9.8 (Bollobás 1965). *Let* $A_1, \ldots, A_m$ *and* $B_1, \ldots, B_m$ *be two sequences of sets such that* $A_i \cap B_j = \emptyset$ *if and only if* $i = j$. *Then*

$$\sum_{i=1}^{m} \binom{a_i + b_i}{a_i}^{-1} \leqslant 1, \tag{9.2}$$

where $a_i = |A_i|$ *and* $b_i = |B_i|$.

As we already mentioned, due to its importance, there are several different proofs of this theorem. We present two of them.

First proof. Let X be the union of all sets $A_i \cup B_i$. We argue by induction on $n = |X|$. For $n = 1$ the claim is obvious, so assume it holds for $n - 1$ and prove it for n. For every point $x \in X$, consider the family of pairs

$$\mathcal{F}_x \coloneqq \{(A_i, B_i - \{x\}) : x \notin A_i\}.$$

Since each of these families $\mathcal{F}_x$ has less than n points, we can apply the induction hypothesis for each of them, and sum the corresponding inequalities (9.2). The resulting sum counts $n - a_i - b_i$ times the term $\binom{a_i + b_i}{a_i}^{-1}$, corresponding to points $x \notin A_i \cup B_i$, and b_i times the term $\binom{a_i + b_i - 1}{a_i}^{-1}$, corresponding to points $x \in B_i$; the total is $\leqslant n$. Hence we obtain that

$$\sum_{i=1}^{m} (n - a_i - b_i) \binom{a_i + b_i}{a_i}^{-1} + b_i \binom{a_i + b_i - 1}{a_i}^{-1} \leqslant n.$$

Since $\binom{k-1}{l} = \frac{k-l}{k}\binom{k}{l}$, the ith term of this sum is equal to $n \cdot \binom{a_i + b_i}{a_i}^{-1}$. Dividing both sides by n we get the result. □

Second proof. Lubell's method of counting permutations. Let, as before, X be the union of all sets $A_i \cup B_i$. If A and B are disjoint subsets of X then we say that a permutation $(x_1, x_2, \ldots, x_n)$ of X *separates* the pair (A, B) if no element of B precedes an element of A, i.e., if $x_k \in A$ and $x_l \in B$ imply $k < l$.

Each of the $n!$ permutations can separate at most one of the pairs (A_i, B_i), $i = 1, \ldots, m$. Indeed, suppose that $(x_1, x_2, \ldots, x_n)$ separates two pairs (A_i, B_i) and (A_j, B_j) with $i \neq j$, and assume that $\max\{k : x_k \in A_i\} \leqslant \max\{k : x_k \in A_j\}$. Since the permutation separates the pair (A_j, B_j),

$$\min\{l : x_l \in B_j\} > \max\{k : x_k \in A_j\} \geqslant \max\{k : x_k \in A_i\}$$

which implies that $A_i \cap B_j = \emptyset$, contradicting the assumption.

We now estimate the number of permutations separating one fixed pair. If $|A| = a$ and $|B| = b$ and A and B are disjoint then the pair (A, B) is separated by exactly

$$\binom{n}{a + b} a! b! (n - a - b)! = n! \binom{a + b}{a}^{-1}$$

permutations. Here $\binom{n}{a+b}$ counts the number of choices for the positions of $A \cup B$ in the permutation; having chosen these positions, A has to occupy the first a places, giving $a!$ choices for the order of A, and $b!$ choices for the order of B; the remaining elements can be chosen in $(n - a - b)!$ ways.

Since no permutation can separate two different pairs (A_i, B_i), summing up over all m pairs we get all permutations at most once

$$\sum_{i=1}^{m} n! \binom{a_i + b_i}{a_i}^{-1} \leqslant n!$$

and the desired bound (9.2) follows. □

Tuza (1984) observed that Bollobás's theorem implies both Sperner's theorem and the LYM inequality. Let $A_1, \ldots, A_m$ be an antichain over a set X. Take the complements $B_i = X - A_i$ and let $a_i = |A_i|$ for $i = 1, \ldots, m$. Then $b_i = n - a_i$ and by (9.2)

$$\sum_{k=1}^{m} \binom{n}{|A_i|}^{-1} = \sum_{i=1}^{m} \binom{a_i + b_i}{a_i}^{-1} \leqslant 1.$$

Due to its importance, the theorem of Bollobás was extended in several ways.

Theorem 9.9 (Tuza 1985). *Let $A_1, \ldots, A_m$ and $B_1, \ldots, B_m$ be collections of sets such that $A_i \cap B_i = \emptyset$ and for all $i \neq j$ either $A_i \cap B_j \neq \emptyset$ or $A_j \cap B_i \neq \emptyset$ (or both) holds. Then for any real number $0 < p < 1$, we have*

$$\sum_{i=1}^{m} p^{|A_i|} (1 - p)^{|B_i|} \leqslant 1.$$

Proof. Let X be the union of all sets $A_i \cup B_i$. Choose a subset $\mathbf{Y} \subseteq X$ at random in such a way that each element $x \in X$ is included in $\mathbf{Y}$ independently and with the same probability p. Let E_i be the event that $A_i \subseteq \mathbf{Y} \subseteq X - B_i$. Then for their probabilities we have $\mathrm{Prob}\,(E_i) = p^{|A_i|}(1 - p)^{|B_i|}$ for every $i = 1, \ldots, m$ (see Exercise 9.4). We claim that, for $i \neq j$, the events E_i and E_j cannot occur at the same time. Indeed, otherwise we would have $A_i \cup A_j \subseteq \mathbf{Y} \subseteq X - (B_i \cup B_j)$, implying $A_i \cap B_j = A_j \cap B_i = \emptyset$, which contradicts our assumption.

Since the events $E_1, \ldots, E_m$ are mutually disjoint, we conclude that $\mathrm{Prob}\,(E_1) + \cdots + \mathrm{Prob}\,(E_m) = \mathrm{Prob}\,(E_1 \cup \cdots \cup E_m) \leqslant 1$, as desired. $\square$

The theorem of Bollobás also has other important extensions. We do not intend to give a complete account here; we only mention some of these results without proof. More information about Bollobás-type results can be found, for example, in a survey by Tuza (1994).

A typical generalization of Bollobás's theorem is its following "skew version." This result was proved by Frankl (1982) by modifying an argument of Lovász (1977) and was also proved in an equivalent form by Kalai (1984). The argument itself employs the *linear algebra method*, and we will give this proof in Sect. 14.3.4.

Theorem 9.10. *Let $A_1, \ldots, A_m$ and $B_1, \ldots, B_m$ be finite sets such that $A_i \cap B_i = \emptyset$ and $A_i \cap B_j \neq \emptyset$ if $i < j$. Also suppose that $|A_i| \leqslant a$ and $|B_i| \leqslant b$. Then $m \leqslant \binom{a+b}{a}$.*

We also have the following "threshold version" of Bollobás's theorem.

Theorem 9.11 (Füredi 1984). *Let $A_1, \ldots, A_m$ be a collection of a-sets and $B_1, \ldots, B_m$ be a collection of b-sets such that $|A_i \cap B_i| \leqslant s$ and $|A_i \cap B_j| > s$ for every $i \neq j$. Then $m \leqslant \binom{a+b-2s}{a-s}$.*

Another line of research in that direction was to consider more that two sequences of sets.

Theorem 9.12 (Tuza 1989b). *Let k be an integer, $k \geqslant 2$, and let $p_1, \ldots, p_k$ be positive reals with $p_1 + \ldots + p_k = 1$. Suppose we have a collection*

$$
\begin{array}{cccc}
A_{1,1} & A_{1,2} & \cdots & A_{1,k} \\
A_{2,1} & A_{2,2} & \cdots & A_{2,k} \\
\vdots & \vdots & \cdots & \vdots \\
A_{m,1} & A_{m,2} & \cdots & A_{m,k}
\end{array}
$$

of k-tuples of finite sets satisfying the following two properties: (i) *for every i and every $j \neq j'$, $A_{i,j} \cap A_{i,j'} = \emptyset$, and* (ii) *for every $i \neq i'$ there exist $j \neq j'$ for which $A_{i,j} \cap A_{i',j'} \neq \emptyset$. Then $\sum_{i=1}^{m} \prod_{a=1}^{k} p_a^{|A_i^a|} \leqslant 1$.*

9.2.3 Strong systems of distinct representatives

Recall that a system of distinct representatives for the sets $S_1, S_2, \ldots, S_k$ is a k-tuple $(x_1, x_2, \ldots, x_k)$ where the elements x_i are distinct and $x_i \in S_i$ for all $i = 1, 2, \ldots, k$. Such a system is *strong* if we additionally have $x_i \notin S_j$ for all $i \neq j$.

Theorem 9.13 (Füredi–Tuza 1985). *In any family of more than $\binom{r+k}{k}$ sets of cardinality at most r, at least $k + 2$ of its members have a strong system of distinct representatives.*

Proof. Let $\mathcal{F} = \{A_1, \ldots, A_m\}$ be a family of sets, each of cardinality at most r. Suppose that *no $k + 2$ of these sets have a strong system of distinct representatives.* We will apply the theorem of Bollobás to prove that then $m \leqslant \binom{r+k}{k}$. Let us make an additional assumption that our sets form an antichain, i.e., that no of them is a subset of another one. By Theorem 9.8 it is enough to prove that, for every $i = 1, \ldots, m$ there exists a set B_i, such that $|B_i| \leqslant k$, $B_i \cap A_i = \emptyset$ and $B_i \cap A_j \neq \emptyset$ for all $j \neq i$.

Fix an i and let $B_i = \{x_1, \ldots, x_t\}$ be a minimal set which intersects all the sets $A_j - A_i$, $j = 1, \ldots, m$, $j \neq i$. (Such a set exists because none of these differences is empty.) By the minimality of B_i, for every $\nu = 1, \ldots, t$ there exists a set $S_\nu \in \mathcal{F}$ such that $B_i \cap S_\nu = \{x_\nu\}$. Fix an arbitrary element $y_i \in A_i$. Then $(y_i, x_1, \ldots, x_t)$ is a strong system of distinct representatives for $t + 1$ sets $A_i, S_1, \ldots, S_t$. By the indirect assumption, we can have at most $k + 1$ such sets. Therefore, $|B_i| = t \leqslant k$, as desired.

In the case when our family $\mathcal{F}$ is not an antichain, it is enough to order the sets so that $A_i \not\subseteq A_j$ for $i < j$, and apply the skew version of Bollobás's theorem. □

9.2.4 Union-free families

A family of sets $\mathcal{F}$ is called r-*union-free* if $A_0 \not\subseteq A_1 \cup A_2 \cup \cdots \cup A_r$ holds for all distinct $A_0, A_1, \ldots, A_r \in \mathcal{F}$. Thus, antichains are r-union-free for $r = 1$.

Let $T(n, r)$ denote the maximum cardinality of an r-union-free family $\mathcal{F}$ over an n-element underlying set. This notion was introduced by Kautz and Singleton (1964). They proved that

$$\Omega(1/r^2) \leqslant \frac{\log_2 T(n, r)}{n} \leqslant O(1/r).$$

This result was rediscovered several times in information theory, in combinatorics by Erdős, Frankl, and Füredi (1985), and in group testing by Hwang and Sós (1987). Dyachkov and Rykov (1982) obtained, with a rather involved proof, that

$$\frac{\log_2 T(n, r)}{n} \leqslant O(\log_2 r / r^2).$$

Recently, Ruszinkó (1994) gave a purely combinatorial proof of this upper bound. Shortly after, Füredi (1996) found a very elegant argument, and we present it below.

Theorem 9.14 (Füredi 1996). *Let $\mathcal{F}$ be a family of subsets of an n-element underlying set X, and $r \geqslant 2$. If $\mathcal{F}$ is r-union-free then $|\mathcal{F}| \leqslant r + \binom{n}{t}$ where*

$$t = \left\lceil (n - r) / \binom{r + 1}{2} \right\rceil.$$

That is,

$$\log_2 |\mathcal{F}| / n \leqslant O\left(\log_2 r / r^2\right).$$

Proof. Let $\mathcal{F}_t$ be the family of all members of $\mathcal{F}$ having their *own* t-subset. That is, $\mathcal{F}_t$ contains all those members $A \in \mathcal{F}$ for which there exists a t-element subset $T \subseteq A$ such that $T \not\subseteq A'$ for every other $A' \in \mathcal{F}$. Let $\mathcal{T}_t$ be the family of these t-subsets; hence $|\mathcal{T}_t| = |\mathcal{F}_t|$. Let $\mathcal{F}_0 = \{A \in \mathcal{F} : |A| < t\}$, and let $\mathcal{T}_0$ be the family of *all* t-subsets of X containing a member of $\mathcal{F}_0$, i.e.,

$$\mathcal{T}_0 = \{T : T \subseteq X, |T| = t \text{ and } T \supset A \text{ for some } A \in \mathcal{F}_0\}.$$

The family $\mathcal{F}$ is an antichain. This implies that $\mathcal{T}_t$ and $\mathcal{T}_0$ are disjoint. The family $\mathcal{F}_0$ is also an antichain, and since $t < n/2$, we know from Exercise 5.9 that $|\mathcal{F}_0| \leqslant |\mathcal{T}_0|$. Therefore,

$$|\mathcal{F}_0 \cup \mathcal{F}_t| \leqslant |\mathcal{T}_t| + |\mathcal{T}_0| \leqslant \binom{n}{t}. \tag{9.3}$$

It remains to show that the family

$$\mathcal{F}' = \mathcal{F} - (\mathcal{F}_0 \cup \mathcal{F}_t)$$

has at most r members. Note that $A \in \mathcal{F}'$ if and only if $A \in \mathcal{F}$, $|A| \geqslant t$ and for every t-subset $T \subseteq A$ there is an $A' \in \mathcal{F}$ such that $A' \neq A$ and $A' \supseteq T$.

We will use this property to prove that $A \in \mathcal{F}'$, $A_1, A_2, \ldots, A_i \in \mathcal{F}$ $(i \leqslant r)$ imply

$$|A - (A_1 \cup \cdots \cup A_i)| \geqslant t(r - i) + 1. \tag{9.4}$$

To show this, assume the opposite. Then the set $A - (A_1 \cup \cdots \cup A_i)$ can be written as the union of some $(r - i)$ t-element sets $T_{i+1}, \ldots T_r$. Therefore, A lies entirely in the union of $A_1, \ldots, A_i$ and these sets $T_{i+1}, \ldots, T_r$. But, by the choice of A, each of the sets T_j lies in some other set $A_j \in \mathcal{F}$ different from A. Therefore, $A \subseteq A_1 \cup \cdots \cup A_r$, a contradiction.

Now suppose that $\mathcal{F}'$ has more than r members, and take any $r + 1$ of them $A_0, A_1, \ldots, A_r \in \mathcal{F}'$. Applying (9.4) we obtain

$$\left| \bigcup_{i=0}^{r} A_i \right| = |A_0| + |A_1 - A_0| + |A_2 - (A_0 \cup A_1)| + \cdots$$

$$+ |A_r - (A_0 \cup A_1 \cup \cdots \cup A_{r-1})|$$

$$\geqslant (tr + 1) + (t(r - 1) + 1) + (t(r - 2) + 1) + \cdots + (t \cdot 0 + 1)$$

$$= t \cdot \frac{r(r + 1)}{2} + r + 1 = t\binom{r + 1}{2} + r + 1.$$

By the choice of t, the right-hand side exceeds the total number of points n, which is impossible. Therefore, $\mathcal{F}'$ cannot have more than r distinct members. Together with (9.3), this yields the desired upper bound on $|\mathcal{F}|$. $\square$

Exercises

9.1.⁻ Let $\mathcal{F}$ be an antichain consisting of sets of size at most $k \leqslant n/2$. Show that $|\mathcal{F}| \leqslant \binom{n}{k}$.

9.2. Derive from Bollobás's theorem the following weaker version of Theorem 9.11. Let $A_1, \ldots, A_m$ be a collection of a-element sets and $B_1, \ldots, B_m$ be a collection of b-element sets such that $|A_i \cap B_i| = t$ for all i, and $|A_i \cap B_j| > t$ for $i \neq j$. Then $m \leqslant \binom{a+b-t}{a-t}$.

9.3. Show that the upper bounds in Bollobás's and Füredi's theorems (Theorems 9.7 and 9.11) are tight.

  Hint: Take two disjoint sets X and S of respective sizes $a + b - 2s$ and s. Arrange the s-element subsets of X in any order: $Y_1, Y_2, \ldots$. Let $A_i = S \cup Y_i$ and $B_i = S \cup (X - Y_i)$.

9.4. Use the binomial theorem to prove the following. Let $0 < p < 1$ be a real number, and $C \subset D$ be any two fixed subsets of $\{1, \ldots, n\}$. Then the sum of $p^{|A|}(1-p)^{n-|A|}$ over all sets A such that $C \subseteq A \subseteq D$, equals $p^{|C|}(1-p)^{n-|D|}$.

9.5.[(!)] (Frankl 1986). Let $\mathcal{F}$ be a k-uniform family, and suppose that it is *intersection free*, i.e., that $A \cap B \not\subseteq C$ for any three sets A, B and C of $\mathcal{F}$. Prove that $|\mathcal{F}| \leqslant 1 + \binom{k}{\lfloor k/2 \rfloor}$.

Hint: Fix a set $B_0 \in \mathcal{F}$, and observe that $\{A \cap B_0 \ : \ A \in \mathcal{F}, A \neq B_0\}$ is an antichain over B_0.

9.6.[+] Let $A_1, \ldots, A_m$ be a family of subsets of an n-element set, and suppose that it is *convex* in the following sense: if $A_i \subseteq B \subseteq A_j$ for some i, j, then B belongs to the family. Prove that the absolute value of the sum $\sum_{i=1}^m (-1)^{|A_i|}$ does not exceed $\binom{n}{\lfloor n/2 \rfloor}$.

Hint: Use the chain decomposition theorem. Observe that the contribution to the sum from each of the chains is of the form $\pm(1 - 1 + 1 - 1 \ldots)$, and so this contribution is $1, -1$ or 0.

9.7.[(!)] Let $x_1, \ldots, x_n$ be real numbers, $x_i \geqslant 1$ for each i, and let S be the set of all numbers, which can be obtained as a linear combinations $\alpha_1 x_1 + \ldots + \alpha_n x_n$ with $\alpha_i \in \{-1, +1\}$. Let $I = [a, b)$ be any interval (in the real line) of length $b - a = 2$. Show that $|I \cap S| \leqslant \binom{n}{\lfloor n/2 \rfloor}$.

Hint: Associate with each such sum $\xi = \alpha_1 x_1 + \ldots + \alpha_n x_n$ the corresponding set $A_\xi = \{i \ : \ \alpha_i = +1\}$ of indices i for which $\alpha_i = +1$. Show that the family of sets A_ξ for which $\xi \in I$, forms an antichain and apply Sperner's theorem.

Note: Erdős (1945) proved a more general result that if $b - a = 2t$ then $|I \cap S|$ is less than or equal to the sum of the t largest binomial coefficients $\binom{n}{i}$.

9.8. Let P be a finite poset and suppose that the largest chain in it has size r. We know (see Theorem 9.1) that P can be partitioned into r antichains. Show that the following argument also gives the desired decomposition: let A_1 be the set of all maximal elements in P; remove this set from P, and let A_2 be the set of all maximal elements in the reduced set $P - A_1$, etc.

9.9.[−] Let $\mathcal{F} = \{A_1, \ldots, A_m\}$ and suppose that

$$|A_i \cap A_j| < \frac{1}{r} \min\{|A_i|, |A_j|\} \quad \text{for all } i \neq j.$$

Show that $\mathcal{F}$ is r-union-free.

9.10.[−] Let $\mathcal{F} = \{A_1, \ldots, A_m\}$ be an r-union-free family. Show that then $\bigcup_{i \in I} A_i \neq \bigcup_{j \in J} A_j$ for any two distinct non-empty subsets I, J of size at most r.

10. Blocking Sets and the Duality

In this chapter we will consider one of the most basic properties of set systems — their duality.

10.1 Duality

A *blocking set* of a family $\mathcal{F}$ is a set T that intersects (blocks) every member of $\mathcal{F}$. A blocking set of $\mathcal{F}$ is *minimal* if none of its proper subset is such. (Such sets are also called *transversals* of $\mathcal{F}$.) The family of all minimal blocking sets of $\mathcal{F}$ is called its *dual* and is denoted by $b(\mathcal{F})$.

Proposition 10.1. *If $\mathcal{F}$ is an antichain then $b(b(\mathcal{F})) = \mathcal{F}$.*

Proof. Take a set $B \in b(b(\mathcal{F}))$. If $B \notin \mathcal{F}$ then for each set $A \in \mathcal{F}$ there is a point $x_A \in A - B$. The set $\{x_A : A \in \mathcal{F}\}$ of all such points is a blocking set of $\mathcal{F}$, and hence, contains at least one minimal blocking set $T \in b(\mathcal{F})$. But this is impossible, because then B must intersect T.

For the other direction just observe that if some set of $\mathcal{F}$ did not belong to $b(b(\mathcal{F}))$, then this set would avoid at least one set from $b(\mathcal{F})$, which is impossible. $\qquad\square$

Let us consider the following problem of "keys of the safe" (Berge 1989). An administrative council is composed of a set X of individuals. Each of them carries a certain weight in decisions, and it is required that every subset $A \subseteq X$ carrying a total weight greater than some threshold fixed in advance, should have access to documents kept in a safe with multiply locks. The minimal "coalitions" which can open the safe constitute an antichain $\mathcal{F}$. The problem consists in determining the minimal number of locks necessary so that by giving one or more keys to every individual, the safe can be opened if and only if at least one of the coalitions of $\mathcal{F}$ is present.

Proposition 10.2. *For every family $\mathcal{F}$ of minimal coalitions, $\ell = |b(\mathcal{F})|$ locks are enough.*

Proof. Let $b(\mathcal{F}) = \{T_1, \ldots, T_\ell\}$. Then give the key of the ith lock to all the members of T_i. It is clear that then every coalition $A \in \mathcal{F}$ will have the keys to all ℓ locks, and hence, will be able to open the safe. On the other hand, if some set B of individuals does not include a coalition then, by Proposition 10.1, the set B is not a blocking set of $b(\mathcal{F})$, that is, $B \cap T_i = \emptyset$ for some i. But this means that people in B lack the ith key, as desired.   □

A family $\mathcal{F}$ is called *self-dual* if $b(\mathcal{F}) = \mathcal{F}$.

For example, the family of all k-element subsets of a $(2k-1)$-element set is self-dual. Another example is the family of $r+1$ sets, one of which has r elements and the remaining r sets have 2 elements (see Fig. 10.1).

Fig. 10.1. Example of a self-dual family

What other families are self-dual? Our nearest goal is to show that a family is self-dual if and only if it is intersecting and not 2-colorable. Let us first recall the definition of these two concepts.

A family is *intersecting* if any two of its sets have a non-empty intersection. The *chromatic number* $\chi(\mathcal{F})$ of $\mathcal{F} \subseteq 2^X$ is the smallest number of colors necessary to color the points in X so that no set of $\mathcal{F}$ of cardinality > 1 is monochromatic. It is clear that $\chi(\mathcal{F}) \geqslant 2$ (as long as $\mathcal{F}$ is non-trivial, i.e., contains at least one set with more than one element).

The families with $\chi(\mathcal{F}) = 2$ are of special interest and are called *2-colorable*. In other words, $\mathcal{F}$ is 2-colorable iff there is a subset S such that neither S nor its complement $X - S$ contain a member of $\mathcal{F}$. We have already considered this concept in Chap. 6. It turns out that $\chi(\mathcal{F}) > 2$ is a necessary condition for a family $\mathcal{F}$ to be self-dual.

For families of sets $\mathcal{F}$ and $\mathcal{G}$, we write $\mathcal{F} \succ \mathcal{G}$ if every member of $\mathcal{F}$ contains at least one member of $\mathcal{G}$.

Proposition 10.3. (i) *A family $\mathcal{F}$ is intersecting if and only if $\mathcal{F} \succ b(\mathcal{F})$.*
(ii) *If $\mathcal{F}$ is an antichain then $\chi(\mathcal{F}) > 2$ if and only if $b(\mathcal{F}) \succ \mathcal{F}$.*

Proof. (i) If $\mathcal{F}$ is intersecting then every $A \in \mathcal{F}$ is also a blocking set of $\mathcal{F}$, and hence, contains at least one minimal blocking set. Conversely, if $\mathcal{F} \succ b(\mathcal{F})$ then every set of $\mathcal{F}$ contains a blocking set of $\mathcal{F}$, and hence, intersects all other sets of $\mathcal{F}$.

(ii) Let us prove that $\chi(\mathcal{F}) > 2$ implies $b(\mathcal{F}) \succ \mathcal{F}$. If not, then there must be a blocking set T of $\mathcal{F}$ which contains no set of $\mathcal{F}$. But its complement $X - T$ also contains no set of $\mathcal{F}$, since otherwise T would not block all the members of $\mathcal{F}$. Thus $(T, X - T)$ is a 2-coloring of $\mathcal{F}$ with no monochromatic set, a contradiction with $\chi(\mathcal{F}) > 2$.

For the other direction, assume that $b(\mathcal{F}) \succ \mathcal{F}$ but $\chi(\mathcal{F}) = 2$. By the definition of $\chi(\mathcal{F})$ there exists a set S such that neither S nor $X - S$ contain a set of $\mathcal{F}$. This, in particular, means that S is a blocking set of $\mathcal{F}$ which together with $b(\mathcal{F}) \succ \mathcal{F}$ implies that $S \supseteq A$ for some $A \in \mathcal{F}$, a contradiction. $\qquad\square$

Corollary 10.4. *Let $\mathcal{F}$ be an antichain. Then the following three conditions are equivalent:*

(1) $b(\mathcal{F}) = \mathcal{F}$;
(2) $\mathcal{F}$ *is intersecting and* $\chi(\mathcal{F}) > 2$;
(3) *both $\mathcal{F}$ and $b(\mathcal{F})$ are intersecting.*

Proof. Equivalence of (1) and (2) follows directly from Proposition 10.3. Equivalence of (1) and (3) follows from the fact that both $\mathcal{F}$ and $b(\mathcal{F})$ are antichains. $\qquad\square$

10.2 The blocking number

Recall that the *blocking number* $\tau(\mathcal{F})$ of a family $\mathcal{F}$ is the minimum number of elements in a blocking set of $\mathcal{F}$, i.e.,

$$\tau(\mathcal{F}) = \min\{|T| \ : \ T \cap A \neq \emptyset \text{ for every } A \in \mathcal{F}\}.$$

We make two observations concerning this characteristic:

If $\mathcal{F}$ contains a k-*matching*, i.e., k mutually disjoint sets, then $\tau(\mathcal{F}) \geqslant k$;
If $\mathcal{F}$ is intersecting, then $\tau(\mathcal{F}) \leqslant \min_{A \in \mathcal{F}} |A|$.

A family $\mathcal{F}$ can have many smallest blocking sets, i.e., blocking sets of size $\tau(\mathcal{F})$. The following result says how many. The *rank* of a family $\mathcal{F}$ is the maximum cardinality of a set in $\mathcal{F}$.

Theorem 10.5 (Gyárfás 1987). *Let $\mathcal{F}$ be a family of rank r, and let $\tau = \tau(\mathcal{F})$. Then the number of blocking sets of $\mathcal{F}$ with τ elements is at most r^τ.*

Proof. We will prove by backward induction on i that every i-element set I is contained in at most $r^{\tau-i}$ τ-element blocking sets. It is obvious for $i = \tau$ and the case $i = 0$ gives the theorem. If $i < \tau$ then there exists a set $A \in \mathcal{F}$ such that $A \cap I = \emptyset$ (because $|I| < \tau(\mathcal{F})$). Now apply the induction hypothesis for the sets $I \cup \{x\}$, $x \in A$. Observe that every blocking set T of $\mathcal{F}$, containing the set I, must contain at least one of the extended sets $I \cup \{x\}$, with $x \in A$ (because $I \cap A = \emptyset$ whereas $T \cap A \neq \emptyset$). By the induction hypothesis, each of the sets $I \cup \{x\}$ with $x \in A$, is contained in at most $r^{\tau-(i+1)}$ τ-element blocking sets of $\mathcal{F}$. Thus, the set I itself is contained in at most $|A| \cdot r^{\tau-i-1} \leqslant r^{\tau-i}$ τ-element blocking sets, as desired. $\qquad\square$

Considering τ pairwise disjoint sets of size r shows that Theorem 10.5 is best possible. An important corollary of this theorem is the following result due to Erdős and Lovász (1975).

Theorem 10.6. *Let $\mathcal{F}$ be an intersecting r-uniform family with $\tau(\mathcal{F}) = r$. Then $|\mathcal{F}| \leqslant r^r$.*

Proof. Each $A \in \mathcal{F}$ is a blocking set of size r.   □

Gyárfás's theorem has the following immediate generalization (we will use a version of it later in Sect. 10.6.1 when dealing with lower bounds for monotone boolean circuits).

Lemma 10.7. *Let $\mathcal{F}$ be a family of rank r, and let $\mathcal{B} \subseteq b(\mathcal{F})$ be a family of its minimal blocking sets of size at least t each. Then, for every $1 \leqslant s \leqslant t$, there exists an s-uniform family $\mathcal{H}_s$ such that $|\mathcal{H}_s| \leqslant r^s$ and every $B \in \mathcal{B}$ contains at least one $H \in \mathcal{H}_s$.*

Proof. Let $\mathcal{F} = \{A_1, \ldots, A_m\}$, and fix this order of sets. Construct the desired family $\mathcal{H}_s$ by induction on s. For $s = 1$ we can take as $\mathcal{H}_1$ the family of all single element sets $\{x\}$ with $x \in A_1$.

Assume now that $\mathcal{H}_{s-1}$ is already constructed. We may assume w.l.o.g. that every member of $\mathcal{H}_{s-1}$ is contained in at least one blocking set from $\mathcal{B}$ (if not, we simply omit the redundant sets from $\mathcal{H}_{s-1}$). For each set $H \in \mathcal{H}_{s-1}$ choose the first index i such that $H \cap A_i = \emptyset$; such an i exists since otherwise H would be a blocking set for $\mathcal{F}$ of size $|H| \leqslant s - 1 < s$, which is impossible since H is a part of at least one minimal blocking set of size $\geqslant s$. Put in $\mathcal{H}_s$ all the sets $H \cup \{x\}$ with $x \in A_i$. Each of these sets has $|H \cup \{x\}| = |H| + 1 = s$ elements, and $|\mathcal{H}_s| \leqslant (\max_i |A_i|) \cdot |\mathcal{H}_{s-1}| \leqslant r \cdot r^{s-1} = r^s$, as desired.   □

Is not is obvious

using flower lemma?

Corollary 10.8. *Let $\mathcal{F}$ be a family of rank r. Then for any $s \geqslant 1$, the family $\mathcal{F}$ has at most r^s minimal blocking sets of size s.*

10.3 Generalized Helly theorems

In terms of the blocking number τ, the simplest Helly-type result for families of sets (Theorem 8.6) says that if $\mathcal{F}$ is r-uniform and each set of $\leqslant r+1$ of its members intersect then $\tau(\mathcal{F}) = 1$. This result can be generalized as follows.

Theorem 10.9 (Lovász 1979). *Let $\mathcal{F}$ be r-uniform. If each collection of k members $(k \geqslant 2)$ of $\mathcal{F}$ intersect then $\tau(\mathcal{F}) \leqslant (r-1)/(k-1) + 1$.*

Proof. By construction. For $j = 1, \ldots, k$ we will select j sets $A_1, \ldots, A_j$ in $\mathcal{F}$ such that

$$1 \leqslant |A_1 \cap \cdots \cap A_j| \leqslant r - (j-1)(\tau(\mathcal{F}) - 1), \tag{10.1}$$

which for $j = k$ gives the desired upper bound on $\tau = \tau(\mathcal{F})$.

For $j = 1$ take $A_1 \in \mathcal{F}$ arbitrarily.

Assume $A_1, \ldots, A_j$ have been selected $(j \leqslant k - 1)$. The set $A_1 \cap \cdots \cap A_j$ intersects every set of $\mathcal{F}$ (why?), thus $|A_1 \cap \cdots \cap A_j| \geqslant \tau(\mathcal{F})$. Take a subset $S \subseteq A_1 \cap \cdots \cap A_j$ with $|S| = \tau - 1$. Since $|S| < \tau(\mathcal{F})$, there must be a set

$A_{j+1} \in \mathcal{F}$ such that $S \cap A_{j+1} = \emptyset$ (this set is different from $A_1, \ldots, A_j$ since S intersects all of them). Thus

$$|A_1 \cap \cdots \cap A_j \cap A_{j+1}| \leqslant |A_1 \cap \cdots \cap A_j \cap \overline{S}| = |A_1 \cap \cdots \cap A_j - S|$$
$$= |A_1 \cap \cdots \cap A_j| - (\tau - 1) \leqslant r - (j-1)(\tau - 1) - \tau + 1 = r - j(\tau - 1).$$

$\square$

For graphs (i.e., for 2-uniform families) Helly's theorem (Theorem 8.6) says that, if in a finite graph any three edges share a common vertex, then this graph is a star. Erdős, Hajnal, and Moon (1964) generalized this easy observation about graphs in a different direction. A set of vertices S *covers* a set of edges $F \subseteq E$ of a graph $G = (V, E)$ if every edge in F has at least one of its endpoints in S.

Theorem 10.10. *If each family of at most $\binom{s+2}{2}$ edges of a graph can be covered by s vertices, then all edges can.*

The complete graph on $s + 2$ vertices shows that this bound is best possible. Graphs are 2-uniform families. The question was how to generalize the result to r-uniform families for arbitrary r. The conjecture was easy to formulate: the formula $\binom{s+r}{r}$. This appears to be the correct answer.

Theorem 10.11 (Bollobás 1965). *If each family of at most $\binom{s+r}{r}$ members of an r-uniform family can be blocked by s points then all members can.*

Proof. Let $\mathcal{F}$ be an r-uniform family, satisfying the assumption of the theorem, and suppose that $\tau(\mathcal{F}) \geqslant s + 1$. Then there is a subfamily $\mathcal{F}' = \{A_1, \ldots, A_m\} \subseteq \mathcal{F}$ such that $\tau(\mathcal{F}') = s + 1$ and $\mathcal{F}'$ is *τ-critical*, that is,

$$\tau(\mathcal{F}' - \{A_i\}) \leqslant s$$

for all $i = 1, \ldots, m$. Our goal is to show that $m \leqslant \binom{r+s}{s}$, contradicting the assumption (that every subfamily with so few members *can* be blocked by s points).

Since $\tau(\mathcal{F}') = s + 1$ and $\mathcal{F}'$ is τ-critical, for each $i = 1, \ldots, m$, the family $\mathcal{F}' - \{A_i\}$ has a blocking set B_i of size s. Hence, $A_j \cap B_i \neq \emptyset$ for all $j \neq i$. Moreover, $A_i \cap B_i = \emptyset$ since B_i has too few elements to intersect all the members of $\mathcal{F}'$. Thus, we can apply the Bolobás theorem (Theorem 9.8) with $a_1 = \ldots = a_m = r$ and $b_1 = \ldots = b_m = s$, which yields

$$m \cdot \binom{s+r}{s}^{-1} = \sum_{i=1}^{m} \binom{a_i + b_i}{a_i}^{-1} \leqslant 1,$$

and the desired upper bound on m follows. $\square$

10.4 Decision trees

Blocking sets play an important role in the theory of boolean functions. In this and the next sections we will present some results in that direction.

Let $f : \{0,1\}^n \to \{0,1\}$ be a boolean function. A (deterministic) *decision tree* for f is a binary tree whose internal nodes have labels from $\{1,\ldots,n\}$ and whose leaves have labels from $\{0,1\}$. If a node has label i then the test performed at that node is to examine the ith bit of the input. If the result is 0, one descends into the left subtree, whereas if the result is 1, one descends into the right subtree. The label of the leaf so reached is the value of the function (on that particular input). The *depth* of a decision tree is the number of edges in a longest path from the root to a leaf, or equivalently, the maximum number of bits tested on such a path. Let $DT(f)$ denote the minimum depth of a decision tree computing f.

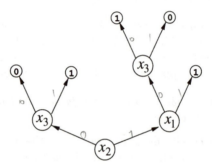

Fig. 10.2. A decision tree of depth 3; on input $(0,1,0)$ it outputs 1.

Given an input $a = (a_1,\ldots,a_n)$ from $\{0,1\}^n$, we would like to know whether $f(a) = 1$ or $f(a) = 0$. How many bits of a must we see in order to answer this question? It is clear that seeing $DT(f)$ bits is always enough: just look at those bits of a which are tested along the (unique!) path from the root to a leaf. In a decision tree all the tests are made in a prescribed order independent of individual inputs. Can we do better if we relax this and allow for each input a to choose its own smallest set of bits to be tested? The question can be formalized as follows.

A *certificate* of a is a set of variables $Y = \{x_{i_1},\ldots,x_{i_k}\}$ such that assigning only these k values already forces the function output the value $f(a)$, independent of the values of remaining variables. That is, $f(b) = f(a)$ for any input b such that $b_{i_1} = a_{i_1}, \ldots, b_{i_k} = a_{i_k}$.

The connection with blocking sets is clear: if $f(a) = 0$ and Y is a certificate of a then Y must intersect the certificates of all the inputs b for which $f(b) = 1$, and *vice versa* (otherwise we could force the function to take both values 0 and 1 on the same input, which, of course, is impossible).

By $C(f, a)$ we denote the minimum size of a certificate for a. The *certificate complexity* of f is $C(f) = \max\{C_0(f), C_1(f)\}$, where

$$C_0(f) = \max_a \{C(f, a) : f(a) = 0\}, \ C_1(f) = \max_a \{C(f, a) : f(a) = 1\}.$$

It is clear that $C(f) \leqslant DT(f)$, i.e., for every input a, seeing its $DT(f)$ bits is enough to determine the value $f(a)$. Is this upper bound optimal? The following example shows that this may be not the case: there are boolean functions f for which $C(f) \leqslant \sqrt{DT(f)}$.

To see this, consider the following function $f(X)$ on $n = m^2$ boolean variables:

$$f = \bigwedge_{i=1}^{m} \bigvee_{j=1}^{m} x_{ij}.$$

Then $C(f) = m$, whereas $DT(f) = m^2$ (see Exercise 10.7). Thus, for this function, $DT(f) = C_0(f) \cdot C_1(f)$.

10.4.1 Depth versus certificate complexity

It turns out that the example given above is, in fact, the worst case. This important result has been re-discovered by many authors in different contexts: Blum and Impagliazzo (1987), Hartmanis and Hemachandra (1991), and Tardos (1989).

Theorem 10.12. *For every boolean function f, $DT(f) \leqslant C_0(f) \cdot C_1(f)$.*

Proof. Induction on the number of variables n. If $n = 1$ then the inequality is trivial.

Let (say) $f(0, \ldots, 0) = 0$; then some set Y of $k \leqslant C_0(f)$ variables can be chosen such that by fixing their value to 0, the function is 0 independently of the other variables. We can assume w.l.o.g. that the set

$$Y = \{x_1, \ldots, x_k\}$$

of the first k variables has this property.

Take a complete decision tree T_0 of depth k on these k variables. Each of its leaves corresponds to a unique input $a = (a_1, \ldots, a_k) \in \{0, 1\}^k$ reaching this leaf. Replace such a leaf by a minimal depth decision tree T_a for the sub-function

$$f_a \rightleftharpoons f(a_1, \ldots, a_k, x_{k+1}, \ldots, x_n).$$

Obviously, $C_0(f_a) \leqslant C_0(f)$ and $C_1(f_a) \leqslant C_1(f)$. We claim that the latter inequality can be strengthened:

$$C_1(f_a) \leqslant C_1(f) - 1 \tag{10.2}$$

To prove this, take an arbitrary input $(a_{k+1}, \ldots, a_n)$ of f_a for which $f_a(a_{k+1}, \ldots, a_n) = 1$. Together with the bits $a = (a_1, \ldots, a_k)$, this gives an input of the whole function f with $f(a_1, \ldots, a_n) = 1$. According to the definition of the quantity $C_1(f)$, there must be a set $Z = \{x_{i_1}, \ldots, x_{i_m}\}$ of $m \leqslant C_1(f)$ variables such that fixing them to the corresponding values $x_{i_1} = a_{i_1}, \ldots, x_{i_m} = a_{i_m}$, the value of f becomes 1 independently of the other variables. A simple (but crucial) observation is that

$$Y \cap Z \neq \emptyset. \tag{10.3}$$

Indeed, if $Y \cap Z = \emptyset$ then the value of $f(0, \ldots, 0, a_{k+1}, \ldots, a_n)$ should be 0 because fixing the variables in Y to 0 forces f to be 0, but should be 1, because fixing the variables in Z to the corresponding values of a_i forces f to be 1, a contradiction.

By (10.3), only $|Z - Y| \leqslant m - 1$ of the bits of $(a_{k+1}, \ldots, a_n)$ must be fixed to force the sub-function f_a to obtain the constant function 1. This completes the proof of (10.2).

Applying the induction hypothesis to each of the sub-functions f_a with $a \in \{0, 1\}^k$, we obtain

$$DT(f_a) \leqslant C_0(f_a)C_1(f_a) \leqslant C_0(f)(C_1(f) - 1).$$

Altogether,

$$DT(f) \leqslant k + \max_a DT(f_a) \leqslant C_0(f) + C_0(f)(C_1(f) - 1) = C_0(f)C_1(f).$$

$\square$

10.4.2 Block sensitivity

Theorem 10.12 gives the following relation between the decision tree depth of boolean functions and their certificate complexity:

$$C(f) \leqslant DT(f) \leqslant C(f)^2.$$

A similar relation also exists between certificate complexity and another important measure of boolean functions – their "block sensitivity."

For a vector $a \in \{0, 1\}^n$ and a subset of indices $S \subseteq \{1, \ldots, n\}$, let $a[S]$ denote the vector a, with *all* bits in S flipped. That is, $a[S]$ differs from a exactly on the bits in S. For example, if $a = (0, 1, 1, 0, 1)$ and $S = \{1, 3, 4\}$, then $a[S] = (1, 1, 0, 1, 1)$. We say that f is *sensitive* to S on a if

$$f(a[S]) \neq f(a).$$

The *block sensitivity* of f on a, denoted $B(f, a)$, is the largest number t for which there exist t disjoint sets $S_1, \ldots, S_t$ such that f is sensitive on a to each of these sets, i.e., $f(a[S_i]) \neq f(a)$ for all $i = 1, \ldots, t$. Let

$$B(f) = \max \{B(f, a) : a \in \{0, 1\}^n\}.$$

Lemma 10.13 (Nisan 1989). *For every boolean function f,*

$$B(f) \leqslant C(f) \leqslant B(f)^2.$$

Proof. The upper bound $B(f) \leqslant C(f)$ follows from the fact that for any input a, any certificate of a must include at least one variable from each set which f is sensitive to on this input a.

The lower bound: $B(f) \geqslant \sqrt{C(f)}$. Take an input $a \in \{0,1\}^n$ achieving certificate complexity, i.e., every certificate for a has size at least $C(f)$. Let S_1 be a minimal set of indices for which $f(a[S_1]) \neq f(a)$, let S_2 be another minimal set disjoint from S_1, and such that $f(a[S_2]) \neq f(a)$, etc. Continue until no such set exists. We have, say, t disjoint sets $S_1, \ldots, S_t$ to each of which the function f is sensitive on a.

The union $I = S_1 \cup \cdots \cup S_t$ must be a certificate of a since otherwise we could pick yet another set S_{t+1} for which $f(a[S_{t+1}]) \neq f(a)$. Thus

$$\sum_{i=1}^{t} |S_i| = |I| \geqslant C(f).$$

If $t \geqslant \sqrt{C(f)}$ then we are done since $B(f) \geqslant t$. If not, then, by the averaging principle (see Sect. 1.5),

$$|S| \geqslant |I|/t \geqslant C(f)/t \geqslant \sqrt{C(f)}$$

for at least one set $S \in \{S_1, \ldots, S_t\}$. Since S is a minimal set for which $f(a[S]) \neq f(a)$, this means that $f(a[S - \{i\}]) = f(a)$ for every $i \in S$. Thus, on the vector $b = a[S]$, the function f is sensitive to each single coordinate $i \in S$; hence

$$B(f) \geqslant B(f, b) \geqslant |S| \geqslant \sqrt{C(f)}.$$

$\square$

10.5 The switching lemma

A boolean function f is a *t-And-Or* function if it can be written as an And of an arbitrary number of clauses, each being an Or of at most t literals, that is,

$$f = C_1 \wedge C_2 \wedge \cdots \wedge C_m,$$

where $C_i = y_{i1} \vee \cdots \vee y_{ia_i}$, $a_i \leqslant t$ and each y_{ij} is a literal. Dually, we call a boolean function f an *s-Or-And* if it can be written as an Or

$$f = M_1 \vee M_2 \vee \cdots \vee M_m$$

of an arbitrary number of monomials, each being an And of at most s literals.

Suppose we have a t-And-Or function f. Our goal is to find its dual representation as an s-Or-And with s as small as possible. If we just multiply the clauses we can get very long monomials, much longer than s. So, the function f itself may not be an s-Or-And. We can try to assign constants 0 and 1 to some variables and "kill off" all long monomials (i.e., evaluate them to 0). If we set some variable x_i, say, to 1, then two things will happen: the literal $\bar{x}_i$ gets value 0 and disappears from all the remaining clauses, and all the clauses C_i containing the literal x_i disappear (they get value 1).

Of course, if we set all variables to constants, then we are done – there will remain no monomials at all. The question becomes interesting if we must leave some given number ℓ of variables not assigned.

The Switching Problem. Given a t-And-Or function f on n variables and a number $\ell < n$, is it possible to assign constants to at most $n - \ell$ variables so that the resulting function is an s-Or-And function? If so – in how many ways we can do this?

This problem is solved by a basic result in complexity theory, known as the *Switching Lemma*. Different versions of this lemma were proved by Ajtai (1983), Furst, Saxe, and Sipser (1984), and Yao (1985). In the most general form (given below) this lemma was proved by Håstad (1986, 1989).

A *restriction* is a map ρ of the set of indices $\{1, \ldots, n\}$ to the set $\{0, 1, *\}$. The restriction ρ can be applied to a function $f = f(x_1, \ldots, x_n)$, then we get the function $f\restriction_\rho$ (called a *subfunction* of f) where the variables are set according to ρ, and $\rho(i) = *$ means that x_i is left unassigned. For example, if $f = (x_1 \vee x_2 \vee x_3) \wedge (\bar{x}_1 \vee x_2) \wedge (x_1 \vee \bar{x}_3)$ and $\rho(1) = 1$, $\rho(2) = \rho(3) = *$ then $f\restriction_\rho = x_2$. We denote by $\mathcal{R}^\ell$ the set of all restrictions assigning exactly ℓ stars. Hence

$$|\mathcal{R}^\ell| = \binom{n}{\ell} 2^{n-\ell}.$$

A *minterm* of f is a restriction ρ such that $f\restriction_\rho \equiv 1$ and which is minimal in the sense that unspecifying every single value $\rho(i) \in \{0, 1\}$ already violates this property. The *length* of a minterm is the number $|\rho^{-1}(*)|$ of unassigned variables.

Switching Lemma. *Let f be a t-And-Or boolean function on n variables, and let ρ be a random restriction in $\mathcal{R}^\ell$. Then*

$$\text{Prob}\left(f\restriction_\rho \text{ is an } s\text{-Or-And }\right) \geq 1 - \left(\frac{\gamma t \ell}{n}\right)^s \tag{10.4}$$

for a fixed constant $\gamma < 5$.

[handwritten: $t \geq 2$ ($t=1$ is trivial) (to be checked)]

[handwritten: for $\ell \geq \frac{n}{t}$ is NOT interesting]

[handwritten: So, we assume $\ell \leq \frac{n}{t} \leq \frac{n}{2}$]

All previous proofs of this lemma used quite involved probabilistic arguments. Razborov (1995) found a new purely combinatorial proof, eliminating the probabilistic argument altogether. Another advantage of this proof is that it also gives us an idea of how to *construct* the desired restriction ρ.

[handwritten: derandomize?]

Let $\min(f)$ be the length of the longest minterm of f, and let

$$\text{Bad}(f, s) = \{\rho \in \mathcal{R}^\ell : \min(f\restriction_\rho) > s\}.$$

In particular, $\text{Bad}^\ell(f, s)$ contains all the restrictions $\rho \in \mathcal{R}^\ell$ for which $f\restriction_\rho$ is not an s-Or-And. [handwritten: ¿ Nothing else?]

[handwritten Arabic/Persian annotation: افرایش‌ها‌ی DT(f|ρ)]

Lemma 10.14 (Razborov 1995). *Let f be a t-And-Or boolean function on n variables. Then, for any $1 \leqslant s \leqslant \ell \leqslant n$,*

$$|\text{Bad}^\ell(s,t)| \leqslant |\mathcal{R}^{\ell-s}| \cdot (4t)^s. \tag{10.5}$$

Before we merge into the proof of Lemma 10.14, let us show that it implies the switching lemma (with a slightly worse constant): take a random restriction ρ in $\mathcal{R}^\ell$. Then, by Lemma 10.14, the probability that $f\!\restriction_\rho$ is *not* an s-Or-And function, is at most

$$\frac{|\text{Bad}^\ell(s,t)|}{|\mathcal{R}^\ell|} \leqslant \frac{\binom{n}{\ell-s}2^{n-\ell+s}(4t)^s}{\binom{n}{\ell}2^{n-\ell}} \leqslant \left(\frac{8t\ell}{n-\ell}\right)^s \leqslant \left(\frac{\gamma t\ell}{n}\right)^s$$

for some constant $\gamma \leqslant 8$; here we used the inequality $\binom{n}{\ell-s}/\binom{n}{\ell} \leqslant \left(\frac{\ell}{n-\ell}\right)^s$.

Remark. Note that f is an s-Or-And if $\min(f) \leqslant s$, but the opposite may be not true! (See Exercise 10.8 for a counterexample.) Therefore, $\text{Bad}^\ell(s,t)$ *may also contain good restrictions*, i.e., restrictions ρ such that $f\!\restriction_\rho$ is an s-Or-And, despite the fact that some of its minterms are long, i.e., despite the fact that $\min(f\!\restriction_\rho) > s$. In this sense Razborov's version of the switching lemma is somewhat stronger.

Razborov's argument gives a very interesting realization of the following *coding principle*, which may be of independent interest:

In order to prove that some set A is not very large try to construct an encoding $F\colon A \to B$ of elements of A with elements of some set B which is a priori known to be small, and *give a way how to retrieve* the element $a \in A$ from its code $F(a)$. Then $|A| \leqslant |B|$.

Proof of Lemma 10.14. Let f be a t-And-Or, that is, f can be written as

$$f = C_1 \wedge C_2 \wedge \cdots \wedge C_m,$$

where each C_i is a clause with at most t literals. Fix this order of clauses and fix an order of literals in each clause. If ρ and π are restrictions on disjoint domains, then $\rho\pi$ will denote the combined restriction, which coincides with ρ and π on their domains.

Now suppose that ρ is a bad restriction, i.e., $\rho \in \text{Bad}^\ell(s,t)$. Then there must be a minterm of $f\!\restriction_\rho$ whose length is at least s. Fix such a minterm π and split it into $k \leqslant s$ sub-restrictions $\pi_1, \ldots, \pi_k$ inductively as follows. Assume that we have already found the restrictions $\pi_1, \pi_2, \ldots, \pi_{i-1}$ to some of the variables in $\rho^{-1}(*)$ such that $\pi_1\pi_2 \cdots \pi_{i-1} \neq \pi$. Apply the restriction $\rho\pi_1 \cdots \pi_{i-1}$ to the function $f = C_1 \wedge C_2 \wedge \cdots \wedge C_m$, and let C_{ν_i} be the *first* (this is important!) clause which is not satisfied by this restriction, i.e.,

$$C_{\nu_i}\!\restriction_{\rho\pi_1 \cdots \pi_{i-1}} \not\equiv 1.$$

Such a clause C_{ν_i} must exist since $\pi_1 \cdots \pi_{i-1}$ is not a minterm of $f\!\restriction_\rho$. Let X_i be the collection of variables of C_{ν_i}, and let Y_i be those variables from X_i

how can it happen?

which are set by π but not by $\pi_1 \cdots \pi_{i-1}$. Note that $Y_i \neq \emptyset$ since $f\lceil_{\rho\pi} \equiv 1$ and thus $C_{\nu_i}\lceil_{\rho\pi} \equiv 1$. We define the ith sub-restriction π_i by: $\pi_i = \pi|_{Y_i}$, that is, on variables in Y_i, the restriction π_i coincides with the restriction π, and assigns $*$'s elsewhere. This way we obtain three sequences:

$$\cdots \ C_{\nu_1} \ \cdots \ C_{\nu_2} \ \cdots \ C_{\nu_i} \ \cdots \ C_{\nu_k} \ \cdots \quad \text{clauses;}$$
$$\cdots \ Y_1 \ \cdots \ Y_2 \ \cdots \ Y_i \ \cdots \ Y_k \ \cdots \quad \text{subsets of their variables:}$$
$$\cdots \ \pi_1 \ \cdots \ \pi_2 \ \cdots \ \pi_i \ \cdots \ \pi_k \ \cdots \quad \text{sub-restrictions.}$$

Now, let k be minimal with the property that $\pi_1, \pi_2, \ldots, \pi_k$ together assign at least s variables. We trim π_k in an arbitrary way so that it still sets the clause C_{ν_k} to 1 and $\pi_1, \pi_2, \ldots, \pi_k$ assigns *exactly* s variables. → *But it happen*

Let $\overline{\pi}_i$ be the uniquely determined restriction which has the same domain Y_i as π_i and does *not* set the clause C_{ν_i} to 1. That is, $\overline{\pi}_i$ evaluates all the literals "touched" by π_i to 0. We let

$$F_1(\rho) = \rho\overline{\pi}_1 \cdots \overline{\pi}_{k-1}\overline{\pi}_k;$$

note that $F_1(\rho) \in \mathcal{R}^{\ell-s}$.

Knowing $F_1(\rho)$ alone is not enough to retrieve the initial restriction ρ. We will therefore put some auxiliary information into a special string, so that by knowing the restriction $F_1(\rho)$ *and* this string will be able to retrieve ρ. To define the desired string, recall that the order of literals in the clauses is fixed. Associate with each clause C_{ν_i} the string $\mathbf{v}_i = (v_{i1}, \ldots, v_{it})$ over the alphabet $\{\circ, \bullet, \star\}$, whose jth letter v_{ij} $(1 \leqslant j \leqslant t)$ is defined as follows. If the jth literal in the clause C_{ν_i} is x_l or $\overline{x}_l$ then:

$$v_{ij} = \begin{cases} \star & \text{if } x_l \text{ is } \textit{not} \text{ set by } \pi_i \text{ or } C_{\nu_i} \text{ has fewer than } j \text{ literals;} \\ \circ & \text{if } x_l \text{ is set by } \pi_i \text{ and } \overline{\pi}_i(l) = \pi_i(l); \\ \bullet & \text{if } x_l \text{ is set by } \pi_i \text{ and } \overline{\pi}_i(l) \neq \pi_i(l). \end{cases}$$

Here is a typical example for the string $\mathbf{v}_i$:

$$C_{\nu_i} = x_3 \vee \overline{x}_4 \vee x_6 \vee x_7 \vee x_{12}$$

$\pi_i =$	0	1	$*$	1	0
$\overline{\pi}_i =$	0	1	$*$	0	0

$\mathbf{v}_i =$	$\circ$	$\circ$	$\star$	$\bullet$	$\circ$

It is easy to see that, knowing the clause C_{ν_i} and the string $\mathbf{v}_i = (v_{i1}, \ldots, v_{it})$, we can uniquely recover both restrictions π_i and $\overline{\pi}_i$. Indeed, the non-$\star$ positions of $\mathbf{v}_i$ tell us which of the literals in C_{ν_i} are set to 0 by the restriction $\overline{\pi}_i$; knowing this restriction, the $\bullet$-positions tell us which of the values in $\overline{\pi}_i$ must be toggled to get π_i.

The constructed string $F_2(\rho) = (\mathbf{v}_1, \ldots, \mathbf{v}_k)$ has the following properties:

(1) each substring $\mathbf{v}_i$ has length t;
(2) each $\mathbf{v}_i$ contains at least one $\bullet$;

why not exactly?

(3) the whole string $F_2(\rho)$ has at most s positions that are non-$\star$'s.

This way we have defined a mapping

$$F : \mathrm{Bad}^\ell(s,t) \to \mathcal{R}^{\ell-s} \times \mathrm{Code}(t,s)$$

which takes ρ to $(F_1(\rho), F_2(\rho))$, and where $\mathrm{Code}(t,s)$ is the set of strings over the alphabet $\{\circ, \bullet, \star\}$, satisfying the three conditions above. To finish the proof it is enough to show that $\mathrm{Code}(t,s)$ contains at most $(4t)^s$ strings and that the mapping F is injective.

How many strings does $\mathrm{Code}(t,s)$ have? By Properties (1)–(3), we can think of every string $\mathbf{v}$ in $\mathrm{Code}(t,s)$ as a sequence of at most $s+1$ substrings of consecutive $\star$'s, interrupted by at most s non-$\star$ values. Properties (1) and (2) ensure that each substring of consecutive $\star$'s has length at most $2t$. There are at most $2t$ (in fact, at most t for this first block) possibilities to put the first non-$\star$ in $\mathbf{v}$; at most $2t$ possibilities to put the second non-$\star$, etc. Since each non-$\star$ has two possibilities (of being $\circ$ or $\bullet$) the total number of such strings is $|\mathrm{Code}(t,s)| \leq (2t)^s 2^s = (4t)^{s+1}$.

In order to show that the mapping F is injective it is enough to show how to retrieve the original restriction ρ from the restriction

needs (+1)?

$$F_1(\rho) = \rho \bar{\pi}_1 \cdots \bar{\pi}_{k-1} \bar{\pi}_k$$

and the string

$$F_2(\rho) = (\mathbf{v}_1, \ldots, \mathbf{v}_k).$$

It is sufficient to retrieve all the pairs $(\pi_i, \bar{\pi}_i)$ $(1 \leq i \leq k)$.

We do this by induction on i. Assume that we already know all the pairs of sub-restrictions $(\bar{\pi}_1, \pi_1), \ldots, (\bar{\pi}_{i-1}, \pi_{i-1})$. Then we also know the restriction $\rho \pi_1 \ldots \pi_{i-1} \bar{\pi}_i \ldots \bar{\pi}_k$. The crucial observation is that

ν_i is the *minimal* index ν such that $C_\nu \lceil_{\rho \pi_1 \cdots \pi_{i-1} \bar{\pi}_i \cdots \bar{\pi}_k} \not\equiv 1$.

Indeed, $C_{\nu_i} \lceil_{\rho \pi_1 \cdots \pi_{i-1}} \not\equiv 1$, $C_{\nu_i} \lceil_{\bar{\pi}_i} \not\equiv 1$ from definitions, and restrictions $\bar{\pi}_{i+1}$, $\ldots$, $\bar{\pi}_k$ do not assign variables from C_{ν_i} at all. Hence

$$C_{\nu_i} \lceil_{\rho \pi_1 \cdots \pi_{i-1} \bar{\pi}_i \cdots \bar{\pi}_k} \not\equiv 1.$$

On the other hand, all C_ν with $\nu < \nu_i$ are already set to 1 by $\rho \pi_1 \ldots \pi_{i-1}$ and hence by $\rho \pi_1 \ldots \pi_{i-1} \bar{\pi}_i \ldots \bar{\pi}_k$.

Now, when we know the clause C_{ν_i}, the rest is easy. Having this clause, we have the set X_i of variables in it. From the string $\mathbf{v}_i$ and X_i we know the domain of $\bar{\pi}_i$, then we retrieve from $F_2(\rho)$ the actual value of $\bar{\pi}_i$ and consult the string $\mathbf{v}_i$ again to get π_i. $\qquad \square$

10.6 Monotone circuits

A boolean function $f(x_1, \ldots, x_n)$ is *monotone* if $f(x_1, \ldots, x_n) = 1$ and $x_i \leq y_i$ for all i, imply $f(y_1, \ldots, y_n) = 1$. A *monotone circuit* is a sequence $f_1, \ldots, f_t$

of monotone boolean functions, called *gates*, each of which is either one of the variables $x_1, \ldots, x_n$ or is obtained from some previous gates via an And or Or operations. That is, each gate f_i has either the form $f_i = x_l$ for some $1 \leqslant l \leqslant n$, or one of the forms $f = g \vee h$ or $f = g \wedge h$ for some $g, h \in \{0, 1, f_1, \ldots, f_{i-1}\}$. The *size* of a circuit is the number t of gates in it. The function *computed* by such a circuit is the last function f_t.

The problem (known as the *lower bounds problem*) is, given an explicit boolean function, to prove that it cannot be computed by a circuit of small size. It is clear that every function can be computed by a circuit of size exponential in the number of variables. However, even in the case of monotone circuits, it is difficult to show that some function is indeed hard, i.e., requires many gates. Until 1985, the largest known lower bound on the size of such circuits for an explicit boolean function of n variables was only $4n$. This was considerably improved in the fundamental paper by Razborov (1985), where a bound of $n^{\Omega(\log n)}$ to the Clique$_n$ function (which outputs 1 iff a given graph on n vertices contains a clique of size $k = \Theta(\log n)$) was established. Shortly afterwards, Andreev (1985) and Alon and Boppana (1987) used similar methods to obtain exponential lower bounds.

All these proofs are based on Razborov's idea to approximate intermediate results of a computation by k-Or-Ands for particular values of k. To get a nontrivial lower bound on the number of gates in a circuit, these approximations must be as tight as possible. The desired tightness was achieved via nontrivial applications of the sunflower lemma of Erdős and Rado (see Sect. 7.1) and its modification, due to Füredi (Lemma 7.2).

In this section we will show that, using some combinatorial properties of blocking sets, one may obtain exponential lower bounds in a relatively easy and direct way. The idea is to approximate intermediate results not only by k-Or-Ands but also by their "duals," k-And-Ors. The idea was first (implicitly) employed by Haken (1995), Haken and Cook (1999), and then extended in Jukna (1997, 1999) to a general combinatorial lower bounds *criterion* for monotone circuits, including circuits with unbounded fan-in gates and circuits with non-decreasing real-valued functions as gates. Here we will demonstrate the idea for the simplest case of monotone circuits with the usual (fan-in 2) And and Or gates.

10.6.1 The lower bounds criterion

A *monotone k-CNF* (conjunctive normal form) is a boolean function which can be written as an And of an arbitrary number of monotone *clauses*, each being an Or of k variables. Dually, a *monotone k-DNF* is a boolean function which can be written as an Or of an arbitrary number of *monomials*, each being an And of k variables.

Note that k-CNF and k-DNF differ from k-And-Or and k-Or-And functions (considered in Sects. 7.3.1 and 10.5) in that now we require that the clauses and monomials have *exactly* k distinct variables.

For two boolean functions f and g in n variables, we write $f \leqslant g$ if $f(x) \leqslant g(x)$ for all $x \in \{0,1\}^n$.

Definition 10.15. Let $f(x_1, \ldots, x_n)$ be a monotone boolean function. We say that f is t-*simple* if for every pair of integers $2 \leqslant s, r \leqslant n$ there exists a monotone s-CNF C consisting of at most $t \cdot (r-1)^s$ clauses, a monotone r-DNF D consisting of at most $t \cdot (s-1)^r$ monomials, and a subset $I \subseteq \{1, \ldots, n\}$ of $|I| \leqslant s - 1$ indices, so that either

$$C \leqslant f \quad \text{or} \quad f \leqslant D \vee \left(\bigvee_{i \in I} x_i \right)$$

(or both) hold.

We have the following general lower bounds criterion for monotone boolean circuits. In fact, the same criterion holds for more general circuits, where we can use as gates not just boolean And and Or but also any *real-valued* functions $\varphi : \mathbb{R}^2 \to \mathbb{R}$ such that $\varphi(x_1, x_2) \leqslant \varphi(y_1, y_2)$ whenever $x_1 \leqslant y_1$ and $x_2 \leqslant y_2$. A similar criterion holds also with unbounded fan-in boolean gates (see Jukna 1999 for details).

Theorem 10.16. *If a monotone boolean function can be computed by a monotone circuit of size t, then it is t-simple.*

Proof. Given a monotone circuit, the idea is to approximate every intermediate gate (more exactly – the function computed at the gate) by an $(s-1)$-And-Or and an $(r-1)$-Or-And, and to show that doing so we do not introduce too many errors. If the function computed by the whole circuit is not t-simple, then it cannot be approximated well by such an And-Or/Or-And pair meaning that every such pair must make many errors. Since the number of errors introduced at each separate gate is small, the total number of gates must be large.

To make as few errors at each gate as possible we will use the following simple fact which allows us to convert an $(s-1)$-And-Or into a small $(r-1)$-Or-And and vice versa (cf. also Lemmas 7.4 and 10.7).

Claim 10.17. *For every $(s-1)$-And-Or f_0 there is an $(r-1)$-Or-And f_1 and an r-DNF D such that $|D| \leqslant (s-1)^r$ and*

$$f_1 \leqslant f_0 \leqslant f_1 \vee D.$$

Dually, for every $(r-1)$-Or-And f_1 there is an $(s-1)$-And-Or f_0 and an s-CNF C such that $|C| \leqslant (r-1)^s$ and

$$f_0 \wedge C \leqslant f_1 \leqslant f_0.$$

Proof. We prove the first claim (the second is dual). Let $f_0 = S_1 \wedge S_2 \wedge \cdots \wedge S_m$ be an $(s-1)$-And-Or. We associate with f_0 the following tree T of fan-out at most $s - 1$. The first node of T corresponds to the first clause S_1, and the outgoing $|S_1|$ edges are labeled by the variables from S_1. Suppose we have reached a node v, and let M be the monomial consisting of the labels of edges from the root to v. If M intersects all the clauses of f_0, then v is a leaf. Otherwise, let S_i be the *first* clause such that $M \cap S_i = \emptyset$. Then the node v has $|S_i|$ outgoing edges labeled by the variables in S_i.

Each path from the root to a leaf of T gives us a monomial M which intersects all the clauses of f_0. The Or of paths in T from the root to leafs of height at most $r - 1$ give us the desired $(r - 1)$-Or-And f_1. It is clear that $f_1 \leqslant f_0$ because each monomial of f_1 intersects all the clauses of f_0. The OR of paths to the nodes of height r give us the desired r-DNF D. This DNF has at most $(s - 1)^r$ monomials since every node of T has fan-out at most $s - 1$.

$\square$

We now turn to the actual proof of Theorem 10.16. Let $F(x_1, \ldots, x_n)$ be a monotone boolean function, and suppose that F can be computed by a monotone circuit of size t. Our goal is to show that then the function F is t-simple. To do this, fix an arbitrary pair of integers $2 \leqslant s, r \leqslant n$.

Let $f = g * h$ be a gate in our circuit computing F. By an *approximator* of this gate we will mean a pair f_0, f_1, where f_0 is an $(s - 1)$-And-Or (a *left* approximator of f) and f_1 is an $(r - 1)$-Or-And (a *right* approximator of f) such that $f_1 \leqslant f_0$.

We say that such an approximator f_0, f_1 of f introduces a new error on input $x \in \{0, 1\}^n$ if the approximators of g and of h did not make an error on x, but the approximator of f does. That is, $g_0(x) = g_1(x) = g(x)$ and $h_0(x) = h_1(x) = h(x)$, but either $f_0(x) \neq f(x)$ or $f_1(x) \neq f(x)$.

We define approximators inductively as follows.

Case 1: f is an input variable, say, $f = x_i$.

In this case we take $f_0 = f_1 \rightleftharpoons x_i$. It is clear that this approximator introduces no errors.

Case 2: f is an And gate, $f = g \wedge h$.

In this case we take $f_0 \rightleftharpoons g_0 \wedge h_0$ as the left approximator of f; hence, f_0 introduces no new errors. To define the right approximator of f we use Claim 10.17 to convert f_0 into an $(r - 1)$-Or-And f_1; hence, $f_1 \leqslant f_0$. Let E_f be the set of inputs on which f_1 introduces a new error, i.e.,

$$E_f \rightleftharpoons \{x : f(x) = f_0(x) = 1 \text{ but } f_1(x) = 0\}.$$

By Claim 10.17, all these errors can be "corrected" by adding a relatively small r-DNF: there is an r-DNF D such that $|D| \leqslant (s - 1)^r$ and $D(x) = 1$ for all $x \in E_f$.

Case 3: f is an Or gate, $f = g \vee h$.

In this case we take $f_1 \rightleftharpoons g_1 \vee h_1$ as the right approximator of f; hence, f_1 introduces no new errors. To define the left approximator of f we use Claim 10.17 to convert f_1 into an $(s - 1)$-And-Or f_0; hence, $f_1 \leqslant f_0$. Let E_f be the set of inputs on which f_0 introduces a new error, i.e.,

$$E_f \rightleftharpoons \{x : f(x) = f_1(x) = 0 \text{ but } f_0(x) = 1\}.$$

By Claim 10.17, all these errors can be "corrected" by adding a relatively small s-CNF: there is an r-CNF C such that $|C| \leqslant (r-1)^s$ and $C(x) = 0$ for all $x \in E_f$.

Proceeding in this way we will reach the last gate of our circuit computing the given function F. Let F_0, F_1 be its approximator, and let E be the set of all inputs $x \in \{0,1\}^n$ on which F differs from at least of one of the functions F_0 or F_1. Since at input gates (= variables) no error was made, for every such input $x \in E$, the corresponding error should be introduced at some intermediate gate. That is, for every $x \in E$ there is a gate f such that $x \in E_f$ (approximator of f introduces an error on x for the first time). But we have shown that, for each gate, all these errors can be corrected by adding an s-CNF of size at most $(r-1)^s$ or an r-DNF of size at most $(s-1)^r$. Since we have only t gates, all such errors $x \in E$ can be corrected by adding an s-CNF C of size at most $t \cdot (r-1)^s$ and an r-DNF D of size at most $t \cdot (s-1)^r$, that is, for all inputs $x \in \{0,1\}^n$, we have

$$C(x) \wedge F_0(x) \leqslant F(x) \leqslant F_1(x) \vee D(x).$$

This already implies that the function F is t-simple. Indeed, if the CNF F_0 is empty (i.e., if $F_0 \equiv 1$) then $C \leqslant F$, and we are done. Otherwise, F_0 must contain some clause S of length at most $s-1$, say, $S = \bigvee_{i \in I} x_i$ for some I of size $|I| \leqslant s-1$. Since $F_0 \leqslant S$, the condition $F_1 \leqslant F_0$ implies $F \leqslant F_1 \vee D \leqslant F_0 \vee D \leqslant S \vee D$, as desired. This completes the proof of Theorem 10.16.   □

10.6.2 Explicit lower bounds

In order to show that a given boolean function cannot be computed by a monotone circuit of size at most t, it is enough, by Theorem 10.16, to show that the function is not t-simple for at least one(!) choice of parameters s and r. In the following three sections we demonstrate how this can be used to derive exponential lower bounds for concrete boolean functions.

In applications, boolean functions f are usually defined as set-theoretic predicates. In this case we say that f accepts a set $S \subseteq \{1, \ldots, n\}$ if and only if f accepts its incidence vector. It is therefore convenient to also have a set-theoretic definition of t-simplicity, which we give now.

A set S is a *positive input* for f if $f(S) = 1$, and a *negative input* if $f(\overline{S}) = 0$, where $\overline{S}$ is the complement of S. Put otherwise, a positive (negative) input is a set of variables which, if assigned the value 1 (0), forces the function to take the value 1 (0) regardless of the values assigned to the remaining variables. Note that one set S can be both positive and negative input! For example, if $f(x_1, x_2, x_3)$ outputs 1 iff $x_1 + x_2 + x_3 \geqslant 2$, then $S = \{1, 2\}$ is both positive and negative input for f, because $f(1, 1, x_3) = 1$ and $f(0, 0, x_3) = 0$.

The following is a set-theoretic equivalent of Definition 10.15 (show this):

Definition 10.18. A function $f(x_1, \ldots, x_n)$ is *t-simple* if for every pair of integers $2 \leqslant s, r \leqslant n$ there exists an $(s-1)$-element set I, a system of s-element subsets $S_1, \ldots, S_K$ and a system of r-element subsets $R_1, \ldots, R_L$ of $\{1, \ldots, n\}$ such that $K \leqslant t(r-1)^s$, $L \leqslant t(s-1)^r$, and at least one of the following two conditions hold:

1. Every negative input of f contains at least one of the sets $S_1, \ldots, S_K$.
2. Every positive input of f either intersects the set I or contains at least one of the sets $R_1, \ldots, R_L$.

We begin with the simplest example. We will also present two more respectable applications, but this special case already demonstrates the common way of reasoning pretty well.

Let us consider a monotone boolean function Δ_m, whose input is an undirected graph on m vertices, represented by $n = \binom{m}{2}$ variables, one for each possible edge. The value of the function is 1 if and only if the graph contains a triangle (three incident vertices). Clearly, there is a monotone circuit of size $O(m^3)$ computing this function: just test whether any of $\binom{m}{3}$ triangles is present in the graph. Thus, the following theorem is tight, up to a poly-logarithmic factor.

Proposition 10.19. *Any monotone circuit computing Δ_m has size at least $\Omega\left(m^3/\log^4 m\right)$.*

Proof. Let t be the minimal number for which Δ_m is t-simple. By Theorem 10.16, it is enough to show that $t \geqslant \Omega\left(m^3/\log^4 m\right)$. For this proof we take $s = \lfloor 5\log^2 m \rfloor$ and $r = 2$. According to Definition 10.18 we have only two possibilities.

Case 1: Every negative input for Δ_m contains at least one of $t(r-1)^s = t$ s-element sets of edges $S_1, \ldots, S_t$.

In this case we consider the graphs $E = E_1 \cup E_2$ consisting of two disjoint non-empty cliques E_1 and E_2 (we look at graphs as sets of their edges). Each such graph E is a negative input for Δ_m, because its complement is a bipartite graph, and hence, has no triangles. We have $2^{m-1} - 1$ such graphs, and each of them must contain at least one of the sets $S_1, \ldots, S_t$. Every of these sets of edges S_i is incident to at least $\sqrt{2s}$ vertices, and if $E \supseteq S_i$ then all these vertices must belong to one of the cliques E_1 or E_2. Thus, at most $2^{m-\sqrt{2s}} - 1$ of our negative inputs E can contain one fixed set S_i, implying that, in this case,

$$ t \geqslant \frac{2^{m-1} - 1}{2^{m-\sqrt{2s}} - 1} \geqslant 2^{\sqrt{2s}-1} \geqslant 2^{3\log m} \geqslant m^3. $$

Case 2: Every positive input for Δ_m either intersects a fixed set I of $s-1$ edges, or contains at least one of $L \leqslant t(s-1)^r = t(s-1)^2$ 2-element sets of edges $R_1, \ldots, R_L$.

As positive inputs for Δ_m we take all triangles, i.e., graphs on m vertices with exactly one triangle; we have $\binom{m}{3}$ such graphs. At most $(s-1)(m-2)$ of them will have an edge in I. Each of the remaining triangles must contain one of $t(s-1)^2$ given pairs of edges R_i. Since two edges can lie in at most one triangle, we conclude that, in this case,

$$t \geqslant \frac{\binom{m}{3} - (s-1)(m-2)}{(s-1)^2} \geqslant \Omega\left(m^3/\log^4 m\right).$$

Thus, in both cases, $t \geqslant \Omega\left(m^3/\log^4 m\right)$, and we are done.   □

The following monotone boolean function was introduced by Andreev (1995). Let $q \geqslant 2$ be a prime power, and consider bipartite graphs $G = (U, V, E)$ with $U = \mathbb{F}_q$ and $V = \mathbb{F}_q$. Since the parts U and V are fixed, we will identify the graph G with the set $E \subseteq \mathbb{F}_q \times \mathbb{F}_q$ of its edges. Given a polynomial $h : \mathbb{F}_q \to \mathbb{F}_q$, its *graph* is the bipartite graph $E_h = \{(i, h(i)) : i \in \mathbb{F}_q\}$.

Define POLY(q, d) to be a monotone boolean function, whose input is a bipartite graph $E \subseteq \mathbb{F}_q \times \mathbb{F}_q$ represented by $n = q^2$ variables $x_{i,j}$, one for each possible edge. The value of the function is 1 if and only if the graph contains the graph E_h of at least one polynomial h over $\mathbb{F}_q$ of degree at most $d-1$. That is,

$$\mathrm{POLY}(q, d) = \bigvee_h \bigwedge_{i \in \mathbb{F}_q} x_{i, h(i)}$$

where the Or is over all q^d polynomials h of degree at most $d-1$.

We see that the POLY(q, d) function can be computed by a monotone boolean circuit of size at most q^{d+1}. Alon and Boppana (1987), using Razborov's method, have proved that this bound is almost optimal. Here we will show that this result can also be derived from Theorem 10.16 via direct computations.

Theorem 10.20 (Alon–Boppana 1987). *If $d \leqslant (q/\ln q)^{1/2}/2$ then any monotone circuit computing* POLY(q, d) *has size at least $q^{\Omega(d)}$.*

Proof. Let $f = \mathrm{POLY}(q, d)$ and take a minimal t for which this function is t-simple. By Theorem 10.16, it is enough to show that $t \geqslant q^{\Omega(d)}$. For this proof we take $s \rightleftharpoons \lceil d \ln q \rceil$ and $r \rightleftharpoons d$. According to Definition 10.18 we have only two possibilities.

Case 1: Every negative input for f contains at least one of $K \leqslant t(r-1)^s$ s-element sets of edges $S_1, \ldots, S_K$.

Let **E** be a random bipartite graph, with each edge appearing in **E** independently with probability $\gamma \rightleftharpoons (2d \ln q)/q$. Since there are only q^d polynomials of degree at most $d-1$, the probability that the complement of **E** will contain the graph of at least one of them does not exceed $q^d(1 - \gamma)^q \leqslant q^{-d}$, by our choice of γ. Hence, with probability at least $1 - q^{-d}$, the graph **E** is a negative input for f. On the other hand, each of the sets S_i is contained in **E** with probability $\gamma^{|S_i|} = \gamma^s$. Thus, in this case,

$$t \geqslant \frac{1 - q^{-d}}{(r-1)^s \gamma^s} \geqslant \left(\frac{q}{2d^2 \ln q} \right)^{\Omega(s)} \geqslant q^{\Omega(d)}.$$

Case 2: Every positive input for f either intersects a fixed set I of at most $s - 1$ edges, or contains at least one of $L \leqslant t(s-1)^r$ r-element sets of edges $R_1, \ldots, R_L$.

Graphs of polynomials of degree at most $d - 1$ are positive inputs for f. Each set of l $(1 \leqslant l \leqslant d)$ edges is contained in either 0 or precisely q^{d-l} of such graphs. Hence, at most $(s-1)q^{d-1}$ of these graphs can contain an edge in I, and at most q^{d-r} of them can contain any of the given graphs R_i. Therefore, in this case we again have

$$t \geqslant \left(1 - \frac{s-1}{q} \right) \frac{q^d}{(s-1)^r \cdot q^{d-r}} \geqslant \left(\frac{q}{s} \right)^{\Omega(r)} \geqslant q^{\Omega(d)}.$$

We have proved that the function f can be t-simple only if $t \geqslant q^{\Omega(d)}$. By Theorem 10.16, this function cannot be computed by monotone circuits of size smaller than $q^{\Omega(d)}$. □

Our last example is a monotone boolean function PALEY(n, k), associated with bipartite Paley graphs. Although the calculations for this function are a bit more complicated, we include this example for two reasons: the argument explores nice structural properties of Paley graphs, and – as we will see in Sects. 15.2.2 and 16.5 – the function itself is a standard example of a hard function for different models of monotone computations.

Let n be an odd prime power, congruent to 1 modulo 4. A *bipartite Paley graph* is a bipartite graph $G_n = (V_1, V_2, E)$ with parts $V_1 = V_2 = \mathbb{F}_n$, where two nodes $x \in V_1$ and $y \in V_2$ are joined by an edge if and only if $x - y$ is a non-zero square modulo n (see Sect. 11.3.2). Since exactly half of the non-zero elements of $\mathbb{F}_n$ are squares (see the proof of Theorem 13.7), this immediately tells us that the graph G_n is d-regular with $d = (n-1)/2$, that is, each node has degree $(n-1)/2$. Moreover, it can be shown (see Theorem 11.9) that this graph also has the following interesting property: for every two disjoint sets A, B of nodes in the first part V_1 or in the second part V_2, with $|A| + |B| = k < \log n$, the number $N(k)$ of nodes (in the opposite part) adjacent to all nodes in A and nonadjacent to every node in B is very close to $n/2^k$, namely

$$|N(k) - 2^{-k}n| \leqslant k\sqrt{n}. \tag{10.6}$$

Define PALEY(n, k) to be the function of $2n$ boolean variables $x_1, \ldots, x_{2n}$ representing the nodes in $V_1 \cup V_2$, which accepts a set of nodes if and only if this set contains some k-element subset $A \subseteq V_1$ together with the set of its common neighbors $\Gamma(A) \rightleftharpoons \{y \in V_2 : (x, y) \in E \text{ for all } x \in A\}$.

We will consider the function PALEY(n, k) with $k = \Theta(\log n)$. It is clear that in this case PALEY(n, k) can be computed by a trivial monotone circuit using at most $2n|\mathcal{F}| = 2n\binom{n}{k} \leqslant n^{O(\log n)}$ fanin-2 And and Or gates. On the other hand, using the general lower bound, given by Theorem 10.16, we can show that, in fact, we cannot expect to do much better: this trivial upper bound is almost optimal.

Theorem 10.21. *Let $k = \lfloor (\log n)/5 \rfloor$. Then any monotone circuit computing* PALEY(n, k) *has size at least $n^{\Omega(\log n)}$.*

Proof. By the definition, all the members of the family

$$\mathcal{F} = \{A \cup \Gamma(A) : A \subseteq V_1, |A| = k\}$$

are positive inputs for $f = \text{PALEY}(n, k)$. On the other hand, if $k \leqslant (\log n)/5$ and n is sufficiently large then, by (10.6),

$$N(2k) \geqslant \frac{n}{4^k} - 2k\sqrt{n} > 0.$$

This, in particularly, implies that the family $\mathcal{F}$ is intersecting, and hence, all its members are also negative inputs for f.

Take now a minimal t for which f is t-simple. By Theorem 10.16, it is enough to show that $t \geqslant n^{\Omega(\log n)}$. For this proof we take $s = r = k$. According to Definition 10.18 we only have two possibilities, depending on which of its items holds.

Case 1: Every member of $\mathcal{F}$ contains at least one of $K \leqslant t(r-1)^s$ s-element sets of nodes $S_1, \ldots, S_K$.

Let $A_1 \cup \Gamma(A_1), \ldots, A_m \cup \Gamma(A_m)$ be the members of $\mathcal{F}$ containing a fixed s-element subset S of $V_1 \cup V_2$. Let $S_1 = S \cap V_1$, $S_2 = S \cap V_2$ and observe that all the sets $A_1, \ldots, A_m$

(i)   contain the set S_1, and
(ii) are contained in $\Gamma(S_2)$.

If less than half of the nodes of S lie in V_2 then $|S_1| \geqslant s/2$, and hence,

$$m \leqslant \binom{n - |S_1|}{k - |S_1|} \leqslant n^{k - |S_1|} \leqslant n^{k/2}.$$

Otherwise, $|S_2| \geqslant s/2$ and in this case we have $m \leqslant \binom{N(s/2)}{k} \leqslant N(k/2)^k$, where, by (10.6),

$$N(k/2) \leqslant n2^{-k/2} + (k\sqrt{n})/2 \leqslant n^{1-\epsilon},$$

for a small (but absolute) constant $\epsilon > 0$. Thus, in both cases one s-element set S can be contained in at most $n^{(1-\epsilon)k}$ members of $\mathcal{F}$. Since $|\mathcal{F}| = \binom{n}{k} \geqslant (n/k)^k$, we need $K \geqslant (n^\epsilon/k)^k$ such sets S, implying that

$$t \geqslant \frac{K}{(r-1)^s} \geqslant \left(\frac{n^\epsilon}{k}\right)^k \cdot r^{-s} = \left(\frac{n^\epsilon}{k^2}\right)^k \geqslant n^{\Omega(\log n)}.$$

Case 2: Every member of $\mathcal{F}$ either intersects a fixed set I of at most $s - 1$ nodes, or contains at least one of $L \leqslant t(s-1)^r$ r-element sets of edges $R_1, \ldots, R_L$.

The same argument as in Case 1 yields that one r-element set of nodes can be contained in at most $\binom{N(r/2)}{k} \leqslant n^{(1-\epsilon)k}$ members of $\mathcal{F}$. So, it remains to estimate how many of these members avoid the set I.

To estimate this number, let $A_1 \cup \Gamma(A_1), \ldots, A_m \cup \Gamma(A_m)$ be the members of $\mathcal{F}$ containing a fixed node $x \in I$. If $x \in V_1$ then x belongs to all the sets $A_1, \ldots, A_m$, implying that $m \leqslant \binom{n-1}{k-1} = \frac{k}{n}\binom{n}{k}$. If $x \in V_2$ then x belongs to all the sets $\Gamma(A_1), \ldots, \Gamma(A_m)$, and hence, all the nodes from $A_1 \cup \cdots \cup A_m$ are adjacent to x, implying that their number cannot exceed the degree $(n-1)/2$ of x. Therefore, in this case $m \leqslant \binom{(n-1)/2}{k} = \binom{n-b}{k}$, with $b = (n+1)/2$. Since $\binom{n-b}{k}/\binom{n}{k} \leqslant \left(\frac{n-k}{n}\right)^b \leqslant e^{-(k/n)b}$, in both cases at least $1 - (s-1) \cdot e^{-k/2} \geqslant 1 - o(1)$ fraction of all $\binom{n}{k}$ sets in $\mathcal{F}$ avoid the set I, and hence, must contain at least one of the sets $R_1, \ldots, R_L$. Since none of the sets R_i is contained in more than $n^{k(1-\epsilon)}$ of the members of $\mathcal{F}$, we conclude that also in this case

$$ t \geqslant \frac{L}{(s-1)^r} \geqslant \frac{(1-o(1))\binom{n}{k}}{(s-1)^r \cdot n^{(1-\epsilon)k}} \geqslant \frac{(1-o(1))n^k}{k^{2k} \cdot n^{(1-\epsilon)k}} \geqslant n^{\Omega(\log n)}. $$

$\square$

Exercises

10.1.[−] The *independence number* $\alpha(\mathcal{F})$ of a family $\mathcal{F} \subseteq 2^X$ is defined as the maximum cardinality $|S|$ of a set $S \subseteq X$ which does not contain any member of $\mathcal{F}$. Prove that $\alpha(\mathcal{F}) = |X| - \tau(\mathcal{F})$.

10.2.[−] Let T be a minimal blocking set of a family $\mathcal{F}$. Show that, for every $x \in T$, there exists an $A \in \mathcal{F}$ such that $T \cap A = \{x\}$.

10.3. Let $\mathcal{F}$ be an r-uniform family and suppose that $\tau(\mathcal{F} - \{A\}) < \tau(\mathcal{F})$ for all $A \in \mathcal{F}$. Prove that $|\mathcal{F}| \leqslant \binom{r+\tau(\mathcal{F})-1}{r}$.

Hint: Observe that, for each $A \in \mathcal{F}$, there is a set B of size $\tau(\mathcal{F}) - 1$ which is disjoint from A but intersects all other members of $\mathcal{F}$; apply the Bollobás theorem (Theorem 9.7).

10.4.[(1)] Let $\mathcal{F}$ and $\mathcal{H}$ be antichains over some set X. Prove that:

(i) $\mathcal{H} = b(\mathcal{F})$ if and only if for every coloring of the points in X in Red and in Blue, *either* $\mathcal{F}$ has a Red set (i.e., all points in this set are red), *or* (exclusive) $\mathcal{H}$ has a Blue set.

(ii) $\mathcal{F} \succ \mathcal{H}$ if and only if $b(\mathcal{H}) \succ b(\mathcal{F})$.

10.5. Consider the following family $\mathcal{F}$. Take k disjoint sets $V_1, \ldots, V_k$ such that $|V_i| = i$ for $i = 1, \ldots, k$. The members of $\mathcal{F}$ are all the sets of the form $V_i \cup T$, where T is any set such that $|T| = k - i$ and $|T \cap V_j| = 1$ for all $j = i+1, \ldots, k$. Show that this family is self-dual, i.e., that $\mathcal{F} = b(\mathcal{F})$. (This construction is due to Erdös and Lovász.)

10.6. A pair of sets (A, B) *separates* a pair of elements (x, y) if $x \in A - B$ and $y \in B - A$. A family $\mathcal{F} = \{A_1, \ldots, A_m\}$ of subsets of $X = \{x_1, \ldots, x_n\}$ is a *complete separator* if every pair of elements in X is separated by at least one pair of sets in $\mathcal{F}$. Let $\mathcal{F}^*$ be the family of all non-empty sets $X_i \rightleftharpoons \{j : x_i \in A_j\}$. Prove that $\mathcal{F}$ is a complete separator if and only if $\mathcal{F}^*$ is an antichain.

Hint: $X_i \nsubseteq X_j$ means that there exists k such that $k \in X_i$ and $k \notin X_j$, i.e., that $x_i \in A_k$ and $x_j \notin A_k$.

10.7.⁻ Prove that any decision tree for the function $f = \bigwedge_{i=1}^{m} \bigvee_{j=1}^{m} x_{ij}$ has depth m^2.

Sketch: Take an arbitrary decision tree for f and construct a path from the root by the following "adversary" rule. Suppose we have reached a node v labeled by x_{ij}. Then follow the outgoing edge marked by 1 if and only if *all* the variables x_{il} with $l \neq j$ were already tested before we reached the node v.

10.8. The *storage access function* is a boolean function $f(x, y)$ on $n + k$ variables $x = (x_0, \ldots, x_{n-1})$ and $y = (y_0, \ldots, y_{k-1})$ where $n = 2^k$, and is defined as follows: $f(x, y) \rightleftharpoons x_{int(y)}$, where $int(y) \rightleftharpoons \sum_{i=0}^{k-1} y_i 2^i$ is the integer whose binary representation is the vector y. Prove that f is a $(k+1)$-Or-And function although some of its minterms have length 2^k.

Hint: For the first claim observe that the value of f only depends on $k + 1$ bits $y_0, \ldots, y_{k-1}$ and $x_{bin(y)}$. For the lower bound, consider the monomial $x_0 x_1 \cdots x_{n-1}$ and show that it is a minterm of f.

10.9.⁺ A *partial b-(n, k, λ) design* is a family $\mathcal{F}$ of k-element subsets of $\{1, \ldots, n\}$ such that any b-element set is contained in at most λ of its members. We can associate with each such design $\mathcal{F}$ a monotone boolean function $f_{\mathcal{F}}$ such that $f_{\mathcal{F}}(S) = 1$ if and only if $S \supseteq F$ for at least one $F \in \mathcal{F}$. Assume that $\ln |\mathcal{F}| < k - 1$ and that each element belongs to at most N members of $\mathcal{F}$. Use Theorem 10.16 to show that for every integer $a \geqslant 2$, every monotone circuit computing $f_{\mathcal{F}}$ has size at least

$$\ell \rightleftharpoons \min \left\{ \frac{1}{2} \left(\frac{k}{2b \ln |\mathcal{F}|} \right)^a, \ \frac{|\mathcal{F}| - a \cdot N}{\lambda \cdot a^b}, \right\}.$$

Hint: Take $s = a$, $r = b$ and show that under this choice of parameters, the function $f_{\mathcal{F}}$ can be t-simple only if $t \geqslant \ell$. When doing this, note that the members of $\mathcal{F}$ are positive inputs for $f_{\mathcal{F}}$. To handle the case of negative inputs, take a random subset in which each element appears independently with probability $p = (1 + \ln |\mathcal{F}|)/k$, and show that its complement can contain a member of $\mathcal{F}$ with probability at most $|\mathcal{F}|(1 - p)^k \leqslant e^{-1}$.

10.10.⁻ Derive Theorem 10.20 from the previous exercise.

Hint: Observe that the family of all q^d graphs of polynomials of degree at most $d - 1$ over $\mathbb{F}_q$ forms a partial b-(n, k, λ) design with parameters $n = q^2$, $k = q$ and $\lambda = q^{d-b}$.

10.11.⁻ Andreev (1987) has shown how, for any prime power $q \geqslant 2$ and $d \leqslant q$, to construct an explicit family $\mathcal{D}$ of subsets of $\{1, \ldots, n\}$ which, for every $b \leqslant d + 1$, forms a partial b–(n, k, λ) design with parameters $n = q^3$, $k = q^2$, $\lambda = q^{2d+1-b}$ and $|\mathcal{F}_v| = q^{2d+1}$. Use Exercise 10.9 to show that the corresponding boolean function $f_{\mathcal{D}}$ requires monotone circuits of size exponential in $\Omega\left(n^{1/3-o(1)}\right)$.

10.12⁽ᐟ⁾ (due to Berkowitz). A k-threshold is a monotone boolean function $T_k^n(x_1, \ldots, x_n)$ which outputs 1 if and only if the input vector $x = (x_1, \ldots, x_n)$ has weight at least k, i.e., if $|x| = x_1 + \cdots + x_n \geqslant k$. Show that

$$T_k^{n-1}(x_1, \ldots, x_{i-1}, x_{i+1}, \ldots, x_n) = \overline{x}_i,$$

for all inputs $(x_1, \ldots, x_n)$ such that $x_1 + \cdots + x_n = k$.

10.13.⁻ A boolean function f is a *slice* function if there is some $0 \leqslant k \leqslant n$ such that for every input $x \in \{0, 1\}^n$,

$$f(x) = \begin{cases} 0 & \text{if } |x| < k; \\ 1 & \text{if } |x| > k. \end{cases}$$

That is, f can be non-trivial only on inputs with exactly k 1's; in this case we also say that f is the k-*slice* function. Use the previous exercise and the fact that the threshold function T_k^n has a monotone circuit of size $O(n^2)$ to prove that for such functions, using the negations cannot help much. Namely, prove that if a slice function f has a non-monotone circuit of size ℓ, then f can also be computed by a monotone circuit of size at most $\ell + O(n^3)$.

10.14. Given a vector $x = (x_1, \ldots, x_n)$ in $\{0, 1\}^n$, associate with it the following two integers $h_+(x) = |x| \cdot 2^n + b(x)$ and $h_-(x) = |x| \cdot 2^n - b(x)$, where $b(x) = \sum_{i=1}^n x_i 2^{i-1}$. Prove that for any two vectors $x \neq y$,

(i)   if $|x| < |y|$, then $h_+(x) < h_+(y)$ and $h_-(x) < h_-(y)$;
(ii) if $|x| = |y|$, then $h_+(x) \leqslant h_+(y)$ if and only if $h_-(x) \geqslant h_-(y)$.

10.15 (Rosenbloom 1997). Let $f(x_1, \ldots, x_n)$ be a k-slice function, $0 \leqslant k \leqslant n$. Use the previous exercise to show that f can be computed by a circuit with $O(n)$ monotone real-valued functions as gates.

Hint: As the last gate take a monotone function $\varphi : \mathbb{R}^2 \to \{0, 1\}$ such that $\varphi(h_+(x), h_-(x)) = f(x)$ for all inputs x of weight $|x| = k$.

11. Density and Universality

In many applications (testing logical circuits, construction of k-wise independent random variables, etc.), vector sets $A \subseteq \{0,1\}^n$ with the following property play an important role:

> For any subset of k coordinates $S = \{i_1, \ldots, i_k\}$ the projection of A onto the indices in S contains all possible 2^k configurations.

Such sets are called (n,k)-*universal*. If the same holds not for all but only for *at least one* subset S of k indices, then A is called (n,k)-*dense*. The minimal number k, for which A is (n,k)-dense, is also known as the *Vapnik–Chervonenkis dimension* of A.

Given n and k, the problem is to find a universal (or dense) set A with as few vectors as possible. In this chapter we will discuss several approaches to its solution.

11.1 Dense sets

Given a vector $v = (v_1, \ldots, v_n)$, its *projection* onto a set of coordinates $S = \{i_1, \ldots, i_k\}$ is the vector $v|_S \rightleftharpoons (v_{i_1}, \ldots, v_{i_k})$. The projection of a set of vectors $A \subseteq \{0,1\}^n$ onto S is the set of vectors $A|_S \rightleftharpoons \{v|_S : v \in A\}$. Thus, A is (n,k)-dense iff $A|_S = \{0,1\}^k$ for at least one subset of k coordinates S.

It is clear that every (n,k)-dense set must contain at least 2^k vectors. On the other hand, if A is the set of all vectors in $\{0,1\}^n$ with less than k ones, then A has

$$H(n,k) \rightleftharpoons \sum_{i=0}^{k-1} \binom{n}{i}$$

vectors but is *not* (n,k)-dense. It turns out, however, that every larger set already *is* (n,k)-dense! This interesting fact, whose applications range from probability theory to computational learning theory, was discovered independently by three sets of authors in remarkable simultaneity: Perles and Shelah (see Shelah 1972), Sauer (1972), and Vapnik and Chervonenkis (1971). No less remarkable is the range of contexts in which the results arose: logic, set theory, and probability theory.

Theorem 11.1. *If $A \subseteq \{0,1\}^n$ and $|A| > H(n,k)$ then A is (n,k)-dense.*

Proof. Induction on n and k. If $k = 1$ then A has at least two different vectors and hence is $(n,1)$-dense. For the induction step take an arbitrary set $A \subseteq \{0,1\}^n$ of size $|A| > H(n,k)$. Let B be the projection of A onto the first $n-1$ coordinates, and C be the set of all vectors u in $\{0,1\}^{n-1}$ for which *both* vectors $(u,0)$ and $(u,1)$ belong to A. A simple but crucial observation is that

$$|A| = |B| + |C|.$$

Now, if $|B| > H(n-1,k)$ then the set B is $(n-1,k)$-dense by induction, and hence the whole set A is also (n,k)-dense. If $|B| \leqslant H(n-1,k)$ then, using the identity $\binom{n}{i} - \binom{n-1}{i} = \binom{n-1}{i-1}$ (see Proposition 1.2) we obtain

$$|C| = |A| - |B| > H(n,k) - H(n-1,k)$$
$$= \sum_{i=0}^{k-1} \binom{n}{i} - \sum_{i=0}^{k-1} \binom{n-1}{i} = \sum_{i=0}^{k-2} \binom{n-1}{i}$$
$$= H(n-1,k-1).$$

By the induction hypothesis, the set C is $(n-1,k-1)$-dense, and since $C \times \{0,1\}$ lies in A, the whole set A is also (n,k)-dense. $\square$

11.2 Hereditary sets

Alon (1983) and Frankl (1983) have independently made an intriguing observation that for results like Theorem 11.1, we can safely restrict our attention to sets with a very special structure.

A set $A \subseteq \{0,1\}^n$ is *monotone decreasing* (or *hereditary*) if $v \in A$ and $u \leqslant v$ implies $u \in A$. (Here, as usual, $u \leqslant v$ means that $u_i \leqslant v_i$ for all i.) Thus, being hereditary means that we can arbitrarily switch 1's to 0's, and the resulting vectors will still belong to the set.

For a set $S \subseteq \{1, \ldots, n\}$ of coordinates, let $t_S(A)$ denote the number of vectors in the projection $A\vert_S$. If v is a vector and i is any of its coordinates, then the *ith neighbor* of v is the vector $v_{i \to 0}$ obtained from v by switching its ith bit to 0; if this bit is 0 then we let $v_{i \to 0} = v$.

Theorem 11.2. *For every subset A of the n-cube $\{0,1\}^n$ there is a hereditary subset B such that $|B| = |A|$ and $t_S(B) \leqslant t_S(A)$ for all sets S of coordinates.*

Before we prove this result, observe that it immediately implies Theorem 11.1: if B is hereditary and $|B| > H(n,k)$, then B must contain a vector v with at least k 1's, and so, must contain all the 2^k vectors obtained from v by changing any subset of these ones to zeroes.

Proof. If A itself is hereditary, there is nothing to do. Otherwise, we have some "bad" coordinates, i.e., coordinates i such that $v_{i \to 0} \notin A$ for some $v \in A$. To correct the situation, we will apply for each such bad coordinate i, the following transformation T_i. Take a vector $v \in A$ with $v_i = 1$, and see if $v_{i \to 0}$ belongs to A. If so, do nothing; otherwise, replace the vector v in A by $v_{i \to 0}$. Apply this transformation as long as possible, and let B denote the resulting set. It is clear that $|B| = |A|$. We also claim that $t_S(B) \leqslant t_S(A)$ for every $S \subseteq \{1, \dots, n\}$.

Indeed, if $i \notin S$ then $t_S(B) = t_S(A)$, and we are done. Suppose that $i \in S$ and let $S' = S - \{i\}$. Assume, for notational convenience, that i was the first coordinate, i.e., that $i = 1$. Now, if $t_S(B) \geqslant t_S(A) + 1$, this can happen only when A has two vectors $x = (1, u, w_1)$ and $y = (1, u, w_2)$ with $u \in \{0, 1\}^{S'}$ and $w_1 \neq w_2$, and such that *exactly one* of them, say x, was altered by T_i. That is, the S-projection of B contains both vectors $(1, u)$ and $(0, u)$, whereas $(0, u)$ does not appear in the S-projection of A. But this is impossible because the fact that the other vector $y = (1, u, w_2)$ was *not* altered by T_i means that its ith neighbor $(0, u, w_2)$ belongs to A, and hence vector $(0, u)$ must appear among the vectors in the S-projection of A. This contradiction proves that $t_S(B) \leqslant t_S(A)$.

Thus, starting with A, we can apply the transformations T_i along all n coordinates $i = 1, \dots, n$, and obtain the set $B = T_n(T_{n-1}(\cdots T_1(A) \cdots))$, which is hereditary, has the same number of vectors as the original set A and satisfies the condition $t_S(B) \leqslant t_S(A)$ for all S. $\square$

Frankl (1983) observed that this result also has other interesting consequences. For a set $A \subseteq \{0, 1\}^n$, let $t_s(A) = \max t_S(A)$ over all $S \subseteq \{1, \dots, n\}$ with $|S| = s$; hence, $t_n(A) = |A|$.

Theorem 11.3 (Bondy 1972). *If $|A| \leqslant n$ then $t_{n-1}(A) = |A|$.*

Proof. We will give a direct proof of this result in Sect. 12.1; here we show that it is a consequence of Theorem 11.2.

By this theorem we may assume that A is hereditary. If A is empty, there is nothing to prove. Otherwise, A must contain the all-0 vector. Hence, at least one of n unit vectors

$$e_i = (0, \dots, 0, 1, 0, \dots, 0),$$

with the 1 in the ith coordinate, does not belong to A. As A is hereditary, this implies that $|A| = t_n(A) = t_S(A)$ for $S = \{1, \dots, n\} - \{i\}$. $\square$

Bollobás (see Lovász 1979, Problem 13.10) extended this result to larger sets.

Theorem 11.4. *If $|A| \leqslant \lceil \frac{3}{2} n \rceil$ then $t_{n-1}(A) \geqslant |A| - 1$.*

Proof. By Theorem 11.2 we may assume that A is hereditary. If there is an i such that $e_i \notin A$, then again $t_{n-1}(A) = |A|$, and we are done. Otherwise, A contains the all-0 vector and all unit vectors $e_1, \dots, e_n$. Let A' be the set of all vectors in A with precisely two 1's. Each such vector covers only two of the unit vectors. Therefore, some e_i must remain uncovered, for otherwise we would have $|A| \geqslant 1 + n + \lceil n/2 \rceil > \lceil \frac{3}{2} n \rceil$. But this means that e_i is the only vector in A with 1 in the ith coordinate, implying that for $S = \{1, \ldots, n\} - \{i\}$, $t_S(A) = |A - \{e_i\}| = |A| - 1$. $\qquad\square$

Combining Theorem 11.2 with the deep Kruskal–Katona theorem about the shadows of arbitrary families of sets, Frankl (1983) derived the following general result, which is the best possible whenever t divides n (see Exercise 11.9). We state it without proof.

Theorem 11.5 (Frankl 1983). *If $A \subseteq \{0,1\}^n$ and $|A| \leqslant \lceil n(2^t - 1)/t \rceil$, then $t_{n-1}(A) \geqslant |A| - 2^{t-1} + 1$.*

The following result concerning the intersection of hereditary sets, due to Kleitman, has many generalizations and applications (see, for example, Exercise 11.8 and Theorem 8.5):

Theorem 11.6 (Kleitman 1966). *Let A, B be monotone decreasing subsets of $\{0,1\}^n$. Then*

$$|A \cap B| \geqslant \frac{|A| \cdot |B|}{2^n}.$$

Proof. Apply induction on n, the case $n = 0$ being trivial. For $\epsilon \in \{0,1\}$, set $c_\epsilon = |A_\epsilon|$ and $d_\epsilon = |B_\epsilon|$, where

$$A_\epsilon \rightleftharpoons \{(a_1, \ldots, a_{n-1}) : (a_1, \ldots, a_{n-1}, \epsilon) \in A\}$$

and

$$B_\epsilon \rightleftharpoons \{(b_1, \ldots, b_{n-1}) : (b_1, \ldots, b_{n-1}, \epsilon) \in B\}.$$

Then

$$
\begin{aligned}
|A \cap B| &= |A_0 \cap B_0| + |A_1 \cap B_1| \\
&\geqslant (c_0 d_0 + c_1 d_1)/2^{n-1} \quad \text{(by induction)} \\
&= (c_0 + c_1)(d_0 + d_1)/2^n + (c_0 - c_1)(d_0 - d_1)/2^n.
\end{aligned}
$$

Since sets A, B are monotone decreasing, we have $A_1 \subseteq A_0$ and $B_1 \subseteq B_0$, implying that $(c_0 - c_1)(d_0 - d_1) \geqslant 0$. $\qquad\square$

11.3 Universal sets

The (n, k)-density of a set of vectors means that its projection on *at least one* set of k coordinates gives the whole binary k-cube. We now consider a stronger property – (n, k)-universality – where we require that the same holds for *all* subsets of k coordinates.

Of course, the whole cube $\{0,1\}^n$ is (n,k)-universal for every $k \leqslant n$. This is the trivial case. Does there exist smaller universal sets? Note that 2^k is a trivial lower bound.

Using the probabilistic argument it can be shown that there *exist* (n,k)-universal sets of size only $k2^k \log n$ (see Theorem 18.5).

This results tells us only that small universal sets exist, but gives us no idea of how to construct them. In this section we will show how to construct explicit sets in $\{0,1\}^n$ which only have size n and are (n,k)-universal as long as $k2^k < \sqrt{n}$. The construction employs some nice combinatorial properties of so-called Paley graphs. We will also use these graphs later to construct boolean functions, which are hard for monotone formulas (Sect. 15.2.2), monotone circuits (Sect. 10.6.2), and monotone span programs (Sect. 16.5).

11.3.1 Isolated neighbor condition

In this section we introduce one property of (bipartite) graphs which is equivalent to the universality property of 0-1 vectors. In the next section we will describe an explicit construction of such graphs based on the famous theorem of Weil (1948) regarding character sums.

By a bipartite graph with parts of size n we will mean a bipartite graph $G = (V_1, V_2, E)$ with $|V_1| = |V_2| = n$. We say that a node $y \in V_2$ is a *common neighbor* for a set of nodes $A \subseteq V_1$ if y is joined to *each* node of A. Dually, a node $y \in V_2$ is a *common non-neighbor* for a set of nodes $B \subseteq V_1$ if y is joined to *no* node of B. Given two disjoint subsets A and B of V_1, we denote by $v(A,B)$ the number of nodes in V_2 which are common neighbors for A, and at the same time are common non-neighbors for B. That is, $v(A,B)$ is the number of nodes in V_2 joined to each node of A and to no node of B.

Definition 11.7 (Gál 1998). A bipartite graph $G = (V_1, V_2, E)$ satisfies the *isolated neighbor condition for k* if $v(A,B) > 0$ for any two disjoint subsets $A, B \subseteq V_1$ such that $|A| + |B| = k$.

Such graphs immediately yield (n,k)-universal sets of 0-1 strings:

Proposition 11.8. *Let G be a bipartite graph with parts of size n and C be the set of columns of its incidence matrix. If G satisfies the isolated neighbor condition for k then C is (n,k)-universal.*

Proof. Let $G = (V_1, V_2, E)$ and $M = (m_{x,y})$ be the incidence matrix of G. That is, M has n rows labeled by nodes x from V_1, n columns labeled by nodes y from V_2, and $m_{x,y} = 1$ if and only if $(x,y) \in E$.

Let $S = \{i_1, \ldots, i_k\}$ be an arbitrary subset of k rows of M and $v = (v_{i_1}, \ldots, v_{i_k})$ be an arbitrary (column) vector in $\{0,1\}^k$. Each row of M corresponds to a node in V_1. Let A be the set of nodes in V_1 corresponding to the 1-coordinates of v, and B be the set of nodes corresponding to the 0-coordinates of v. Since $|A| + |B| = |S| = k$ and our graph satisfies the isolated neighbor condition for k, there must be a node $y \in V_2$ which is joined to each node of A and to no node of B. But this means that the values of the y-th column of M at rows from S coincide with the corresponding values of the vector v, as desired.   □

11.3.2 Paley graphs

Here we will show how to construct explicit bipartite graphs satisfying the isolated neighbor condition for k close to $\log n$.

A *bipartite Paley graph* is a bipartite graph $G_q = (V_1, V_2, E)$ with parts $V_1 = V_2 = \mathbb{F}_q$ for q odd prime congruent to 1 modulo 4; two nodes, $x \in V_1$ and $y \in V_2$, are joined by an edge if and only if $x - y$ is a non-zero square in $\mathbb{F}_q$, i.e., if $x - y = z^2 \pmod{q}$ for some $z \in \mathbb{F}_q$, $z \neq 0$. The condition $q \equiv 1 \pmod 4$ is only to ensure that -1 is a square in the field (see Exercise 11.7), so that the resulting graph is undirected.

Given two disjoint sets of nodes $A, B \subseteq V_1$, let $v(A, B)$, as before, denote the number of nodes in V_2 joined to each node of A and to no node of B. It appears that for $|A| + |B| = k < (\log q)/3$, this number is very close to $q/2^k$, independent of what the sets A, B actually are.

Theorem 11.9. *Let $G_q = (V_1, V_2, E)$ be a bipartite Paley graph with $q \geqslant 9$, and A, B be disjoint sets of nodes in V_1 such that $|A| + |B| = k$. Then*

$$\left| v(A, B) - 2^{-k} q \right| \leqslant k\sqrt{q}. \tag{11.1}$$

In particular, $v(A, B) > 0$ as long as $k2^k < \sqrt{q}$.

This result is a slight modification of a similar result of Bollobás and Thomason (1981) about general (non-bipartite) Paley graphs; essentially the same result was proved earlier by Graham and Spencer (1971). The proof is based on the theorem of Weil (1948) regarding character sums. Its special case states the following.

Let χ be the *quadratic residue character* in $\mathbb{F}_q$: $\chi(x) = x^{(q-1)/2}$. That is, $\chi(x) = 1$ if x is a non-zero square in $\mathbb{F}_q$, $\chi(x) = -1$ if x is non-square, and $\chi(0) = 0$. Also, $\chi(x \cdot y) = \chi(x) \cdot \chi(y)$.

Theorem 11.10 (Weil 1948). *Let $f(t)$ be a polynomial over $\mathbb{F}_q$ which is not the square of another polynomial and has precisely s distinct zeros. Then*

$$\left| \sum_{x \in \mathbb{F}_q} \chi(f(x)) \right| \leqslant (s - 1)\sqrt{q}.$$

We omit the proof of this important result. Weil's original proof relied heavily on several ideas from algebraic geometry. Since then other (but still complicated) proofs have been found; the interested reader can find the details in (Schmidt 1976).

With Weil's result, the above stated property of Paley graphs can be derived by easy computations.

Proof of Theorem 11.9. Recall that (x, y) is an edge in G_q if and only if $\chi(x - y) = 1$. Say that a node $x \in V_2$ is a *copy* of a node $y \in V_1$ if both these nodes correspond to the same element of $\mathbb{F}_q$; hence, each node of V_2 is a copy of precisely one node in V_1. Moreover, no x is joined to its copy y since then $\chi(x - y) = \chi(0) = 0$.

Let A' and B' be the set of all copies of nodes in A and, respectively, in B. Also let $U = V_2 - (A' \cup B')$. Define

$$g(x) = \prod_{a \in A} (1 + \chi(x - a)) \prod_{b \in B} (1 - \chi(x - b))$$

and observe that, for each node $x \in U$, $g(x)$ is non-zero if and only if x is joined to every node in A and to no node in B, in which case it is precisely 2^k. Hence,

$$\sum_{x \in U} g(x) = 2^k \cdot v^*(A, B), \tag{11.2}$$

where $v^*(A, B)$ is the set of those nodes in U which are joined to each node of A and to no node of B.

Expanding the expression for $g(x)$ and using the fact that $\chi(x \cdot y) = \chi(x) \cdot \chi(y)$, we obtain

$$g(x) = 1 + \sum_C (-1)^{|C \cap B|} \chi(f_C(x)),$$

where $f_C(x)$ denotes the polynomial $\prod_{c \in C}(x - c)$, and the sum is over all *non-empty* subsets C of $A \cup B$. By Weil's theorem,

$$\left| \sum_{x \in \mathbb{F}_q} \chi(f_C(x)) \right| \leqslant (|C| - 1)\sqrt{q}.$$

Hence,

$$\left| \sum_{x \in \mathbb{F}_q} g(x) - q \right| \leqslant \sum_C (|C| - 1)\sqrt{q} = \sqrt{q} \sum_{s=2}^{k} \binom{k}{s} (s - 1)$$
$$= \sqrt{q}((k - 2)2^{k-1} + 1).$$

Here the last equality follows from the identity $\sum_{s=1}^{k} s \binom{k}{s} = k2^{k-1}$ (see Exercise 1.5).

The summation above is over all nodes $x \in V_2 = \mathbb{F}_q$. However, for every node $x \in A' \cup B'$, $g(x) \leqslant 2^{k-1}$, and the nodes of $A' \cup B'$ can contribute at most

$$\left| \sum_{x \in A' \cup B'} g(x) \right| \leqslant k \cdot 2^{k-1}.$$

Therefore,

$$\left| \sum_{x \in U} g(x) - q \right| \leqslant \sqrt{q}((k-2)2^{k-1} + 1) + k \cdot 2^{k-1}.$$

Dividing both sides by 2^k and using (11.2), together with the obvious estimate $v(A, B) - v^*(A, B) \leqslant |A' \cup B'| = k$, we conclude that

$$\left| v(A, B) - 2^{-k} q \right| \leqslant \frac{k\sqrt{q}}{2} - \sqrt{q} + \frac{\sqrt{q}}{2^k} + \frac{k}{2} + k, \tag{11.3}$$

which does not exceed $k\sqrt{q}$ as long as $q \geqslant 9$. □

Theorem 11.9 together with Proposition 11.8 give us, for infinitely many values of n, and for every k such that $k2^k < \sqrt{n}$, an explicit construction of (n, k)-universal sets of size n. In Sect. 14.4.1 we will show how to construct such sets of size $n^{O(k)}$ for arbitrary k using some elementary properties of linear codes.

11.4 Full graphs

We have seen that universal sets of 0-1 strings correspond to bipartite graphs satisfying the isolated neighbor condition. Let us now ask a slightly different question: how many vertices a graph must have in order to contain *every k-vertex* graph as an induced subgraph? Such graphs are called *k-full*. That is, given k, we are looking for graphs of small order (the order of a graph is the number of its vertices) which contain every graph of order k as an induced subgraph.

Note that if G is a k-full graph of order n then $\binom{n}{k}$ is at least the number of non-isomorphic graphs of order k, so

$$\binom{n}{k} \geqslant 2^{\binom{k}{2}}/k!$$

and thus

$$\cdot\ n \geqslant 2^{(k-1)/2}.$$

On the other hand, for every k it is possible to exhibit a k-full graph of order $n = 2^k$. This nice construction is due to Bollobás and Thomason (1981).

Let P_k be a graph of order $n = 2^k$ whose vertices are subsets of $\{1, \ldots, k\}$, and where two distinct vertices A and B are joined if and only if $|A \cap B|$ is even; if one of the vertices, say A, is an empty set then we join B to A if and only if $|B|$ is even. Note that the resulting graph is regular: each vertex has degree $2^{k-1} - 1$.

Theorem 11.11 (Bollobás–Thomason 1981). *The graph P_k is k-full.*

Proof. Let G be a graph with vertex set $\{v_1, v_2, \ldots, v_k\}$. We claim that there are sets $A_1, A_2, \ldots, A_k$ uniquely determined by G, such that

$A_i \subseteq \{1, \ldots, i\}, \quad i \in A_i,$

and, for $i \neq j$,

$|A_i \cap A_j|$ is even if and only if v_i and v_j are joined in G.

Indeed, suppose we have already chosen the sets $A_1, A_2, \ldots, A_{j-1}$. Our goal is to choose the next set A_j which is properly joined to all the sets $A_1, A_2, \ldots, A_{j-1}$, that is, $|A_j \cap A_i|$ must be even precisely when v_j is joined to v_i in G. We will obtain A_j as the last set in a sequence $B_1 \subseteq B_2 \subseteq \ldots \subseteq B_{j-1} = A_j$, where, for each $1 \leqslant i < j$, B_i is a set properly joined to all sets $A_1, A_2, \ldots, A_i$.

As the first set B_1 we take either $\{j\}$ or $\{1, j\}$ depending on whether v_j is joined to v_1 or not. Having the sets $B_1, \ldots, B_{i-1}$ we want to choose a set B_i. If v_j is joined to v_i then we set $B_i = B_{i-1}$ or $B_i = B_{i-1} \cup \{i\}$ depending on whether $|B_{i-1} \cap A_i|$ is even or odd. If v_j is not joined to v_i then we act dually. Observe that our choice of whether i is in B_i will effect $|B_i \cap A_i|$ (since $i \in A_i$) but none of $|B_i \cap A_l|$, $l < i$ (since $A_l \subseteq \{1, \ldots, l\}$). After $j - 1$ steps we will obtain the desired set $B_{j-1} = A_j$. □

Exercises

11.1. − Let $A \subseteq \{0, 1\}^n$ be (n, k)-dense and suppose that no vector in A has more than r 1's. Prove that some two vectors in A have at most $r - k$ 1's in common.

11.2 (Alon 1986). Let A be a 0-1 matrix of 2^n rows and n columns, the ith row being the binary representation of $i - 1$ ($1 \leqslant i \leqslant 2^n$). Show that for any choice of k distinct columns of A and any choice of k bits, there are exactly 2^{n-k} rows of A that have the jth chosen bit in the jth chosen column.

11.3. Let $A \subseteq \{0, 1\}^n$, $|A| = n$. By induction on k prove that, for every $k = 1, 2, \ldots, n - 1$, there exist k coordinates such that the projection of A onto these coordinates has more than k vectors. For $k = n - 1$ this is the well-known Bondy's theorem (Theorem 12.1).

11.4 (Chandra et al. 1983). Try to prove the following (n, k)-universality criterion for the case $k = 2$. Given a set $A \subseteq \{0, 1\}^n$ of $m = |A|$ vectors, look at it as an $m \times n$ matrix, whose rows are the vectors of A. Let $v_1, \ldots, v_n \in \{0, 1\}^m$ be the *columns* of this matrix, and let $\overline{v}_1, \ldots, \overline{v}_n$ be their complements, i.e., $\overline{v}_i$ is obtained from v_i by switching all its bits to the opposite values. Prove that A is $(n, 2)$-universal if and only if all the vectors $v_1, \ldots, v_n, \overline{v}_1, \ldots, \overline{v}_n$ are different and form an antichain in $\{0, 1\}^m$, i.e., are mutually incomparable.

11.5. Let $A \subseteq \{0,1\}^n$, $|A| = m$. Look at A as an $m \times n$ matrix, and let $\mathcal{F}_A$ be the family of those subsets of $\{1, \ldots, m\}$, whose incidence vectors are columns of this matrix. Show that A is (n, k)-universal if and only if the family $\mathcal{F}_A$ is *k-independent* in the following sense: for every k distinct members $S_1, \ldots, S_k$ of $\mathcal{F}_A$ all 2^k intersections $\bigcap_{i=1}^k T_i$ are non-empty, where each T_i can be either S_i or its complement $\overline{S}_i$.

11.6.⁻ Show that the converse of Proposition 11.8 also holds: if the set of rows of the incidence matrix of a given bipartite graph is (n, k)-universal then the graph satisfies the isolated neighbor condition for k.

11.7. Let p be a prime with $p \equiv 1 \pmod 4$. Show that -1 is a square in the field $\mathbb{F}_p$.

Sketch: Let P be the product of all nonzero elements of $\mathbb{F}_p$. If -1 is not a square, then $x^2 = -1$ has no solutions; so, the set of all $p - 1$ nonzero elements of $\mathbb{F}_p$ can be divided into $(p-1)/2$ pairs such that the product of the elements in each pair is -1; hence $P = 1$. On the other hand, for any $x \neq \pm 1$ there exists exactly one $y \neq x$ with $xy = 1$, so all the elements of $\mathbb{F}_p - \{-1, 0, +1\}$ can be divided into pairs so that the product of elements in each pair is 1; hence, $P = -1$, a contradiction.

11.8.⁽!⁾ Recall that a set $A \subseteq \{0,1\}^n$ of vectors is monotone decreasing if $v \in A$ and $u \leqslant v$ implies $u \in A$. Similarly, say that a set is monotone increasing if $v \in A$ and $u \geqslant v$ implies $u \in A$. Show that Kleitman's theorem (Theorem 11.6) implies the following: Let A, B be monotone increasing and C monotone decreasing subsets of $\{0,1\}^n$. Then

$$|A \cap B| \geqslant \frac{|A| \cdot |B|}{2^n}$$

and

$$|A \cap C| \leqslant \frac{|A| \cdot |C|}{2^n}.$$

Hint: For the first inequality, apply Kleitman's theorem to the complements of A and B. For the second inequality, take $B = \{0,1\}^n - C$, and apply the first inequality to the pair A, B to get

$$|A| - |A \cap C| = |A \cap B| \geqslant 2^{-n} |A| (2^n - |C|).$$

11.9.⁻ Show that the lower bound $t_{n-1}(A) \geqslant |A| - 2^{t-1} + 1$ given in Theorem 11.5 is the best possible whenever t divides n.

Hint: Split $\{1, \ldots, n\}$ into n/t disjoint subsets $S_1, \ldots, S_{n/t}$ with $|S_i| = t$ and define $A = \{v : \exists i \text{ such that } v_j = 0 \text{ for all } j \notin S_i\}$.

12. Witness Sets and Isolation

Given a set A of distinct 0-1 vectors and a vector u in A, how many bits of u must we know in order to distinguish it from the other vectors in A? Such a set of bits is a *witness* for the fact that $u \notin A - \{u\}$. In this chapter we will give some basic estimates on the size of these witnesses. We will also consider a related problem of how to *isolate* an object within a given universum according to its weight. Finally, we will describe the so-called "dictator paradox" saying that, if the society fulfills some simple "democracy axioms," then there will always be an individual (a dictator?) whose options prevail against all options.

12.1 Bondy's theorem

Let $A \subseteq \{0,1\}^n$ be a set of m distinct 0-1 vectors of length n. A set $S \subseteq \{1,\ldots,n\}$ of coordinates is a *witness* for a vector u in A if for every other $v \in A$ there exists a coordinate in S on which u differs from v. We may also say that *exposing* the entries of u corresponding to S uniquely determines u among vectors in A. The minimum size of a witness for u in A is denoted by $w_A(u)$ (or by $w(u)$, if the underlying set A is clear from the context).

It is easy to show that every set of m vectors contains a vector whose witness has size at most $\log_2 m$ (see Exercise 12.2). On the other hand, it is obvious that $w_A(u) \leqslant |A| - 1$ for any A and $u \in A$, and a simple example shows that this is tight: if A consists of the all-0 vector 0^n and the n vectors with precisely one 1, then $w_A(0^n) = n$.

The following result, due to Bondy (1972), shows that if we take only $m \leqslant n$ vectors, then all the vectors will already have *one and the same* witness of size at most $m - 1$. The *projection* of a vector $v = (v_1,\ldots,v_k)$ onto a set of coordinates $S = \{i_1,\ldots,i_k\}$ is the vector $v\!\restriction_S \; = (v_{i_1},\ldots,v_{i_k})$. The projection of a set of vectors A is the set $A\!\restriction_S = \{v\!\restriction_S \; : \; v \in A\}$.

Theorem 12.1 (Bondy 1972). *For every set A of 0-1 vectors there exists a set S of at most $|A| - 1$ coordinates such that all the vectors $\{v\!\restriction_S \; : \; v \in A\}$ are distinct.*

Proof. Suppose that A is a counterexample, that is, $|A\!\restriction_S| < |A|$ for every set S of at most $|A| - 1$ coordinates. Let S be a *maximal* set of coordinates for which $|A\restriction_S| \geqslant |S| + 1$. Since $|A\restriction_S| \leqslant |A| - 1$, at least two vectors $u \neq v \in A$ must coincide on S. Take a coordinate $i \notin S$ on which these two vectors differ, and set $T := S \cup \{i\}$. Since the vectors u, v coincide on S but differ on T, the projection $A\restriction_T$ must have at least one more vector than $A\restriction_S$; hence,

$$|A\restriction_T| \geqslant |A\restriction_S| + 1 \geqslant |S| + 2 = |T| + 1,$$

a contradiction with the maximality of S.   □

Given k, how large must a set A be in order to be sure that at least one of its vectors will have no witness of size $\leqslant k$? It is clear that any such set A must have more than 2^k vectors; this is a trivial lower bound. A trivial upper bound is 2^n. The following simple observation shows that much fewer vectors are enough.

Proposition 12.2. *In every set of more than $2^k \binom{n}{k}$ 0-1 vectors of length n there is a vector which has no witness of size k.*

Proof. Let A be a set of 0-1 vectors of length n, and assume that every vector in it has a witness of size k. Then each vector $u \in A$ has its own set S_u of k coordinates on which this vector differs from all other vectors in A. That is, we can assign to each vector $u \in A$ its "pattern" – a set S_u of k bits and the projection $u\restriction_S$ of u onto this set – so that different vectors will receive different patterns, i.e., if $u \neq v$ then either $S_u \neq S_v$ or $S_u = S_v$ but u and v differ on some coordinate in S_u. There are $\binom{n}{k}$ possible subsets of k coordinates and, on each of these sets, vectors can take no more than 2^k possible values. Thus, there are at most $\binom{n}{k}2^k$ possible patterns and, since each vector in A must have its own pattern, we conclude that $|A| \leqslant \binom{n}{k}2^k$.   □

12.2 Average witnesses

Since the worst-case witness sets may have to be large, it is natural to consider the *average* witness size:

$$w_{ave}(A) := \frac{1}{|A|} \sum_{u \in A} w_A(u).$$

The same example, as in the previous section, shows that the gap between the worst-case witness size and the average witness size may be exponential: if A is the set of $n+1$ vectors with at most one 1, then $w_{ave}(A) = 2n/(n+1) \leqslant 2$, but in the all-0 vector all n bits must be exposed.

How large can $w_{ave}(A)$ be as a function of $|A|$? The following result of Kushilevitz, Linial, Rabinovitch, and Saks (1996) says that the average witness size of *any* set does not exceed the square root of its size, and that this bound is almost optimal.

Theorem 12.3. *For every set A of m 0-1 vectors, $\mathrm{w}_{\text{ave}}(A) \leqslant 2m^{1/2}$. On the other hand, for infinitely many numbers m, there exists a set A of m 0-1 vectors such that $\mathrm{w}_{\text{ave}}(A) \geqslant \frac{1}{2\sqrt{2}}m^{1/2}$.*

Proof. Upper bound. Take an arbitrary set A of m vectors and order its vectors $u_1, u_2, \ldots, u_m$ by decreasing value of their smallest witness size: $\mathrm{w}(u_1) \geqslant \mathrm{w}(u_2) \geqslant \cdots \geqslant \mathrm{w}(u_m)$. Consider the sum of the first k largest values $\sum_{i=1}^{k} \mathrm{w}(u_i)$ for a value k soon to be set. Find a set T of at most $k-1$ coordinates as guaranteed by Bondy's theorem applied to the set $\{u_1, \ldots, u_k\}$ and expose the T-coordinates in *all* vectors of A. By the property of T, vectors $u_1, \ldots, u_k$ are already mutually distinguished. The T-coordinates of every vector u_j with $j > k$, distinguish u_j from all $u_1, \ldots, u_k$, except, perhaps, *one* u_i (because no two of the vectors $u_1, \ldots, u_k$ coincide on T). It is possible to expose a single additional bit in u_i to distinguish u_i from u_j. Apply this step for every u_j, $j > k$. Consequently, each of $u_1, \ldots, u_k$ is distinguished from every other vector in A. No more than $m-k$ additional bits get exposed in this process, so:

$$\sum_{i=1}^{k} \mathrm{w}(u_i) \leqslant k(k-1) + m - k = k^2 - 2k + m. \tag{12.1}$$

In particular, it follows that $\mathrm{w}(u_k) \leqslant k - 2 + m/k$.

Putting these two observations together we get

$$\sum_{i=1}^{m} \mathrm{w}(u_i) = \sum_{i=1}^{k} \mathrm{w}(u_i) + \sum_{i=k+1}^{m} \mathrm{w}(u_i)$$

$$\leqslant (k^2 - 2k + m) + (m - k)\left(k - 2 + \frac{m}{k}\right).$$

Pick $k = m^{1/2}$; the above inequality then yields $\sum_{i=1}^{m} \mathrm{w}(u_i) \leqslant 2m^{3/2}$, which means that $\mathrm{w}_{\text{ave}}(A) \leqslant 2m^{1/2}$, as desired.

Lower bound. We will explicitly construct a set $A \subseteq \{0,1\}^n$ which achieves the lower bound. Let p be a prime and consider a projective plane $\mathrm{PG}(2,p)$ of order p (see Sect. 13.4). Such a plane consists of $n = p^2 + p + 1$ points $P = \{1, \ldots, n\}$ and n subsets of points $L_1, \ldots, L_n \subseteq P$ (called *lines*) satisfying the following three conditions: (i) each line has exactly $p+1$ points; (ii) every two lines intersect in exactly one point, and (iii) exactly $p+1$ lines meet in one point.

We consider n-dimensional vectors where the coordinates correspond to points of P, and define $A \subseteq \{0,1\}^n$ to be the family of $m = 2n$ binary vectors, of which n are the incidence vectors of lines of $\mathrm{PG}(2,p)$, and another n are all unit vectors, i.e., incidence vectors of all singletons $\{i\}$, $i \in P$.

For a vector $u \in A$, corresponding to a line L, $\mathrm{w}(u) = 2$, since it suffices to expose the coordinates corresponding to any two points on L. Such a pair distinguishes u from all singletons, and since distinct lines share exactly one point, this pair also distinguishes u from the incidence vectors of other lines.

On the other hand, $w(u) = p + 2$ if $u = (0, \ldots, 0, 1, 0, \ldots, 0)$ corresponds to a singleton point $i \in P$. To distinguish u from the incidence vector of a line L containing i, a zero in u should be exposed in a coordinate that corresponds to a point on L other than i. There are $p+1$ lines, whose pairwise intersection is $\{i\}$, so to distinguish u from all of them, at least $p + 1$ distinct 0-entries should be exposed. To distinguish u from other singletons, the 1-entry should be exposed as well (the alternative being to expose all $p^2 + p$ 0-entries).

Putting things together, we get

$$W_{\text{ave}}(A) = \frac{1}{|A|} \sum_{u \in A} w(u) = \frac{1}{2n}(2n + (p+2)n) = \frac{p+4}{2} \geqslant \frac{n^{1/2}}{2\sqrt{2}}.$$

□

The next natural problem concerning 0-1 vectors is the following question about the *distribution* of their witness sizes:

> Given an integer t, $1 \leqslant t \leqslant m$, and a set of m vectors, how many of its vectors have a witness of size at least (or at most) t?

If we know nothing more about the set except for its size, the question turns out to be difficult. Still, Kushilevitz et al. (1996) have found the following interesting partial solutions (see also Example 12.3):

Lemma 12.4. *Let A be a set of m distinct 0-1 vectors. Then*

(a) *for any $t \leqslant m$ at most t of vectors in A have a minimal witness of size at least $t + m/t - 2$;*

(b) *for any $t \leqslant \sqrt{m}$ at least $t^2 - t$ of vectors in A have a witness of size at most $2t + \log_2 m$.*

Proof. The first claim (a) follows from the proof of the upper bound in Theorem 12.3: let k be the number of vectors $u \in A$ for which $w(u) \geqslant t + m/t - 2$, and apply (12.1).

To prove the second claim (b), reorder the vectors in A as follows: split the vectors into two groups according to their first coordinate, and let the vectors of the smaller group (i.e., of the group containing at most half of the vectors) precede those in the larger. Expose the first coordinate in all vectors of the smaller group. Proceed recursively in the same manner on each group separately (by looking at next coordinates), and so on, until each group reduces to a single vector (see Fig. 12.1). Observe that:

(i)   each vector is distinguished from all those following it (but not necessarily from those preceding it);

(ii) no vector has more than $\log_2 m$ bits exposed (since each time one bit is exposed in at most one-half of the vectors of a current group).

Let B be the set of the first t^2 vectors. Applying the first claim (a) to this set, we conclude that at most t of its vectors have a witness of size at least $2t$. Therefore, at least $t^2 - t$ of the vectors in B can be distinguished

```
1  0   0 0
1  0   1 0
1  1   1 1
1  1   0 1
1  1   0 0
0  1   0 1
0  1   1 0
0  1   1 1
0  0   1 1
0  0   1 0
0  0   0 0
0  0   0 1
```

Fig. 12.1. Exposed bits are in boldface; a vector u *follows* vector v if u is below v.

from other members of B at the cost of exposing at most $2t$ additional bits in each of them. We call these vectors *good*. By (i) and (ii), at the cost of exposing at most $\log_2 m$ bits, each good vector v is already distinguished from all the vectors in A following it. On the other hand, all the vectors preceding v belong to B, and hence, v is distinguished also from them by at most $2t$ additional bits. Thus, we have at least $t^2 - t$ good vectors v and for each of them, $\mathrm{w}_A(v) \leqslant 2t + \log_2 m$. □

12.3 The isolation lemma

Let X be some set of n points, and $\mathcal{F}$ be a family of subsets of X. Let us assign a weight $w(x)$ to each point $x \in X$ and let us define the weight of a set E to be $w(E) = \sum_{x \in E} w(x)$. It may happen that several sets of $\mathcal{F}$ will have the minimal weight. If this is not the case, i.e., if $\min_{E \in \mathcal{F}} w(E)$ is achieved by a unique $E \in \mathcal{F}$, then we say that w is *isolating for $\mathcal{F}$*.

The following lemma, due to K. Mulmuley, U. Vazirani, and V. Vazirani (1987), says that – independent of what our family $\mathcal{F}$ actually is – a randomly chosen w is isolating for $\mathcal{F}$ with large probability.

Lemma 12.5. *Let $\mathcal{F}$ be a family of subsets of an n-element set X. Let $\mathbf{w} : X \to \{1, \ldots, N\}$ be a random function, each $\mathbf{w}(x)$ independently and uniformly chosen over the range. Then*

$$\mathrm{Prob}\,(\mathbf{w} \text{ is isolating for } \mathcal{F}) \geqslant 1 - \frac{n}{N}.$$

Proof (Spencer 1995). For a point $x \in X$, set

$$\alpha(x) = \min_{E \in \mathcal{F}; \, x \notin E} \mathbf{w}(E) - \min_{E \in \mathcal{F}; \, x \in E} \mathbf{w}(E - \{x\}).$$

A crucial observation is that evaluation of $\alpha(x)$ does not require knowledge of $\mathbf{w}(x)$. As $\mathbf{w}(x)$ is selected uniformly from $\{1, \ldots, N\}$,

$$\mathrm{Prob}\,(\mathbf{w}(x) = \alpha(x)) \leqslant 1/N,$$

so that

$$\text{Prob}\,(\mathbf{w}(x) = \alpha(x) \text{ for some } x \in X) \leqslant n/N.$$

But if $\mathbf{w}$ had two minimal sets $A, B \in \mathcal{F}$ and $x \in A - B$, then

$$\min_{E \in \mathcal{F}; x \notin E} \mathbf{w}(E) = \mathbf{w}(B),$$

$$\min_{E \in \mathcal{F}; x \in E} \mathbf{w}(E - \{x\}) = \mathbf{w}(A) - \mathbf{w}(x),$$

so $\mathbf{w}(x) = \alpha(x)$. Thus, if $\mathbf{w}$ is *not* isolating for $\mathcal{F}$ then $\mathbf{w}(x) = \alpha(x)$ for some $x \in X$, and we have already established that the last event can happen with probability at most n/N.   □

This lemma has many applications in the theory of computing. In particular, Mulmuley et al. (1987) used it to give an efficient randomized algorithm for finding a perfect matching in a graph. This result is a standard demonstration of the isolation lemma. Below we describe an application of different type: we use this lemma to show that, in the model of switching networks, counting is not weaker than nondeterminism. (Comparing the power of different modes of computation is one of the main problems in the theory of computing.)

A (switching-and-rectifier) *network* is a directed acyclic graph $G = (V, E)$ with two specified vertices $s, t \in V$, *some* of whose edges are labeled by variables x_i or their negations $\overline{x}_i$. The size of G is defined as the number of vertices. Each input $a = (a_1, \ldots, a_n) \in \{0,1\}^n$ defines a subgraph $G(a)$ of G obtained by deleting all edges whose labels are evaluated by a to 0, and removing the labels from the remaining edges. Let $|G(a)|$ denote the number of s-t paths in $G(a)$. A network G computes a boolean function in a natural way: it accepts the input a if and only if $|G(a)| > 0$. This is a *nondeterministic* mode of computation: we accept the input if and only if the labels of at least one s-t path in G are consistent with it. A *parity network* is a network with a *counting* mode of computation: we accept the input a if and only if the number of s-t paths consistent with a is odd, i.e., iff $|G(a)| = 1 \pmod 2$.

Using the isolation lemma one can show that, at the cost of a slight increase of size, every (nondeterministic) network may be simulated by a parity network.

Theorem 12.6 (Wigderson 1994). *If a boolean function in n variables can be computed by a network of size L, then it can also be computed by a parity network of size at most $n \cdot L^c$, where $c \leqslant 10$.*

Proof. Given a graph $G = (V, E)$, a weight function $w : E \to \{1, \ldots, 2 \cdot |E|\}$ and an integer l, define the (unweighted, layered) version $G_w^l = (V', E')$ of G as follows. Replace every vertex $u \in V$ by $l + 1$ new vertices $u_0, u_1, \ldots, u_l$ in V' (i.e., V' consists of $l + 1$ copies of V, arranged in layers). For every edge (u, v) in E and every $0 \leqslant i \leqslant l - w(e)$ we put an edge $\big(u_i, v_{i+w(e)}\big)$

in E' (see Fig. 12.2). Let $d_w(G)$ denote the weight of the shortest s-t path in G (the weight of a path is the sum of weights of its edges; a path is *shortest* if its weight is minimal); hence, $d_w(G) \leqslant M = 2|V| \cdot |E| \leqslant |V|^3$ and $|V'| \leqslant (1+l)|V|$.

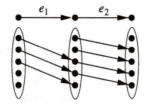

Fig. 12.2. $l = 4$, $w(e_1) = 2$ and $w(e_2) = 1$

It can be shown (do this!) that the graphs G_w^l have the following properties:

(i) if G has no s-t path, then for every w and l, G_w^l has no s_0-t_l path;

(ii) if G has an s-t path and $l = d_w(G)$, then G_w^l has an s_0-t_l path. Moreover, the later path is unique if the shortest s-t path in G is unique.

Now let $G = (V, E)$ be a network computing a given boolean function $f(x_1, \ldots, x_n)$. Say that a weight function w is *good* for an input $a \in \{0,1\}^n$ if either $G(a)$ has no s-t paths or the shortest s-t path in $G(a)$ is unique. For each input $a \in \{0,1\}^n$, taking the family $\mathcal{F}$ to be all s-t paths in the graph $G(a)$, the isolation lemma (Lemma 12.5) implies that at least one-half of all weight functions w are good for a. By a standard counting argument, there exists a set W of $|W| \leqslant \log_2(2^n) = n$ weight functions such that at least one $w \in W$ is good for every input. If w is good for a, then the graph $G_w^l(a)$ with $l = d_w(G(a))$ has the properties (i) and (ii). For different inputs a, the corresponding values of l may be different, but they all lie in the interval $1, \ldots, M$. Thus, there exist $m \leqslant n \cdot M$ networks $H_1, \ldots, H_m$ (with each $H_j = G_w^l$ for some $w \in W$ and $1 \leqslant l \leqslant M$) such that, for every input $a \in \{0,1\}^n$, the following holds:

(iii) if $|G(a)| = 0$, then $|H_j(a)| = 0$ for all j;

(iv) if $|G(a)| > 0$, then $|H_j(a)| = 1$ for at least one j.

Let s_j, t_j be the specified vertices in H_j, $j = 1, \ldots, m$. We construct the desired parity network H as follows: to each H_j add the unlabeled edge (s_j, t_j), identify t_j and s_{j+1} for every $j < m$, and add the unlabeled edge (s_1, t_m) (see Fig. 12.3).

It is easy to see that, for every input $a \in \{0,1\}^n$, $|H(a)| = 1 \,(\mathrm{mod}\,2)$ if and only if $|G(a)| > 0$. Indeed, if $|G(a)| = 0$, then by (iii), $H(a)$ has precisely

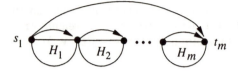

Fig. 12.3. Construction of the parity network H

two s_1-t_m paths (formed by added unlabeled edges). On the other hand, if $|G(a)| > 0$, then by (iv), at least one $H_j(a)$ has precisely one s_j-t_j path, implying that the total number of s_1-t_m paths in $H(a)$ is odd. Thus, H is a parity network computing the same boolean function f.   □

For the sake of completeness, let us mention (without proof) the following interesting "parity-type" isolation lemma proved by Valiant and Vazirani (1986). View the cube $\{0,1\}^n$ as n-dimensional vector space $\mathbb{F}_2^n$, and let $\langle u, v \rangle = \sum_{i=1}^n u_i v_i \pmod 2$ denote the scalar product over $\mathbb{F}_2$.

Lemma 12.7. *Let $S \subseteq \{0,1\}^n$, $|S| \geqslant 2$. Let $\mathbf{w}_1, \ldots, \mathbf{w}_n$ be chosen independently from $\{0,1\}^n$ at random. Then, with probability at least $1/4$, there is an i such that $\langle v, \mathbf{w}_1 \rangle = \ldots = \langle v, \mathbf{w}_i \rangle = 0$ for precisely one vector $v \in S$.*

12.4 Isolation in politics: the dictator paradox

One of the problems of politics involves averaging out individual preferences to reach decisions acceptable to society as a whole. In this section we will prove one isolation-type result due to Arrow (1950) which shows that, under some simple "democracy axioms" this is indeed a difficult task.

The simple process of voting can lead to surprisingly counterintuitive paradoxes. For example, if three people vote for three candidates, giving the rankings $x < y < z$, $y < z < x$, $z < x < y$, then a majority prefers y to x ($x < y$), x to z ($z < x$) but also z to y ($y < z$). In general, we have the following situation.

Suppose that $I = \{1, \ldots, n\}$ is a society consisting of a set of n individuals. These individuals are to be offered a choice among a set X of options, for example, by a referendum. We assume that each individual i has made her/his mind up about the relative worth of the options. We can describe this by a total order $<_i$ on X, for each $i \in I$, where $x <_i y$ means that the individual i prefers option y to option x. So, after a referendum we have a set $R = \{<_1, \ldots, <_n\}$ of total orders on X. A *social choice function* F takes such a set of total orders as input and comes up with a "social preference" on X, i.e., with some total order $<$ on X. Being total means, in particular, that the order $<$ is transitive: if $x < y$ and $y < z$ then $x < z$.

Given a social choice function F, a *dictator* is an individual $i_0 \in I$ such that for every referendum, the resulting social preference $<$ coincides with the preference $<_{i_0}$ of this individual. That is, for any given set of total orders $R = \{<_1, \ldots, <_n\}$, the social choice function will output the order $<_{i_0}$, independent of preferences $<_i$ made by other individuals $i \neq i_0$.

Arrow's theorem asserts that, if the choice function F fulfills some natural "democracy axioms," then there will *always* be a dictator! That is, once we fix some social choice function F, then there will be an individual (a dictator?) whose options prevail against all options.

Let us consider the following three natural *democracy axioms*:

(A1) If $x < y$ (in the social preference), then the same remains true if the individual preferences are changed in y's favour.

(A2) If $Y \subseteq X$ is a set of options and if during two referendums no individual changes his/her mind about the options in Y then the society also don't changes its mind about these options.

(A3) For any distinct options $x, y \in X$, there is some system of individual preferences for which the corresponding social preference has $x < y$. That is, it should be possible for society to prefer y to x if enough individuals do so.

Theorem 12.8. *If $|X| \geqslant 3$ then for every social choice function, satisfying the three democracy axioms above, there is a dictator.*

Proof. We follow the elegant argument from Cameron (1994). Suppose that we have a social choice function. If (x, y) is an ordered pair of distinct options, we say that a set J of individuals is (x, y)-*decisive* if, whenever all members of J prefer y to x, then so does the social order; formally, if $x <_i y$ for all $i \in J$, then $x < y$. Further, we say that J is *decisive* if it is (x, y)-decisive for some distinct $x, y \in X$.

Let J be a minimal decisive set. It follows from (A1)–(A3) that, for any distinct options $x, y \in X$, if every individual prefers y to x then so does the social order. Hence, $J \neq \emptyset$. Suppose that J is (x, y)-decisive, and let i_0 be a member of J.

Claim 12.9. $J = \{i_0\}$.

To prove the claim, suppose the opposite and let $J' \rightleftharpoons J - \{i_0\}$ and $K \rightleftharpoons I - J$. Let v be an option in X different from x and y (remember that $|X| \geqslant 3$). Consider the individual preferences $<_i, i \in I$ for which

$$x <_{i_0} y <_{i_0} v$$
$$v <_i x <_i y \quad \text{for all } i \in J'$$
$$y <_j v <_j x \quad \text{for all } j \in K$$

Then $x < y$, since all members of the (x, y)-decisive set J think so, and $y < v$, since if $v < y$ then J' would be (v, y)-decisive (nobody outside J' thinks so), contradicting the minimality of J. Hence $x < v$. But then $\{i_0\}$ is (x, v)-decisive, since nobody else agrees with this order. By minimality of J, we have $J = \{i_0\}$, as desired.

Claim 12.10. i_0 *is a dictator.*

We have to prove that $\{i_0\}$ is (u,v)-decisive for any pair of different options $u \neq v$. The case when $u = x$ is covered by the (proof of) Claim 12.9, and we are left with two possible situations: either $v = x$ or neither $v = x$ nor $u = x$. The argument in both cases is similar.

Case 1: $u \neq x$ and $v \neq x$.

Consider individual preferences in which

$$u <_{i_0} x <_{i_0} v$$
$$v <_j u <_j x \quad \text{for all } j \neq i_0$$

Then $u < x$ (because everybody thinks so) and $x < v$ (because i_0 thinks so and, by Claim 12.9, is (x,v)-decisive for any $v \neq x$); hence $u < v$, and $\{i_0\}$ is (u,v)-decisive because nobody else agrees with this order.

Case 2: $v = x$.

Take $z \notin \{u,x\}$ and consider individual preferences in which

$$u <_{i_0} z <_{i_0} x$$
$$z <_j x <_j u \quad \text{for all } j \neq i_0$$

Then $u < z$ (because i_0 thinks so and both u, z differ from x) and $z < x$ (because everybody thinks so); hence $u < x$, and $\{i_0\}$ is (u,x)-decisive.

This completes the proof of the claim, and thus, the proof of the theorem. □

Exercises

12.1.⁻ Bondy's theorem (Theorem 12.1) implies that, if we take n binary vectors of length n, then all these vectors differ on some set of $n-1$ bits. Does this hold for $n+1$ vectors?

12.2.⁻ Prove that every set of m vectors contains a vector whose witness has size at most $\log_2 m$.

12.3.⁺ Generalize Lemma 12.4 as follows. Let A be a set of m 0-1 vectors. For an integer l, $1 \leqslant l \leqslant m$, let

$$f(m,l) = \max\{k : k \geqslant 1 \text{ and } k + m/k \geqslant l + 2\}.$$

Prove that:

(a) at most $f(m,l)$ vectors in A have a minimal witness of size at least l;
(b) for any $k \leqslant m$, at least $k - f(k, l - \log_2 m)$ vectors in A have witness of size at most l.

12.4. Lemma 12.5 isolates the unique set with the minimal weight. With what probability will there be a unique set with the *maximal* weight?

12.5. Prove that Lemma 12.5 also holds when the weight of a set is defined to be the *product* of the weights of its elements.

13. Designs

The use of combinatorial objects, called designs, originates from statistical applications. Let us assume that we wish to compare v varieties of wines. In order to make the testing procedure as fair as possible it is natural to require that:

(a) each participating person tastes the same number (say k) of varieties so that each person's opinion has the same weight;
(b) each pair of varieties is compared by the same number (say λ) of persons so that each variety gets the same treatment.

One possibility would be to let everyone taste all the varieties. But if v is large, this is very impractical (if not dangerous, as in the case of wines), and the comparisons become rather unreliable. Thus, we should try to design the experiment so that $k < v$.

Definition 13.1. Let $X = \{1,\ldots,v\}$ be a set of *points* (or *varieties*). A (v, k, λ) *design* over X is a collection $\mathcal{D}$ of distinct subsets of X (called *blocks*) such that the following properties are satisfied:

(1) each set in $\mathcal{D}$ contains exactly k points;
(2) every pair of distinct points is contained in exactly λ blocks.

The number of blocks is usually denoted by b. If we replace (2) by the following property:

(2') every t-element subset of X is contained in exactly λ blocks,

then the corresponding family is called a t-(v, k, λ) *design*. A *Steiner system* $S(t, k, v)$ is a t-(v, k, λ)-design with $\lambda = 1$. A design, in which $b = v$ (i.e., the number of blocks and points is the same) is often called *symmetric*.

13.1 Regularity

In every design, every pair of points lies in the same number of blocks. It is easy to show that then the same also holds for every single point. A family of sets $\mathcal{F}$ is *r-regular* if every point lies in exactly r sets; r is the *replication number* of $\mathcal{F}$.

Theorem 13.2. *Let $\mathcal{D}$ be a (v, k, λ) design containing b blocks. Then $\mathcal{D}$ is r-regular with the replication number r satisfying the equations*

$$r(k - 1) = \lambda(v - 1). \tag{13.1}$$

and

$$bk = vr. \tag{13.2}$$

Proof. Let $a \in X$ be fixed and assume that a occurs in r_a blocks. We count in two ways the cardinality of the set

$$\{(a, x, B) \ : \ B \in \mathcal{D}; a, x \in B; x \neq a\}.$$

For each of the $v - 1$ possibilities for x ($x \neq a$) there are exactly λ blocks B containing both a and x. The cardinality of the set is therefore $(v-1)\lambda$. On the other hand, for each of the r_a blocks B containing a, the element $x \in B - \{a\}$ can be chosen in $|B| - 1 = k - 1$ ways. Hence $(v - 1)\lambda = r_a(k - 1)$. This shows that r_a is independent of the choice of a and proves (13.1).

To prove the second claim we count in two ways the cardinality of the set

$$\{(x, B) \ : \ B \in \mathcal{D}, \ x \in B\}.$$

For each $x \in X$ the block B can be chosen in r ways. On the other hand, for each of the b blocks B the element $x \in B$ can be chosen in k ways. Hence $vr = bk$, as desired.   □

Thus, every design is an r-regular family with the parameter r satisfying both equations (13.1) and (13.2). It turns out that for regularity the second condition (13.2) is also sufficient. The proof presented here is due to David Billington (see Cameron 1994).

Theorem 13.3. *Let $k < v$ and $b \leqslant \binom{v}{k}$. If $bk = vr$ then there is an r-regular family $\mathcal{F}$ of k-subsets of $\{1, \ldots, v\}$ with $|\mathcal{F}| = b$.*

Proof. There is a simple way to make a k-uniform family $\mathcal{F}$ "more regular." (We have already used a similar argument in the proof of Theorem 11.2 to make a given set of binary vectors "more hereditary.")

Let r_x be the *replication number* of x, the number of sets of $\mathcal{F}$ which contain x. (In our previous notation this is the *degree* $d(x)$ of a point in the family $\mathcal{F}$. Here we follow the notation which is usual in the design theory.)

If $r_x > r_y$, then there must exist a $(k - 1)$-set A, containing neither x nor y, such that $\{x\} \cup A \in \mathcal{F}$ and $\{y\} \cup A \notin \mathcal{F}$. Now form a new family $\mathcal{F}'$ by removing $\{x\} \cup A$ from $\mathcal{F}$ and including $\{y\} \cup A$ in its place. In the new family, $r'_x = r_x - 1$, $r'_y = r_y + 1$, and all other replication numbers are unaltered. Starting with *any* family of k-sets, we reach by this process a family in which all the replication numbers differ by at most 1 (an *almost regular* family), containing the same number of sets as the original family. By double counting, the average replication number is

$$\frac{1}{v}\sum r_x = \frac{1}{v}\sum_{A \in \mathcal{F}} |A| = \frac{bk}{v};$$

and an almost regular family whose average replication number is an integer must be regular. $\square$

13.2 Finite linear spaces

Sometimes it is possible to show that a design has at least as many blocks as it has points. The well-known Fisher's Inequality (see Theorem 14.6) implies that if $\mathcal{D}$ is a (v, k, λ) design then $|\mathcal{D}| \geqslant v$. Many generalizations exist. For example, the Petrenjuk–Ray-Chaudhuri–Wilson Inequality (Petrenjuk 1968, Ray-Chaudhuri, and Wilson 1975) states that, if $\mathcal{D}$ is a $2s$-(v, k, λ) design with $v \geqslant k + s$ then $|\mathcal{D}| \geqslant \binom{v}{s}$. Both results can be obtained using the linear algebra method (cf. Exercise 14.13).

Some of these results, however, may be proved by direct double counting. Such, for example, is the argument due to Conway for the case of "finite linear spaces." (Do not confuse these linear spaces with those from Analysis.)

A (finite) *linear space* over a set X is a family $\mathcal{L}$ of its subsets, called *lines*, such that:

- every line contains at least two points, and
- any two points are on exactly one line.

Theorem 13.4 (De Bruijn–Erdős 1948). *If $\mathcal{L}$ is a linear space over X then $|\mathcal{L}| \geqslant |X|$, with equality iff any two lines share exactly one point.*

Proof (due to J. Conway). Let $b = |\mathcal{L}| \geqslant 2$ and $v = |X|$. For a point $x \in X$, let r_x, as above, be its replication number, i.e., the number of lines in $\mathcal{L}$ containing x. If $x \notin L$ then $r_x \geqslant |L|$ because there are $|L|$ lines joining x to the points on L. Suppose $b \leqslant v$. So, for $x \notin L$, we have $b(v - |L|) \geqslant v(b - r_x)$. Hence

$$b = \sum_{L \in \mathcal{L}} 1 = \sum_{L \in \mathcal{L}} \sum_{x:x\notin L} \frac{1}{v - |L|} \leqslant \frac{b}{v} \sum_{L \in \mathcal{L}} \sum_{x:x\notin L} \frac{1}{b - r_x}$$

$$= \frac{b}{v} \sum_{x \in X} \sum_{L:x\notin L} \frac{1}{b - r_x} = \frac{b}{v} \sum_{x \in X} 1 = b,$$

and this implies that all inequalities are equalities so that $b = v$, and $r_x = |L|$ whenever $x \notin L$. $\square$

There are several methods to construct (symmetric) designs. In the next two sections we will study two of them: one comes from "difference sets" in abelian groups, and the other from "finite geometries." The third important construction, which arises from Hadamard matrices, will be described in Chap. 14 (see Theorem 15.6).

13.3 Difference sets

Let $\mathbb{Z}_v$ be an additive abelian group of integers modulo v. We can look at $\mathbb{Z}_v$ as the set of integers $\{0, 1, \ldots, v - 1\}$ where the sum is modulo v.

Definition 13.5. Let $2 \leqslant k < v$ and $\lambda \geqslant 1$. A (v, k, λ) *difference set* is a k-element subset $D = \{d_1, d_2, \ldots, d_k\} \subseteq \mathbb{Z}_v$ such that the collection of values $d_i - d_j$ $(i \neq j)$ contains every element in $\mathbb{Z}_v - \{0\}$ exactly λ times.

Since the number of pairs (i, j) with $i \neq j$ equals $k(k - 1)$ and these give each of the $v - 1$ nonzero elements λ times as a difference, it follows that

$$\lambda(v - 1) = k(k - 1). \tag{13.3}$$

If D is a difference set, we call the set

$$a + D = \{a + d_1, a + d_2, \ldots, a + d_k\}$$

a *translate* of D. Notice that our assumption $k < v$ together with (13.3) implies that all the translates of a difference set are different. Indeed, if $a + D = D$ for some $a \neq 0$, then there is a permutation π of $\{1, \ldots, k\}$ so that $\pi(i) \neq i$ and $a + d_i = d_{\pi(i)}$ for all i. Hence, a can be expressed as a difference $d_{\pi(i)} - d_i$ in k ways; but $\lambda < k$ by (13.3) and our assumption that $k < v$.

Theorem 13.6. *If $D = \{d_1, d_2, \ldots, d_k\}$ is a (v, k, λ) difference set then the translates*

$$D, 1 + D, \ldots, (v - 1) + D$$

are the blocks of a symmetric (v, k, λ) design.

Proof. We have v blocks over v points. Since, clearly, every one of the translates contains k points, it is sufficient to show that every pair of points is contained in exactly λ blocks. Let $x, y \in \mathbb{Z}_v$, $x \neq y$. Suppose that $x, y \in a + D$ for some $a \in \mathbb{Z}_v$. Then $x = a + d_i$ and $y = a + d_j$ for some pair $i \neq j$. Also, we have $d_i - d_j = x - y = d$. Now, there are exactly λ pairs $i \neq j$ such that $d_i - d_j = d$, and for each such pairs, there is exactly one a for which $x, y \in a + D$, namely, $a = x - d_i = y - d_j$. $\quad\square$

Let us now describe one construction of difference sets. *Squares* (or *quadratic residues*) in $\mathbb{Z}_v$ are the elements a^2 for $a \in \mathbb{Z}_v$.

Theorem 13.7. *If v is a prime power and $v \equiv 3 \,(\mathrm{mod}\, 4)$, then the nonzero squares in $\mathbb{Z}_v$ form a (v, k, λ) difference set with $k = (v - 1)/2$ and $\lambda = (v - 3)/4$.*

The condition $v \equiv 3 \pmod 4$ is only used to ensure that -1 is not a square in $\mathbb{Z}_v$, i.e., that $-1 \not\equiv a^2 \pmod v$ for all $a \in \mathbb{Z}_v$. This fact follows from elementary group theory, and we omit its proof here.

Proof. Let D be the set of all nonzero squares, and $k = |D|$. First, observe that $k = (v - 1)/2$. Indeed, the nonzero squares in $\mathbb{Z}_v$ are the elements a^2 for

$a \in \mathbb{Z}_v - \{0\}$. But for every such a the equation $x^2 = a^2$ has two different solutions $x = \pm a$. So, every pair $(+a, -a)$ gives rise to only one square. This means that exactly half of the nonzero elements in $\mathbb{Z}_v$ are squares, and hence $k = (v-1)/2$.

By the remark above, -1 is not a square in $\mathbb{Z}_v$. Hence, if S is the set of all nonzero squares then $-S = \{-s : s \in S\}$ is exactly the set of nonsquares. For any $s \in S$, the pair $(x, y) \in S \times S$ satisfies the equation $x - y = 1$ if and only if the pair $(sx, sy) \in S \times S$ satisfies the equation $sx - sy = s$, or equivalently, if and only if the pair $(sy, sx) \in S \times S$ satisfies the equation $sy - sx = -s$. This shows that all nonzero squares $s \in S$ and all nonsquares $-s \in -S$ have the same number λ of representations as a difference of two nonzero squares. We can compute λ from the equation (13.3), which gives $\lambda = k(k-1)/(v-1) = (v-3)/4$.   □

13.4 Projective planes

Let $\mathcal{L} \subseteq 2^X$ be a linear space with $|\mathcal{L}| = b$ and $|X| = v$. By Theorem 13.4, $b \geqslant v$. In this section we will consider linear spaces with $b = v$ and with an additional requirement that every line has the same number, say $q + 1$, of points. Then $\mathcal{L}$ turns into a symmetric (v, k, λ) design with $\lambda = 1$ and $k = q + 1$. Such a design is known as a *projective plane of order q*. (The reason for taking the block size of the form $k = q + 1$ is that, for any prime power q, such a design has a very transparent construction using the Galois field $\mathbb{F}_q$; we will give this construction below.) By Theorem 13.2, we have $v = b = q^2 + q + 1$.

Projective planes have many applications. They are particularly useful to show that some bounds in Extremal Set Theory are optimal (cf., for example, Lemma 2.1 and Theorems 7.1, 12.3). Due to their importance, projective planes deserve a separate definition.

Definition 13.8. A *projective plane* of order q consists of a set X of $q^2 + q + 1$ elements called *points*, and a family $\mathcal{L}$ of subsets of X called *lines*, having the following properties:

(P1) Every line has $q + 1$ points.
(P2) Any two points lie on a unique line.

The only possible projective plane of order $q = 1$ is a triangle. For $q = 2$, the unique plane of order q is the famous *Fano plane* (see Fig. 13.1).

Additional properties of projective planes are summarized as follows:

Proposition 13.9. *A projective plane of order q has the properties*

(P3) *Any point lies on $q + 1$ lines.*
(P4) *There are $q^2 + q + 1$ lines.*
(P5) *Any two lines meet in a unique point.*

Fig. 13.1. The Fano plane with 7 lines and 3 points on a line

Proof. (P3) Take a point x. There are $q(q+1)$ points different from x; each line through x contains q further points, and there are no overlaps between these lines (apart from x). So, there must be $q+1$ lines through x.

(P4) Counting in two ways the pairs (x, L) with $x \in L$, we obtain $|\mathcal{L}| \cdot (q+1) = (q^2 + q + 1) \cdot (q+1)$, so $|\mathcal{L}| = q^2 + q + 1$.

(P5) Let L_1 and L_2 be lines, and x a point of L_1. Then the $q+1$ points of L_2 are joined to x by different lines; since there are only $q+1$ lines through x, they all meet L_2 in a point; in particular, L_1 meets L_2. □

A nice property of projective planes is their duality. Let $(X, \mathcal{L})$ be a projective plane of order q, and let $M = (m_{x,L})$ be its incidence matrix. That is, M is n by n 0-1 matrix, the rows and columns of which correspond to points and lines, and $m_{x,L} = 1$ iff $x \in L$. Each row and column of M has exactly $q+1$ 1's, and any two rows and any two columns share exactly one 1. Thus, the transpose of M leaves the matrix unchanged.

13.4.1 The construction

The standard construction for projective planes of any prime order $q \geqslant 2$ is the following.

Let V be the set of all vectors (x_0, x_1, x_2) of elements of $\mathbb{F}_q$, where x_0, x_1, x_2 are not all zero. We identify the vectors that can be obtained from each other by multiplying by a nonzero element of $\mathbb{F}_q$, and call each such collection of vectors a *point*. That is, points of our plane are *sets*

$$[x_0, x_1, x_2] = \{(cx_0, cx_1, cx_2) : c \in \mathbb{F}_q, \ c \neq 0\}$$

of $q-1$ vectors in V. There are $(q^3 - 1)/(q - 1) = q^2 + q + 1$ such sets, and hence, so many points. The *line* $L(a_0, a_1, a_2)$, where $(a_0, a_1, a_2) \in V$, is defined to be the set of all those points $[x_0, x_1, x_2]$ for which

$$a_0 x_0 + a_1 x_1 + a_2 x_2 = 0. \tag{13.4}$$

How many points does such a line $L(a_0, a_1, a_2)$ have?

Because $(a_0, a_1, a_2) \in V$, this vector has at least one nonzero component; say $a_0 \neq 0$. Therefore, the equation (13.4) has exactly $q^2 - 1$ solutions $(x_0, x_1, x_2) \in V$: for arbitrary x_1, x_2, not both zero, this equation uniquely determines x_0. Since each $[x_0, x_1, x_2]$ consists of $q-1$ vectors, there are exactly $(q^2 - 1)/(q - 1) = q + 1$ points $[x_0, x_1, x_2]$ satisfying (13.4). In other words: there are exactly $q + 1$ points on each line. So, it remains to verify that *any two points lie on a unique line.*

To show this, let $[x_0, x_1, x_2]$ and $[y_0, y_1, y_2]$ be two distinct points. How many lines contain both these points? For each such line $L(a_0, a_1, a_2)$,

$$a_0 x_0 + a_1 x_1 + a_2 x_2 = 0,$$
$$a_0 y_0 + a_1 y_1 + a_2 y_2 = 0.$$

Without loss of generality $x_0 \neq 0$. Then $a_0 = -a_1 x_1/x_0 - a_2 x_2/x_0$, and we can replace the second equation by

$$a_1 \left(y_1 - \frac{y_0}{x_0} x_1 \right) + a_2 \left(y_2 - \frac{y_0}{x_0} x_2 \right) = 0. \tag{13.5}$$

If

$$y_1 - \frac{y_0}{x_0} x_1 = y_2 - \frac{y_0}{x_0} x_2 = 0$$

then $(y_0, y_1, y_2) = (cx_0, cx_1, cx_2)$ with $c = y_0/x_0$, and hence, $[y_0, y_1, y_2] = [x_0, x_1, x_2]$, which is impossible since we consider distinct points. Therefore, at least one of them, say $y_1 - (y_0/x_0)x_1$, is nonzero. Then for arbitrary nonzero a_2, both a_1 and a_0 are uniquely determined by (13.5) and the first equation; and if (a_0, a_1, a_2) is a solution then (ca_0, ca_1, ca_2) for $c \neq 0$ are all the solutions. Consequently, every two different points $[x_0, x_1, x_2]$ and $[y_0, y_1, y_2]$ are contained in a unique line, as desired.

The constructed projective plane is usually denoted by $PG(2, q)$.

13.4.2 Bruen's theorem

A *blocking set* in a projective plane is a set of points which intersects every line. The smallest (with respect to the set–theoretic inclusion) blocking sets are just the lines (show this!). This is why blocking sets containing a line are called *trivial.*

What can be said about the size of non-trivial blocking sets? Lines themselves have $q + 1$ points, and these are trivial blocking sets. Can we find a non-trivial blocking set with, say $q + 2$ or $q + 3$ points? The fundamental result due to Bruen (1970) says that any non-trivial blocking set in a projective plane of order q must have at least $q + \sqrt{q} + 1$ points, and this lower bound is tight when q is a square (that is, for square q blocking sets of this size exist). For the prime order q, Blokhuis (1995) improved Bruen's bound to $3(q + 1)/2$ (which is also optimal).

Theorem 13.10 (Bruen 1970). *Let B be a non-trivial blocking set in a projective plane of order q. Then $|B| \geqslant q + \sqrt{q} + 1$.*

This result captures very interesting property of projective planes: if we take *any* set of at most $q + \sqrt{q}$ points, then either it contains a line or avoids a line (the third is impossible!).

Proof. The proof employs the double counting argument (introduced in Chap. 1). Let $m = |B|$ and consider the following set of triples of points

$$J = \{(y, x, x') : \text{there is a line } L \text{ such that } y \in L - B \text{ and } x, x' \in L \cap B\}.$$

Counting the triples in two different ways, we will prove that

$$|J| \geqslant m(m - 1)(2q + 1 - m) \tag{13.6}$$

and

$$|J| \leqslant (q^2 + q + 1 - m)(m - q - 1)(m - q). \tag{13.7}$$

Before we prove these two estimates, let us see how they imply the theorem. From (13.6) and (13.7) we obtain (after simple calculations)

$$\begin{aligned}
0 &\leqslant m^2 - 2q(q + 1)^2 m + q^2 + q + 1 \\
&= [m - (q + \sqrt{q} + 1)] \cdot [m - (q - \sqrt{q} + 1)].
\end{aligned}$$

If the first term is negative, then the second should also be negative, meaning that $|B| = m < q - \sqrt{q} + 1 \leqslant q$, which is impossible since no set with q points can intersect all the lines. Thus, the first term must be positive, which may happen only if $m \geqslant q + \sqrt{q} + 1$, as desired.

Let us now turn to the proof of estimates (13.6) and (13.7).

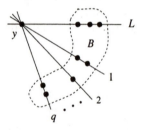

Fig. 13.2. A typical situation for the upper and lower bounds

Lower bound (13.6). Take a pair (x, x') of distinct points in B. These points belong to a unique line L. Since the blocking set B is non-trivial, there is a point $y \in L - B$ (see Fig 13.2). Each of the remaining q lines, containing y, must have a point from B, and these points must be different for different lines (since these q lines intersect only in one point y). Thus $m \geqslant |B \cap L| + q$, and hence

$$|L - B| \geqslant q + 1 - |B \cap L| \geqslant 2q + 1 - m.$$

Thus, for each pair (x, x') of distinct points in B there are at least $2q + 1 - m$ possibilities to choose a point y so that $(y, x, x') \in J$. Since there are exactly $m(m - 1)$ possibilities to choose a (ordered!) pair (x, x') of different points in B, we are done.

Upper bound (13.7). Fix a point $y \notin B$ and let $L_1, \ldots, L_{q+1}$ be the lines containing this point. Set $l_i \rightleftharpoons |B \cap L_i|$. Since these lines cover all the points, and are mutually disjoint outside the point y, we have

$$l_1 + \cdots + l_{q+1} = m.$$

For each line L_i we have $l_i(l_i - 1)$ possibilities to choose a pair of points (x, x') with $x, x' \in B \cap L_i$. Thus, for each $y \notin B$, the number of pairs (x, x') with $(y, x, x') \in J$, is at most

$$\sum_{i=1}^{q+1} l_i(l_i - 1) = \sum_{i=1}^{q+1} (l_i^2 - 2l_i + 1) + \sum_{i=1}^{q+1} l_i - (q+1)$$

$$= \sum_{i=1}^{q+1} (l_i - 1)^2 + m - (q+1)$$

$$\leqslant \left(\sum_{i=1}^{q+1} (l_i - 1) \right)^2 + m - q - 1$$

$$= \left(\sum_{i=1}^{q+1} l_i - (q+1) \right)^2 + m - q - 1$$

$$= (m - q - 1)^2 + m - q - 1 = (m - q - 1)(m - q).$$

Since there are only $q^2 + q + 1 - m$ possibilities to choose such a point y, the desired upper bound in (13.7) follows.   □

13.5 Resolvable designs

Suppose $\mathcal{D}$ is a (v, k, λ) design over a set X. A *parallel class* in $\mathcal{D}$ is a subset of disjoint blocks from $\mathcal{D}$ whose union is X. Observe that a parallel class contains v/k blocks, and every point of X appears in exactly one of these blocks. Moreover, by (13.1) and (13.2), we have

$$r = |\mathcal{D}| \, v/k = \lambda(v-1)/(k-1)$$

such classes, where r is the replication number of $\mathcal{D}$ (the number of blocks containing a given point). A partition of $\mathcal{D}$ into r parallel classes is called a *resolution*, and a design is said to be *resolvable* if it has at least one resolution.

Let us consider the following example from Anderson and Honkala (1997). We have a football league of $2n$ teams and each team plays exactly once against every other team. We wish to arrange the league schedule so that all the matches are played during $2n - 1$ days, and on each day every team plays one match. Is this possible?

What we are looking for is a resolvable $(2n, 2, 1)$ design. For convenience, let our ground set (of teams) be $X = \{*, 1, \ldots, 2n - 1\}$, where $*$ is some symbol different from $1, \ldots, 2n - 1$. Since by (13.1), the replication number equals

$$r = \lambda(v - 1)/(k - 1) = 1 \cdot (2n - 1)/(2 - 1) = 2n - 1,$$

we have to show how to partite the collection $\mathcal{D}$ of all 2-element subsets of X into $2n - 1$ parallel classes $\mathcal{D}_1, \ldots, \mathcal{D}_{2n-1}$; the ith class $\mathcal{D}_i$ gives us the set of matches played at the ith day.

Define $\{i, *\} \in \mathcal{D}_i$ for all $i \in X - \{*\}$, and $\{a, b\} \in \mathcal{D}_i$, if

$$a + b \equiv 2i \pmod{2n - 1}$$

for $a, b \in X - \{*\}$. Since $2n - 1$ is odd, each 2-element subset of X belongs to a unique $\mathcal{D}_i$; and the unique block in $\mathcal{D}_i$ containing an element $a \in X$ is $\{a, b\}$ where $b \equiv 2i - a \pmod{2n - 1}$ if $a \neq i$, and $\{i, *\}$ if $a = i$.

13.5.1 Affine planes

An *affine plane* $AG(2, q)$ of order q is a $(q^2, q, 1)$ design. By (13.1), each point of this plane belongs to $r = (q^2 - 1)/(q - 1) = q + 1$ lines, and by (13.2), we have $b = vr/k = q^2 + q$ lines altogether. Put otherwise, an affine plane of order q has q^2 points and satisfies the following conditions:

(A1) every line has q points;
(A2) any two points lie on a unique line;
(A3) any point lies on $q + 1$ lines;
(A4) there are $q^2 + q$ lines.

Hence, the main difference from projective planes is that now we can have "parallel" lines, i.e., lines which do not meet each other.

There are two basic constructions of affine planes.

Construction 1. An affine plane can be obtained from a projective plane by removing any one of its lines. Let $(X, \mathcal{L})$ be a projective plane of order q. Fix one of its lines $L_0 \in L$ and consider the design $(X', \mathcal{L}')$ where

$$X' = X - L_0$$

and

$$\mathcal{L}' = \{L - L_0 : L \in \mathcal{L}, L \neq L_0\}$$

(see Fig. 13.3). It is easy to verify (Exercise 13.10) that the obtained design $(X', \mathcal{L}')$ is an affine plane of order q. The line L_0 is called the *line at infinity*. For each line $L' \in \mathcal{L}'$ of the affine plane there is a unique point $x \in L_0$ such that $L' \cup \{x\} \in \mathcal{L}$; this point is called the *infinite point* of L'.

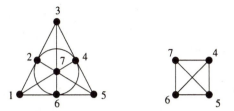

Fig. 13.3. Construction 1 applied to the Fano plane; $L_0 = \{1, 2, 3\}$ is the removed line (the line at infinity).

Construction 2. Let q be a prime power, and consider the set of points $X = \mathbb{F}_q \times \mathbb{F}_q$. Let $\mathcal{D}$ be the set of all blocks of the form

$$L(a, b) \rightleftharpoons \{(x, y) \in X \,:\, y = ax + b\}$$

and

$$L(c) \rightleftharpoons \{(c, y) \,:\, y \in \mathbb{F}_q\},$$

where $a, b, c \in \mathbb{F}_q$. We will show that $\mathcal{D}$ is a $(q^2, q, 1)$ design. Clearly, there are q^2 points and each block contains exactly q of them. Hence, we only need to show that every pair of points $(x_1, y_1), (x_2, y_2) \in \mathbb{F}_q \times \mathbb{F}_q$ is contained in a unique block. If $x_1 = x_2$ then the unique block containing this pair is $L(x_1)$. If $x_1 \neq x_2$ then the system of equations $y_1 = ax_1 + b$, $y_2 = ax_2 + b$ has a unique solution (a, b); hence, the unique block containing that pair is $L(a, b)$, and we are done.

It can be shown that affine planes are resolvable designs (see Exercise 13.11): the parallel classes are $\{L - \{x\} \,:\, x \in L\}$ for $x \in L_0$, in Construction 1, and $\{L(c) \,:\, c \in \mathbb{F}_q\}$ together with $\{L(a, b) \,:\, b \in \mathbb{F}_q\}$ for all $a \in \mathbb{F}_q$, in Construction 2.

13.5.2 Blocking sets in affine planes

A *blocking set* in an affine plane is a set of points which intersects every line. What can be said about the size of these sets?

If q is the order of our affine plane then we can get a blocking set of size $2q - 1$ just by taking a union of any two non-parallel lines. Indeed, take any two non-parallel (i.e., intersecting) lines L_1 and L_2, and let $B = L_1 \cup L_2$. Then $|B| = |L_1| + |L_2| - 1 = 2q - 1$. If B avoided some line L then all three lines L_1, L_2 and L would have the same infinite point (because all the lines of the corresponding projective plane intersect); but then lines L_1 and L_2 would be parallel, a contradiction. Hence, B is a blocking set.

Are there blocking sets with less than $2q - 1$ points? The negative answer was given by Jamison (1977) and Brower and Schrijev (1978). In the case when q is a prime number, this fact can be derived from a classical result, known as the Chevalley–Warning theorem about the number of solutions of a system of multivariate polynomials over a finite field. Its proof is very simple, and we present one from Alon (1995). The *degree* of a multivariate polynomial $f(x_1, \ldots, x_m)$ is the maximum exponent of any variable when f is written as a sum of monomials (i.e., products of the variables).

Theorem 13.11. *Let p be a prime. For $j = 1, \ldots, n$ let $f_j(x_1, \ldots, x_m)$ be a polynomial of degree d_j over $\mathbb{F}_p$, where p is a prime. If $d_1 + \cdots + d_n \leqslant m - 1$ then the number N of common zeros of $f_1, \ldots, f_n$ is divisible by p. In particular, if there is one common zero, then there is another one.*

Proof. By Fermat's Little Theorem (see Exercise 1.12), $a^{p-1} \equiv 1 \pmod{p}$ for all $a \in \mathbb{F}_p$, $a \neq 0$. Hence (in $\mathbb{F}_p$),

$$N = \sum_{x_1, \ldots, x_m \in \mathbb{F}_p} \prod_{j=1}^{n} \left(1 - f_j(x_1, \ldots, x_m)^{p-1} \right). \tag{13.8}$$

By expanding the right-hand side we get a linear combination of monomials of the form

$$\prod_{i=1}^{m} x_i^{t_i} \quad \text{with} \quad \sum_{i=1}^{m} t_i \leqslant (p-1) \sum_{j=1}^{n} d_i < (p-1)m.$$

Hence, in each such monomial there is an i with $t_i < p - 1$. But then (see Exercise 13.12)

$$\sum_{x_i \in \mathbb{F}_p} x_i^{t_i} = 0 \quad (\text{in } \mathbb{F}_p),$$

implying that the contribution of each monomial to the sum (13.8) is 0 modulo p, completing the proof of the theorem □

Theorem 13.12 (Jamison 1977; Brower–Schrijev 1978). *Let q be a prime. If B is a blocking set in an affine plane of order q then $|B| \geqslant 2q - 1$.*

Proof (due to Alon 1995). Since q is a prime, our plane is the 2-dimensional affine space over the field $\mathbb{F}_q$ (see Construction 2 above). Suppose B is a blocking set, i.e., intersects all the lines of this space. We may assume that $(0, 0) \in B$, and define $A = B - \{(0, 0)\}$. Then A intersects all lines not through $(0, 0)$.

   A line not through $(0, 0)$ has the form $L(a', b')$ for some $b' \neq 0$ or the form $L(c)$ for some $c \neq 0$. Therefore, each such line is determined by an equation $ax + by = 1$ for some $a, b \in \mathbb{F}_q$, not both zero ($a = -a'/b'$, $b = 1/b'$ in the first case, and $a = 1/c$, $b = 0$ in the second case). Hence, for every $(0, 0) \neq (a, b) \in \mathbb{F}_q^2$ there exists a point $(u, v) \in A$ such that $au + bv = 1$. Therefore, if we let

$$f(x, y) = \prod_{(u, v) \in A} (1 - ux - vy)$$

then $f(a, b) = 0$ for all $(a, b) \neq (0, 0)$, and $f(0, 0) = 1$. Consider the following polynomial equation in the $2(q - 1)$ variables x_i, y_i, $1 \leqslant i \leqslant q - 1$:

$$\sum_{i=1}^{q-1} f(x_i, y_i) - q + 1 = 0.$$

Obviously, the only zero of this equation is the trivial solution $x_i = y_i = 0$ for all $i = 1, \ldots, q - 1$. By the Chevalley–Waring theorem (see Theorem 13.11), this implies that the degree of the above polynomial, which is $|A|$, is at least the number of variables, which is $2(q - 1)$. Hence

$$|B| = |A| + 1 \geqslant 2(q - 1) + 1 = 2q - 1,$$

completing the proof of the theorem. $\square$

Exercises

13.1. Let $\mathcal{D} \subseteq 2^X$ be a (v, k, λ) design with b blocks, and let r be its replication number (i.e., each element occurs in r blocks). Prove that its complement $\overline{\mathcal{D}} = \{X - B : B \in \mathcal{D}\}$ is a $(v, v - k, b - 2r + \lambda)$ design provided that $b - 2r + \lambda > 0$.

Hint: A pair of elements $x \neq y$ is contained in $X - B$ if and only if B contains neither x nor y. The number of blocks of $\mathcal{D}$ containing neither x nor y is $b - 2r + \lambda$ by the principle of inclusion and exclusion.

13.2.⁻ Show that the number b of blocks in a t-(v, k, λ) design is given by $b = \lambda \binom{v}{t} / \binom{k}{t}$. *Hint*: Count in two ways the number of pairs (T, B) where T is a t-element set of points and B is a block.

13.3. Construct a projective plane $\mathrm{PG}(2, q)$ of order $q = 3$.

13.4.⁻ Show that, in a projective plane of order q, its lines are the only blocking sets of size $q + 1$. *Hint*: See Exercise 8.6.

13.5. From the previous example we know that no set of q points intersects all the lines. We can ask the dual question: are there sets of size q that intersect every *non-trivial* blocking set? Show that there are no such sets and that the only sets of size $q + 1$ that intersect every blocking set are the lines.

Sketch (due to Blokhuis): Take any set S of q points, and start with a line L_0 not intersecting this set. Delete one (suitable) point $x \in L_0$ from that line, and add a point $y_i \in L_i - S$ on every other line L_i through the deleted point x. The resulting set $(L_0 - \{x\}) \cup \{y_1, \ldots, y_q\}$ is blocking and is disjoint from S. This blocking set might be trivial (i.e., contain a line), but it can be made non-trivial by deleting some unnecessary point of the line we started with.

13.6 (due to Bruen). Let S be a nontrivial blocking set in a projective plane of order q. Show that:

(i) $|S| \leqslant q^2 - \sqrt{q}$, and
(ii) no line contains more than $|S| - q$ points of S.

> *Hint:* To (i): observe that the complement of S is also a nontrivial blocking set, and apply Bruen's theorem. To (ii): take a line L and a point $x \in L-S$; there are q other lines through x and these lines intersect S; argue that then $|L \cap S| + q \leqslant |S|$.

13.7.⁻ Let S be a set of $q + 2$ points in a projective plane of order q. Prove that every line, that meets S, meets it twice.

13.8. Let S be a set of points in a projective plane of order q. Suppose that no three points of S are colinear (i.e., lie on a line). Prove that then $|S| \leqslant q+1$ if q is odd, and $|S| \leqslant q + 2$ if q is even.

> *Sketch:* Fix a point $x \in S$; for each other point $y \in S$ the pair x, y lies on one of $q + 1$ lines (containing x), and these lines must be different for different y. This proves the odd case. For the even case show that, if $|S| = q + 2$ then every line that meets S, meets it twice.

13.9. Color some q points of a projective plane of order q in red, and the rest in blue. Prove that, for any two different sets $A \neq B$ of red points, there is a set C of q blue points such that $A \cup C$ is a blocking set but $B \cup C$ avoids at least one line. *Hint:* Take a point $x \in A - B$, and show that some two lines L_1 and L_2 meet in the (red) point x and have no more red points; take $C = L_1 - \{x\}$.

13.10.⁻ Take a projective plane of order q, i.e., a design satisfying the conditions (P1)–(P5), and apply the first construction from Sect. 13.5.1 to it. Show that the resulting design satisfies the conditions (A1)–(A4).

13.11.⁻ Consider a $(q^2, q, 1)$ design, i.e., an affine plane of order q. Show that this design is resolvable. More generally, let a *parallel class* be a set of mutually disjoint lines, and show the following:

(i)   each parallel class contains q lines;
(ii) there are $q + 1$ such classes;
(iii) any two lines from different classes meet in a point;
(iv) lines of each parallel class cover the whole point set.

> *Hint:* Each parallel class contains exactly one line through any point; so, the $q+1$ lines through a point x contain representatives of all the classes.

13.12. Let $p \geqslant 2$ be a prime and consider the field $\mathbb{F}_p$. From Algebra we know that there is an $a \in \mathbb{F}_p$ such that $\mathbb{F}_p - \{0\} = \{a^i : i = 0, 1, \ldots, p-2\}$. Use this fact together with Fermat's Little Theorem to prove that, for every $t \leqslant p - 2$, $\sum_{x \in \mathbb{F}_p} x^t = 0$ in $\mathbb{F}_p$. *Hint:* $\sum_{i=0}^{n} z^i = (z^{n+1} - 1)/(z - 1)$.

13.13.⁺ Research problem: find a combinatorial proof of Jamison–Brower–Schrijev's theorem about the size of blocking sets for affine planes (in a similar spirit to that of Bruen's theorem for projective planes).

Part III

The Linear Algebra Method

14. The Basic Method

The general frame for the *linear algebra method* in combinatorics is the following: if we want to come up with an upper bound on the size of a set of objects, associate them with elements in a vector space V of relatively low dimension, and show that these elements are linearly independent; hence, we cannot have more objects in our set than the dimension of V.

14.1 The linear algebra background

In this book we will not use the whole power of abstract linear algebra – for our purposes the following two special types of linear spaces will be enough.

Let $\mathbb{F}$ be a field. By a *linear space* (or *vector space*) over the field $\mathbb{F}$ we will mean either the set $\mathbb{F}^n$ of all n-tuples $v = (v_1, \ldots, v_n)$ with $v_i \in \mathbb{F}$, or a set of all n-tuples $(v_1, \ldots, v_n)$, where each $v_i : \mathbb{F}^m \to \mathbb{F}$ is a (multivariate) polynomial over $\mathbb{F}$. We have two basic operations in such a space: the component wise addition $u + v = (u_1 + v_1, \ldots, u_n + v_n)$ and multiplication by scalars $\lambda v = (\lambda v_1, \ldots, \lambda v_n)$, $\lambda \in \mathbb{F}$.

A *linear combination* of the vectors $v_1, \ldots, v_m$ is a vector of the form $\lambda_1 v_1 + \ldots + \lambda_m v_m$ with $\lambda_i \in \mathbb{F}$. A *subspace* of V is a nonempty subset of V, closed under linear combinations. The *span* of $v_1, \ldots, v_m$, denoted by span $\{v_1, \ldots, v_m\}$, is the set of all linear combinations of these vectors. A vector u *depends* on the vectors $v_1, \ldots, v_m$ if $u \in$ span $\{v_1, \ldots, v_m\}$. The vectors $v_1, \ldots, v_m$ are *linearly independent* if none of them is dependent on the rest. Equivalently, the (in)dependence can be defined as follows.

A *linear relation* among the vectors $v_1, \ldots, v_m$ is a linear combination that gives the zero vector:

$$\lambda_1 v_1 + \ldots + \lambda_m v_m = 0.$$

This relation is nontrivial if $\lambda_i \neq 0$ for at least one i. It is easy to see that the vectors $v_1, \ldots, v_m$ are linearly independent if and only if no nontrivial relation exists between them. A *basis* of V is a set of independent vectors which spans V. A fundamental fact in linear algebra says that *any two bases of V have the same cardinality*; this number is called the *dimension* dim V of V.

A further basic fact is the so-called *linear algebra bound* (see any standard linear algebra book for the proof):

Proposition 14.1. *If $v_1, \ldots, v_k$ are linearly independent vectors in a vector space of dimension m then $k \leqslant m$.*

An important operation in vector spaces is the scalar product of two vectors. Given two vectors $u = (u_1, \ldots, u_n)$ and $v = (v_1, \ldots, v_n)$, their *scalar product* $\langle u, v \rangle$ (also called *inner product* and denoted $u \cdot v$) is defined by:

$$\langle u, v \rangle = u \cdot v = u_1 v_1 + \cdots + u_n v_n.$$

Vectors u and v are *orthogonal* if $u \cdot v = 0$. If $U \subseteq V$ is a subspace of V then the *dual* (or *orthogonal complement*) is the subspace

$$U^\perp = \{v \in V : u \cdot v = 0 \text{ for all } u \in U\}.$$

The following useful inequality connects the dimensions of two orthogonal subspaces.

Proposition 14.2. *Let V be a finite dimensional linear space, and $U \subseteq V$ be a subspace. Then $\dim U + \dim U^\perp = \dim V$.*

If $A = \{a_{ij}\}$ is an $m \times n$ matrix over some field $\mathbb{F}$ and v is a vector in $\mathbb{F}^m$, then $v \cdot A$ is the vector in $\mathbb{F}^n$ whose jth coordinate is the scalar product of v with the jth column of A. Thus, the rows of A are linearly independent if and only if $v \cdot A \neq 0$ for all $v \neq 0$. Similarly, if $u \in \mathbb{F}^n$, then $A \cdot u$ is the vector in $\mathbb{F}^m$ whose ith coordinate is the scalar product of u with the ith row of A. When writing $v \cdot A$ or $A \cdot u$ we always assume that the dimensions "match", and do not bother distinguishing between row and column vectors.

The *column rank* of a matrix A is the dimension of the vector space spanned by its columns. The *row rank* of A is the dimension of the vector space spanned by its rows. One of the first nontrivial results in matrix theory asserts that the *row and column ranks are equal*; this common value is the rank of A, denoted by $\mathrm{rk}(A)$ or $\mathrm{rk}A$ (or $\mathrm{rk}_\mathbb{F}(A)$ if the field needs to be specified). The following inequalities hold for the rank:

$$\mathrm{rk}(A) - \mathrm{rk}(B) \leqslant \mathrm{rk}(A + B) \leqslant \mathrm{rk}(A) + \mathrm{rk}(B); \tag{14.1}$$

$$\mathrm{rk}(AB) \leqslant \min\{\mathrm{rk}(A), \mathrm{rk}(B)\}. \tag{14.2}$$

If $x = (x_1, \ldots, x_n)$ denotes the vector of indeterminates, and $b = (b_1, \ldots, b_m) \in \mathbb{F}^m$, then the matrix equation $A \cdot x = b$ is a concise form of writing a system of m linear equations in variables $x_1, \ldots, x_n$:

$$a_{i1} x_1 + a_{i2} x_2 + \cdots + a_{in} x_n = b_i \qquad (i = 1, \ldots, m).$$

We have the following useful criterion for such a system being solvable. Let $a_1, \ldots, a_n \in \mathbb{F}^m$ denote the columns of A. Observe that $A \cdot x = x_1 a_1 + x_2 a_2 + \cdots + x_n a_n$. It follows that the set $\{A \cdot x : x \in \mathbb{F}^n\}$ is a *columns space* of A, i.e., the set of all vectors spanned by the columns of A. The system $A \cdot x = b$ is thus solvable if and only if $b \in \operatorname{span}\{a_1, \ldots, a_n\}$, or equivalently, if and only if $\operatorname{rk}A = \operatorname{rk}[A|b]$, where $[A|b]$ denotes the $m \times (n+1)$ matrix obtained by adding the column b to A. A system $A \cdot x = b$ is *homogeneous* if $b = 0$. The set of solutions of $A \cdot x = 0$ is clearly a subspace (of all vectors, orthogonal to all the rows of A) and, by Proposition 14.2, its dimension is $n - \operatorname{rk}(A)$. We summarize this important result:

Proposition 14.3. *Let A be an $m \times n$ matrix over a field $\mathbb{F}$. Then the set of solutions of the system of linear equations $A \cdot x = 0$ is a linear subspace of dimension $n - \operatorname{rk}(A)$ of the space $\mathbb{F}^n$.*

The *norm* (or *length*) of a vector $v = (v_1, \ldots, v_n)$ is the number

$$\|v\| = \langle v, v \rangle^{1/2} = \left(\sum_{i=1}^{n} v_i^2 \right)^{1/2}.$$

The following basic inequality, known as the *Cauchy–Schwarz inequality*, estimates the scalar product of two vectors in terms of their norms (we have already used it in previous sections; now we will prove it):

Proposition 14.4. *For any real vectors $u, v \in \mathbb{R}^n$,*

$$\langle u, v \rangle \leqslant \|u\| \cdot \|v\|.$$

That is,

$$\left(\sum_{i=1}^{n} v_i u_i \right)^2 \leqslant \left(\sum_{i=1}^{n} v_i^2 \right) \left(\sum_{i=1}^{n} u_i^2 \right). \tag{14.3}$$

Proof. For any constant $\lambda \in \mathbb{R}$ we have

$$0 \leqslant \langle \lambda u - v, \lambda u - v \rangle = \langle \lambda u, \lambda u - v \rangle - \langle v, \lambda u - v \rangle$$
$$= \lambda^2 \langle u, u \rangle - 2\lambda \langle u, v \rangle + \langle v, v \rangle.$$

Substituting $\lambda = \frac{\langle u, v \rangle}{\langle u, u \rangle}$ we get

$$0 \leqslant \frac{\langle u, v \rangle^2}{\langle u, u \rangle^2} \langle u, u \rangle - 2\frac{\langle u, v \rangle^2}{\langle u, u \rangle} + \langle v, v \rangle = \langle v, v \rangle - \frac{\langle u, v \rangle^2}{\langle u, u \rangle}$$

Rearranging the last inequality, we get $\langle u, v \rangle^2 \leqslant \langle u, u \rangle \langle v, v \rangle = \|u\| \cdot \|v\|$. $\quad\square$

If a linear space V has dimension n, then we can find many subsets of n independent vectors in it (the bases for V). But not every n-element subset of V is such. If, however, some subset of V has this property (that every n vectors of it are linearly independent) then one says that this set consists of vectors in "general position." Namely, we say that a set of vectors $W \subseteq V$ is in *general position*, if any $n = \dim V$ of the vectors of W are linearly independent.

In cases when $V = \mathbb{F}^n$ for some field $\mathbb{F}$, there is a particularly elegant and explicit choice of $|\mathbb{F}|$ points in general position, using the so-called *moment curve*. This curve is defined as the range of the function

$$\mathbb{F} \ni a \mapsto m(a) \rightleftharpoons (1, a, a^2, \ldots, a^{n-1}) \in \mathbb{F}^n.$$

Lemma 14.5. *Let $V = \mathbb{F}^n$ and $|\mathbb{F}| \geqslant n$. Then the set of $|\mathbb{F}|$ vectors $m(a)$, $a \in \mathbb{F}$, is in general position.*

Proof. For n distinct elements $a_1, \ldots, a_n \in \mathbb{F}$, consider the determinant of the corresponding $n \times n$ matrix with rows $m(a_i)$. This determinant is known as a Vandermonde determinant (cf. Exercise 14.11); its value is

$$\prod_{1 \leqslant i < j \leqslant n} (a_j - a_i) \neq 0.$$

Therefore the rows $m(a_i)$ are linearly independent (consult Exercise 14.12 for this last conclusion). □

Let us now look how the linear algebra argument works in concrete situations.

14.2 Spaces of incidence vectors

Suppose we are given a family $\mathcal{F}$ of sets satisfying some conditions. We want to know how many sets such a family can have. In some situations it is sufficient to associate sets to their incidence vectors and show that these vectors are linearly independent.

14.2.1 Fisher's inequality

Suppose that each two sets of our family share the same number of elements. How large can such a family be? The answer is given by a fundamental result of design theory – known as *Fisher's inequality*.

Theorem 14.6 (Fisher's inequality). *Let $A_1, \ldots, A_m$ be distinct subsets of $\{1, \ldots, n\}$ such that $|A_i \cap A_j| = k$ for some fixed $1 \leqslant k \leqslant n$ and every $i \neq j$. Then $m \leqslant n$.*

Proof. Let $v_1, \ldots, v_m \in \{0, 1\}^n$ be incidence vectors of $A_1, \ldots, A_m$. By the linear algebra bound (Proposition 14.1) it is enough to show that these vectors are linearly independent over the reals. Assume the contrary, i.e., that the linear relation $\sum_{i=1}^{m} \lambda_i v_i = 0$ exists, with not all coefficients being zero. Obviously, $\langle v_i, v_j \rangle = |A_i|$ if $j = i$, and $\langle v_i, v_j \rangle = k$ if $j \neq i$. Consequently,

$$0 = \left(\sum_{i=1}^{m} \lambda_i v_i\right)\left(\sum_{j=1}^{m} \lambda_j v_j\right) = \sum_{i=1}^{m} \lambda_i^2 \langle v_i, v_i \rangle + \sum_{1 \leqslant i \neq j \leqslant m} \lambda_i \lambda_j \langle v_i, v_j \rangle$$

$$= \sum_{i=1}^{m} \lambda_i^2 |A_i| + \sum_{1 \leqslant i \neq j \leqslant m} \lambda_i \lambda_j k = \sum_{i=1}^{m} \lambda_i^2 (|A_i| - k) + k \cdot \left(\sum_{i=1}^{m} \lambda_i\right)^2.$$

Clearly, $|A_i| \geqslant k$ for all i and $|A_i| = k$ for at most one i, since otherwise the intersection condition would not be satisfied. But then the right-hand is greater than 0 (because the last sum can vanish only if at least two of the coefficients λ_i are nonzero), a contradiction.   □

This theorem was first proved by the statistician R. A. Fisher in 1940 for the case when $k = 1$ and all sets A_i have the same size (such configurations are known as balanced incomplete block designs). In 1948, de Bruijn and Erdős relaxed the uniformity condition for the sets A_i (see Theorem 13.4). This was generalized by R. C. Bose in 1949, and later by several other authors. But it was the two-page paper of Bose where the linear argument was first applied to solve a combinatorial problem. The general version, stated above, was first proved by Majumdar (1953); the proof we presented is a variation of a simplified argument found in Babai and Frankl (1992).

14.2.2 Inclusion matrices

In Chap. 11 we have proved a fundamental result concerning the Vapnik–Chervonenkis dimension (see Theorem 11.1). Let us restate this result in terms of set systems.

Let $\mathcal{F}$ be a family of subsets of an n-element set X. Such a family is (n, k)-*dense* if there is a subset $Y \subseteq X$ of cardinality $|Y| = k$ such that every subset Z of Y occurs as $Z = E \cap Y$ for some $E \in \mathcal{F}$.

Theorem 14.7. *Let $\mathcal{F}$ be a family of subsets of an n-element set X. If $\mathcal{F}$ has more than $\sum_{i=0}^{k-1} \binom{n}{i}$ members, then $\mathcal{F}$ is (n, k)-dense.*

We presented two proofs of this result. One was by induction on n and k, whereas the other was based on a striking observation made by Alon (1983) and Frankl (1983) that for results like this, it is enough to consider only those families which are downwards closed. Here we will give one more proof, based on linear independence. The advantage of this argument is that it can be easily modified to yield a similar result for uniform families (see Exercise 14.14).

Proof (due to Frankl and Pach 1983). Let $Y_1, Y_2, \ldots, Y_r$, $r = \sum_{i=0}^{k-1} \binom{n}{i}$, be an enumeration of all subsets of X of size at most $k - 1$, and let $E_1, E_2, \ldots, E_s$ denote the members of $\mathcal{F}$. Define an $s \times r$ 0-1 matrix $M = (m_{ij})$ by

$$m_{ij} = 1 \text{ if and only if } E_i \supseteq Y_j.$$

Assume that $s > r$. Then the rows of M cannot be linearly independent. That is, there exist real coefficients $\lambda_1, \lambda_2, \ldots, \lambda_s$, not all zero, such that

$$g(Y_j) = \sum_{i=1}^{s} \lambda_i m_{ij} = \sum_{E_i \supseteq Y_j} \lambda_i = 0 \qquad (14.4)$$

for all $j = 1, \ldots, r$. Let $Y \subseteq X$ be a set of minimum cardinality such that $g(Y) \neq 0$. (Note that such a set exists, because not all of the coefficients λ_i are zero.) By (14.4), $|Y| \geqslant k$. On the other hand, using the inclusion-exclusion principle, it can be shown (see Exercise 3.6) that for any subset $Z \subseteq Y$,

$$\sum_{E_i \cap Y = Z} \lambda_i = \sum_{Z \subseteq S \subseteq Y} (-1)^{|S-Z|} g(S).$$

By the minimal choice of Y we see that all the terms on the right-hand sum vanish, except for the term corresponding to $S = Y$, and hence, both of the sums are equal to $\pm g(Y)$, which is non-zero. The sum on the left-hand side being non-zero means, in particular, that its range cannot be empty. This yields, in particular, that for every $Z \subseteq Y$ there exists $E_i \in \mathcal{F}$ such that $E_i \cap Y = Z$. □

Our next example concerns the approximation of boolean functions by low degree polynomials over $\mathbb{F}_2$. The celebrated result, due to Razborov (1987), says that the majority function cannot be computed by constant depth circuits of polynomial size, even if we allow unbounded fanin And, Or and Parity functions as gates. This result was obtained in two steps: (i) show that functions, computable by small circuits, can be approximated by low degree polynomials, and (ii) prove that the majority function is hard to approximate by such polynomials. The proof of (i) is probabilistic, and we will present it later (see Lemma 20.5). The proof of (ii) employs the linear algebra argument, and we present it below.

The k-threshold function is a boolean function T_k^n which outputs 1 if and only if at least k of the bits in the input vector are 1.

Lemma 14.8 (Razborov 1987). *Let $n/2 \leqslant k \leqslant n$. Every polynomial of degree at most $2k - n - 1$ over $\mathbb{F}_2$ differs from the k-threshold function on at least $\binom{n}{k}$ inputs.*

Proof (Lovász–Shmoys–Tardos 1995). Let g be a polynomial of degree $d \leqslant 2k - n - 1$ over $\mathbb{F}_2$ and let U denote the set of all vectors where it differs from T_k^n. Let A denote the set of all 0-1 vectors of length n containing exactly k 1's.

Consider the 0-1 matrix $M = (m_{a,u})$ whose rows are indexed by the members of A, columns are indexed by the members of U, and $m_{a,u} = 1$ if and only if $a \geqslant u$. For two vectors a and b we denote by $a \wedge b$ the coordinatewise And of these vectors. Our goal is to prove that the columns of M span the whole linear space; since the dimension of this space is $|A| = \binom{n}{k}$, this will mean that we must have $|U| \geqslant \binom{n}{k}$ columns.

The fact that the columns of M span the whole linear space follows directly from the following claim saying that every unit vector lies in the span.

Claim 14.9. *Let $a \in A$ and $U_a = \{u \in U : m_{a,u} = 1\}$. Then*

$$\sum_{u \in U_a} m_{b,u} = \begin{cases} 1 & \text{if } b = a; \\ 0 & \text{if } b \neq a. \end{cases}$$

Proof. By the definition of U_a, we have (all sums are over $\mathbb{F}_2$):

$$\sum_{u \in U_a} m_{b,u} = \sum_{\substack{u \in U \\ u \leqslant a \wedge b}} 1 = \sum_{x \leqslant a \wedge b} (T_k^n(x) + g(x)) = \sum_{x \leqslant a \wedge b} T_k^n(x) + \sum_{x \leqslant a \wedge b} g(x).$$

The second term of this last expression is 0, since $a \wedge b$ has at least $d + 1$ 1's (Exercise 14.16). The first term is also 0 except if $a = b$.

This completes the proof of the claim, and thus, the proof of the lemma. $\square$

14.2.3 Disjointness matrices

Let $k \leqslant n$ be natural numbers, and X be a set of n elements. A *k-disjointness matrix* over X is a 0-1 matrix $D = D(n, k)$ whose rows and columns are labeled by subsets of X of size at most k; the entry $D_{A,B}$ in the A-th row and B-th column is defined by:

$$D_{A,B} = \begin{cases} 0 & \text{if } A \cap B \neq \emptyset, \\ 1 & \text{if } A \cap B = \emptyset. \end{cases}$$

This matrix plays an important role in computational complexity (we will use it in Sects. 15.2.2 and 16.4). Its importance stems from the fact that it has full rank over $\mathbb{F}_2$, i.e., all its $\sum_{i=0}^{k} \binom{n}{i}$ rows are linearly independent.

Theorem 14.10. *The k-disjointness matrix $D = D(n, k)$ has full rank over $\mathbb{F}_2$, that is,*

$$\mathrm{rk}_{\mathbb{F}_2}(D) = \sum_{i=0}^{k} \binom{n}{i}.$$

There are several proofs of this result. Usually, it is derived from more general facts about Möbius inversion or general intersection matrices. Here we present one particularly simple and direct proof due to Razborov (1987).

Proof. Let $N = \sum_{i=0}^{k} \binom{n}{i}$. We must show that the rows of D are linearly independent over $\mathbb{F}_2$, i.e., that for any non-zero vector $\lambda = (\lambda_{I_1}, \lambda_{I_2}, \ldots, \lambda_{I_N})$ in $\mathbb{F}_2^N$ we have $\lambda \cdot D \neq 0$. For this, consider the following polynomial:

$$f(x_1, \ldots, x_n) = \sum_{|I| \leqslant k} \lambda_I \prod_{i \in I} x_i.$$

Since $\lambda \neq 0$, at least one of the coefficients λ_I is nonzero, and we can find some I_0 such that $\lambda_{I_0} \neq 0$ and I_0 is *maximal* in that $\lambda_I = 0$ for all $I \supset I_0$. Assume w.l.o.g. that $I_0 = \{1, \ldots, t\}$, and make in the polynomial f the substitution $x_i := 1$ for all $i \notin I_0$. After this substitution has been made, a *non-zero* polynomial over the first t variables $x_1, \ldots, x_t$ remains such that the term $x_1 x_2 \cdots x_t$ is left untouched (here we use the maximality of I_0). Hence, after the substitution we obtain a polynomial which is 1 for some assignment $(a_1, \ldots, a_t)$ to its variables. But this means that the polynomial f itself takes the value 1 on the assignment $b = (a_1, \ldots, a_t, 1, \ldots, 1)$. Hence,

$$1 = f(b) = \sum_{|I| \leqslant k} \lambda_I \prod_{i \in I} b_i.$$

Let $J_0 = \{i : a_i = 0\}$. Then $|J_0| \leqslant k$ and, moreover, $\prod_{i \in I} b_i = 1$ if and only if $I \cap J_0 = \emptyset$, which is equivalent to $D_{I,J_0} = 1$. Thus,

$$\sum_{|I| \leqslant k} \lambda_I D_{I,J_0} = 1,$$

meaning that the J_0-th coordinate of the vector $\lambda \cdot D$ is non-zero. $\square$

14.3 Spaces of polynomials

In order to apply the linear algebra method, in many situations it is particularly useful to associate sets not to their incidence vectors but to some (multivariate) polynomials $f(x_1, \ldots, x_n)$ and show that these polynomials are linearly independent as a members of the corresponding functions space. This idea, know as the *polynomial technique*, has found many applications. We will present only few of them. All these applications are based on the following simple and powerful lemma connecting algebra to linear algebra.

Lemma 14.11. *For $i = 1, \ldots, m$ let $f_i : \Omega \to \mathbb{F}$ be functions and $v_i \in \Omega$ elements such that*

(a) $f_i(v_i) \neq 0$ *for all $1 \leqslant i \leqslant m$;*

(b) $f_i(v_j) = 0$ *for all $1 \leqslant j < i \leqslant m$.*

Then $f_1, \ldots, f_m$ are linearly independent members of the space $\mathbb{F}^\Omega$.

Proof. By contradiction: Suppose there is a nontrivial linear relation

$$\lambda_1 f_1 + \lambda_2 f_2 + \cdots + \lambda_m f_m = 0$$

between the f_i's. Take the smallest i for which $\lambda_i \neq 0$. Substitute v_i for the variables. By the assumption, all but the ith term vanish. What remains is $\lambda_i f_i(v_i) = 0$, which implies $\lambda_i = 0$ because $f_i(v_i) \neq 0$, a contradiction. $\square$

14.3.1 Two-distance sets

Our first illustration of the independence criterion is pregnant with surprisingly powerful applications, which we will consider in Sects. 14.3.2 and 14.3.3.

Let $a_1, \ldots, a_m$ be points in the n-dimensional Euclidean space $\mathbb{R}^n$. If the pairwise distances of the a_i are all equal then $m \leqslant n + 1$. (Show this!) But what happens if we relax the condition and require only that the pairwise distances between the a_i take *two* values? Such a set is called *two-distance set*.

We shall see that then m is about $n^2/2$. Indeed, it is easy to construct a two-distance set in $\mathbb{R}^n$ with $\binom{n}{2}$ points (Exercise 14.18). On the other hand, we have the following upper bound.

Theorem 14.12 (Larman–Rogers–Seidel 1977). *Every two-distance set in* $\mathbb{R}^n$ *has at most* $\binom{n}{2} + 3n + 2$ *points.*

Proof. Let $a_1, \ldots, a_m$ be a two-distance set of distinct points in $\mathbb{R}^n$. Using the notation

$$\|x\| = \Big(\sum_{i=1}^{n} x_i^2\Big)^{1/2}$$

for the Euclidean norm of a point $x = (x_1, \ldots, x_n)$ in $\mathbb{R}^n$, the distance between two points x, y in $\mathbb{R}^n$ is $\|x - y\|$. Since for our set of points $a_1, \ldots, a_m$ this distance can take only one of two values d_1 or d_2, no of which is zerro (why?), it is natural to associate with each point a_i the following polynomial in n real variables $x \in \mathbb{R}^n$:

$$f_i(x) \rightleftharpoons (\|x - a_i\|^2 - d_1^2) \cdot (\|x - a_i\|^2 - d_2^2).$$

Then $f_i(a_i) = (d_1 d_2)^2 \neq 0$, but $f_i(a_j) = 0$ for every $j \neq i$. By Lemma 14.11, these polynomials are linear independent (as members of the space of all functions $f : \mathbb{R}^n \to \mathbb{R}$). What is a vector space in which they reside? It is easy to see that every such polynomial is an appropriate linear combination of the following polynomials

$$\Big(\sum_{l=1}^{n} x_i^2\Big)^2, \ \Big(\sum_{l=1}^{n} x_i^2\Big) x_j, \ x_i x_j, \ x_i, \ 1, \ \text{ for } i, j = 1, \ldots, n;$$

their number is $1 + n + \big(\binom{n}{2} + n\big) + n + 1 = \binom{n}{2} + 3n + 2$. Thus, the polynomials $f_1, \ldots, f_m$ belong to a linear space of dimension at most $\binom{n}{2} + 3n + 2$. As they are linearly independent, their number m cannot exceed the dimension, completing the proof of the theorem. □

We can rewrite the upper bound in Theorem 14.12 as

$$m \leqslant \binom{n}{2} + 3n + 2 = \binom{n+2}{2} + n + 1.$$

A significant improvement was achieved by Blokhuis (1981) who showed that the second term $n + 1$ here is redundant. His trick was to show that the polynomials $f_1, \ldots, f_m$ *together* with the polynomials $x_1, \ldots, x_n, 1$ are linearly independent. This nice idea was later employed to derive more impressing results (cf. Exercise 14.22).

14.3.2 Sets with few intersection sizes

In this section we demonstrate how the polynomial technique can be used to obtain far reaching extensions of Fisher's inequality.

Let $\mathcal{F}$ be a family of subsets of some n-element set, and let $L \subseteq \{0, 1, \ldots\}$ be a finite set of integers. We say that $\mathcal{F}$ is *L-intersecting* if $|A \cap B| \in L$ for every pair A, B of distinct members of $\mathcal{F}$.

Suppose we know only the size of L. What can be then said about the number of sets in $\mathcal{F}$? Fisher's inequality tells us that $|\mathcal{F}| \leqslant n$ when $|L| = 1$. In the case of uniform families, the celebrated result of Ray-Chaudhuri and Wilson (1975) gives the upper bound $|\mathcal{F}| \leqslant \binom{n}{|L|}$. The non-uniform version of this result was proved by Frankl and Wilson (1981).

Theorem 14.13 (Frankl–Wilson 1981). *If $\mathcal{F}$ is an L-intersecting family of subsets of a set of n elements, then $|\mathcal{F}| \leqslant \sum_{i=0}^{|L|} \binom{n}{i}$.*

Both these results are best possible: take the family of all subsets of an n-element set with s elements (with at most s elements, respectively).

The original proof of these theorems used the machinery of higher incidence matrices. Fortunately, these results now admit conceptually simpler proofs using linear spaces of multivariate polynomials.

Proof of Theorem 14.13 (due to Babai 1988). Let $\mathcal{F} = \{A_1, \ldots, A_m\}$ where $|A_1| \leqslant \ldots \leqslant |A_m|$. Let $L = \{l_1, \ldots, l_s\}$ be the set of all possible intersection sizes. That is, for every $i \neq j$ there is a k such that $|A_i \cap A_j| = l_k$. With each set A_i we associate its incidence vector $v_i = (v_{i1}, \ldots, v_{in})$, where $v_{ij} = 1$ if $j \in A_i$; otherwise $v_{ij} = 0$. For $x, y \in \mathbb{R}^n$, let (as before) $\langle x, y \rangle = \sum_{i=1}^n x_i y_i$ denote their standard scalar product. Clearly, $\langle v_i, v_j \rangle = |A_i \cap A_j|$.

For $i = 1, \ldots, m$, let us define the polynomial f_i in n variables by

$$f_i(x) = \prod_{k \, : \, l_k < |A_i|} (\langle v_i, x \rangle - l_k) \qquad (x \in \mathbb{R}^n).$$

Observe that $f_i(v_j) = 0$ for all $1 \leqslant j < i \leqslant m$, and $f_i(v_i) \neq 0$ for all $1 \leqslant i \leqslant m$. By Lemma 14.11, the polynomials $f_1, \ldots, f_m$ are linear independent over $\mathbb{R}$. What is a small vector space in which these polynomials can reside? The f_i's are polynomials of degree at most s, but we can do better. The domain being $\{0, 1\}^n$ implies that $x_i^2 = x_i$ for each variable x_i. Thus, pure monomials of degree $\leqslant s$ form a basis (where a pure monomial has at most one occurrence of each variable), and we have only $\sum_{i=0}^s \binom{n}{i}$ of them. $\square$

Using essentially the same argument, we can also prove the following "modular" version of this theorem (we leave the proof as Exercise 14.21). Write $r \in L \pmod{p}$ if $r = l \pmod{p}$ for at least one $l \in L$.

Theorem 14.14 (Deza–Frankl–Singhi 1983). *Let L be a set of integers and p be a prime number. Assume $\mathcal{F} = \{A_1, \ldots, A_m\}$ is a family of subsets of a set of n elements such that*

(a) $|A_i| \notin L \pmod{p}$ $(1 \leqslant i \leqslant m)$;

(b) $|A_i \cap A_j| \in L \pmod{p}$ $(1 \leqslant j < i \leqslant m)$.

Then $|\mathcal{F}| \leqslant \sum_{i=0}^{|L|} \binom{n}{i}$.

These theorems and their modifications have found many striking applications in combinatorics and geometry. An excellent exposition is given in the book by Babai and Frankl (1992).

Roughly, the main idea of all these applications is the following. If we identify the members of our family $\mathcal{F}$ with vertices and join two members if and only if their intersection has a particular size, then the theorems above ensure that the graph cannot have a large clique or a large independent set (or both). To demonstrate the idea, we use it to construct so-called Ramsey graphs.

14.3.3 Constructive Ramsey graphs

Recall that a *clique* of size t in a graph is a set of t of its vertices, each pair of which is joined by an edge. Similarly, an *independent set* of size t is a set of t vertices with no edge between them. A graph is a *Ramsey graph* (with respect to t) if it has no clique and no independent set of size t.

Given t, we are interested in the largest possible number n for which such a graph (on n vertices) exists. The existence of Ramsey graphs of size $n = 2^{t/2}$ is known: this was proved by Erdős (1947) using the *probabilistic method* (see Theorem 18.1). The theorem states only the mere *existence* of a graph, and gives no way to find it.

For many years, only an easy construction of a Ramsey graph of size $n = (t-1)^2$ was known: take the disjoint union of $t-1$ cliques of size $t-1$ each. The first non-trivial construction of Ramsey graphs on $n = \Omega(t^3)$ vertices was given by Zsigmond Nagy in 1972 (see Exercise 14.20).

Substantial progress in that direction was made by Frankl (1977) who was able to construct Ramsey graphs of super-polynomial size

$$n = t^{\Omega(\ln t / \ln \ln t)}.$$

Subsequently, a simpler proof was found by Frankl and Wilson (1981), using their result about families with one missing intersection size modulo a prime power. To be self-contained, here we present a slightly weaker version whose proof relies only on Theorems 14.13 and 14.14.

The desired graph is defined as follows. Let p be a prime number, and $v = p^3$. Let G_p be a graph whose vertices are subsets of $\{1, \ldots, v\}$ of cardinality $p^2 - 1$, and where two vertices A and B are joined by an edge if and only if

$$|A \cap B| \neq -1 (\bmod p).$$

Theorem 14.15 (Frankl 1977, Frankl–Wilson 1981). *The graph G_p has $n = \binom{v}{p^2-1}$ vertices and has neither a clique nor an independent set on more than $t = \sum_{i=0}^{p-1} \binom{v}{i}$ vertices.*

Proof. If $A_1, \ldots, A_r$ is a clique in G_p then $|A_i \cap A_j| \neq -1 (\bmod p)$ for every $1 \leqslant i < j \leqslant r$, implying that $|A_i \cap A_j| \in L (\bmod p)$ for $L = \{0, 1, \ldots, p-2\}$. On the other hand, each of the sets A_i has size $p^2 - 1 = -1 (\bmod p)$, and hence, $|A_i| \notin L (\bmod p)$. Theorem 14.14 implies that in this case $r \leqslant t$.

Now suppose that $A_1, \ldots, A_r$ forms an independent set in G_p. Then $|A_i \cap A_j| \in L$, for every $1 \leqslant i < j \leqslant r$, where $L = \{p-1, 2p-1, \ldots, p^2-p-1\}$. Theorem 14.13 again yields that in this case $r \leqslant t$.   □

To get the desired lower bound $n \geqslant t^{\Omega(\ln t / \ln \ln t)}$, we just have to select p properly. The exact computation is somewhat tedious, and we leave them as an exercise. We only sketch the way these computations should proceed. By the density of primes, there is always a prime between N and $2N$, for any positive integer N; so we may pretend that p is an integer rather than a prime. Since $v = p^3$, $t = \sum_{i=0}^{p-1} \binom{v}{i} \leqslant p^{O(p)}$ whereas $n = \binom{v}{p^2-1} \geqslant p^{\Omega(p^2)}$, which is at least $t^{\Omega(\ln t / \ln \ln t)}$ for $p = \Omega(\ln t / \ln \ln t)$.

14.3.4 Bollobás theorem – another proof

Our last application of the polynomial technique is a new proof of Bollobás's theorem. The original proofs of this theorem were combinatorial (we presented two of them in Sect. 9.2.2 devoted entirely to this fundamental result). The following proof uses linear algebra, and is due to Lovász (1977b). His original proof employs a machinery of multilinear algebra. We follow a simplified variant due Frankl (1982) (see also Babai and Frankl 1992).

Theorem 14.16. *Let $A_1, \ldots, A_m$ be sets of size r and $B_1, \ldots, B_m$ be sets of size s such that*

(a) A_i and B_i *are disjoint for* $i = 1, \ldots, m$;

(b) A_i and B_j *intersect whenever* $i \neq j$.

Then

$$m \leqslant \binom{r+s}{s}.$$

Proof. Let X be the union of all sets $A_i \cup B_i$. Take a set $V \subseteq \mathbb{R}^{r+1}$ of $|V| = |X|$ vectors $v = (v_0, v_1, \ldots, v_r)$ in *general position*, i.e., every $r+1$ of these vectors are linearly independent (we may, for instance, select them from the moment curve, cf. Lemma 14.5). Identify the elements of X with vectors in this set; hence, we look $A_1, \ldots, A_m$ as r-element subsets and $B_1, \ldots, B_m$ as s-element subsets of vectors in V. Associate with each set B_j the following polynomial in $r + 1$ variables $x = (x_0, x_1, \ldots, x_r)$:

$$f_j(x) = \prod_{v \in B_j} \langle v, x \rangle = \prod_{v \in B_j} (v_0 x_0 + v_1 x_1 + \cdots + v_r x_r)$$

Notice that $f_j(x) = 0$ if and only if $\langle v, x \rangle = 0$ for some $v \in B_j$.

The vectors corresponding to elements in A_i generate a subspace span A_i. Since each set A_i consists of r linearly independent vectors in $\mathbb{R}^{r+1}$, its span has dimension r, and by Proposition 14.2, (span $A_i)^\perp$ has dimension 1. We can, therefore, select for each $i = 1, \ldots, m$, a nonzero vector $a_i \in (\text{span } A_i)^\perp$ so that, for every $v \in V$, $\langle v, a_i \rangle = 0$ if and only if $v \in \text{span } A_i$. On the other hand, the general position constraint means that for $v \in V$, $v \in \text{span } A_i$ if and only if $v \in A_i$ (because $|A_i \cup \{v\}| \leqslant r+1$ and any $r+1$ vectors in V are linearly independent). In other words, for $v \in V$,

$$\langle v, a_i \rangle = 0 \quad \text{if and only if} \quad v \in A_i.$$

This implies that $f_j(a_i) = 0$ if $A_i \cap B_j \neq \emptyset$, and $f_j(a_i) \neq 0$ if $A_i \cap B_j = \emptyset$. But by our assumption, the intersection $A_i \cap B_j$ is non-empty if and only if $i \neq j$. Thus, $f_j(a_i) \neq 0$ precisely when $i = j$. By Lemma 14.11, the polynomials $f_1, \ldots, f_m$ are linear independent, and it remains to give an upper bound on the dimension of the corresponding linear space.

Each polynomial $f_j(x)$ is homogeneous of degree $s = |B_j|$ in $r+1$ variables. The small vector space in which they reside is generated by monomials of the form $x_1^{i_1} \cdots x_{r+1}^{i_{r+1}}$ with $i_1 + \cdots + i_{r+1} = s$. By Proposition 1.4, the number of such monomials, and hence the dimension of the corresponding space, is $\binom{r+s}{s}$, and we are done. □

Lovász's proof has the advantage that it can be generalized. Consider the matrix M, where $M_{ij} = f_j(a_i)$. In the proof above we have shown that this matrix is diagonal. However, to apply Lemma 14.11 we need only the matrix be lower (or upper) triangular. Therefore, the condition $A_i \cap B_j \neq \emptyset$ if $i \neq j$ can be changed to $A_i \cap B_j \neq \emptyset$ if $i < j$ (cf. Theorem 9.10).

14.4 Combinatorics of linear spaces

So far, we have used *algebraic* properties of linear spaces to derive some results in combinatorics. Next two results demonstrate that these spaces themselves have interesting *combinatorial* properties as well.

14.4.1 Universal sets from linear codes

A set of 0-1 strings $A \subseteq \{0,1\}^n$ is (n,k)-*universal* if, for any subset of k coordinates $S = \{i_1, \ldots, i_k\}$, the projection of A onto the indices in S contains all possible 2^k configurations.

In Sect. 11.3 we have shown how to construct explicit (n,k)-universal sets of size about n, when $k \leqslant (\log n)/3$. This construction was based on Paley graphs. Here we will show how to construct such sets of size $n^{O(k)}$ for arbitrary k. The construction is based on some elementary properties of linear codes.

A (binary) *linear code* of length n is just a set of vectors $C \subseteq \{0,1\}^n$ which forms a linear subspace of $\mathbb{F}_2^n$. A Hamming distance between two vectors is the number of coordinates where these two vectors have different values. The *minimal distance* of a code C is a minimal Hamming distance between any pair of distinct vectors in C.

It can be easily shown that the minimal distance of C coincides with the minimum weight of (i.e., the number of 1's in) a non-zero vector from C (see Exercise 14.2). This simple property of linear codes implies that their duals are universal.

Proposition 14.17. *If C is a linear code of length n and its dual $C^\perp$ has minimal distance $k+1$ then the code itself is (n,k)-universal.*

Proof. Take a set $I \subseteq \{1, \ldots, n\}$ with $|I| \leqslant k$. The set of all projections of strings in C onto I is a linear subspace in $\{0,1\}^I$, and (by Proposition 14.2) this subspace is proper if and only if all vectors $v \in C$ satisfy a non-trivial linear relation $\sum_i \lambda_i v_i = 0 \bmod 2$ whose support $\{i : \lambda_i = 1\}$ is contained in I. But, by definition, $C^\perp$ consists exactly of all relations $(\lambda_1, \ldots, \lambda_n)$ satisfied by C, and its minimal distance is exactly the minimal possible cardinality of a set I for which the projection of C onto I is proper. $\qquad\square$

It is known (see, for example, MacWilliams and Sloane (1977)) that the dual of a binary BCH code of minimal distance k has only $O(n^{\lfloor k/2 \rfloor})$ vectors. By Proposition 14.17, these codes give us explicit (n,k)-universal sets consisting of only so many vectors.

One of the best known explicit constructions of (n,k)-universal sets of size only $2^{O(k^4)} \log n$ is due to Alon (1986). His construction is based on a Justensen-type code constructed by Friedman (1984).

14.4.2 Short linear combinations

Let A be a set of vectors in $\{0,1\}^n$ and consider its span over the field $\mathbb{F}_2$. Each vector in span A is a linear combination of some vectors from A. Some of these combinations may be short, but some may be long. Given A, we are interested in the smallest number k such that every vector from span A is a sum (over $\mathbb{F}_2$) of at most k vectors of A; we call this k the *spanning diameter* of A.

Of course, the answer depends on how large the span is, compared with the set itself. It is easy to show that if $|A| > |\text{span } A| / 2$, then the spanning diameter of A is at most 2 (see Exercise 14.26). But what if A is a smaller fraction of span A, say, an α-fraction for some $\alpha > 1/4$? It appears that then the spanning diameter does not exceed 4. In general, we have the following upper bound on k.

Theorem 14.18. *Let $A \subseteq \{0,1\}^n$. If $|A| \geqslant \alpha \cdot |\text{span } A|$ for some $0 < \alpha \leqslant 1$, then every vector from span A is a sum of at most k vectors from A, where k is the maximal number satisfying the inequality*

$$k - \lfloor \log_2 k \rfloor - 1 \leqslant \log_2(1/\alpha) \tag{14.5}$$

Theorem 14.18 can be derived from known bounds on the covering radius of binary linear codes (see, for example, Cohen et al. 1997; Theorem 8.1.21). Here we present a direct argument due to Pavel Pudlák.

Proof. Let v_0 be an arbitrary vector in span A. Let

$$v_0 = v_1 + v_2 + \ldots + v_k, \tag{14.6}$$

where $v_1, v_2, \ldots, v_k \in A$ and k is the *minimal* number such that v_0 can be represented in this way. (Here and through the proof all the sums are over $\mathbb{F}_2$.) Our goal is to show that then k must satisfy (14.5). Since $\alpha \leqslant 1$, the cases $k = 1$ and $k = 2$ are trivial. So, assume that $k \geqslant 3$.

We will need a lower bound on the size of distance-3 codes. Such codes can be obtained by shortening the Hamming code (see, for example, MacWilliams and Sloane (1977)); Exercise 14.29 sketches a way about how this can be done.

Claim 14.19. *There exists a set $C \subseteq \mathbb{F}_2^k$ such that any two vectors of C differ in at least 3 coordinates and $\log_2 |C| \geqslant k - \lfloor \log_2 k \rfloor - 1$.*

Fix such a set C, and let B be the set of all those vectors u from span A which can be represented in the form

$$u = c_1 v_1 + c_2 v_2 + \cdots + c_k v_k$$

for $c = (c_1, \ldots, c_k) \in C$. The key point is that all the translates

$$u + A \rightleftharpoons \{u + v : v \in A\},$$

with $u \in B$, are mutually disjoint.

Claim 14.20. *For every pair u, u' of distinct vectors from B, the sets $u + A$ and $u' + A$ are disjoint.*

Proof of Claim 14.20. Suppose not. Then for some $v, v' \in A$ we have $u + v = u' + v'$, and hence, $v + v' = u + u'$. Let c, c' be the vectors from C for which $u = c_1 v_1 + \cdots + c_k v_k$ and $u' = c'_1 v_1 + \cdots + c'_k v_k$. Then

$$v + v' = u + u' = (c_1 + c'_1)v_1 + (c_2 + c'_2)v_2 + \cdots + (c_k + c'_k)v_k.$$

Since vectors c and c' differ in at least three coordinates, we have on the right-hand side the sum of at least three vectors, say $v_{i_1} + \cdots + v_{i_l}$, with $l \geqslant 3$. But then in the equation (14.6) we can replace these three (or more) vectors $v_{i_1}, \ldots, v_{i_l}$ by two vectors v, v', which contradicts the minimality of k. $\square$

The same argument also implies that no two distinct vectors $c, c' \in C$ can lead to one and the same vector $u \in B$, which means that $|B| = |C|$.

This, together with Claim 14.20, implies

$$|A| \cdot |C| = |A| \cdot |B| = \sum_{u \in B} |u + A| = \left| \bigcup_{u \in B} (u + A) \right| \leqslant |\text{span } A|.$$

Hence, $\log_2 |C| \leqslant \log_2(1/\alpha)$ which, together with Claim 14.19, yields the desired upper bound (14.5) on k. $\square$

14.5 The flipping cards game

There are situations in theory of computing, where switching to the linear algebra *language* alone can lead to interesting results. In particular, linear combination and/or scalar product can often be used to encode some useful information about the input vectors which, in its turn, can lead to surprisingly efficient algorithms.

Suppose that we have two 0-1 vectors $u = (u_1, \ldots, u_n)$ and $v = (v_1, \ldots, v_n)$ of length n. We want to decide whether $u = v$, but our access to the bits is very limited: at any moment we can see at most one bit of each pair of the bits u_i and v_i. We can imagine the corresponding bits to be written on two sides of a card, so that we can see all the cards, but only one side of each card:

A *probe* consists in flipping of one or more of the cards. After every probe we can write down some information but the memory is not reusable – after the next probe we have to use new memory (i.e., we cannot wipe it out). Moreover, this is the only memory for us: seeing the information written here (but not the cards themselves), we ask to flip some of the cards; seeing the actual values of the cards and using the current information from the memory, we either give an answer or we write some additional bits of information in the memory; after that the cards are closed for us, and we make the next probe.

Suppose we are charged for every bit of memory that we use but not for the number of probes. The goal is to decide if both sides of all cards are the same using as little of memory as possible. Of course, n bits of memory are always enough: simply write u in the memory, and flip all the cards to see v. Can we do better? To enjoy the next two results, the reader is invited to stop for a moment and try to imagine a protocol which uses less than n bits of memory.

Theorem 14.21. *Let* $n = r^2$ *for some* $r \geqslant 1$. *It is possible to test the equality of two vectors in* $\{0,1\}^n$ *using only* $r + 1$ *probes and writing down only* r *bits in the memory.*

Proof. The following protocol is due to J. Edmonds and R. Impagliazzo. Split the given vectors u and v into r pieces of length r: $u = (u^1, \ldots, u^r)$ and $v = (v^1, \ldots, v^r)$. In the first probe look at vector u and compute the vector

$$w_0 \rightleftharpoons u^1 + u^2 + \cdots + u^r,$$

where the sum is over $\mathbb{F}_2$. Write down this vector w_0 in the memory (using r bits), and make subsequent r probes as follows. During the ith probe flip the cards of the ith piece; compute the vector

$$w_i \rightleftharpoons u^1 + \cdots + u^{i-1} + v^i + u^{i+1} + \cdots + u^r$$

and just test if the obtained vector w_i coincides with the vector w_0 (written in the memory). Answer "$u = v$" if all the vectors $w_1, \ldots, w_r$ coincide with w_0, and "$u \neq v$" otherwise. If we answer "$u = v$", we know that, after the first probe, $u^1 + u^2 + \cdots + u^r = v^1 + u^2 + \cdots + u^r$ and hence $u^1 = v^1$; the same argument is valid for other probes, hence $u = v$ and the protocol is correct. $\square$

Using the language of scalar products, Pudlák and Sgall (1997) have shown that, in fact, $O((\log n)^2)$ bits are enough.

Theorem 14.22. *It is possible to test the equality of two vectors in* $\{0,1\}^n$ *using only* $O(\log n)$ *probes and writing down only* $O(\log n)$ *bits in the memory about each probe.*

Proof. Each probe corresponds to a subset $I \subseteq \{1, \ldots, n\}$; after this probe we see n bits: $|I|$ bits $\{u_i : i \in I\}$ of u and $n - |I|$ bits $\{v_i : i \notin I\}$ of v. We think of u and v as 0-1 vectors in real vector space $\mathbb{R}^n$. The idea is to compute (a square of) the Euclidean distance

$$\|u - v\|^2 = \langle u, u \rangle + \langle v, v \rangle - 2 \langle u, v \rangle$$

$$= \sum_{i=1}^{n} u_i^2 + \sum_{i=1}^{n} v_i^2 - 2 \left[\left(\sum_{i=1}^{n} u_i \right) \cdot \left(\sum_{i=1}^{n} v_i \right) - \sum_{i \neq j} u_i v_j \right]$$

of u and v, and check if it is 0. We compute $\langle u, u \rangle$ and $\langle v, v \rangle$ each using one probe (probe $I = \{1, \ldots, n\}$ for $\langle u, u \rangle$ and probe $I = \emptyset$ for $\langle v, v \rangle$) and $\lceil \log(n+1) \rceil$ bits of memory (to write down these two numbers between 0 and n). It remains to compute the product $\langle u, v \rangle = \sum_{i=1}^{n} u_i v_i$.

To do this, we first compute the product $N = \left(\sum_{i=1}^{n} u_i \right) \left(\sum_{i=1}^{n} v_i \right)$ using the same probes and additional $2\lceil \log(n+1) \rceil$ bits of the memory (to write the value of this product which lies between 0 and n^2). To compute the desired product $\langle u, v \rangle$ we need to subtract from N the sum $\sum_{i \neq j} u_i v_j$ of cross-terms. This is easily done using $2\lceil \log n \rceil$ probes: choose them so that each of the cross-terms can be computed by one of them, and for each probe sum all these terms assigned to it. After each of these probes we write the resulting partial sum using $O(\log n)$ bits of memory. $\square$

Exercises

14.1.$^{-}$ Prove the Pythagoras theorem: if the vectors u, v are orthogonal, then $\|u + v\|^2 = \|u\|^2 + \|v\|^2$.

14.2.$^{-}$ Show that the minimal distance of a linear code coincides with the minimum weight of its non-zero vector. *Hint*: Every linear code contains the zero vector.

14.3. Prove the following stronger version of Proposition 14.17. Let C be a linear code of length n and minimal distance $k + 1$ and let $C^{\perp}$ be its dual. Then for every subset S of $l \leqslant k$ coordinates, every 0-1 string of length l appears as a projection of $C^{\perp}$ onto S one and the same number of times.

> *Hint*: Take a matrix whose rows form a basis of $C^{\perp}$, observe that every k columns of this matrix are linearly independent and use Proposition 14.3.

14.4.$^{-}$ Let $V \subseteq \mathbb{F}_2^n$ be a subspace of dimension d. Show that $|V| = 2^d$.

14.5.$^{-}$ Let $\mathcal{F}$ be a family of subsets of an n-element set such that: (i) every set of $\mathcal{F}$ has an *even* number of elements, and (ii) each pair of sets share an *even* number of elements. Construct such a family with at least $2^{\lfloor n/2 \rfloor}$ sets.

14.6 (Babai–Frankl 1992). Show that the upper bound $2^{\lfloor n/2 \rfloor}$ in the previous exercise cannot be improved.

> *Hint*: Let S be the set of incidence vectors of all sets in $\mathcal{F}$, and let U the span of this set (over $\mathbb{F}_2$). Argue that the rules (i) and (ii) imply that U is a subspace of $U^{\perp}$, and apply Proposition 14.2.

14.7.$^{(!)}$ Prove the following "Oddtown Theorem" (see Babai and Frankl (1992) for the explanation of this name). Let $\mathcal{F}$ be a family of subsets of an n-element set such that: (i) every set of $\mathcal{F}$ has an *odd* number of elements, and (ii) each pair of sets share an *even* number of elements. Prove that then $|\mathcal{F}| \leqslant n$. Compare this with Exercise 14.5.

> *Hint*: The incidence vectors of sets in $\mathcal{F}$ are linear independent over $\mathbb{F}_2$.

14.8. $^-$ Show that the Euclidean distance between any two 0-1 vectors is the square root of their Hamming distance.

14.9. $^-$ Show that the mutual orthogonality of $(+1, -1)$-vectors implies their linear independence (over the reals).

14.10. $^-$ Using the Cauchy–Schwarz inequality show that if $u = (u_1, \ldots, u_n)$ is a vector in $\mathbb{R}^n$ then $|u| \leqslant \sqrt{n} \cdot \|u\|$, where $|u| \rightleftharpoons |u_1| + \ldots + |u_n|$ and $|u_i|$ is the absolute value of u_i.

Hint: Take a vector $v = (v_1, \ldots, v_n)$ with $v_i = 1$ if $u_i > 0$ and $v_i = -1$, otherwise. Observe that $|u| = \langle u, v \rangle$ and $\|v\| = \sqrt{n}$.

14.11. Recall that the *determinant* $\det(A)$ of an $n \times n$ matrix $A = (a_{ij})$ is a sum of $n!$ signed products $\pm a_{1i_1} a_{2i_2} \cdots a_{ni_n}$, where $(i_1, i_2, \ldots, i_n)$ is a permutation of $(1, 2, \ldots, n)$, the sign being $+1$ or -1, according as the number of *inversions* of $(i_1, i_2, \ldots, i_n)$ is even or odd; an inversion occurs when $i_r > i_s$ but $r < s$. The *Vandermonde* matrix is the $n \times n$ matrix X_n whose ith row is $(1, x_i, x_i^2, \ldots, x_i^{n-1})$. Prove that $\det(X_n) = \prod_{1 \leqslant i < j \leqslant n} (x_j - x_i)$.

Hint: Argue by induction on n. Multiply each column by x_1 and subtract it from the next column on the right, starting from the right–hand side; this yields $\det(X_n) = (x_n - x_1) \cdots (x_2 - x_1) \det(X_{n-1})$.

14.12. Let A be an $n \times n$ matrix with rows $v_1, \ldots, v_n$. Prove the following "independence condition": if $\det(A) \neq 0$ then vectors $v_1, \ldots, v_n$ are linearly independent.

Hint: Prove the other direction: if these vectors are linear dependent then $\det(A) = 0$. Take a linear combination $\sum_i \lambda_i v_i = 0$ where some coefficient $\lambda_k \neq 0$. Dividing this equation by λ_k we get the row v_k as a linear combination of the remaining rows: $v_k = \sum_{i \neq k} \alpha_i v_i$ where $\alpha_i = -\lambda_i / \lambda_k$. Use the linearity of determinant (check this property!) to get

$$\det(v_1, \ldots, v_k, \ldots, v_n) = \sum_{i \neq k} \alpha_i \det(v_1, \ldots, v_i, \ldots, v_n).$$

In the sum on the right, each determinant has the ith row equal to the kth row (with $k \neq i$) and is therefore equal to 0.

14.13. Let $\mathcal{D} = \{B_1, \ldots, B_b\}$ be a (v, k, λ) design, i.e., a family of k-element subsets of a v-element set of points $X = \{x_1, \ldots, x_v\}$ such that every two points belong to exactly λ sets. Use Fisher's inequality to show that $b \geqslant v$. *Hint*: Take $A_i \rightleftharpoons \{j : x_i \in B_j\}$.

14.14 (Frankl–Pach 1984). Prove the following analog of Theorem 14.7 for uniform families. Let n, l, k be natural numbers, $n \geqslant l \geqslant k$, and let $\mathcal{F}$ be an l-uniform family of subsets of an n-element set. If $|\mathcal{F}| > \binom{n}{k-1}$ then $\mathcal{F}$ is (n, k)-dense.

Hint: Consider the inclusion matrix whose columns are labeled by $(k-1)$-element subsets only, and argue that the minimal set Y, for which $g(Y) \neq 0$, must still have at least k elements.

14.15.⁻ Let $f(x_1, \ldots, x_n)$ be a polynomial over $\mathbb{F}_2$ of degree $d < n$ which is not identically 1. Show that then $f(v) = 0$ for at least one non-zero vector v with at most $d + 1$ 1's.

14.16.⁻ Let $h = \prod_{i \in S} x_i$ be a monomial of degree $d = |S| \leqslant n - 1$, and let a be a 0-1 vector with at least $d+1$ 1's. Show that then, over $\mathbb{F}_2$, $\sum_{b \leqslant a} h(b) = 0$.

14.17 (Babai et al. 1991). Let $\mathbb{F}$ be a field, $H_1, \ldots, H_m \subseteq \mathbb{F}$ and $H = H_1 \times \cdots \times H_m$. Prove that, for any function $f : H \to \mathbb{F}$ there exists a polynomial $\tilde{f}$ in m variables over $\mathbb{F}$ such that:

(i) $\tilde{f}$ has degree $\leqslant |H_i|$ in its ith variable, and
(ii) $\tilde{f}$, restricted to H, agrees with f.

Is such a polynomial unique?

Hint: Associate with each vector $u = (u_1, \ldots, u_m)$ in H a polynomial

$$g_u(x) = \prod_{i=1}^{m} \prod_{h \in H_i - \{u_i\}} (x_i - h)$$

and show that every function $f : H \to \mathbb{F}$ is a linear combination of the g_u's, restricted to H.

14.18.⁻ Construct a two-distance set in $\mathbb{R}^n$ of size $\binom{n}{2}$.

14.19. Prove the following generalization of Theorem 14.12 for *s-distance* sets. Let $a_1, \ldots, a_m$ be points in $\mathbb{R}^n$ and suppose that the pairwise distances between them take at most s values. Prove that $m \leqslant \binom{n+s+1}{s}$.

Sketch: Let $d_1, \ldots, d_s$ be the distances permitted, and consider the polynomials $f_i(x) = \prod_{i=1}^{s}(\|x - a_i\|^2 - d_i^2)$. To estimate the dimension of the subspace containing all of them, expand the norm-square expression in each factor, replace the sum $\sum_{i=1}^{n} x_i^2$ by a new variable z, and multiply the constant terms by a new variable t. Observe that then each f_i becomes a homogeneous polynomial of degree s in $n + 2$ variables $x_1, \ldots, x_n, z, t$, and apply Proposition 1.4.

14.20 (Nagy 1972). Let G be a graph whose vertices are 3-element subsets of $\{1, \ldots, t\}$, and where two vertices A and B are joined by an edge if and only if $|A \cap B| = 1$. Use Exercise 14.7 and Fisher's inequality to show that this graph has neither a clique nor an independent set of size $t + 1$.

14.21. Write down a complete proof of Theorem 14.14.

Hint: Work in the finite field $\mathbb{F}_p$ instead of that of real numbers $\mathbb{R}$. This time, due to the condition $|A_i| \notin L \pmod{p}$, we can take $f_i(x) = \prod_{l \in L}(v_i \cdot x - l)$, i.e., we do not need the condition $l < |A_i|$.

14.22[(!)] (Ray-Chaudhuri–Wilson 1975). Prove the following uniform version of Theorem 14.13: if $A_1, \ldots, A_m$ is a k-uniform L-intersecting family of subsets of an n-element set, then $m \leqslant \binom{n}{s}$, where $s = |L|$.

Sketch (Alon–Babai–Suzuki 1991): Start as in the proof of Theorem 14.13 and define the same polynomials $f_1, \ldots, f_m$ of degree at most s. Associate with each subset I of $\{1, \ldots, n\}$ of cardinality $|I| \leqslant s - 1$ the following polynomial of degree at most s:

$$g_I(x) = \left(\left(\sum_{j=1}^{n} x_j \right) - k \right) \prod_{i \in I} x_i,$$

and observe that for any subset $S \subseteq \{1, \ldots, n\}$, $g_I(S) \neq 0$ if and only if $|S| \neq k$ and $S \supseteq I$. Use this property to show that the polynomials g_I *together* with the polynomials f_i are linearly independent. For this, assume

$$\sum_{i=1}^{m} \lambda_i f_i + \sum_{|I| \leqslant s-1} \mu_I g_I = 0$$

for some $\lambda_i, \mu_I \in \mathbb{R}$. Substitute A_j's for the variables in this equation to show that $\lambda_j = 0$ for every $j = 1, \ldots, m$. What remains is a relation among the g_I. To show that this relation must be also trivial, assume the opposite and re-write this relation as $\mu_1 g_{I_1} + \cdots + \mu_t g_{I_t} = 0$ with all $\mu_i \neq 0$ and $|I_1| \geqslant |I_j|$ for all $j > 1$. Show that then $g_{I_1}(I_1) \neq 0$ and $g_{I_1}(I_j) = 0$ for all $j > 1$.

14.23. Let $A_1, \ldots, A_m$ and $B_1, \ldots, B_m$ be subsets of an n-element set such that $|A_i \cap B_i|$ is odd for all $1 \leqslant i \leqslant m$, and $|A_i \cap B_j|$ is even for all $1 \leqslant i < j \leqslant m$. Show that then $m \leqslant n$.

14.24[(!)] (Frankl–Wilson 1981). Let p be a prime, and $n = 4p - 1$. Consider the graph $G = (V, E)$ whose vertex set V consists of all 0-1 vectors of length n with precisely $2p - 1$ 1's each; two vectors are adjacent if and only if the Euclidean distance between them is $\sqrt{2p}$. Show that G has no independent set of size larger than $\sum_{i=0}^{p-1} \binom{n}{i}$.

Hint: Use Exercise 14.8 to show that two vectors from V are adjacent in G precisely when they share $p - 1$ 1's in common, and apply Theorem 14.14.

Comment: This construction was used by Frankl and Wilson (1981) to resolve an old problem proposed by H. Hadwiger in 1944: how many colors do we need in order to color the points of the n-dimensional Euclidean space $\mathbb{R}^n$ so that each monochromatic set of points misses some distance? A set is said to *miss distance* d if no two of its points are at distance d apart from each other. Larman and Rogers (1972) proved that Hadwiger's problem reduces to the estimating the minimum number of colors $\chi(n)$ necessary to color the points of $\mathbb{R}^n$ such that pairs of points of unit distance are colored differently. The graph G we just constructed shows that $\chi(n) \geqslant 2^{\Omega(n)}$ (see the next exercise). Kahn and Kalai (1993) used a similar construction to disprove another 60 years old and widely believed conjecture of K. Borsuk (1933) that every set of diameter one in n-dimensional real space R^n can be partitioned in at most $n + 1$ disjoint pieces of smaller diameter. Kahn and Kalai presented an infinite sequence of examples where the minimum number of pieces grew as an exponential function of $\sqrt{n}$, rather than just as a linear function $n + 1$, as conjectured. Interested reader can find these surprising solutions in the book of Babai and Frankl (1992).

14.25. The *unit distance graph* on $\mathbb{R}^n$ has the infinite set $\mathbb{R}^n$ as its vertex set, and two points are adjacent if their (Euclidean) distance is 1. Let $\chi(n)$ be the minimum number of colors necessary to color the points of the Euclidean space $\mathbb{R}^n$ such that pairs of points of unit distance are colored differently. Use the graph from the previous exercise to show that $\chi(n) \geqslant 2^{\Omega(n)}$.

Hint: Observe that $\chi(G) \geqslant |V|/\alpha(G)$ and replace each 0-1 vector v by the vector ϵv, where $\epsilon = 1/\sqrt{2p}$. How does this change the distance?

14.26.$^{-}$ Let $A \subseteq \mathbb{F}_2^n$ and suppose that $|A| > |\text{span } A|\,/2$. Prove that every vector in span A is a sum of at most 2 vectors from A.

Hint: Show that, for every $v \in \mathbb{F}_2^n$, the set $A \cap (v + A)$ has at least one vector.

14.27. Theorem 14.18 gives an *upper* bound on the spanning diameter of sets A in terms of their density $\alpha = |A|/|\text{span } A|$. Show that for infinitely many values of k, the bound (14.5) is optimal, that is, exhibit sets A whose spanning diameter is the maximal number satisfying (14.5).

Hint: Consider the set consisting of all-0 vector and k vectors with precisely one 1; its density is $\alpha = (k+1)/2^k$.

14.28 (*Hamming code*). Let r be a positive integer, and let $k = 2^r - 1$. Consider the $r \times k$ matrix H whose columns are all the distinct nonzero vectors of $\{0,1\}^r$. Let $C \subseteq \mathbb{F}_2^k$ be the set of vectors, each of which is orthogonal (over $\mathbb{F}_2$) to all the rows of H. Prove that C is a linear code of minimal distance 3 and has precisely 2^{k-r} code words,

Hint: Show that no vector of weight 1 or 2 can be orthogonal to all the rows of H, and use Exercise 14.2.

14.29. Prove the Claim 14.19.

Sketch: If k has the form $k = 2^r - 1$, then we can take C to be a Hamming code (see previous exercise). Otherwise, take r such that $k = 2^r + x$ for some integer $0 \leqslant x < 2^r - 1$, and let C be a Hamming code of length $K = 2^{r+1} - 1$. By fixing the last $K - k$ of coordinates to appropriate constants, it is possible to obtain from C a set of vectors $C' \subseteq \{0,1\}^k$ of size $|C'| \geqslant |C|/2^{K-k} = 2^{k-r-1}$, such that any two of its vectors still differ in at least 3 coordinates. The obtained code C' may be not linear, but we do not require that.

14.30.$^{-}$ Let $x_1, \ldots, x_n$ be real numbers, and $\sigma : [n] \to [n]$ a permutation of $[n] = \{1, \ldots, n\}$. Show that then $\sum_{i=1}^n x_i \cdot x_{\sigma(i)} \leqslant \sum_{i=1}^n x_i^2$.

Hint: Use the Cauchy–Schwarz inequality.

14.31. Prove that among any $2^{k-1} + 1$ vectors in $\mathbb{F}_2^n$ some k of them must be linearly independent. *Hint:* Take a maximal subset of linearly independent vectors and form all possible sums (over $\mathbb{F}_2$).

15. Orthogonality and Rank Arguments

Linear independence is one of the most basic concepts in linear algebra. Not less important are also the concepts of orthogonality and rank. In this chapter we consider some combinatorial applications of these two concepts.

15.1 Orthogonality

Recall that two vectors u, v are *orthogonal* if their scalar product is zero, $\langle u, v \rangle = 0$. Intuitively, the orthogonality (like the linear independence) tells us that the vectors are "quite different." In this section we demonstrate how this information can be used to solve some combinatorial problems.

15.1.1 Orthogonal coding

Linear independence is not the only way to obtain good upper bounds. If the members of a family $\mathcal{F}$ can be *injectively* associated with the elements of $\mathbb{F}_q^m$, then $|\mathcal{F}| \leqslant q^m$. If we are lucky, the associated "code-vectors" will be orthogonal to some subspace of dimension d, which (due to Proposition 14.2) immediately improves our bound to $|\mathcal{F}| \leqslant q^{m-d}$. We demonstrate this idea by the following result.

Let $\mathcal{A}$ and $\mathcal{B}$ be two families of subsets of an n-element set. We are interested in how large their Cartesian product can be, i.e., we would like to estimate the number $|\mathcal{A} \times \mathcal{B}| = |\mathcal{A}| \cdot |\mathcal{B}|$ of all pairs (A, B) with $A \in \mathcal{A}$ and $B \in \mathcal{B}$. If we know nothing about the families, then this number can be as large as 2^{2n}. If we know that both families are monotone increasing (or monotone decreasing), Kleitman's theorem (Theorem 11.6) gives a non-trivial upper bound:

$$|\mathcal{A}| \cdot |\mathcal{B}| \leqslant 2^n \cdot |\mathcal{A} \cap \mathcal{B}|.$$

If we know that all the intersections $A \cap B$ with $A \in \mathcal{A}$ and $B \in \mathcal{B}$, have the same size modulo 2, then we can get even better bound.

Theorem 15.1 (Ahlswede–El Gamal–Pang 1984). *Let $\mathcal{A}$ and $\mathcal{B}$ be two families of subsets of an n-element set with the property that $|A \cap B|$ is even for all $A \in \mathcal{A}$ and $B \in \mathcal{B}$. Then $|\mathcal{A}| \cdot |\mathcal{B}| \leqslant 2^n$.*

The same holds also with "even" replaced by "odd." In this case the bound is slightly better: $|\mathcal{A}| \cdot |\mathcal{B}| \leqslant 2^{n-1}$ (see Exercise 15.1).

Proof (due to Delsarte and Piret 1985). With each subset of X associate its incidence vector, and look at these vectors as elements of the n-dimensional vector space $\mathbb{F}_2^n$. Let U and V be the sets of incidence vectors of $\mathcal{A}$ and $\mathcal{B}$, respectively. Fix a vector $v_0 \in V$ and let $V_0 \rightleftharpoons \{v_0 + v : v \in V\}$. Moreover, let U' and V_0' be the subspaces spanned by U and V_0, respectively. Then

$$|\mathcal{A}| \cdot |\mathcal{B}| = |U| \cdot |V| = |U| \cdot |V_0| \leqslant |U'| \cdot |V_0'| \leqslant 2^{\dim U' + \dim V_0'}. \tag{15.1}$$

The key point is that (in $\mathbb{F}_2$) $\langle u, w \rangle = 0$ for all $u \in U$ and $w \in V_0$. This (with $w = v_0 + v$, $v \in V$) follows from the fact that $\langle u, v \rangle$ is exactly the parity of points in the intersection of corresponding sets, and from our assumption that all these intersections have the same parity: $\langle u, w \rangle = \langle u, v_0 \rangle + \langle u, v \rangle = 0$.

Therefore, by Proposition 14.2, we obtain that $\dim U' \leqslant n - \dim V_0'$. Putting this estimate in (15.1) we get the desired upper bound $|\mathcal{A}| \cdot |\mathcal{B}| \leqslant 2^n$. $\qquad \square$

15.1.2 A bribery party

Let us imagine the following situation. A town has n inhabitants, n is even. The members of the city council would like to keep their jobs as long as possible. So, each year they organize an (open) vote by asking each inhabitant whether he/she was happy this year or not. Assume also that the common opinion is "stable" in the following (somewhat artificial) sense: between any pair of (not necessarily subsequent) votes exactly half of inhabitants change their opinion.

After k years the council counts the votes, given by each single inhabitant (the vote was open), chooses those of them who, during these k years, voted YES the same number of times as NO, and invites these inhabitants to a party (a "bribery party"?). How many inhabitants will come to that party? (We assume that every invited person comes.) Of course, if k is odd, nobody will come. But what about an even k? Using elementary linear algebra it can be shown that, for even k, at least n/k of citizens will not come.

The question can be formalized as follows. Let $s(k)$ be the number of inhabitants who will *not* come to the party. Associate with the ith voting a $(-1, +1)$-vector $v_i = (v_{i1}, \ldots, v_{in})$ where $v_{ij} = +1$ if and only if the jth inhabitant voted YES during the ith voting. Then $s(k)$ is exactly the number of nonzero entries in the vector $v_1 + v_2 + \cdots + v_k$. Moreover, the "stability" of votings ensures that the vectors $v_1, \ldots, v_k$ are mutually orthogonal, i.e., $\langle v_i, v_j \rangle = 0$ for $i \neq j$.

The desired lower bound $s(k) \geqslant n/k$ on the number of persons who will not come to the party follows from the following more general result (in our case $c_1 = \ldots = c_k = 1$):

Lemma 15.2 (Alon 1990a). *Let $v_i = (v_{i1}, v_{i2}, \ldots, v_{in})$, $1 \leqslant i \leqslant k$, be k mutually orthogonal vectors, where $v_{ij} \in \{-1, +1\}$, and let $c_1, c_2, \ldots, c_k$ be k reals, not all zero. Then the vector $y = c_1 v_1 + c_2 v_2 + \cdots + c_k v_k$ has at least n/k nonzero entries.*

Proof. Assume, without loss of generality, that $|c_1| = max_{1 \leqslant i \leqslant k} |c_i|$. Put $y = (y_1, y_2, \ldots, y_n)$ and let s be the number of nonzero entries in y, i.e., $s = |S|$ where $S = \{j : y_j \neq 0\}$. Also let $|y|$ stand for the vector $(|y_1|, \ldots, |y_n|)$ of the absolute values of entries of y. Since vectors $v_1, \ldots, v_k$ are mutually orthogonal, we have

$$
kc_1^2 n \geqslant \sum_{i=1}^{k} c_i^2 n = \sum_{i=1}^{k} \langle c_i v_i, c_i v_i \rangle = \left\langle \sum_{i=1}^{k} c_i v_i, \sum_{i=1}^{k} c_i v_i \right\rangle = \langle y, y \rangle
$$
$$
= \sum_{j=1}^{n} y_j^2 = \sum_{j \in S} |y_j|^2 = \frac{1}{s} \left(\sum_{j \in S} 1 \right) \left(\sum_{j \in S} |y_j|^2 \right) \geqslant \frac{1}{s} \left(\sum_{j \in S} |y_j| \right)^2,
$$

where the last inequality follows from Cauchy–Schwarz inequality (14.3). On the other hand, since v_1 is orthogonal to all the vectors $v_2, \ldots, v_k$,

$$
\sum_{j=1}^{n} |y_j| \geqslant \sum_{j=1}^{n} y_j v_{1j} = \sum_{j=1}^{n} \sum_{i=1}^{k} c_i v_{ij} v_{1j}
$$
$$
= \sum_{i=1}^{k} c_i \sum_{j=1}^{n} v_{ij} v_{1j} = \sum_{i=1}^{k} c_i \langle v_i, v_1 \rangle = c_1 \langle v_1, v_1 \rangle = c_1 \cdot n.
$$

Substituting this estimate into the previous one we obtain $s \geqslant n/k$, as desired. □

This result gives us some new information about Hadamard matrices. A Hadamard matrix is a square $n \times n$ matrix H with entries in $\{-1, +1\}$ and with row vectors mutually orthogonal (and hence with column vectors mutually orthogonal). Thus, H has a maximal rank n. (Recall that the rank of a matrix is the minimal number of linearly independent rows.)

The following corollary from Alon's lemma says that not only the Hadamard matrix itself but also each of its large enough submatrices has maximal rank.

Corollary 15.3. *If $t > (1 - 1/r)n$, then every $r \times t$ sub-matrix H' of an $n \times n$ Hadamard matrix H has rank r (over the reals).*

Proof. Suppose this is false. Then there is a real nontrivial linear combination of the rows of H' that vanishes. But by Lemma 15.2 this combination, taken with the corresponding rows of H, has at least n/r nonzero entries, and at least one of these must appear in a column of H', a contradiction. □

Changing some entries of a real matrix by appropriate reals we can always reduce its rank. The *rigidity* of a matrix M is the function $R_M(r)$, which for a given r, gives the minimum number of entries of M which one has to change in order to reduce its rank to r or less. Due to its importance, the rigidity of Hadamard matrices deserves continuous attention. For an $n \times n$ Hadamard matrix H, Pudlák, Razborov, and Savický (1988) proved that $R_H(r) \geqslant \frac{n^2}{r^3 \log r}$. Alon's lemma implies that $R_H(r) \geqslant \frac{n^2}{r^2}$.

Corollary 15.4. *If less than $(n/r)^2$ entries of an $n \times n$ Hadamard matrix H are changed (over the reals) then the rank of the resulting matrix remains at least r.*

Proof. Split H into n/r submatrices with r rows in each. Since less than $(n/r)(n/r)$ of the entries of H are changed, in at least one of these $r \times n$ submatrices strictly less than n/r changes are made. Thus, there is an $r \times t$ submatrix, with $t > n - n/r$, in which no change has taken place. The result now follows from Corollary 15.3.   □

15.1.3 Hadamard matrices

In the previous section we demonstrated how, using an elementary linear algebra, one can prove some non-trivial facts about the rank and rigidity of Hadamard matrices. In this section we will establish several other important properties of these matrices. Recall that a Hadamard matrix of order n is an $n \times n$ matrix with entries in $\{-1, +1\}$ whose rows (columns) are mutually orthogonal.

The first property is the *Lindsey Lemma*. Its proof can be found in Erdős and Spencer (1974). We present the neat proof given in Babai, Frankl and Simon (1986).

Lemma 15.5 (J. H. Lindsey). *Let H be an $n \times n$ Hadamard matrix and T be an arbitrary $a \times b$ sub-matrix of H. Then the difference between the number of $+1's$ and $-1's$ in T is at most $\sqrt{abn}$.*

Proof. Let $v_i = (v_{i1}, \ldots, v_{in})$ denote the ith row of H. Assume that T consists of its first a rows and b columns, and let

$$\alpha = \sum_{i=1}^{a} \sum_{j=1}^{b} v_{ij}.$$

We want to prove that $\alpha \leqslant \sqrt{abn}$.

Set $x = (1^b 0^{n-b})$ and consider the vector $y = (y_1, \ldots, y_n) = \sum_{i=1}^{a} v_i$. By the Cauchy–Schwarz inequality (14.3),

$$\alpha^2 = \langle x, y \rangle^2 \leqslant \|x\|^2 \|y\|^2 = b \cdot \|y\|^2.$$

Since H is Hadamard, the vectors v_i are orthogonal; hence

$$\|y\|^2 = \langle y, y \rangle = \left\langle \sum_{i=1}^{a} v_i, \sum_{i=1}^{a} v_i \right\rangle = \sum_{i=1}^{a} \sum_{j=1}^{a} \langle v_i, v_j \rangle = \sum_{i=1}^{a} \langle v_i, v_i \rangle = an.$$

Thus, $\alpha^2 \leqslant b \cdot \|y\|^2 = abn$, as desired. $\square$

We can multiply any rows and columns of a Hadamard matrix by -1 to obtain other Hadamard matrices. In particular, starting from an arbitrary Hadamard matrix, we can reduce it to the form where the first row or the first column (or both) consist entirely of 1's. In this case the matrix is called *normalized*.

Theorem 15.6. *If H is a Hadamard matrix of order n and its first row (column) consists entirely of 1's, then every other row (column) has $n/2$ positive and $n/2$ negative entries. If $n > 2$ then any two rows (columns) other than the first have exactly $n/4$ 1's in common.*

Proof. The first statement immediately follows from the fact that the scalar product of any row with the first row is 0.

To prove the second statement, let u and v be two rows other than first, and let a (resp. b) be the number of places where they both have 1's (resp. -1's). Because u has the same number $n/2$ of 1's and -1's, we get the following picture:

$$
\begin{array}{ccccc}
u & +1+1\dots+1 & +1+1\dots+1 & -1-1\dots-1 & -1-1\dots-1 \\
v & +1+1\dots+1 & -1-1\dots-1 & +1+1\dots+1 & -1-1\dots-1 \\
\\
 & a & n/2-a & n/2-b & b
\end{array}
$$

Since the total number of $+1$'s in v is $n/2$, we have $a + (n/2 - b) = n/2$, and hence, $a = b$. The orthogonality of u and v then implies that $a - (n/2 - a) - (n/2 - b) + b = 0$, i.e., that $a = n/4$. $\square$

Let H be a Hadamard matrix of order n. Take all the rows of H and $-H$, and change all -1's to 0. This way we obtain a set of $2n$ binary vectors of length n called the *Hadamard code* C_n.

Theorem 15.7. *Every two codewords in C_n differ in at least $n/2$ coordinates.*

Proof. Take any $x, y \in C_n$, $x \neq y$. If these two vectors have been obtained from the ith rows of H and $-H$ respectively, then they disagree in all n coordinates. Otherwise, there are two different rows u and v in H such that x is obtained (by changing -1's to 0's) from u or $-u$, and y from v or $-v$. In all cases, x and y differ in $n/2$ coordinates, because $\pm u$ and $\pm v$ are orthogonal. $\square$

Hadamard matrices can also be used to construct combinatorial designs with good parameters. Recall that a (v, k, λ) *design* is a k-uniform family of subsets (called also *blocks*) of a v-element set such that every pair of distinct points is contained in exactly λ of these subsets; if the number of blocks is the same as the number v of points, then the design is *symmetric* (see Chap. 13).

By Theorem 15.6, we have that, if there is a Hadamard matrix of order n, then $n = 2$ or n is divisible by 4. It is conjectured that Hadamard matrices exist for *all* orders that are divisible by 4.

Theorem 15.8. *If there exists a Hadamard matrix of order $4n$ then there exists a symmetric $(4n - 1, 2n - 1, n - 1)$ design.*

Proof. Let H be a Hadamard matrix of order $4n$, and assume that it is normalized, i.e., the first row and the first column consist entirely of 1's. Form a $(4n - 1) \times (4n - 1)$ 0-1 matrix M by deleting the first column and the first row in H, and changing -1's to 0's. This is an incidence matrix of a symmetric $(4n - 1, 2n - 1, n - 1)$ design, because by Theorem 15.6, each row of M has $2n - 1$ 1's and any two columns of M have exactly $n - 1$ 1's in common. □

15.2 Rank arguments

In this section we consider several applications based on the sub-additivity of the rank of 0-1 matrices and its relation to the dimension.

15.2.1 Balanced families

A family $A_1, \ldots, A_m$ of distinct sets is *balanced* if there exist two disjoint and non-empty subsets of indices I and J such that

$$\bigcup_{i \in I} A_i = \bigcup_{j \in J} A_j \quad \text{and} \quad \bigcap_{i \in I} A_i = \bigcap_{j \in J} A_j.$$

Theorem 15.9 (Lindstrom 1993). *Every family of $m \geqslant n + 2$ distinct subsets of an n-element set is balanced.*

Proof. With each subset A of $\{1, \ldots, n\}$ we can associate the incidence vector $v = (x_1, y_1, x_2, y_2, \ldots, x_n, y_n)$ of the pair $(A, \overline{A})$ in the usual way: $x_i = 1$ iff $i \in A$, and $y_i \rightleftharpoons 1 - x_i$. These vectors belong to the vector space V (over $\mathbb{R}$) of all vectors v for which $x_1 + y_1 = \cdots = x_n + y_n$.

Claim 15.10. *The dimension of V is $n + 1$.*

To prove the claim, observe that for any vector $v = (x_1, y_1, x_2, y_2, \ldots, x_n, y_n)$ in V, the knowledge of $n+1$ coordinates $x_1, \ldots, x_n, y_1$ is enough to reconstruct the whole vector v; namely $y_i = x_1 + y_1 - x_i$. So, our space V is the set of solutions $v \in \mathbb{R}^{2n}$ of the system of linear equations $M \cdot v = 0$, where M is the $(n - 1) \times (2n)$ matrix

$$\begin{pmatrix} 1 & 1 & -1 & -1 & 0 & 0 & \cdots & 0 & 0 \\ 1 & 1 & 0 & 0 & -1 & -1 & \cdots & 0 & 0 \\ \vdots & \vdots & \vdots & \vdots & \vdots & \vdots & \cdots & \vdots & \vdots \\ 1 & 1 & 0 & 0 & 0 & 0 & \cdots & -1 & -1 \end{pmatrix}$$

By Proposition 14.3, $\dim V = 2n - \mathrm{rk}(M) = 2n - (n-1) = n+1$, as desired.

Now let $v_i = (v_{i,1}, \ldots, v_{i,2n})$ be the vector corresponding to the ith set A_i, $i = 1, \ldots, m$. By the assumption, the vectors $v_1, \ldots, v_m$ are distinct and all belong to the subspace V. Since $m \geqslant n+2 > n+1 = \dim V$, there must be a nontrivial linear relation between these vectors, which we can write as

$$\sum_{i \in I} \alpha_i v_i = \sum_{j \in J} \beta_j v_j,$$

where I and J are non-empty, $I \cap J = \emptyset$, and $\alpha_i, \beta_j > 0$ for all $i \in I$ and $j \in J$. But this means that

$$\bigcup_{i \in I} A_i = \bigcup_{j \in J} A_j \quad \text{and} \quad \bigcup_{i \in I} \overline{A}_i = \bigcup_{j \in J} \overline{A}_j,$$

where (using the identity $\overline{A} \cup \overline{B} = \overline{A \cap B}$) the last equality is equivalent to $\bigcap_{i \in I} A_i = \bigcap_{j \in J} A_j$. □

15.2.2 Lower bounds for boolean formulas

A *DeMorgan formula* (or just a *formula*) is a circuit with And, Or, and Not gates, whose underlying graph is a tree. Such formulas can be defined inductively as follows:

(1) Every boolean variable x_i and its negation $\overline{x}_i$ is a formula of size 1 (these formulas are called *leaves*).
(2) If F_1 and F_2 are formulas of size l_1 and l_2, then both $F_1 \wedge F_2$ and $F_1 \vee F_2$ are formulas of size $l_1 + l_2$.

Note that the size of F is exactly the number of leaves in F.

Given a boolean function f, how it can be shown that it is hard, i.e., that it cannot be computed by a formula of small size? Easy counting shows that almost all boolean functions in n variables require formulas of size exponential in n. Still, for a *concrete* boolean function f, the largest remains the lower bound $n^{3-o(1)}$ proved by Håstad (1993).

The main difficulty here is that we allow negated variables $\overline{x}_i$ as leaves. It is therefore natural to look at what happens if we forbid this and require that our formulas are *monotone* in that they do not have negated leaves. Of course, not every boolean function $f(x_1, \ldots, x_n)$ can be computed by such a formula – the function itself must be also *monotone*: if $f(x_1, \ldots, x_n) = 1$

and $x_i \leqslant y_i$ for all i, then $f(y_1, \ldots, y_n) = 1$. Under this restriction progress is substantial: we are able to prove that some explicit monotone functions require monotone formulas of super-polynomial size. This fact is not new to us – in Sect. 10.6 we have already proved such lower bounds for a more general model of monotone *circuits*. However, the proof for formulas employs a very elegant rank argument, and hence, is much in the spirit of this chapter.

Reduction to set-covering. Let A and B be two disjoint subsets of $\{0, 1\}^n$. A boolean formula F *separates* A and B if $F(a) = 1$ for all $a \in A$ and $F(b) = 0$ for all $b \in B$.

Definition 15.11. A *rectangle* is a subset $R = A' \times B'$ of $A \times B$ such that A' and B' are separated by a variable x_i or its negation $\overline{x}_i$, that is, there must be is a coordinate $i \in \{1, \ldots, n\}$ such that $a(i) \neq b(i)$ for all vectors $a \in A'$ and $b \in B'$. If we have a stronger condition that $a(i) = 1$ and $b(i) = 0$ for all $a \in A'$ and $b \in B'$ (i.e., if we do not allow negations $\overline{x}_i$), then the rectangle is *monotone*.

The following simple lemma reduces the (computational) problem of proving a lower bound on the size of a DeMorgan formulas separating a pair A, B to a (combinatorial) problem of proving a lower bound on the number of mutually disjoint rectangles covering the Cartesian product $A \times B$.

Lemma 15.12 (Rychkov 1985). *If A and B can be separated by a (monotone) DeMorgan formula of size t then the set $A \times B$ can be covered by t mutually disjoint (monotone) rectangles.*

Proof. Let F be an optimal DeMorgan formula which separates the pair A, B, i.e., $A \subseteq F^{-1}(1)$ and $B \subseteq F^{-1}(0)$. Let $t = size(F)$. We argue by the induction on t.

Base case. If $size(F) = 1$ then F is just a single variable x_i or its negation. In that case the Cartesian product $A \times B$ is the rectangle itself, and we are done.

Induction step. Assume that the theorem holds for all formulas smaller than F, and suppose that $F = F_1 \wedge F_2$ (the case $F = F_1 \vee F_2$ is similar). Let $t_i = size(F_i)$, hence $t = t_1 + t_2$. Define $B_1 \rightleftharpoons \{b \in B : F_1(b) = 0\}$ and $B_2 \rightleftharpoons B - B_1$. Notice that F_i separates A and B_i for $i = 1, 2$. Applying the induction hypothesis to the subformula F_i yields that the product $A \times B_i$ can be covered by t_i mutually disjoint rectangles, for both $i = 1, 2$. Since $A \times B = A \times B_1 \cup A \times B_2$, we have that the set $A \times B$ can be covered by $t_1 + t_2 = t$ rectangles, as desired. □

We can use Rychkov's lemma to derive the well-known lower bound due to Khrapchenko (1971). Given two disjoint subsets A and B of $\{0, 1\}^n$, let $N(A, B)$ denote the set of all pairs (a, b) where $a \in A$, $b \in B$ and these vectors differ in exactly one coordinate; in this case we also say that a and b are *neighbors*.

Intuitively, if $N(A, B)$ is large, then every formula separating A and B must be large, since the formula must distinguish many pairs of "very similar" inputs. The following theorem of Khrapchenko makes this intuition precise.

Theorem 15.13 (Khrapchenko 1971). *Let F be a DeMorgan formula that separates A and B. Then F has size at least*

$$\frac{|N(A, B)|^2}{|A| \cdot |B|}.$$

Proof. Consider a partition of $A \times B$ into rectangles as in Rychkov's lemma: $A \times B = \bigcup_{i=1}^{t} A_i \times B_i$. Let

$$R_i \rightleftharpoons N(A, B) \cap (A_i \times B_i).$$

Then in the rectangle $A_i \times B_i$, each row and column can contain at most one element of R_i. This is because each element of A_i differs from each element in B_i in one particular position j, whereas (a, b) is in $N(A, B)$ only if a and b differ in exactly one position. Hence, for any given $a \in A_i$, the only possible $b \in B_i$ for which $(a, b) \in R_i$ is one which differs from a exactly in position j. As a result, we have $|R_i| \leqslant |A_i|$ and $|R_i| \leqslant |B_i|$, which implies

$$|R_i|^2 \leqslant |A_i| \cdot |B_i|.$$

Furthermore, $N(A, B) = R_1 \cup R_2 \cup \cdots \cup R_t$, with all of the R_i's disjoint. Since each R_i includes the neighbors in the ith rectangle, together they include all neighbors. Since the R_i's are disjoint, $|N(A, B)| = \sum_{i=1}^{t} |R_i|$, and hence, $|N(A, B)|^2 = \left(\sum_{i=1}^{t} |R_i| \right)^2$ which, by Cauchy–Schwarz inequality (14.3), does not exceed

$$t \cdot \sum |R_i|^2 \leqslant t \cdot \sum |A_i| \cdot |B_i| = t|A| \cdot |B|.$$

$\square$

Khrapchenko's theorem can be used to show that some explicit boolean functions require formulas of quadratic size. Consider, for example, the parity function $f = x_1 \oplus \cdots \oplus x_n$. Taking $A = f^{-1}(1)$ and $B = f^{-1}(0)$ we see that $|N(A, B)| = n|A| = n|B|$, and hence, f requires DeMorgan formulas of size at least n^2.

Unfortunately, this (quadratic) lower bound is the best that we can achieve using this theorem (Exercise 15.9).

The rank lower bound. Every set $\mathcal{R}$ of rectangles, whose union gives the whole set $A \times B$, is its *cover*. The *canonical* monotone cover $\mathcal{R}_{\mathrm{mon}}(A, B)$ consists of n rectangles $R_1, R_2, \ldots, R_n$, where

$$R_i \rightleftharpoons \{(a, b) \in A \times B : a(i) = 1 \text{ and } b(i) = 0\}.$$

By a *matrix over* A, B we mean a matrix M over a field $\mathbb{F}$ whose rows are indexed by elements of the set A and columns by elements of the set B. Given a rectangle $R \subseteq A \times B$, we denote by M_R the corresponding submatrix of M. By $\widehat{M}_R$ we denote the matrix (over A, B) which is obtained from the matrix M by changing all its entries $m_{u,v}$ with $(u, v) \notin R$, to 0.

The following theorem gives a lower bound on the size of a DeMorgan formula separating A and B, in terms of the rank of these matrices. By $L_+(f)$ we denote the minimal size of a monotone DeMorgan formula computing f.

Theorem 15.14 (Razborov 1990). *For any $A, B \subseteq \{0, 1\}^n$ and a monotone boolean function f such that $f(A) = 1$, $f(B) = 0$ and any non-zero matrix M over A, B (over an arbitrary field $\mathbb{F}$), we have*

$$L_+(f) \geqslant \frac{\mathrm{rk}(M)}{\max_R \mathrm{rk}(M_R)} \qquad (15.2)$$

where the maximum is over all rectangles $R \in \mathcal{R}_{\mathrm{mon}}(A, B)$.

Proof. Let $t = L_+(f)$. By Rychkov's lemma we know that there exists a set $\mathcal{R}$ of $|\mathcal{R}| \leqslant t$ mutually disjoint (monotone) rectangles which cover the set $A \times B$. So, $M = \sum_{R \in \mathcal{R}} \widehat{M}_R$, and hence, by the sub-additivity of the rank (see (14.1)),

$$\mathrm{rk}(M) = \mathrm{rk}\left(\sum_{R \in \mathcal{R}} \widehat{M}_R\right) \leqslant \sum_{R \in \mathcal{R}} \mathrm{rk}\left(\widehat{M}_R\right).$$

On the other hand, for every $R \in \mathcal{R}$ there is a rectangle $R' \in \mathcal{R}_{\mathrm{mon}}(A, B)$ such that $R \subseteq R'$. Therefore, $\mathrm{rk}(M_R) = \mathrm{rk}\left(\widehat{M}_R\right) \leqslant \mathrm{rk}\left(\widehat{M}_{R'}\right)$, and

$$\mathrm{rk}(M) \leqslant |\mathcal{R}| \cdot \max_{R' \in \mathcal{R}_{\mathrm{mon}}} \mathrm{rk}\left(\widehat{M}_{R'}\right),$$

implying the desired lower bound on $|\mathcal{R}|$ and hence, on $t = L(f)$. $\qquad \square$

It is clear that the same lower bound (15.2) also holds for *non-monotone* formulas, if we extend the set $\mathcal{R}_{\mathrm{mon}}(A, B)$ by adding n "dual" rectangles $R'_1, \ldots, R'_n$, where

$$R'_i \rightleftharpoons \{(a, b) \in A \times B : a(i) = 0 \text{ and } b(i) = 1\}.$$

However, Razborov (1992) has proved that in this case the result is useless: for any boolean function f in n variables, the fraction on the right-hand side of (15.2) does not exceed $O(n)$ (see Sect. 20.8). Fortunately, in the monotone case, Theorem 15.14 *can* give large lower bounds, and we are going to show this in the next section.

The lower bound for Paley functions. Let us consider bipartite graphs $G = (V_1, V_2, E)$ with $|V_1| = |V_2| = n$. With any such graph we can associate a monotone boolean function $f_{G,k}$ as follows. The function has $2n$ variables, one for each node of G, and accepts a set of nodes $X \subseteq V_1 \cup V_2$ if and only if X contains some subset $S \subseteq V_1$ of size at most k, together with the set of its *common neighbors*

$$\Gamma(S) \rightleftharpoons \{j \in V_2 : (i, j) \in E \text{ for all } i \in S\}.$$

That is, $f_{G,k}$ is the Or of all $\sum_{i=0}^{k} \binom{n}{i}$ monomials $\bigwedge_{i \in S \cup \Gamma(S)} x_i$ where $S \subseteq V_1$ and $|S| \leqslant k$. By $\widehat{\Gamma}(S)$ we will denote the set of all *common non-neighbors* of S, that is,

$$\widehat{\Gamma}(S) = \{j \in V_2 : (i,j) \in E \text{ for no } i \in S\}.$$

By the definition, for every bipartite graph G, the function $f_{G,k}$ is an Or of at most $\sum_{i=0}^{k} \binom{n}{i}$ And's, each of length at most $2n$. It appears that, for graphs satisfying the isolated neighbor condition (see Definition 11.7), this trivial formula is almost optimal.

Recall that a bipartite graph $G = (V_1, V_2, E)$ satisfies the *isolated neighbor condition for k* if for any two disjoint subsets $S, T \subseteq V_1$ such that $|S|+|T| = k$, there is a node $v \in V_2$ which is a common neighbor of all the nodes in S and is isolated from all the nodes in T, i.e., if $\Gamma(S) \cap \widehat{\Gamma}(T) \neq \emptyset$.

Lemma 15.15 (Gál 1998). *If G satisfies the isolated neighbor condition for $2k$, then the function $f_{G,k}$ does not have a monotone DeMorgan formula of size smaller than $\sum_{i=0}^{k} \binom{n}{i}$.*

Proof. If the graph G satisfies the isolated neighbor condition for $2k$, then we have the following intersection property: for any two subsets $S, T \subseteq V_1$ of size at most k,

$$S \cap T = \emptyset \text{ if and only if } \Gamma(S) \cap \widehat{\Gamma}(T) \neq \emptyset. \tag{15.3}$$

Now consider the following 0-1 matrix M. Its rows and columns are labeled by subsets of V_1 of size at most k; the entries are defined by:

$$M_{S,T} = 1 \text{ if and only if } S \cap T = \emptyset.$$

This is a disjointness matrix $D(n, k)$, considered in Sect. 14.2.3, and we already know (see Theorem 14.10) that it has full rank over $\mathbb{F}_2$:

$$\mathrm{rk}_{\mathbb{F}_2}(M) = \sum_{i=0}^{k} \binom{n}{i}.$$

In order to apply Theorem 15.14, we will now label the rows and columns of this matrix by vectors from special subsets of vectors A and B in $\{0,1\}^{2n}$ so that:

(a) $f_{G,k}(A) = 1$ and $f_{G,k}(B) = 0$;
(b) for every $R \in \mathcal{R}_{\mathrm{mon}}(A, B)$, the submatrix M_R has rank 1.

To achieve these goals, do the following. If a row of the intersection matrix M is labeled by a set S, then relabel this row by the incidence vector v_S of $S \cup \Gamma(S)$; hence, v_S is a binary vector of length $2n$ such that, for all coordinates $i \in V_1 \cup V_2$,

$$v_S(i) = 1 \text{ if and only if } i \in S \cup \Gamma(S).$$

If a column of M is labeled by a set T, then relabel this column by the vector u_T such that

$$u_T(i) = 0 \text{ if and only if } i \in T \cup \widehat{\Gamma}(T).$$

Let A be the set of all vectors v_S, and B be the set of all vectors u_T, where S and T range over all subsets of V_1 of size at most k.

To verify the condition (a), note that $f_{G,k}(x) = 1$ if and only if $x \geqslant v_S$ for some S. Hence, $f(a) = 1$ for all $a \in A$. To show that $f(b) = 0$ for all $b \in B$, observe that, by (15.3), each pair of vectors v_S and u_T has a coordinate i on which $0 = u_T(i) < v_S(i) = 1$, and hence, $u_T \not\geqslant v_S$ implying that u_T must be rejected by f, as desired.

Let us now verify the second condition (b).

For each node $i \in V_1 \cup V_2$, let R_i denote the rectangle consisting of all pairs of vectors (v_S, u_T) for which $v_S(i) = 1$ and $u_T(i) = 0$. If $i \in V_1$ then the corresponding entry of the intersection matrix M is 0 because then $S \cap T \neq \emptyset$; if $i \in V_2$ then this entry is 1 because then $\Gamma(S) \cap \widehat{\Gamma}(T) \neq \emptyset$, and by (15.3), $S \cap T = \emptyset$. Thus, each of the rectangles R_i is either 0-monochromatic or 1-monochromatic, and hence, has rank 1.

We have shown that the set of rectangles $\mathcal{R} = \{R_i : v \in V_1 \cup V_2\}$ is a canonical monotone cover of M, and $\text{rk}(R_i) = 1$ for each i because all these rectangles are monochromatic. By Razborov's theorem (Theorem 15.14), we conclude that

$$L^+(f_{G,k}) \geqslant \frac{\text{rk}(M)}{\max_R \text{rk}(M_R)} = \text{rk}(M) \geqslant \sum_{i=0}^{k} \binom{n}{i}.$$

$\square$

Explicit bipartite graphs, satisfying the isolated neighbor condition for $k = \Omega(\log n)$ are known. Such are, for example, *Paley graphs* G_n constructed in Sect. 11.3.2. By Lemma 15.15, the corresponding boolean function $f_{G_n,k}$ requires monotone formula of size at least $\binom{n}{k} = n^{\Omega(\log n)}$.

For a related function PALEY(n, k) we have proved in Sect. 10.6.2 that it requires even monotone *circuits* of the same (super-polynomial) size. The main difference between formulas and circuits is that in a circuit each intermediate result of computation may be used many times, whereas in a formula each query of a previously obtained result requires re-computing it again. Thus (at least potentially) circuits are more powerful than formulas. In the monotone case this fact is even provable: there are explicit monotone boolean functions that can be computed by monotone circuits of polynomial size but require monotone formulas of super-polynomial size (Karchmer–Wigderson 1990). In this context it is interesting to note that in the case of circuits the isolated neighbor condition of Paley graphs alone is not enough – we have employed a more general property of these graphs, given by Theorem 11.9.

Exercises

15.1.⁻ Let $\mathcal{A}$ and $\mathcal{B}$ be families of subsets of an n-element set with the property that $|A \cap B|$ is *odd* for all $A \in \mathcal{A}$ and $B \in \mathcal{B}$. Prove that then $|\mathcal{A}| \cdot |\mathcal{B}| \leqslant 2^{n-1}$.

Hint: Replace A in the proof of Theorem 15.1 by a larger set $A' = A \cup A_0$ where $A_0 = \{u_0 + u : u \in A\}$ for some fixed $u_0 \in A$. Show that $A \cap A_0 = \emptyset$, and argue as in that proof with A' instead of A.

15.2. Define the matrices H_{2m}, $m = 2, 4, 8, \ldots$, inductively as follows:

$$H_2 = \begin{pmatrix} 1 & 1 \\ 1 & -1 \end{pmatrix}, \qquad H_{2m} = \begin{pmatrix} H_m & H_m \\ H_m & -H_m \end{pmatrix}.$$

Show that these matrices are Hadamard.

15.3.⁻ Show that Alon's lemma (Lemma 15.2) is sharp, at least whenever k divides n and there exists a $k \times k$ Hadamard matrix H.

Hint: Take n/k copies of H.

15.4. Take an $n \times n$ matrix over some field $\mathbb{F}$ and suppose that all of its rows are different. Prove the following: if some column is linearly dependent on the others then after the deletion of this column, all the rows in the resulting n by $n-1$ matrix are still different.

15.5[(!)] (Babai–Frankl 1992). Give a linear algebra proof of Bondy's theorem (Theorem 12.1): In any $n \times n$ 0-1 matrix M with all rows being different, there is a column, after deletion of which the resulting n by $n-1$ matrix still has no equal rows.

Sketch: Consider two cases depending on what is the determinant of M. If $\det(M) = 0$ then some column is linearly dependent on the others, and we are in the situation of the previous exercise. If $\det(M) \neq 0$ then take a row v_i of M with the minimal number of 1's, and expand the determinant by this row. Conclude that for some j, the term $v_{ij} \cdot \det M_{ij} \neq 0$, where M_{ij} is the $(n-1)$ by $(n-1)$ minor obtained by deleting the ith row and the jth column. Hence, $v_{ij} = 1$ and no two rows of the minor are identical. Use this to prove that deleting the jth column from the whole matrix M leaves no equal rows.

15.6.⁺ For $n \geqslant 1$, $d \geqslant 0$, $n \equiv d \,(\mathrm{mod}\,2)$, let $K(n,d)$ denote the minimal cardinality of a family V of ± 1 vectors of length n, such that for any ± 1 vector w of length n, there is a $v \in V$ such that the value of the scalar product $\langle v, w \rangle$ (over reals) lies between $-d$ and d. Prove that:

(i) $K(n,0) \leqslant n$ (Knuth 1986).

Hint: Consider ± 1 vectors $v_0, v_1, \ldots, v_n$ of length n, where the ith vector v_i has first i coordinates equal to -1 and the rest equal to $+1$; hence, v_0 has no -1's at all whereas v_n consists entirely of -1's. Observe that $\langle w, v_0 \rangle = -\langle w, v_n \rangle$, while $\langle w, v_i \rangle = \langle w, v_{i+1} \rangle \pm 2$ for each $i = 0, 1, \ldots, n-1$. Note that $\langle w, v_i \rangle$ is even for all i, and apply the pigeonhole principle to show that $\langle w, v_i \rangle = 0$ for some $i \leqslant n - 1$.

(ii) $K(n, d) \leqslant \lceil n/(d+1) \rceil$ (Alon et al. 1988).

> *Hint:* Consider the same vectors as before, and select only the vectors $u_j \rightleftharpoons v_{j\cdot(d+1)+1}$ for $j = 0, 1, \ldots, r$; $r = \lceil n/(d+1) \rceil - 1$. Observe that for any ± 1 vector w and any j, $0 \leqslant j < r$, $\langle w, u_j \rangle = \langle w, u_{j+1} \rangle \pm (2d+2)$.

Note: Alon et al. (1988) have also proved that this upper bound is tight.

15.7 (Alon et al. 1988). Let V be the set of all ± 1 vectors of length n. A vector is *even* if it has an even number of -1's, otherwise it is *odd*. Let $f(x_1, \ldots, x_n)$ be a multilinear polynomial of degree less than $n/2$ over the reals, i.e.,

$$f = \sum_{|S| < n/2} \alpha_S \prod_{i \in S} x_i,$$

where $\alpha_S \in \mathbb{R}$. Suppose that $f(v) = 0$ for every even vector $v \in V$. Prove that then $f \equiv 0$, i.e., $\alpha_S = 0$ for all S. Does the same holds if $f(v) = 0$ for every odd vector $v \in V$?

Sketch: By the hypothesis, for every even subset $T \subseteq N$ we have

$$\sum_{|S| < n/2} \alpha_S (-1)^{|S \cap T|} = 0.$$

It thus suffices (why?) to show that the rows of the matrix

$$A = \left\{ (-1)^{|S \cap T|} : |T| \text{ even and } |S| < n/2 \right\}$$

are linearly independent (over the reals). For this, show that the matrix $M = A^T A$ has non-zero determinant. The (S_1, S_2)-th entry of M is the sum

$$\sum_T (-1)^{|S_1 \cap T| + |S_2 \cap T|} = \sum_T (-1)^{|(S_1 \oplus S_2) \cap T|}$$

over all even T. If $S_1 = S_2$, then this sum is 2^{n-1}. If $S_1 \neq S_2$, then $0 < |S_1 \oplus S_2| < n$; use this to show that in this case the sum is 0.

15.8.⁻ Let $n = 2m+1$, and consider the majority function $\text{MAJ}_n(x_1, \ldots, x_n)$, which outputs 1 iff $x_1 + \ldots + x_n \geqslant m + 1$. The best known upper bound $O(n^{4.57})$ for the formula size of MAJ_n is due to Valiant. Use Khrapchenko's theorem to show that this function requires DeMorgan formulas of size $\Omega(n^2)$.

Hint: Take $A = \{a : |a| = m + 1\}$ and $B = \{b : |b| = m\}$.

15.9.⁻ Show that Khrapchenko's theorem cannot yield larger than quadratic lower bounds. *Hint:* Each vector in $\{0, 1\}^n$ has only n neighbors.

15.10.⁺ Research problem: It is not known if the converse of Rychkov's lemma (Lemma 15.12) holds. Suppose that $A \times B$ can be covered by t mutually disjoint rectangles. Does there then exist a DeMorgan formula which separates A, B and has size at most t^c for some absolute constant c?

16. Span Programs

In 1993 Karchmer and Wigderson introduced an interesting linear algebraic model for computing boolean functions – the *span program*. A span program for a function $f(x_1, \ldots, x_n)$ is presented as a matrix over some field, with rows labeled by variables x_i or their negations $\overline{x}_i$ (one variable can label many rows). The span program accepts an input assignment if and only if the all-1 vector can be obtained as a linear combination of the rows whose labels are satisfied by the input. The size of the span program is the number of rows in the matrix. A span program is *monotone* if only positive literals are used as labels of the rows, i.e. negated variables are not allowed.

The model appears to be quite strong: classical models for computing boolean functions – like switching networks or DeMorgan formulas – can be simulated by span programs without any increase in size. Therefore, proving lower bounds on the size of span programs is a hard task, even for monotone span programs.

In this chapter we will show how this task can be solved using the linear algebra argument.

16.1 The model

We describe the model more precisely.

Let $\mathbb{F}$ be a field. A *span program* over $\mathbb{F}$ is given by a matrix M over $\mathbb{F}$ with its rows labeled by literals $x_1, \ldots, x_n, \overline{x}_1, \ldots, \overline{x}_n$. For an input $a = (a_1, \ldots, a_n) \in \{0, 1\}^n$, let M_a denote the submatrix of M obtained by keeping those rows whose labels are satisfied by a. That is, M_a contains rows labeled by those x_i for which $a_i = 1$ and by those $\overline{x}_i$ for which $a_i = 0$. The program M *accepts* the input a if the all-ones vector $\mathbf{1}$ (or any other, fixed in advance, vector) belongs to the span of the rows of M_a. A span program computes a boolean function f if it accepts exactly those inputs a where $f(a) = 1$. The *size* of a span program is the number of rows in it.

The number of columns is not counted as a part of the size. It is always possible to restrict the matrix of a span program to a set of linearly independent columns without changing the function computed by the program, therefore it is not necessary to use more columns than rows. However, it is usually easier to design a span program with a large number of columns, many of which may be linearly dependent.

16.2 Span programs and switching networks

One of the oldest models for computing boolean functions is that of switching networks. This model includes that of DeMorgan formulas (considered in Sect. 15.2.2) and was intensively studied after C.E. Shannon introduced this model more than 50 years ago.

A *switching network* is a graph $G = (V, E)$ with two specified vertices $s, t \in V$, whose edges are labeled by variables x_i or their negations $\overline{x}_i$. The graph may have multiple edges, i.e., several edges may have the same endpoints. The size of G is defined as the number of edges. Each input $a = (a_1, \ldots, a_n)$ switches the labeled edges on or off by the following rule: the edge, labeled by x_i, is switched on if $a_i = 1$ and is switched off if $a_i = 0$; the edge, labeled by $\overline{x}_i$, is switched on if $a_i = 0$ and is switched off if $a_i = 1$. A switching network G computes a boolean function in a natural way: it accepts the input a if and only if there is a path from s to t along which all edges are switched on by a. That is, each input switches the edges on or off, and we accept that input if and only if after that there is a nonzero conductivity between the vertices s and t. It can be easily shown that DeMorgan formulas correspond to a very special type of switching networks, whose underlying graph consists of parallel-sequential components (see Exercise 16.6).

Theorem 16.1 (Karchmer–Wigderson 1993). *If a boolean function can be computed by a switching network of size s then it can also be computed by a span program of size at most s over any field.*

Proof. Let $G = (V, E)$ be a switching network for a function f, with $s, t \in V$ its special vertices. Fix a field $\mathbb{F}$ and take the standard basis $\{e_i : i \in V\}$ of the $|V|$-dimensional space over $\mathbb{F}$, i.e., e_i is a binary vector of length $|V|$ with exactly one 1 in the ith coordinate.

The span program M is constructed as follows. For every edge $e = \{i, j\}$ in E add the row $e_i - e_j$ to M and label this row by the label of e. It is easy to see that there is an s-t path in G, all whose labeled edges are switched on by a, if and only if the rows of M_a span the vector $e_s - e_t$. Therefore, M computes f, and its size is $|E|$. $\qquad\square$

16.3 Monotone span programs

A span program is called *monotone* if the labels of the rows are only positive literals $x_1, \ldots, x_n$. A boolean function $f(x_1, \ldots, x_n)$ is *monotone* if $f(a_1, \ldots, a_n) \leqslant f(b_1, \ldots, b_n)$ as long as $a_i \leqslant b_i$ for all i. It is clear that such programs can compute only monotone functions; moreover, every monotone boolean function can be computed by a monotone span program (Exercise 16.7).

So far the largest known lower bound for (non-monotone) span program size is $\Omega\left(n^{3/2}/\log n\right)$, which follows from the lower bound on the size of switching networks, due to Nechiporuk (1966). The argument is based on counting the number of subfunctions of a given boolean function, and cannot lead to much larger lower bounds.

In the case of monotone span programs the situation is much better: here we can even prove superpolynomial lower bounds. The bounds of order $n^{\Omega(\log n/\log\log n)}$ were first obtained by Babai et al. (1996) and Babai, Gál, and Wigderson (1999) for explicit boolean functions defined by bipartite graphs with certain properties. Then, using a different property of the underlying bipartite graphs, Gál (1992) simplified and improved these bounds to $n^{\Omega(\log n)}$.

All these proofs were based on a general combinatorial lower bound for such programs, found by Beimel, Gál, and Paterson (1996). An intriguing aspect of their proof is that it reduces the lower bounds problem for the size of monotone span programs to the proof that a particular set of vectors is linearly independent. This criterion, together with its applications for explicit boolean functions, is one of the most interesting recent applications of the linear algebra method in computer science. To acknowledge the power of these results, we first show that some seemingly hard functions can be computed by surprisingly small monotone span programs.

16.3.1 Threshold functions

A *k-threshold function* $T_k^n(x_1,\ldots,x_n)$ outputs 1 iff $x_1+\cdots+x_n\geqslant k$.

Proposition 16.2 (Karchmer–Wigderson 1993). *Let* $\mathbb{F}$ *be a field with at least* $n+1$ *elements. Then, for any* $1\leqslant k\leqslant n$, *the function* T_k^n *can be computed by a monotone span program over* $\mathbb{F}$ *of size* n.

Proof. Since the field $\mathbb{F}$ has more than n elements, we can apply Lemma 14.5 and find a set $\{v_0,v_1,\ldots,v_n\}\subset\mathbb{F}^k$ of $n+1$ vectors in *general position*, i.e., any k of these vectors are linearly independent. Moreover, we may assume w.l.o.g. that $v_0=\mathbf{1}$ (the all-1 vector). This suggests the following span program M over $\mathbb{F}$: M is an $n\times k$ matrix whose ith row ($1\leqslant i\leqslant n$) is v_i and is labeled by x_i. It is now straightforward to check that, for any input $a\in\{0,1\}^n$, the vector $\mathbf{1}$ is spanned by the rows of M_a iff $|a|\geqslant k$. Indeed, if $|a|\geqslant k$ then the vectors $\{v_i:a_i=1\}$ contain a basis for $\mathbb{F}^k$ (because some k of them are linearly independent), and hence, vector $\mathbf{1}$ is a linear combination of them. If $|a|\leqslant k-1$ then all the vectors $\{v_0\}\cup\{v_i:a_i=1\}$ are linearly independent, and hence, $v_0=\mathbf{1}$ cannot be a linear combination of $\{v_i:a_i=1\}$. $\qquad\square$

Taking $\mathbb{F}=\mathrm{GF}(2^l)$ with l the smallest integer $>\log n$ (which corresponds to binary encoding of field elements), it is possible to reduce the constructed span program over $\mathbb{F}$ to a program over the field GF(2) of size $O(n \log n)$ (see Karchmer–Wigderson (1993) for details).

16.3.2 Non-bipartite graphs

Recall that a graph is *bipartite* if its vertex set can be partitioned into two independent sets; a set of vertices is *independent* if there are no edges between them. A bipartite graph is *complete* if every two vertices from different parts are joined by an edge.

Consider the function *Non-Bipartite$_n$*, whose input is an undirected graph on m vertices, represented by $n = \binom{m}{2}$ variables, one for each possible edge. The value of the function is 1 if and only if the graph is not bipartite.

Theorem 16.3 (Beimel–Gál–Paterson 1996). *The function Non-Bipartite$_n$ can be computed over the field $\mathbb{F}_2$ by a monotone span program of size n.*

Proof. We construct a monotone span program accepting exactly the non-bipartite graphs as follows. There will be n rows, each labeled by a variable. There is a column for each possible complete bipartite graph on n vertices. The column for a given complete bipartite graph contains the value 0 in each row that corresponds to an edge of the given graph and contains 1 in every other row.

This program rejects every bipartite graph G. This is because G is contained in some complete bipartite graph, and so there will be a column that contains only 0's in the rows labeled by the edges of G. Therefore the vector **1** is not a linear combination of these rows.

Next, we show that the program accepts every non-bipartite graph. Since the span program is monotone, it is sufficient to show that it accepts every *minimal* non-bipartite graph, i.e., every odd cycle. Let C be an odd cycle. The intersection of any cycle with any complete bipartite graph has an even number of edges. So the odd cycle C has an odd number of edges which are *not* in any given complete bipartite graph. Hence, the sum of the row vectors corresponding to all the edges in C is odd in each column, i.e., gives the vector **1** over $\mathbb{F}_2$, and so C is accepted by the span program. $\square$

16.3.3 Odd factors

A *spanning subgraph* of a graph $G = (V, E)$ is a graph $G' = (V, F)$ where $F \subseteq E$. Note that the set of vertices is the same. A (connected) *component* of a graph is a maximal set of its vertices such that there is a path between any two of them. A graph is *connected* if it consists of just one component. The *degree* $d_E(i)$ of a vertex i is the number of edges of E which are incident to i.

An *odd factor* in a graph is a spanning subgraph with all degrees odd. We need the following property of odd factors.

Lemma 16.4. *If a graph is connected then it has an odd factor if and only if the number of its vertices is even.*

Proof. Suppose that G has an odd factor $G' = (V, F)$. Hence, all degrees $d_F(i)$ are odd. By Euler's theorem (see Theorem 1.7), the sum $\sum_{i \in V} d_F(i)$ equals $2|F|$, and hence, is even. Thus, the number $|V|$ of summands must be even, as claimed.

For the other direction, suppose that the graph $G = (V, E)$ is connected and has an even number of vertices, say $V = \{x_1, \ldots, x_{2m}\}$. For every $i = 1, \ldots, m$, fix any one path $P_i = (V_i, E_i)$ connecting x_i to x_{i+m}. Let F be the set of those edges from E which appear in an odd number of the sets $E_1, \ldots, E_m$ (see Fig. 16.1).

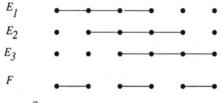

Fig. 16.1. The case $m = 3$

We claim that the subgraph (V, F) is the desired odd factor. Indeed, observe that if a vertex x appears in a path P_i then either $d_{E_i}(x)$ is even or $d_{E_i}(x) = 1$, and this last event happens iff x is a leaf of this path, i.e., if $x = x_i$ or $x = x_{i+m}$. Since each vertex $x \in V$ is a leaf of exactly one of the paths $P_1, \ldots, P_m$, we have that the sum of degrees $D(x) = \sum_{i=1}^{m} d_{E_i}(x)$ is odd. It remains to observe that, by the definition of F, this sum $D(x)$ is congruent modulo 2 to the degree $d_F(x)$ of x in the graph (V, F). □

We now consider the following function $Oddfactor_n$ on $n = m^2$ variables: the input is an $m \times m$ 0-1 matrix representing a bipartite graph X with m vertices in each part; the graph X is accepted if it has an odd factor. Note that, by Lemma 16.4, X is rejected exactly if it has a component with an odd number of vertices.

Theorem 16.5 (Babai–Gál–Wigderson 1996). *$Oddfactor_n$ can be computed over the field $\mathbb{F}_2$ by a monotone span program of size n.*

Proof. We construct the desired span program for $Oddfactor_n$ as follows. Let $V = V_1 \cup V_2$ be the vertex set ($|V_1| = |V_2| = m$), and let $X = \{x_{i,j}\}$ with $i \in V_1$ and $j \in V_2$ be the corresponding set of boolean variables (one for each potential edge). Take the standard basis $\{e_i : i \in V\}$ of the $2m$-dimensional space over $\mathbb{F}_2$, i.e., e_i is a binary vector of length $2m$ with exactly one 1 in the ith coordinate. Let M be the m^2 by $2m$ matrix whose rows are vectors $e_i + e_j$ labeled by the corresponding variables $x_{i,j}$. We claim that this span program computes $Oddfactor_n$. To verify this we have to show that the all-1 vector $\mathbf{1} = (1, \ldots, 1)$ is a sum (over $\mathbb{F}_2$) of vectors of the form $e_i + e_j$ precisely when the corresponding edges (i, j) form an odd factor.

Take an arbitrary graph $E \subseteq V_1 \times V_2$. Suppose that E has an odd factor $F \subseteq E$. Since the degree $d_F(i)$ of each vertex $i \in V$ in the subgraph F is odd, we have

$$\sum_{(i,j)\in F} (e_i + e_j) = \mathbf{1}$$

because for each $i \in V$, the vector e_i occurs exactly $d_F(i)$ times in this sum. Thus, our span program M accepts the graph E, as desired.

Suppose now that E has no odd factors. By Lemma 16.4, the graph E must have a connected component with an odd number of vertices. Take such a component $G' = (A, B, F)$ where $A \subseteq V_1$ and $B \subseteq V_2$; hence, $|A \cup B|$ is odd. The program M must reject the graph E. Assume the opposite, i.e., that some subset of rows, labeled by edges in E, sum up to the all-1 vector $\mathbf{1}$. Since our subgraph G' is a connected component, no vertex from the set $A \cup B$ is incident to a vertex from outside. This means that the 1's in the positions, corresponding to vertices in $A \cup B$, can be obtained only by summing along the edges in that component. Hence, there must be a subset of edges $H \subseteq E \cap (A \times B)$ such that the vector

$$w = \sum_{(i,j)\in H} (e_i + e_j)$$

has 1's in all the coordinates from $A \cup B$. For each $i \in A \cup B$, the number of terms $e_i + e_j$ in this sum is exactly the degree $d_H(i)$ of i in the subgraph H. By Euler's theorem, the sum $\sum_{i \in A \cup B} d_H(i)$ equals $2|H|$, and hence, is even. Since $|A \cup B|$ is odd, there must be at least one $i_0 \in A \cup B$ for which $d_H(i_0) \equiv 0 \,(\mathrm{mod}\,2)$ (the sum of an odd number of odd numbers would be odd). But this means that the vector w has a 0 in the i_0-th coordinate, a contradiction.

Thus, the designed span program M correctly computes $Oddfactor_n$. Since it has only $m^2 = n$ rows, we are done.   □

Despite the simplicity of this span program for $Oddfactor_n$, the function itself has a close affinity to the perfect matching problem, which makes it difficult for monotone boolean models. Namely, Razborov (1985b) has proved an $n^{\Omega(\log n)}$ lower bound on the size of any monotone circuit (with And and Or but no Not gates) that accepts all perfect matchings, and rejects any constant fraction of all 2-colorings.

Note that every perfect matching is an odd factor, and should be accepted. For rejected graphs, identify every 2-coloring of V with the graph of all monochromatic edges. Hence, an odd 2-coloring (in which each color occupies an odd number of vertices) has two odd components, and thus must be rejected. As odd 2-colorings constitute half of all 2-colorings, Razborov's argument yields an $n^{\Omega(\log n)}$ lower bound on the size of any monotone boolean circuit computing $Oddfactor_n$.

Thus, proving that some explicit families require large monotone span programs is a harder task than that for monotone boolean circuits. This task (large lower bounds for the size of span programs) was recently resolved by Beimel et al. (1996), Babai et al. (1996), Babai et al. (1999) and Gál (1998) using the linear algebra argument.

But before we go on to these beautiful results, let us first demonstrate another more direct argument for threshold functions.

16.3.4 A lower bound for threshold functions

Recall that a threshold function T_k^n accepts an input vector if and only if its has at least k 1's. We have shown in Sect. 16.3.1 that this function has small monotone span program over large fields, and have mentioned that over $\mathbb{F}_2$ the function T_k^n can be computed by a monotone span program of size $O(n \log n)$. Let us now prove a lower bound.

Theorem 16.6 (Karchmer–Wigderson 1993). *Any monotone span program computing T_2^n over the field $\mathbb{F}_2$ has size at least $n \log n$.*

Proof. When working in the field $\mathbb{F}_2$, it is often useful to keep in mind the following definition of the rejection, which is based on the following simple observation (see Exercise 16.1): a set of vectors $A \subseteq \mathbb{F}^t$ spans the all-1 vector $\mathbf{1}$ if and only if there is an *odd* vector $r \in \mathbb{F}^t$ (i.e., a vector with an odd number of 1's) such that $A \cdot r \neq \mathbf{0}$, i.e., if r is non-orthogonal to at least one vector in A. Thus, a span program M *rejects* an input a if and only if $M_a \cdot r = \mathbf{0}$ for at least one odd vector r.

Now let M be a monotone span program for T_2^n. Let t be the number of columns in M, and R the set of all odd vectors in $\mathbb{F}_2^t$. Clearly $|R| = 2^{t-1}$. Let X_i be the span of the rows of M labeled by the ith variable, and let $d_i = \dim X_i$. Recall that M rejects an input a if and only if at least one vector from R is orthogonal to all the vectors from those sets X_i for which $a_i = 1$.

For every $i = 1, \ldots, n$, let $R_i = R \cap (X_i)^{\perp}$. Since M rejects vectors of weight one, $\mathbf{1} \notin X_i$, and so for every i, $|R_i| = 2^{t-d_i-1}$ (see Exercise 16.2).

We now claim that $R_i \cap R_j = \emptyset$. To see this, observe that for every pair $i \neq j$ we must have vector $\mathbf{1}$ in the span of $X_i \cup X_j$ (as M accepts the vector with 1's in positions i and j). Therefore, for some $u_i \in X_i$ and $u_j \in X_j$, $u_i \oplus u_j = \mathbf{1}$. If there is a vector $r \in R_i \cap R_j$, then $r \cdot \mathbf{1} = 1$ while $r \cdot u_i = r \cdot u_j = 0$, a contradiction.

The previous two paragraphs imply that

$$\sum_{i=1}^{n} 2^{t-d_i-1} = \sum_{i=1}^{n} |R_i| = \left| \bigcup_{i=1}^{n} R_i \right| \leqslant |R| = 2^{t-1},$$

and hence, $\sum_{i=1}^{n} 2^{-d_i} \leqslant 1$. Applying Jensen's inequality (see Proposition 1.10) with $\lambda_i = 1/n$, and $f(d_i) = 2^{-d_i}$, we obtain

$$\frac{1}{n} \geqslant \frac{1}{n} \sum_i 2^{-d_i} = \sum_i \lambda_i f(d_i) \geqslant f\left(\sum_i \lambda_i d_i\right) = 2^{-(\sum_i d_i)/n},$$

implying that the matrix M must have at least $\sum_i d_i \geqslant n \log n$ rows.   □

16.4 A general lower bound

Recall that a *Sperner system* (or an *antichain*) is a family of sets none of which contains any other member of the family. A *filter* is a family which is closed upwards: if $A \in \mathcal{F}$ and $A \subseteq B$ then $B \in \mathcal{F}$. The minimal members of a filter form a Sperner system; and each Sperner system $\mathcal{F}$ uniquely defines a filter (consisting of all not necessarily proper supersets of its members).

There is a one-to-one correspondence between Sperner systems and monotone boolean functions. A *minterm* of a monotone boolean function is a minimal set of variables which, if assigned the value 1, forces the function to take the value 1 regardless of the values assigned to the remaining variables. It is clear that the set of all minterms of f forms a Sperner system, and that every Sperner system defines (uniquely) a monotone boolean function.

Definition 16.7. A Sperner system $\mathcal{F}$ is *self-avoiding* if one can associate a set $T_H \subseteq H$ with each $H \in \mathcal{F}$, called the *core* of H, such that

1. T_H determines H in the family. That is, no other set in the family $\mathcal{F}$ contains T_H as a subset.
2. For every $H \in \mathcal{F}$ and every subset $Y \subseteq T_H$, the set

$$S(Y) = \bigcup_{A \in \mathcal{F}, A \cap Y \neq \emptyset} A - Y,$$

called the *spread* of Y, does not contain any member of $\mathcal{F}$.

We have the following general combinatorial lower bound on the size of span programs.

Theorem 16.8 (Beimel–Gál–Paterson 1996). *Let f be a monotone boolean function and $\mathcal{F}$ be the family of its minterms. If $\mathcal{F}$ is self-avoiding then for every field $\mathbb{F}$, every monotone span program over $\mathbb{F}$ computing f has size at least $|\mathcal{F}|$.*

Proof. The idea is to show that if a small monotone span program accepts all the minterms of the function f then it must also accept an input that does not contain any minterms, a contradiction.

Let M be the matrix of a span program computing f, and let r be the number of rows of M. Our goal is to prove that $r \geqslant |\mathcal{F}|$.

Any member of $\mathcal{F}$ is accepted by the program. By definition, this means that, for every $H \in \mathcal{F}$, there is some vector $v_H \in \mathbb{F}^r$ such that $v_H \cdot M = 1$, and where v_H has nonzero coordinates only at rows labeled by elements from H. For any given $H \in \mathcal{F}$ there may be several such vectors; we pick one of them and denote it by v_H.

Since vectors v_H are taken from the vector space $\mathbb{F}^r$, its dimension r cannot be smaller than the number of linearly independent vectors among the vectors v_H, $H \in \mathcal{F}$. We will show that, in fact, *all* the vectors v_H for $H \in \mathcal{F}$ must be linearly independent, which directly yields the desired lower bound $r \geqslant |\mathcal{F}|$.

Suppose that this is not the case, i.e., for some $H \in \mathcal{F}$

$$v_H = \sum_{A \in \mathcal{A}} \alpha_A v_A, \qquad (*)$$

where $\alpha_A \in \mathbb{F}$ and $\mathcal{A} = \mathcal{F} - \{H\}$. Our goal is to show that $(*)$ leads to a contradiction.

Fix a core T_H of H, let $t = |T_H|$ and consider the following $(2^t - 1) \times (2^t - 1)$ 0-1 matrix Q. The rows and columns of Q are indexed by the nonempty subsets of T_H, and

$Q(Y, Z) = 1$ if and only if $Y \cap Z \neq \emptyset$.

This matrix has full rank over any field $\mathbb{F}$, i.e., all its columns are linearly independent over $\mathbb{F}$ (see Exercise 16.4).

Notice that the last column of Q indexed by T_H consists of all 1's. To get the desired contradiction, we will now show that, if $(*)$ held, then we could get this last column as a linear combination of the other columns of Q, which is impossible since Q has full rank. Thus, what we need is to find coefficients λ_Z for all $\emptyset \neq Z \subset T_H$, so that, for every nonempty subset Y of T_H,

$$\sum_{\emptyset \neq Z \subset T_H} \lambda_Z \cdot Q(Y, Z) = 1.$$

By Condition 1 of Definition 16.7, for every $A \in \mathcal{A}$, the intersection $A \cap T_H$ is a proper subset of T_H. Hence, taking the coefficients

$$\lambda_Z \rightleftharpoons \sum_{A \in \mathcal{A}, A \cap T_H = Z} \alpha_A,$$

we have for every nonempty subset Y of T_H, that

$$\sum_{\emptyset \neq Z \subset T_H} \lambda_Z \cdot Q(Y, Z) = \sum_{\substack{\emptyset \neq Z \subset T_H \\ Z \cap Y \neq \emptyset}} \left(\sum_{\substack{A \in \mathcal{A} \\ A \cap T_H = Z}} \alpha_A \right) = \sum_{\substack{A \in \mathcal{A} \\ A \cap Y \neq \emptyset}} \alpha_A.$$

Thus, to finish the proof of the theorem, it remains to prove the following claim.

Claim 16.9. *If (∗) holds, then for any nonempty subset $Y \subseteq T_H$,*

$$\sum_{A \in \mathcal{A}, A \cap Y \neq \emptyset} \alpha_A = 1.$$

Proof of Claim 16.9. Suppose that for some $Y \subseteq T_H$ this sum is $\gamma \neq 1$, and consider the vector

$$v = \sum_{A \in \mathcal{A}, A \cap Y \neq \emptyset} \alpha_A v_A - v_H. \tag{16.1}$$

Since $v_H \cdot M = 1$ and $v_A \cdot M = 1$ for all $A \in \mathcal{A}$, we have that $v \cdot M = (\gamma - 1)\mathbf{1}$. Thus $1/(\gamma - 1)v \cdot M = \mathbf{1}$, and the program accepts the set

$$I \rightleftharpoons \{i : v_i \neq 0\}$$

of elements that label the rows corresponding to nonzero coordinates of v.

By Definition 16.7, the spread $S(Y)$ does not contain any member of $\mathcal{F}$, and hence, should be rejected by the program. Thus, to get a desired contradiction it is enough to show that $I \subseteq S(Y)$.

Recall that each v_A has nonzero coordinates only at rows labeled by elements from A. Thus, for any A such that $A \cap Y = \emptyset$, the coordinates of v_A are zero at rows labeled by elements of Y. By (∗),

$$v = \sum_{A \in \mathcal{A}, A \cap Y \neq \emptyset} \alpha_A v_A - \sum_{A \in \mathcal{A}} \alpha_A v_A = 0 - \sum_{A \in \mathcal{A}, A \cap Y = \emptyset} \alpha_A v_A.$$

Therefore, the vector v has zero coordinates at all rows labeled by elements of Y, i.e., $I \cap Y = \emptyset$. On the other hand, by (16.1) all the nonzero coordinates of v are at rows labeled by elements that appear in at least one set $A \in \mathcal{F}$ such that $A \cap Y \neq \emptyset$ (because H also belongs to $\mathcal{F}$ and $Y \subseteq H$). Thus, the set I lies entirely in the spread $S(Y) = \bigcup\{A - Y : A \in \mathcal{F}, A \cap Y \neq \emptyset\}$, a contradiction.

This completes the proof of Claim 16.9, and thus, the proof of Theorem 16.8.   □

16.5 Explicit self-avoiding families

For Theorem 16.8 to yield a large lower bound on the size of monotone span programs we need explicit Sperner systems that are self-avoiding and have many sets. A very elegant construction of such systems, based on a particular property of bipartite Paley graphs was found by Gál (1998). We have already used this property in Sect. 15.2.2 and called it the "isolated neighbor condition" (see Definition 11.7). We recall this condition.

Let $G = (V_1, V_2, E)$ be a bipartite graph. For $A \subseteq V_1$, let

$$\Gamma(A) \rightleftharpoons \{u \in V_2 : (v, u) \in E \text{ for all } v \in A\}$$

be the set of common neighbors of the vertices in A, and let

$$\widehat{\Gamma}(A) = \{u \in V_2 : (v, u) \notin E \text{ for all } v \in A\}$$

be the set of all common non-neighbors of A. The graph G satisfies the *isolated neighbor condition for k* if $\Gamma(A) \cap \widehat{\Gamma}(B) \neq \emptyset$ for any two disjoint sets $A, B \subseteq V_1$ such that $|A| + |B| = k$.

Let $\mathcal{F}_{G,k}$ be the family of all subsets $A \cup \Gamma(A)$ where $A \subseteq V_1$, $|A| = k$. It is clear that $\mathcal{F}_{G,k}$ is a Sperner system. In order to apply the general lower bound, given by Theorem 16.8, we have to prove that this family is self-avoiding.

Lemma 16.10 (Gál 1998). *If the graph G satisfies the isolated neighbor condition for $2k$ ($k \geqslant 1$) then the family $\mathcal{F}_{G,k}$ is self-avoiding.*

Proof. Since G satisfies the isolated neighbor condition for $2k$, then for any two subsets $A, B \subseteq V_1$ of size at most k,

$$A \cap B \neq \emptyset \iff \Gamma(A) \cap \widehat{\Gamma}(B) = \emptyset. \tag{16.2}$$

We will use this property to show that the family $\mathcal{F} = \mathcal{F}_{G,k}$ is self-avoiding.

The sets in $\mathcal{F}$ have the form $H = A \cup \Gamma(A)$ where A is a k-element subset of V_1. Since any two k-element subsets are incomparable, we can take the set A as the core T_H of $H = A \cup \Gamma(A)$. It remains to prove that the spread

$$S(Y) = \bigcup_{|B|=k, B \cap Y \neq \emptyset} (B - Y) \cup \Gamma(B)$$

of no subset $Y \subseteq A$ contains a member of $\mathcal{F}$.

To show that some set S does not contain any member of $\mathcal{F}$ it is enough to show that some blocking set T of $\mathcal{F}$ (i.e., a set which intersects every member of $\mathcal{F}$) lies entirely in the complement $\overline{S}$ of S. We claim that, for $S = S(Y)$, such is the set $T = Y \cup \widehat{\Gamma}(Y)$.

The fact that T is a blocking set for $\mathcal{F}$ follows directly from (16.2): if $T \cap E = \emptyset$ for some member $E = B \cup \Gamma(B)$ of $\mathcal{F}$, then also $Y \cap B = \emptyset$; but then $\widehat{\Gamma}(Y) \cap \Gamma(B) \neq \emptyset$, a contradiction.

The same argument yields that $T \subseteq \overline{S(Y)}$. Indeed, in $S(Y)$ we take the union over all k-element sets B intersecting the set Y. Since we always have $Y \subseteq \overline{B - Y}$, it remains to show that $\widehat{\Gamma}(Y) \subseteq \overline{\Gamma(B)}$. But this again follows directly from (16.2) because $Y \cap B \neq \emptyset$ implies that the sets $\widehat{\Gamma}(Y)$ and $\Gamma(B)$ are disjoint, and hence, $\widehat{\Gamma}(Y)$ lies entirely in the complement of $\Gamma(B)$. This completes the proof of Lemma 16.10.   □

We already know how to construct explicit bipartite graphs, satisfying the isolated neighbor condition for $k = \Omega(\log n)$: such are the bipartite Paley graphs G_n constructed in Sect. 11.3.2.

The function, whose minterm set coincides with $\mathcal{F}_{G_n,k}$, is the monotone boolean function $\text{PALEY}(n, k)$ introduced in Sect. 10.6.2. There we have proved that this function is hard for monotone circuits. Using another argument, we have proved in Sect. 15.2.2 that this function cannot be realized by a small monotone formulas.

Now we can show that this function is also hard for monotone span programs: the bound given in Theorem 11.9 implies that G_n satisfies the isolated neighbor condition for $2k$, as long as $2k4^k < \sqrt{n}$, and hence, is such for $k = \Omega(\log n)$. Applying Theorem 16.8 together with Lemma 16.10 we conclude that any monotone span program computing $\mathrm{PALEY}(n, k)$ must have size at least $|\mathcal{F}_{G_n,k}| = \binom{n}{k} = n^{\Omega(\log n)}$.

We finish this chapter with four open problems about span programs formulated by Babai, Gál, and Wigderson (1999):

1. Do there exist exponential size self-avoiding Sperner systems?
2. Do there exist functions admitting polynomial size monotone span programs which require exponential size monotone circuits?
3. Do there exist functions admitting polynomial size monotone circuits which require superpolynomial size monotone span programs?
4. How much monotone span programs are weaker than non-monotone span programs?

Exercises

16.1.$^-$ Let $A \subseteq \mathbb{F}_2^n$. Show that $\mathbf{1} \notin \mathrm{span}\, A$ if and only if $A \cdot r = \mathbf{0}$ for some odd vector $r \in \mathbb{F}_2^n$. *Hint:* $U \subseteq (\mathrm{span}\,\{1\})^\perp$ iff all the vectors in U are even.

16.2. Let $V \subseteq \mathbb{F}_2^n$ be a linear space and $y \in \mathbb{F}_2^n$ be a vector. Assume that $y \notin V^\perp$. Show that then $v \cdot y = 0$ for precisely one-half of the vectors v in V.

Sketch: Split V into V_0 and V_1 according to whether $v \cdot y = 0$ or $v \cdot y = 1$. Take $x \in V$ such that $x \cdot y = 1$; hence, $x \in V_1$. Show that then $x + V_0 \subseteq V_1$, $x + V_1 \subseteq V_0$, $|x + V_0| = |V_0|$ and $|x + V_1| = |V_1|$.

16.3. The *disjointness matrix* D_n is a $2^n \times 2^n$ 0-1 matrix whose rows and columns are labeled by the subsets of an n-element set, and the (A, B)-th entry is 1 if and only if $A \cap B = \emptyset$. Prove that this matrix has full rank, i.e., that $\mathrm{rk}(D_n) = 2^n$.

Hint: Use the induction on n together with the following recursive construction of D_n:

$$D_1 = \begin{pmatrix} 1 & 1 \\ 1 & 0 \end{pmatrix}, \qquad D_n = \begin{pmatrix} D_{n-1} & D_{n-1} \\ D_{n-1} & 0 \end{pmatrix}$$

16.4. The *intersection matrix* Q_n is a $(2^n - 1) \times (2^n - 1)$ 0-1 matrix whose rows and columns are labeled by the *non-empty* subsets of an n-element set, and the (A, B)-th entry is 1 if and only if $A \cap B \neq \emptyset$. Prove that this matrix also has full rank over any field.

Hint: If we subtract D_n from the all-1 matrix I_n then we get a matrix Q_n with one additional null column and row. Combine this fact with (14.1) and the previous exercise.

16.5 (Beimel–Gál–Paterson 1996). Let $L_1, \ldots, L_n$ be n subsets of $\{1, \ldots, n\}$ such that $|L_i \cap L_j| \leqslant 1$ for all $i \neq j$. For example, the lines of a projective plane can be used. Given these sets, let us consider the following family $\mathcal{F} = \{\{x_i, y_j\} : j \in L_i\}$ of 2-element subsets of the $(2n)$-element set $X = \{x_1, \ldots, x_n, y_1, \ldots, y_n\}$. Prove that the family $\mathcal{F}$ is self-avoiding.

Hint: Define the core T_H to be just the set H itself. If, say, $H = \{x_1, y_1\}$ then its spread is $S(H) = \{x_j : j \in L_1\} \cup \{x_i : 1 \in L_i\} - \{x_1, y_1\}$.

16.6. Recall that a *DeMorgan formula* is defined inductively as follows: every boolean variable x_i and its negation $\overline{x}_i$ is a formula of size 1, and if F_1 and F_2 are formulas of size l_1 and l_2, then both $F_1 \wedge F_2$ and $F_1 \vee F_2$ are formulas of size $l_1 + l_2$. Show that, if a boolean function can be computed by a DeMorgan formula F of size s, then it can be also computed by a switching network of the same size.

Hint: Argue by induction on the size of F. If $F = F_1 \vee F_2$ then connect the corresponding to F_1 and F_2 switching networks in parallel; otherwise (if $F = F_1 \wedge F_2$) connect them sequentially.

16.7.[(!)] Show that every (monotone) boolean function can be computed by a (monotone) span program.

Hint: Every boolean function on n variables can be presented as an Or of monomials of length n, each monomial being an And of variables or their negations. Alternative: use Theorem 16.1.

16.8[(!)] (Karchmer–Wigderson 1993). Let M be a span program computing f over $\mathbb{F}_2$. Such a program is *canonical* if the columns of M are in one-to-one correspondence with the vectors in $f^{-1}(0)$, and for every $b \in f^{-1}(0)$, the column corresponding to b in M_b is an all-0 column. Show that every span program can be converted to a canonical span program of the same size (= the number of rows) and computing the same function.

Hint: Take a vector $b \in f^{-1}(0)$. By Exercise 16.1, there is an odd vector $r = r_b$ for which $M_b \cdot r_b = 0$. Define the column corresponding to b in a new span program M' to be $M_b \cdot r_b$. Do this for all $b \in f^{-1}(0)$. Show that, for every vector $a \in \mathbb{F}_2^n$, the rows of M'_a span the all-1 vector $\mathbf{1}$ if and only if $f(a) = 1$.

16.9.[+] Research problem: Let k be the minimal number for which the following holds: there exist n colorings $c_1, \ldots, c_n$ of the n-cube $\{0,1\}^n$ in k colors $\{1, \ldots, k\}$ such that, for every triple of vectors x, y, z there exists a coordinate i on which not all three vectors agree and the three colors $c_i(x), c_i(y), c_i(z)$ are distinct. Bound the smallest number k of colors for which such a good collection of colorings $c_1, \ldots, c_n$ exists.

Comment: This problem is connected with proving lower bounds on the size of *non-monotone* span programs, see Wigderson (1993).

16.10$^+$ (Wigderson 1993). Consider the version of the problem above where we additionally require that the colorings c_i are *monotone*, i.e., $x <_y y$ implies $c_i(x) \leqslant c_i(y)$. Prove that in this case $k = \Omega(n)$.

16.11$^{(!)}$ (Pudlák–Sgall 1998). A *monotone dependency program* over a field $\mathbb{F}$ is given by a matrix M over $\mathbb{F}$ with its rows labeled by variables $x_1, \ldots, x_n$. For an input $a = (a_1, \ldots, a_n) \in \{0, 1\}^n$, let (as before) M_a denote the submatrix of M obtained by keeping those rows whose labels are satisfied by a. The program M *accepts* the input a if and only if the rows of M_a are linearly dependent (over $\mathbb{F}$). A program computes a boolean function f if it accepts exactly those inputs a where $f(a) = 1$. The *size* of a span program is the number of rows in it.

(i) Suppose that a boolean function $f \not\equiv 1$ is computed by a monotone dependency program M of size smaller than the number of minterms of f. Prove that then there exists a set of minterms $\mathcal{F}$, $|\mathcal{F}| \geqslant 2$, such that for any non-trivial partition $\mathcal{F} = \mathcal{F}_0 \cup \mathcal{F}_1$; $\mathcal{F}_0, \mathcal{F}_1 \neq \emptyset$, $\mathcal{F}_0 \cap \mathcal{F}_1 = \emptyset$, the set

$$S(\mathcal{F}_0, \mathcal{F}_1) = \left(\bigcup_{A \in \mathcal{F}_0} A \right) \cap \left(\bigcup_{B \in \mathcal{F}_1} B \right)$$

contains at least one minterm of f.

> *Sketch:* For every minterm A of f choose some linear dependence v_A of the rows of M, i.e., v_A is a vector such that $v_A \cdot M = 0$, and v_A has a nonzero coordinates only at rows labeled by variables from A. The vectors v_A are linearly dependent (why?). Let $\mathcal{F}$ be a minimal set of minterms such that $\{v_A : A \in \mathcal{F}\}$ are linearly dependent. Thus, $\sum_{A \in \mathcal{F}} \alpha_A v_A = 0$ for some coefficients $\alpha_A \in \mathbb{F}$, $\alpha_A \neq 0$. Observe that for any non-trivial partition $\mathcal{F} = \mathcal{F}_0 \cup \mathcal{F}_1$,
> $$v = \sum_{A \in \mathcal{F}_0} \alpha_A v_A = - \sum_{A \in \mathcal{F}_1} \alpha_A v_A \neq 0.$$
> Let B be the set of variables labeling the rows of M corresponding to nonzero coordinates of v. This set lies in $S(\mathcal{F}_0, \mathcal{F}_1)$ and contains at least one minterm of f.

(ii) Use the previous fact to show that the function

$$f = (x_1 \vee x_2) \wedge (x_3 \vee x_4) \wedge \cdots \wedge (x_{2n-1} \vee x_{2n})$$

cannot be computed by a monotone dependency program of size smaller than 2^n. Show that this function has a small monotone span program.

> *Sketch:* Each minterm A of f has precisely one variable from each of the sets $\{x_{2i-1}, x_{2i}\}$, $i = 1, \ldots, n$; hence, there are 2^n minterms. Suppose that f has a program of size smaller than 2^n, and let $\mathcal{F}$ be the set of minterms guaranteed by (i). Pick i such that both sets of minterms $\mathcal{F}_0 = \{A \in \mathcal{F} : x_{2i-1} \notin A\}$ and $\mathcal{F}_1 = \{A \in \mathcal{F} : x_{2i} \notin A\}$ are non-empty (why this is possible?). By (i), the set $S(\mathcal{F}_0, \mathcal{F}_1)$ must contain at least one minterm B of f. But, by the definition of $\mathcal{F}_0$ and $\mathcal{F}_1$, this minterm can contain neither x_{2i-1} nor x_{2i}, a contradiction.

Part IV

The Probabilistic Method

17. Basic Tools

The probabilistic method is a powerful tool for tackling many problems in discrete mathematics. This is one of the strongest tools in combinatorics and in graph theory. It is also useful in number theory and in combinatorial geometry. More recently, it has been applied in the development of efficient algorithms and in the study of various computational problems.

Roughly speaking, the method works as follows: trying to prove that an object with certain properties exists, one defines an appropriate probability space of objects and shows that a randomly chosen element of this space has the desired properties with a positive probability. A prototype of this method is the following averaging (counting) argument:

If $x_1, \ldots, x_n \in \mathbb{R}$ and

$$\frac{x_1 + \cdots + x_n}{n} \geqslant a \tag{17.1}$$

then for some j

$$x_j \geqslant a. \tag{17.2}$$

The usefulness of the method lies in the fact that the average (17.1) is often easier to compute than to exhibit a specific x_j for which (17.2) can be proved to hold.

We briefly recall some basic definitions of (discrete) probability.

17.1 Probabilistic preliminaries

A *finite probability space* consists of a finite set Ω and a function $\text{Prob} : \Omega \to [0, 1]$, such that $\sum_{x \in \Omega} \text{Prob}(x) = 1$. A probability space is a representation of a random experiment, where we choose a member of Ω at random and $\text{Prob}(x)$ is the probability that x is chosen. Subsets $A \subseteq \Omega$ are called *events*. The probability of an event is defined by $\text{Prob}(A) = \sum_{x \in A} \text{Prob}(x)$, i.e., the probability that a member of A is chosen.

We call Ω the *domain* (or a *sample space*) and we call Prob a *probability distribution*. The most common probability distribution is the *uniform distribution*, which is defined as $\text{Prob}(x) = 1/|\Omega|$ for each $x \in \Omega$; the corresponding sample space is then called *symmetric*.

Some elementary properties follow directly from the definitions. For any two events A and B (here and throughout $\overline{A} = \Omega - A$ stands for the complement of A):

Prob $(A \cup B) = $ Prob $(A) + $ Prob $(B) - $ Prob $(A \cap B)$;

Prob $(\overline{A}) = 1 - $ Prob (A);

Prob $(A \cap B) \geqslant $ Prob $(A) - $ Prob $(\overline{B})$;

If $B_1, \ldots, B_m$ is a partition of Ω then Prob $(A) = \sum_{i=1}^{m}$ Prob $(A \cap B_i)$.

For two events A and B, the *conditional probability of A given B*, denoted Prob $(A|B)$, is the probability that one would assign to A if one knew that B occurs. Formally, we define

$$\text{Prob}\,(A|B) \rightleftharpoons \frac{\text{Prob}\,(A \cap B)}{\text{Prob}\,(B)},$$

when Prob $(B) \neq 0$. For example, if we are choosing a uniform integer from $\{1, \ldots, 6\}$, A is the event that the number is 2 and B is the event that the number is even, then Prob $(A|B) = 1/3$, whereas Prob $(B|A) = 1$.

An event A is *independent* of B if Prob $(A|B) = $ Prob (A). Since this is equivalent to Prob $(A \cap B) = $ Prob $(A) \cdot $ Prob (B), independence is a reflexive relation, and we can say that A and B are independent. It is very important to note that the "independence" has nothing to do with the "disjointness" of the events: if, say, Prob $(A) > 0$, Prob $(B) > 0$ and $A \cap B = \emptyset$ then the events A and B *are* dependent!

An event A is *mutually independent* of the events $B_1, \ldots, B_m$ if

$$\text{Prob}\,(A \mid \cap_{i \in I} B_i) = \text{Prob}\,(A)$$

for every subset $I \subseteq \{1, \ldots, m\}$. Note that A might be independent of *each* of the events $B_1, \ldots, B_m$, but not be mutually independent of them. To see this, consider flipping a fair coin twice and the three events: B_1, B_2, A, where B_i is the event that the ith flip is a head and A is the event that both flips are the same. Then A is independent of B_1 and of B_2 but Prob $(A \mid B_1 B_2) = 1$.

Let Γ be finite set, and $0 \leqslant p \leqslant 1$. A *random subset* $\mathbf{S}$ of Γ is obtained by flipping a coin, with probability p of success, for each element of Γ to determine whether the element is to be included in $\mathbf{S}$; the distribution of $\mathbf{S}$ is the probability distribution on $\Omega = 2^{\Gamma}$ given by Prob $(S) = p^{|S|}(1-p)^{|\Gamma|-|S|}$ for $S \subseteq \Gamma$. We will mainly consider the case when $\mathbf{S}$ is uniformly distributed, that is, when $p = 1/2$. If $\mathcal{F}$ is a family of subsets, then its *random member* $\mathbf{S}$ is a uniformly distributed member; in this case, $\Omega = \mathcal{F}$ and $\mathbf{S}$ has the probability distribution Prob $(S) = 1/|\mathcal{F}|$. Note that, for $p = 1/2$, a random subset of Γ is just a random member of 2^{Γ}.

A *random variable* is a variable defined as a function $X : \Omega \rightarrow \mathbb{R}$ of the domain of a probability space. For example, if X is a uniform integer chosen from $\{1, \ldots, n\}$, then $Y \rightleftharpoons 2X$ and $Z \rightleftharpoons$ "the number of prime divisors of X" are both random variables, and so is X itself. The *distribution* of a random variable X is the function $f : \Omega \rightarrow [0,1]$ defined as $f(i) \rightleftharpoons $ Prob $(X = i)$, where here and in what follows Prob $(X = i)$ denotes the probability of the event $A = \{x \in \Omega : X(x) = i\}$. An *indicator random variable* for the event A is a random 0-1 variable X such that $X = 1$ if and only if A occurs. Two random variables X and Y are *independent* if for every $i, j \in \mathbb{R}$, the events $X^{-1}(i)$ and $Y^{-1}(j)$ are independent.

One of the most basic probabilistic notions is the expected value of a random variable. This is defined for any real-valued random variable X, and intuitively, it is the value that we would expect to obtain if we repeated a random experiment several times and took the average of the outcomes of X. More formally, the *mean* or *expectation* of X is defined as the sum

$$E[X] = \sum_i i \cdot \text{Prob}(X = i)$$

over all numbers i in the range of X. For example, if X is the number of spots on the top face when we roll a fair die, then the expected number of spots is $E[X] = \sum_{i=1}^{6} \frac{i}{6} = 3.5$.

The following properties of expectation can be easily verified (we address them in Exercises). The first fundamental property is known as the *linearity of expectation*. Let $X_1, X_2, \ldots, X_n$ be random variables. Then

$$E[X_1 + X_2 + \cdots + X_n] = E[X_1] + E[X_2] + \cdots + E[X_n].$$

If $X_1, X_2, \ldots, X_n$ are mutually independent then

$$E[X_1 \cdot X_2 \cdot \ldots \cdot X_n] = E[X_1] \cdot E[X_2] \cdot \ldots \cdot E[X_n].$$

The *variance* of a random variable X is defined as

$$\text{Var}[X] = E\left[(X - E[X])^2\right].$$

The variance can be computed by the formula

$$\text{Var}[X] = E\left[X^2\right] - (E[X])^2.$$

If X is a random variable which takes 1 with probability p and is 0 otherwise, then $E[X] = p$. In this case we say that X is a Bernoulli variable with probability of success p. The standard *binomial distribution* $B(n, p)$ is the sum $X = X_1 + \cdots + X_n$ of n independent Bernoulli random variables $X_1, \ldots, X_n$ each with success probability p. That is, $B(n, p)$ is the number of heads obtained by flipping a weighted coin n times, each time getting a head with probability p. The word "binomial" is justified by the fact that, for any $0 \leqslant k \leqslant n$,

$$\text{Prob}(X = k) = \binom{n}{k} p^k (1 - p)^{n-k}.$$

The expectation of such a variable X is $E[X] = \sum_{i=1}^{n} E[X_i] = np$. The variance is also easy to compute (do this!): $\text{Var}[X] = np(1 - p)$.

17.2 Elementary tools

We now briefly mention some of the main weapons of the probabilistic method. In subsequent chapters we will show how some of them work in concrete situations.

Counting Sieve. This simple principle says that the probability of a disjunction of events is at most the sum of the probabilities of the events, $\text{Prob}\left(\bigcup A_i\right) \leqslant \sum \text{Prob}\left(A_i\right)$. Thus, if A_i's are some "bad" events and $\sum \text{Prob}\left(A_i\right) < 1$ then we know that $\text{Prob}\left(\bigcap \overline{A_i}\right) > 0$, i.e., that, with positive probability, *none* of these bad events happens.

Independence Sieve. If the events $A_1, \ldots, A_n$ are mutually independent and all $\text{Prob}\left(A_i\right) < 1$ then $\text{Prob}\left(\bigcap \overline{A_i}\right) > 0$.

Linearity of Expectation. This principle says that if $X_1, \ldots, X_n$ are random variables, then $\text{E}\left[X_1 + \cdots + X_n\right] = \text{E}\left[X_1\right] + \cdots + \text{E}\left[X_n\right]$. The strength of this property lies in the fact that it remains valid even if the X_i's are dependent on each other. The use of expectation in existence proofs is based on the following *pigeonhole property* of expectation: a random variable cannot always be smaller (or always greater) than its expectation.

Markov's Inequality. Let X be a non-negative random variable and λ a real number. Then

$$\text{Prob}\left(X \geqslant \lambda\right) \leqslant \frac{\text{E}\left[X\right]}{\lambda}.$$

Proof:

$$\text{E}\left[X\right] = \sum_x x \cdot \text{Prob}\left(X = x\right) \geqslant \sum_{x \geqslant \lambda} \lambda \cdot \text{Prob}\left(X = x\right) = \lambda \cdot \text{Prob}\left(X \geqslant \lambda\right).$$

Chernoff's Inequalities. These inequalities are special (and hence, more informative) cases of Markov's inequality applied to sums of random variables. Let $Y = Y_1 + \cdots + Y_n$, where $\text{Prob}\left(Y_i = +1\right) = \text{Prob}\left(Y_i = -1\right) = 1/2$ and the Y_i are mutually independent. Then for any $\lambda > 0$,

$$\text{Prob}\left(Y > \lambda\right) < e^{-\lambda^2/2n}, \tag{17.3}$$

and, by symmetry,

$$\text{Prob}\left(|Y| > \lambda\right) < 2e^{-\lambda^2/2n}. \tag{17.4}$$

For sums of random 0-1 variables we have the following estimates. Let $X = X_1 + \cdots + X_n$ be the sum of n random variables, taking their values in $\{0, 1\}$ independently and with probabilities $\text{Prob}\left(X_i = 1\right) = p$ and $\text{Prob}\left(X_i = 0\right) = 1 - p$. Hence, X has a binomial distribution $B(n, p)$, and $\text{E}\left[X\right] = pn$. Taking exponents and applying Markov's inequality, we obtain

$$\text{Prob}\left(X \geqslant m\right) \leqslant \text{Prob}\left(e^{tX} \geqslant e^{tm}\right) \leqslant e^{-tm}\text{E}\left[e^{tX}\right]$$

for any $t \geqslant 0$, where using the inequality $1 + a \leqslant e^a$ and the independence of the X_i's,

$$E\left[e^{tX}\right] = E\left[\prod_{i=1}^{n} e^{tX_i}\right] = \prod_{i=1}^{n} E\left[e^{tX_i}\right] = (pe^t + 1 - p)^n \leqslant e^{pn(e^t-1)}.$$

Putting $t = \ln(m/pn)$ yields the following useful inequality due to Chernoff (1952): if $m \geqslant pn$ then

$$\text{Prob}\,(X \geqslant m) \leqslant \left(\frac{pn}{m}\right)^m \cdot e^{m-pn}. \tag{17.5}$$

An easy consequence of this is the following (see, e.g., Hagerup and Rüb 1989 for simple proofs of these inequalities): for every $0 < \lambda < 1$,

$$\text{Prob}\,(X \geqslant (1 + \lambda)pn) \leqslant e^{-\lambda^2 pn/3} \tag{17.6}$$

and

$$\text{Prob}\,(X \leqslant (1 - \lambda)pn) \leqslant e^{-\lambda^2 pn/2}. \tag{17.7}$$

Chebyshev's Inequality. The basic Chebyshev's inequality states the following. Let X be a random variable. Then for any real $\lambda > 0$,

$$\text{Prob}\,(|X - E\,[X]| \geqslant \lambda) \leqslant \frac{\text{Var}\,[X]}{\lambda^2}.$$

Proof. We define a random variable $Y \rightleftharpoons (X - E\,[X])^2$, and apply Markov's inequality. This yields

$$\text{Prob}\,(|X - E\,[X]| \geqslant \lambda) \leqslant \text{Prob}\,(Y \geqslant \lambda^2) \leqslant E\,[Y]/\lambda^2 = \text{Var}\,[X]/\lambda^2.$$

$\square$

The Second Moment Method. Taking $\lambda = \epsilon \cdot E\,[X]$ in Chebyshev's inequality, we obtain

$$\text{Prob}\,(|X - E\,[X]| \geqslant \epsilon E\,[X]) \leqslant \frac{\text{Var}\,[X]}{\epsilon^2 E\,[X]^2}.$$

If we know that $\text{Var}\,[X] = o(E\,[X]^2)$, then for any fixed ϵ, the above probability is $o(1)$. This is the *second moment method*: if $\text{Var}\,[X] = o(E\,[X]^2)$, then $X \sim E\,[X]$, i.e., X is almost always almost equal to its expectation $E\,[X]$. Thus, if we know the expectation $E\,[X]$, then we know only that $X \geqslant E\,[X]$ for *at least one* realization of X. Additional knowledge that $\text{Var}\,[X] = o(E\,[X]^2)$ gives us additional information that $X \geqslant E\,[X]$ for *many* such realizations.

17.3 Advanced tools

Using the elementary tools mentioned above, one shows the existence of a "good" point (configuration, coloring, tournament, etc.) by showing that the set of bad points has measure less than one. By moving slightly away from the threshold value we usually find that the bad points are negligible in measure so that a randomly selected point will almost surely be good. A method is "advanced" if it enables us to find "rare" points, i.e., if it works even when the set of good points is very small in measure. As Joel Spencer tightly observed: "... advanced methods allow us to find a needle in a haystack, whereas elementary methods just show us the hay." Below we shortly describe the most famous of these advanced methods.

The Deletion Method. This method takes a random configuration which is not "good" but is "bad" in only a few places. An alteration, generally a deletion, is made at each bad place until the object, now generally smaller, is "good," i.e., has the desired properties.

The Lovász Sieve. This is a method which allows some dependence among the events. Let $A_1, \ldots, A_m$ be events. The *dependency graph* for these events is a graph G on vertices $\{1, \ldots, n\}$ such that for each i, $1 \leqslant i \leqslant n$, the event A_i is mutually independent of all the events A_j for which $(i, j) \notin G$. (That is, A_i is independent of any boolean combination of these A_j.) Let d be the maximum degree of a vertex in G, and let $\text{Prob}(A_i) \leqslant p$ for all $i = 1, \ldots, n$. Lovász sieve says: *If $ep(d+1) < 1$ then* $\text{Prob}\left(\bigcap_{i=1}^{n} \overline{A_i}\right) > 0$

The Entropy. Let a random variable Y assume m possible values with probabilities $p_1, \ldots, p_m$. The *entropy* of Y, denoted by $H[Y]$, is given by $H[Y] = \sum_{i=1}^{m} -p_i \log_2 p_i$. Entropy measures the "information content" of Y. If $m = 2^t$ and all $p_i = 2^{-t}$, then $H[Y] = t$. Convexity arguments show that if the p_i's are smaller then the entropy must be larger. In a contrapositive form we may express this as a *concentration property*: if $H[Y] \leqslant t$ then there must be some b for which $\text{Prob}(Y = b) \geqslant 2^{-t}$.

In some applications the *subadditivity property* is also useful. Let $Y = (Y_1, \ldots, Y_n)$ be a random variable taking values in the set $B = \prod_{i=1}^{n} B_i$, where each of the coordinates Y_i of Y is a random variable taking values in B_i. Then $H[Y] \leqslant \sum_{i=1}^{n} H[Y_i]$ (see Chap. 23 for the proof).

Janson's Inequality. Let Ω be a fixed set and $\mathbf{Y} \subseteq \Omega$ a random subset where the events "$y \in \mathbf{Y}$" are mutually independent over all $y \in \Omega$. Let $A_1, \ldots, A_m \subseteq \Omega$. Let B_i be the event "$\mathbf{Y} \supseteq A_i$." Write $i \sim j$ if $i \neq j$ and $A_i \cap A_j \neq \emptyset$. Roughly, $\sim$ represents the dependence of the corresponding events B_i. Let $M = \prod \text{Prob}(\overline{B_i})$. Janson's Inequality is:

$$M \leqslant \text{Prob}\left(\bigcap \overline{B_i}\right) \leqslant M \cdot e^{\frac{\Delta}{2(1-\epsilon)}},$$

where $\Delta = \sum_{i \sim j} \text{Prob}(B_i \cap B_j)$ and $\epsilon = \max \text{Prob}(B_i)$.

Azuma's Inequality. This inequality extends Chernoff's bounds to more general random variables. Let $\Omega = \Omega_1 \times \Omega_2 \times \cdots \times \Omega_m$ be a product probability space and X a random variable on it satisfying the following Lipschitz-type condition: if $y, y' \in \Omega$ differ in only one coordinate then $|X(y) - X(y')| \leqslant 1$. Set $\mu = \mathrm{E}[X]$, the expectation of X. Then

$$\mathrm{Prob}\,(X \geqslant \mu + a) < \mathrm{e}^{-a^2/2m}.$$

The same inequality holds for $\mathrm{Prob}\,(X \leqslant \mu - a)$.

Talagrand's Inequality. This inequality has a similar framework to Azuma. Let $\Omega = \prod_{i=1}^{m} \Omega_i$ be a product probability space. For $A \subseteq \Omega$ and $x = (x_1, \ldots, x_m) \in \Omega$, define a "distance" $\rho(A, x)$ as the least t so that, for any real vector $\alpha = (\alpha_1, \ldots, \alpha_m)$ there exists $y = (y_1, \ldots, y_m) \in A$ for which

$$\sum_{i: x_i \neq y_i} \alpha_i \leqslant t\|\alpha\| = t\left(\sum_{i=1}^{m} \alpha^2\right)^{1/2}.$$

Note critically that y may depend on α. Set $A_t \rightleftharpoons \{x \in \Omega : \rho(A, x) \leqslant t\}$. Talagrand's Inequality is:

$$\mathrm{Prob}\,(A) \cdot (1 - \mathrm{Prob}\,(A_t)) \leqslant \mathrm{e}^{-t^2/4}.$$

This inequality is reminiscent of isoperimetric inequalities. Indeed, suppose $\Omega_i = \{0, 1\}$ with uniform distribution so that Ω is the n-cube $\{0, 1\}^n$ with uniform distribution. If $x \in A_t$ then, taking all $\alpha_i = 1$, x must be within Hamming distance $t\sqrt{m}$ from A_t. However, the choice of arbitrary α_i's make this inequality far stronger.

In the following chapters we will demonstrate how all these (except the last three) tools work in concrete situations. The three exceptions are Azuma's, Janson's, and Talagrand's inequalities whose applications are more involved. The interested reader is invited to consult the book by Alon and Spencer (1992) – a comprehensive guide to the state of the art of the probabilistic methods.

Exercises

17.1. Let $B_1, \ldots, B_t$ be a partition of a sample space Ω into mutually disjoint sets. Adam's Theorem (Erdős and Spencer (1974) call this result Adam's because it has been rediscovered so often) states that then, for every event A, $\mathrm{Prob}\,(A) = \sum_{i=1}^{t} \mathrm{Prob}\,(A \mid B_i) \cdot \mathrm{Prob}\,(B_i)$. Prove this theorem. Using it, show that $\mathrm{Prob}\,(A) \leqslant \max_{1 \leqslant i \leqslant t} \mathrm{Prob}\,(A \mid B_i)$.

17.2. $^-$ Recall that the scalar product of two vectors $u, v \in \mathbb{F}_2^n$ is defined by $\langle u, v \rangle = \sum_{i=1}^{n} u_i v_i \,(\mathrm{mod}\,2)$. Let $\mathbf{u}$ be a random vector uniformly distributed over $\mathbb{F}_2^n$. Show that: $\mathrm{Prob}\,(\langle \mathbf{u}, v \rangle = 1) = 1/2$ for every $v \neq 0$, and $\mathrm{Prob}\,(\langle \mathbf{u}, v \rangle = \langle \mathbf{u}, w \rangle) = 1/2$ for every $v \neq w$.

Hint: Take an i for which $v_i = 1$ and split the space $\mathbb{F}_2^n$ into 2^{n-1} pairs u, u' that differ only in their ith coordinate. For each of these pairs, $\langle u, v \rangle \neq \langle u', v \rangle$.

17.3.⁻ Let $X : \Omega \to \{x_1, \ldots, x_r\}$ and $Y : \Omega \to \{y_1, \ldots, y_s\}$ be random variables. Prove that

1. $\mathrm{E}\,[a \cdot X + b \cdot Y] = a \cdot \mathrm{E}\,[X] + b \cdot \mathrm{E}\,[Y]$ for any constants a and b.
2. If X and Y are independent then $\mathrm{E}\,[X \cdot Y] = \mathrm{E}\,[X] \cdot \mathrm{E}\,[Y]$ and $\mathrm{Var}\,[X + Y] = \mathrm{Var}\,[X] + \mathrm{Var}\,[Y]$.
3. $\mathrm{Var}\,[X] = \mathrm{E}\,[X^2] - (\mathrm{E}\,[X])^2$. *Hint:* $\mathrm{E}\,[X \cdot \mathrm{E}\,[X]] = (\mathrm{E}\,[X])^2$.

17.4.⁻ Let X be a non-negative integer-valued random variable. Show that $\mathrm{E}\,[X^2] \geqslant \mathrm{E}\,[X]$, $\mathrm{Prob}\,(X = 0) \geqslant 1 - \mathrm{E}\,[X]$ and $\mathrm{E}\,[X] = \sum_{x=1}^{\infty} \mathrm{Prob}\,(X \geqslant x)$.

17.5. Use the Cauchy–Schwarz inequality to show that, for any random variable X, $\mathrm{E}\,[X]^2 \leqslant \mathrm{E}\,[X^2]$.

17.6 (The second moment method; Lovász 1979). Let $A_1, \ldots, A_n$ be arbitrary events. Define $a = \sum_{i=1}^{n} \mathrm{Prob}\,(A_i)$ and $b = \sum_{i<j} \mathrm{Prob}\,(A_i \cap A_j)$. Prove that

$$\mathrm{Prob}\,(\overline{A}_1 \cdots \overline{A}_n) \leqslant \frac{a + 2b}{a^2} - 1.$$

Hint: Let X be the number of A_i's that occur. Use Chebyshev's inequality to show that $\mathrm{Prob}\,(X = 0) \leqslant a^{-2}\mathrm{E}\,[(X - a)^2]$, and use the linearity of expectation to expand the right hand expression.

17.7 (Alon–Spencer 1992). Let $\mathcal{F}$ be a family of subsets of a given n-element set X. The *discrepancy* of $\mathcal{F}$ is the minimal number d for which there exists a coloring $c: X \to \{-1, +1\}$ of points in X in two "colors" -1 and $+1$ such that, for every $S \in \mathcal{F}$, the absolute value of $c(S) \rightleftharpoons \sum_{x \in S} c(x)$ does not exceed d. Use Chernoff's bound (17.4) to prove that the discrepancy of every family with at most m members does not exceed $\sqrt{2n \ln(2m)}$.

17.8.⁺ Chebyshev's inequality is particularly useful for sums of independent random variables. Let $X_1, \ldots, X_n$ be pairwise independent random variables with the identical expectation, denoted μ, and identical variance, denoted σ^2. Let $Z = (\sum X_i)/n$. Prove that

$$\mathrm{Prob}\,(|Z - \mu| \geqslant \lambda) \leqslant \frac{\sigma^2}{\lambda^2 n}.$$

Sketch: Consider the random variables $Y_i \rightleftharpoons X_i - \mathrm{E}\,[X_i]$. Note that the Y_i's are pairwise independent, and each has zero expectation. Apply Chebyshev's inequality to the random variable $Z = (\sum X_i)/n$, and use the linearity of expectation to show that $\mathrm{Prob}\,(|Z - \mu| \geqslant \lambda) \leqslant \mathrm{E}\,[(\sum Y_i)^2]/(\lambda^2 \cdot n^2)$. Then (again using the linearity of expectation) show that $\mathrm{E}\,[(\sum Y_i)^2] = n \cdot \sigma^2$.

17.9. Let $X_1, \ldots, X_n$ be n independent 0-1 random variables such that $\mathrm{Prob}\,(X_i = 1) = p_i$ and $\mathrm{Prob}\,(X_i = 0) = 1 - p_i$. Let $X = \sum_{i=1}^{n} X_i \,(\mathrm{mod}\,2)$. Prove that $\mathrm{Prob}\,(X = 1) = \frac{1}{2}\left[1 - \prod_i(1 - 2p_i)\right]$.

Hint (due to T. Hofmeister): Consider the random variable $Y = Y_1 \cdots Y_n$, where $Y_i = 1 - 2X_i$, and observe that $\mathrm{E}\,[Y] = 1 - 2 \cdot \mathrm{Prob}\,(Y = -1)$.

18. Counting Sieve

The probabilistic method is most striking when it is applied to prove theorems whose statement does not seem to suggest the need for probability at all. It is therefore surprising what results may be obtained from such simple principles like the counting sieve:

The probability of a disjunction of events is at most the sum of the probabilities of the events, $\mathrm{Prob}\left(\bigcup A_i\right) \leqslant \sum \mathrm{Prob}\left(A_i\right)$.

The examples given below demonstrate the power of this principle.

18.1 Ramsey numbers

Let K_n denote the clique of size n, i.e., a complete graph on n vertices. We color its edges (not vertices!) in two colors – red and blue – and say that a set S of its vertices is *monochromatic* if the edges of K_S (a clique on the vertices S) are either all red or all blue. The Ramsey number $R(t)$ is the smallest n such that by coloring the edges of K_n red and blue in *any* way there will always be a monochromatic clique of size t. By simple logic, $R(t) > n$ means that we can always color the edges of K_n in such a way that no monochromatic K_t appears.

In Sect. 14.3.3 we have derived (using linear algebra) a constructive lower bound $R(t) \geqslant t^{\Omega(\log t/\log\log t)}$. We have also stated a much better lower bound $R(t) > 2^{t/2}$ obtained by Paul Erdős via probabilistic argument. Let us now look at the proof.

Theorem 18.1 (Erdős 1947). *If* $\binom{n}{t} \cdot 2^{1-\binom{t}{2}} < 1$, *then* $R(t) > n$. *In particular,* $R(t) > 2^{t/2}$ *for all* $t \geqslant 3$.

Proof. Consider a random 2-coloring of the edges of K_n obtained by coloring each edge independently either red or blue, where each color is equally likely. Informally, we have an experiment in which a fair coin is flipped to determine the color of each edge. For any fixed set S of t vertices, let A_S be the event that the induced subgraph of K_n is monochromatic (i.e., that either all its edges are red or they are blue). Then $\mathrm{Prob}\left(A_S\right) = 2 \cdot 2^{-\binom{t}{2}}$, as for A_S to hold all $\binom{t}{2}$ "coin flips" must be the same. The probability that at least one t-clique is monochromatic is $\text{Prob}(\bigcup_S A_S)$. An exact formula for this probability would be difficult to find as the events A_S may have complex interactions. We may, however, use the counting sieve and obtain

$$\text{Prob}\left(\bigcup A_S\right) \leqslant \sum \text{Prob}(A_S) = \binom{n}{t} 2^{1-\binom{t}{2}} < 1.$$

Thus, with positive probability, no event A_S occurs, and hence, there is a point in the probability space for which $\bigcap \overline{A}_S$ holds. This point is a coloring of K_n, and the event $\bigcap \overline{A}_S$ is precisely that under this coloring there is *no* monochromatic K_t, implying that $R(t) > n$. Finally, if $t \geqslant 3$ and we take $n = \lfloor 2^{t/2} \rfloor$, then

$$\binom{n}{t} \cdot 2^{1-\binom{t}{2}} < \frac{n^t}{t!} \cdot \frac{2^{1+t/2}}{2^{t^2/2}} < 1,$$

and hence $R(t) > 2^{t/2}$ for all $t \geqslant 3$. $\qquad\square$

18.2 Van der Waerden's theorem

Let $W(r, k)$ be the least number n such that any coloring of $\{1, 2, \ldots, n\}$ in r colors gives a monochromatic arithmetic progression with k terms, i.e., for any such coloring there exist integers a, b such that all the points

$$a, a + b, a + 2b, \ldots, a + (k-1)b$$

get the same color. The existence of $W(r, k)$ for any r and k is the celebrated theorem of van der Waerden (1927). In Chap. 29 we will show how this theorem can be derived using a very powerful Ramsey-type result due to Hales and Jewett (1963).

How fast does $W(r, k)$ grow? Easy probabilistic argument shows that the growth is exponential, even for $r = 2$.

Theorem 18.2. $W(2, k) > 2^{k/2}$. *That is, the set $\{1, \ldots, n\}$ may be two-colored so that no $2 \log n$-term arithmetic progression is monochromatic.*

Proof. Color $\{1, \ldots, n\}$ randomly, i.e., we again flip a coin n times to determine a color of each point. For each arithmetic progression S with k terms, let A_S be the event that S is monochromatic. Then $\Pr[A_S] = 2 \cdot 2^{-|S|} = 2^{-k+1}$. There are no more than $\binom{n}{2}$ progressions (since each is uniquely determined by its first and second elements); so if $\binom{n}{2} 2^{-k+1} < 1$, we have that $\text{Prob}(\bigcup A_S) \leqslant \sum \text{Prob}(A_S) < 1$, and the desired coloring exists. Therefore $W(2, k)$ must be larger than any n for which $\binom{n}{2} 2^{-k+1} < 1$, e.g., for which $n^2 \leqslant 2^k$. $\qquad\square$

18.3 Tournaments

A *tournament* is an oriented graph $T = (V, E)$ such that $(x, x) \notin E$ for all $x \in V$, and for any two vertices $x \neq y$ exactly one of (x, y) and (y, x) belongs to E. The name tournament is natural, since one can think of the set V as a set of players in which each pair participates in a single match, where $(x, y) \in E$ iff x beats y.

Say that a tournament has the property P_k if for every set of k players there is one who beats them all, i.e., if for any subset $S \subseteq V$ of k players there exists a player $y \notin S$ such that $(y, x) \in E$ for all $x \in S$.

Theorem 18.3 (Erdős 1963). *If $\binom{n}{k}(1 - 2^{-k})^{n-k} < 1$, then there is a tournament of n players that has the property P_k.*

Proof. Consider a random tournament of n players, i.e., the outcome of every game is determined by the flip of fair coin. For a set S of k players, let A_S be the event that *no $y \notin S$ beats all of S*. Each $y \notin S$ has probability 2^{-k} of beating all of S and there are $n - k$ such possible y, all of whose chances are mutually independent. Hence $\mathrm{Prob}\,(A_S) = (1 - 2^{-k})^{n-k}$ and $\mathrm{Prob}\,(\bigcup A_S) \leqslant \binom{n}{k}(1 - 2^{-k})^{n-k} < 1$, i.e., with positive probability no event A_S occurs. Therefore, there is a point in the probability space for which none of the events A_S happens. This point is a tournament T and this tournament has the property P_k. □

Using estimates $\binom{n}{k} < \left(\frac{en}{k}\right)^k$ and $(1 - 2^{-k})^{n-k} < e^{-(n-k)/2^k}$, one may show that the condition of Theorem 18.3 is satisfied if, say, $n \geqslant k^2 \cdot 2^{k+1}$.

18.4 Property B revised

Let $B(k)$ be the minimum possible number of sets in a k-uniform family which is not 2-colorable. Put otherwise, the meaning of $B(k)$ is the following: for any k-uniform family of less than $B(k)$ sets, it is possible to color the points in red and blue so that no set in the family is monochromatic.

We have proved in Chap. 6 that $B(k) \geqslant 2^{k-1}$ (Theorem 6.2) We also proved an upper bound $B(k) \leqslant k^k$ under the additional assumption that our family is intersecting (Theorem 6.3). Using the probabilistic method we can eliminate this assumption and essentially improve the upper bound. This improvement was found by turning the probabilistic argument "on its head": the sets become random and each coloring defines an event.

Theorem 18.4 (Erdős 1964). *If k is sufficiently large, then*

$$B(k) \leqslant (1 + o(1))\frac{e \ln 2}{4}k^2 2^k.$$

Proof. Set $r = \lfloor k^2/2 \rfloor$. Let $\mathbf{A}_1, \mathbf{A}_2, \ldots$ be independent random members of $[r]^k$, that is, $\mathbf{A}_i$ ranges over the set of all $A \subseteq \{1, \ldots, r\}$ with $|A| = k$, and $\mathrm{Prob}\,(\mathbf{A}_i = A) = \binom{r}{k}^{-1}$. Consider the family $\mathcal{F} = \{\mathbf{A}_1, \ldots, \mathbf{A}_b\}$, where b is a parameter to be specified later. Let χ be a coloring of $\{1, \ldots, r\}$ in red and blue, with a red points and $r - a$ blue points. Using Jensen's inequality (see Proposition 1.10), for any such coloring and any i, we have

$$\mathrm{Prob}\,(\mathbf{A}_i \text{ is monochromatic}) = \mathrm{Prob}\,(\mathbf{A}_i \text{ is red}) + \mathrm{Prob}\,(\mathbf{A}_i \text{ is blue})$$
$$= \frac{\binom{a}{k} + \binom{r-a}{k}}{\binom{r}{k}} \geqslant 2\binom{r/2}{k} \Big/ \binom{r}{k} \rightleftharpoons p,$$

where, by the asymptotic formula (1.6) for the binomial coefficients, $p \sim e^{-1}2^{1-k}$. Since the members $\mathbf{A}_i$ of $\mathcal{F}$ are independent, the probability that a given coloring χ is legal for $\mathcal{F}$ equals

$$\prod_{i=1}^{b} (1 - \mathrm{Prob}\,(\mathbf{A}_i \text{ is monochromatic})) \leqslant (1 - p)^b.$$

Hence, the probability that at least one of all 2^r possible colorings will be legal for $\mathcal{F}$ does not exceed $2^r(1 - p)^b \leqslant e^{r \ln 2 - pb}$, which is less than 1 for $b \rightleftharpoons (1 + o(1))k^2 2^{k-2} e \ln 2$. But this means that there must be at least one realization of the (random) family $\mathcal{F}$, which has only b sets and which cannot be colored legally. $\qquad\square$

As for exact values of $B(k)$, only the first two $B(2) = 3$ and $B(3) = 7$ are known. The value $B(2) = 3$ is realized by the graph K_3. We address the inequality $B(3) \leqslant 7$ in Exercise 18.3.

18.5 The existence of small universal sets

A set of 0-1 strings of length n is (n, k)-*universal* if, for any subset of k coordinates $S = \{i_1, \ldots, i_k\}$, the projection

$$A\!\restriction_S \;\rightleftharpoons\; \{(a_{i_1}, \ldots, a_{i_k}) : (a_1, \ldots, a_n) \in A\}$$

of A onto the coordinates in S contains all possible 2^k configurations.

In Sects. 11.3 and 14.4.1 we presented two *explicit* constructions of such sets of size about n, when $k \leqslant (\log n)/3$, and of size $n^{O(k)}$, for arbitrary k. We have also mentioned that it is possible to prove the existence of much smaller universal sets of size only $2^k \log n$ (note that 2^k is a trivial lower bound). We now prove this.

Theorem 18.5 (Kleitman–Spencer 1973). *If* $\binom{n}{k}2^k(1 - 2^{-k})^r < 1$, *then there is an* (n, k)-*universal set of size* r.

Proof. Let $\mathbf{A}$ be a set of r random 0-1 vectors of length n, each entry of which takes values 0 or 1 independently and with equal probability $1/2$. For every fixed set S of k coordinates and for every fixed vector $v \in \{0,1\}^S$,

$$\text{Prob}\,(v \notin \mathbf{A}\lceil_S) = \prod_{a \in \mathbf{A}} \text{Prob}\,(v \neq a\lceil_S) = \prod_{a \in \mathbf{A}} \left(1 - 2^{-|S|}\right) = \left(1 - 2^{-k}\right)^r.$$

Since there are only $\binom{n}{k}2^k$ possibilities to choose a pair (S,v), the set $\mathbf{A}$ is *not* (n,k)-universal with probability at most $\binom{n}{k}2^k(1 - 2^{-k})^r$, which is strictly smaller than 1. Thus, at least one set A of r vectors must be (n,k)-universal, as claimed. $\qquad\square$

By using the fact that $\binom{n}{k} < (en/k)^k$ and $(1 - 2^{-k})^r \leqslant e^{-r/2^k}$, and by a special simple construction for $k \leqslant 2$ (cf., for example, Exercise 11.4), it is easy to derive from the last theorem that for every n and k there is an (n,k)-universal set of size at most $k2^k \log n$.

18.6 Cross-intersecting families

A pair of families $\mathcal{A}, \mathcal{B}$ is *cross-intersecting* if every set in $\mathcal{A}$ intersects every set in $\mathcal{B}$. The *degree* $d_{\mathcal{A}}(x)$ of a point x in $\mathcal{A}$ is the number of sets in $\mathcal{A}$ containing x. The *rank* of $\mathcal{A}$ is the maximum cardinality of a set in $\mathcal{A}$.

If $\mathcal{A}$ has rank a, then, by the pigeonhole principle, each set in $\mathcal{A}$ contains a point x which is "popular" for the members of $\mathcal{B}$ in that $d_{\mathcal{B}}(x) \geqslant |\mathcal{B}|/a$. Similarly, if $\mathcal{B}$ has rank b, then each member of $\mathcal{B}$ contains a point y for which $d_{\mathcal{A}}(y) \geqslant |\mathcal{A}|/b$. However, this alone does not imply that we can find a point which is popular in *both* families $\mathcal{A}$ and $\mathcal{B}$. It turns out that if we relax the "degree of popularity" by one-half, then such a point exists.

Theorem 18.6 (Razborov–Vereshchagin 1999). *Let $\mathcal{A}$ be a family of rank a and $\mathcal{B}$ be a family of rank b. Suppose that the pair $\mathcal{A}, \mathcal{B}$ is cross-intersecting. Then there exists a point x such that*

$$d_{\mathcal{A}}(x) \geqslant \frac{|\mathcal{A}|}{2b} \quad \text{and} \quad d_{\mathcal{B}}(x) \geqslant \frac{|\mathcal{B}|}{2a}.$$

Proof. Assume the contrary and let $\mathbf{A}, \mathbf{B}$ be independent random sets that are uniformly distributed in $\mathcal{A}, \mathcal{B}$ respectively. That is, for each $A \in \mathcal{A}$ and $B \in \mathcal{B}$, $\text{Prob}\,(\mathbf{A} = A) = 1/|\mathcal{A}|$ and $\text{Prob}\,(\mathbf{B} = B) = 1/|\mathcal{B}|$. Since the pair $\mathcal{A}, \mathcal{B}$ is cross-intersecting, the probability of the event "$\exists x(x \in \mathbf{A} \cap \mathbf{B})$" is equal to 1. Since the probability of a disjunction of events is at most the sum of the probabilities of the events, we have

$$\sum_x \text{Prob}\,(x \in \mathbf{A} \cap \mathbf{B}) \geqslant 1.$$

Let X_0 consist of those points x for which

$$\frac{d_{\mathcal{A}}(x)}{|\mathcal{A}|} = \mathrm{Prob}\,(x \in \mathbf{A}) < \frac{1}{2b},$$

and X_1 consist of the remaining points. Note that by our assumption, for any $x \in X_1$,

$$\mathrm{Prob}\,(x \in \mathbf{B}) = \frac{d_{\mathcal{B}}(x)}{|\mathcal{B}|} < \frac{1}{2a}$$

holds. By double counting (see Proposition 1.6), $\sum_x d_{\mathcal{A}}(x) = \sum_{A \in \mathcal{A}} |A|$. Hence,

$$\sum_{x \in X_1} \mathrm{Prob}\,(x \in \mathbf{A} \cap \mathbf{B}) = \sum_{x \in X_1} \mathrm{Prob}\,(x \in \mathbf{A}) \cdot \mathrm{Prob}\,(x \in \mathbf{B})$$

$$< \frac{1}{2a} \cdot \sum_{x \in X_1} \mathrm{Prob}\,(x \in \mathbf{A}) \leqslant \frac{1}{2a} \cdot \sum_x \mathrm{Prob}\,(x \in \mathbf{A})$$

$$= \frac{1}{2a} \cdot \sum_x \frac{d_{\mathcal{A}}(x)}{|\mathcal{A}|} = \frac{1}{2a\,|\mathcal{A}|} \cdot \sum_x d_{\mathcal{A}}(x) = \frac{1}{2a\,|\mathcal{A}|} \cdot \sum_{A \in \mathcal{A}} |A| \leqslant \frac{a\,|\mathcal{A}|}{2a\,|\mathcal{A}|} = \frac{1}{2}.$$

In a similar way we obtain

$$\sum_{x \in X_0} \mathrm{Prob}\,(x \in \mathbf{A} \cap \mathbf{B}) < \frac{1}{2},$$

a contradiction. $\qquad\square$

This theorem has the following application to boolean functions. Recall that a DNF (a disjunctive normal form) is an Or of monomials, each being an And of literals, where a literal is a variable x_i or its negation $\overline{x}_i$. The *size* of a DNF is the number of its monomials, and the *rank* is the maximum length of a monomial in it.

Given a pair F_0, F_1 of DNFs, a decision tree for such a pair is a usual decision tree (see Sect. 10.4) with the exception that this time each leaf is labeled by F_0 or by F_1. Given an input $v \in \{0,1\}^n$ we follow the (unique) path from the root until we reach some leaf. We require that the DNF labeling the leaf so reached must be falsified by v. Clearly, not every pair of DNFs will have a decision tree: we need that, for every input, at least one of the DNFs outputs 0 on it; that is, the formula $F_0 \wedge F_1$ must be not satisfiable; in this case we say that the pair of DNFs is *legal*. As before, the *depth* of a decision tree is the number of edges in a longest path from the root to a leaf. Let $DT(F_0, F_1)$ denote the minimum depth of a decision tree for the pair F_0, F_1.

In particular, if F_1 is a DNF of some boolean function f, and F_0 is a DNF of its negation $\overline{f}$, then their And $F_0 \wedge F_1$ is clearly unsatisfiable, and a decision tree for this pair is precisely the decision tree for f. If, moreover, F_0 has rank a and F_1 has rank b, then we already know (see Theorem 10.12) that $DT(F_0, F_1) \leqslant ab$.

Using Theorem 18.6, we can prove an upper bound which takes into account not only the rank of DNFs but also their size.

Theorem 18.7. *Let F_0 be a DNF of size M and rank a, and F_1 be a DNF of size N and rank b. Suppose that $F_0 \wedge F_1$ is not satisfiable. Then*

$$DT(F_0, F_1) \leqslant 2a \ln N + 2b \ln M + 2.$$

Proof. We interpret a monomial A as the set of its literals, and let $\mathcal{A} = \{A : A \text{ is a monomial in } F_0\}$. Let $\mathcal{B}$ be obtained in the same way using F_1 instead of F_0, but this time we flip all literals, i.e., $x_i \in A \in F_1$ gets replaced by $\overline{x}_i$, and $\overline{x}_i$ gets replaced by x_i. Then the unsatisfiability of $F_0 \wedge F_1$ means that every set in $\mathcal{A}$ intersects with each set in $\mathcal{B}$. Starting from the root, we construct the desired decision tree by testing the most popular variables first.

Applying Theorem 18.6 we find a literal that belongs to many sets from both $\mathcal{A}, \mathcal{B}$ and we probe its underlying variable x_i. Then, for any input $v \in \{0, 1\}^n$, either we learn that at least $M/2b$ monomials of F_0 are false on v or we learn that at least $N/2a$ monomials of F_1 are false on v.

Update F_0, F_1 by deleting false monomials. Again, $F_0 \wedge F_1$ is unsatisfiable, thus we can form new $\mathcal{A}, \mathcal{B}$ and apply Theorem 18.6. Repeat this until one of F_0, F_1 has no monomials. If this is F_0 then it is false on the given assignment. Otherwise, F_1 is false.

We estimate the number of evaluated variables. Let t_0 be the number of times when at least a fraction $1/2b$ of monomials from the current F_0 was deleted, and t_1 be the number of times when at least a fraction $1/2a$ of monomials from the current F_1 was deleted; hence $t_0 + t_1$ is the total number of variables tested on the input v. Then we have $M \left(1 - \frac{1}{2b}\right)^{t_0 - 1} \geqslant 1$ and $N \left(1 - \frac{1}{2a}\right)^{t_1 - 1} \geqslant 1$. Using the estimate $1 - x < e^{-x}$ (see (1.3)), these inequalities reduce to $M \cdot e^{-(t_0 - 1)/2b} \geqslant 1$ and $N \cdot e^{-(t_1 - 1)/2a} \geqslant 1$, implying that $t_0 \leqslant 2b \ln M + 1$ and $t_1 \leqslant 2a \ln N + 1$. $\qquad\square$

For completeness, let us mention (without proof) the following recent result concerning the *size* of cross-intersecting families. Say that a family $\mathcal{A}$ is *non-trivial* if $\bigcap_{A \in \mathcal{A}} A = \emptyset$.

Theorem 18.8 (Frankl–Tokushige 1998). *Let a, b and n be integers with $2 \leqslant a \leqslant b$ and $n \geqslant a + b$. Let $\mathcal{A}$ be a non-trivial a-uniform family and $\mathcal{B}$ a non-trivial b-uniform family of subsets of an n-element set. If the pair $\mathcal{A}, \mathcal{B}$ is cross-intersecting, then*

$$|\mathcal{A}| + |\mathcal{B}| \leqslant 2 + \binom{n}{b} - 2\binom{n - a}{b} + \binom{n - 2a}{b}.$$

This bound is optimal, as the following example shows: let $\mathcal{A} = \{A_0, A_1\}$ be disjoint a-element subsets and let $\mathcal{B}$ be the family of all b-element subsets B such that $B \cap A_0 \neq \emptyset$ and $B \cap A_1 \neq \emptyset$.

Exercises

18.1. Prove a lower bound for the general van der Waerden's function $W(r, k)$. *Hint*: Modify the proof of Theorem 18.2 to the case of more than two colors.

18.2 (Razborov–Vereshchagin 1999). Show that the bound in Theorem 18.6 is tight up to a multiplicative factor of 2.

Hint: Consider the following pair of families

$$\mathcal{A} = \{A_1, \ldots, A_b\}, \quad \text{where} \quad A_i = \{(i, 1), (i, 2), \ldots, (i, a)\},$$
$$\mathcal{B} = \{B_1, \ldots, B_a\}, \quad \text{where} \quad B_j = \{(1, j), (2, j), \ldots, (b, j)\}.$$

18.3.⁻ Prove that $B(3) \leqslant 7$. That is, exhibit a family of seven 3-element sets which is not 2-colorable. *Hint*: Consider the Fano configuration (Fig. 13.1).

18.4 (Razborov 1990). Consider the family of all pairs (A, B) of disjoint k-element subsets of $\{1, \ldots, n\}$. A set Y *separates* the pair (A, B) if $A \subseteq Y$ and $B \cap Y = \emptyset$. Prove that there exist $\ell = 3k4^k \ln n$ sets such that every pair (A, B) is separated by at least one of them.

Hint: Pick subsets $Y_1, \ldots, Y_\ell$ of $\{1, \ldots, n\}$ randomly and independently, each with probability 2^{-n}. Show that the probability that none of them separates a given pair (A, B) is at most $\left(1 - 2^{-2k}\right)^\ell$ and use the counting sieve.

18.5⁺ Let X be a set of $n = kr$ points and consider their colorings $c\colon X \to \{1, \ldots, k\}$ by k colors. Say that such a coloring c is *balanced* if each color is used for the same number of points, i.e., if $\left|c^{-1}(i)\right| = r$ for every color $i = 1, \ldots, k$. Given a k-element set of points, say that it is *differently colored* if no two of its points get the same color. Prove that there exist $\ell = O(ke^k \log n)$ balanced colorings $c_1, \ldots, c_\ell$ such that every k-element subset of X is differently colored by at least one of them.

Hint: Consider independent copies $\mathbf{c}_1, \ldots, \mathbf{c}_\ell$ of a balanced coloring $\mathbf{c}$ selected at random from the set of all $n!/(r!)^k$ such colorings. Show that for every k-element subset S of X, $\mathbf{c}$ colors S differently with probability $p = r^k \cdot \binom{n}{k}^{-1}$. Use the counting sieve to show that, with probability at least $1 - \binom{n}{k}(1 - p)^\ell$, every k-element subset S will be colored differently by at least one of $\mathbf{c}_1, \ldots, \mathbf{c}_\ell$. Recall that $r = n/k$ and use Proposition 1.3 to show that this probability is nonzero for some $\ell = O(ke^k \log n)$

18.6.⁻ (Khasin 1969). Consider the *k-threshold function* $T_k^n(x_1, \ldots, x_n)$ which outputs 1 if and only if $x_1 + \cdots + x_n \geqslant k$. In Sect. 7.3.2 we have shown that any depth-3 *Or-And-Or formula* for T_k^n must have size exponential in k. What about the upper bounds? Use the previous exercise to show that T_k^n can be computed by a monotone *Or-And-Or formula* of size $O(ke^k n \log n)$.

Hint: Each balanced k-coloring c of $\{1, \ldots, n\}$ gives us an And-Or formula $F_c = \bigwedge_{i=1}^{k} \bigvee_{c(j)=i} x_j$. Use the previous exercise to combine them into an Or-And-Or formula for T_k^n.

19. The Lovász Sieve

Let $A_1, \ldots, A_n$ be events in a probability space. In combinatorial applications the A_i are usually "bad" events. We wish to show $\text{Prob}\left(\bigcap \overline{A}_i\right) > 0$, so that there is a point (coloring, tournament, etc) which is "good." By the counting sieve, this holds if $\sum \text{Prob}(A_i) < 1$. In one sense this is best possible: when events A_i are pairwise disjoint, the condition cannot be weakened. At the opposite extreme, if the events A_i are mutually independent, then the only thing we need to ensure $\bigcap \overline{A}_i \neq \emptyset$ is that all $\text{Prob}(A_i) < 1$. The *Lovász sieve* is employed when there is "much independence" among the A_i.

19.1 The local lemma

Let $A_1, \ldots, A_n$ be events. A graph $G = (V, E)$ on the set of vertices $V = \{1, \ldots, n\}$ is said to be a *dependency graph* if, for all i, A_i is mutually independent of all the events A_j such that j is *not* adjacent to i in G, i.e., for which $\{i, j\} \notin E$. We emphasize that A_i must not only be independent of each such A_j individually but also must be independent of any boolean combination of the A_j's.

The following fact is known as the *Lovász Local Lemma*.

Lemma 19.1 (Erdős–Lovász 1975). *Let G be a dependency graph of events $A_1, \ldots, A_n$. Assume that each vertex of G has degree at most d. Assume $\text{Prob}(A_i) \leqslant p$ for all i. If $4pd \leqslant 1$ then $\text{Prob}\left(\overline{A}_1 \overline{A}_2 \cdots \overline{A}_n\right) > 0$.*

Proof (Spencer 1995). We prove by induction on m that for any m events (calling them $A_1, \ldots, A_m$ for convenience only)

$$\text{Prob}\left(A_1 \mid \overline{A}_2 \cdots \overline{A}_m\right) \leqslant 2p.$$

For $m = 1$ this is obvious. Let $2, \ldots, k$ be the vertices from $\{2, \ldots, m\}$ which are adjacent to 1 in the dependency graph G. Using the identity

$$\text{Prob}(A \mid BC) = \frac{\text{Prob}(AB \mid C)}{\text{Prob}(B \mid C)}, \tag{19.1}$$

we can write

$$\text{Prob}\left(A_1 \mid \overline{A}_2 \cdots \overline{A}_m\right) = \frac{\text{Prob}\left(A_1 \overline{A}_2 \cdots \overline{A}_k \mid \overline{A}_{k+1} \cdots \overline{A}_m\right)}{\text{Prob}\left(\overline{A}_2 \cdots \overline{A}_k \mid \overline{A}_{k+1} \cdots \overline{A}_m\right)}. \tag{19.2}$$

We bound the numerator

$$\text{Prob}\left(A_1\overline{A}_2\cdots\overline{A}_k \mid \overline{A}_{k+1}\cdots\overline{A}_m\right) \leqslant \text{Prob}\left(A_1 \mid \overline{A}_{k+1}\cdots\overline{A}_m\right)$$
$$= \text{Prob}\left(A_1\right) \leqslant p$$

since A_1 is mutually independent of $A_{k+1},\ldots,A_m$. The denominator, on the other hand, can be bounded by the induction hypothesis

$$\text{Prob}\left(\overline{A}_2\cdots\overline{A}_k \mid \overline{A}_{k+1}\cdots\overline{A}_m\right) = 1 - \text{Prob}\left(\bigcup_{i=2}^{k} A_i \mid \overline{A}_{k+1}\cdots\overline{A}_m\right)$$
$$\geqslant 1 - \sum_{i=2}^{k}\text{Prob}\left(A_i \mid \overline{A}_{k+1}\cdots\overline{A}_m\right)$$
$$\geqslant 1 - 2p(k-1) \geqslant 1/2,$$

because $k - 1 \leqslant d$ and $2pd \leqslant 1/2$. Thus

$$\text{Prob}\left(A_1 \mid \overline{A}_2\cdots\overline{A}_m\right) \leqslant p/(1/2) = 2p,$$

completing the induction. Finally,

$$\text{Prob}\left(\overline{A}_1\cdots\overline{A}_n\right) = \prod_{i=1}^{n}\text{Prob}\left(\overline{A}_i \mid \overline{A}_1\cdots\overline{A}_{i-1}\right) \geqslant (1 - 2p)^n > 0.$$

$\square$

When the events A_i are not symmetric a more general form of the Lovász sieve is appropriate. This generalization is due to Spencer (1977).

Lemma 19.2. *Let $G = (V, E)$ be a dependency graph of events $A_1,\ldots,A_n$. Suppose there exist real numbers $x_1,\ldots,x_n$, $0 \leqslant x_i < 1$, so that, for all i,*

$$\text{Prob}\left(A_i\right) \leqslant x_i \cdot \prod_{\{i,j\}\in E}(1 - x_j).$$

Then

$$\text{Prob}\left(\overline{A}_1\overline{A}_2\cdots\overline{A}_n\right) \geqslant \prod_{i=1}^{n}(1 - x_i).$$

In particular, with positive probability no event A_i holds.

Proof. The induction hypothesis of the earlier proof is replaced by

$$\text{Prob}\left(A_1 \mid \overline{A}_2\cdots\overline{A}_m\right) \leqslant x_i,$$

and, using the same identity (19.1), the denominator of (19.2) is set equal to

$$\prod_{j=2}^{k}\text{Prob}\left(\overline{A}_j \mid \overline{A}_{j+1}\cdots\overline{A}_m\right),$$

which by the induction hypothesis, is at least

$$\prod_{j=2}^{k}(1 - x_j) = \prod_{\{1,j\}\in E} (1 - x_j).$$

$\square$

19.2 Counting sieve for almost independence

The Lovász sieve works well when we have "much independence" between the events. In a similar vein, there is also an estimate, due to Razborov (1988), which works well if the events are "almost k-wise independent."

Let $A_1, \ldots, A_n$ be events, each of which appears with the same probability $\text{Prob}(A_i) = p$. If *all* these events are mutually independent, then

$$\text{Prob}\left(\bigcup_{i=1}^{n} A_i\right) = 1 - \text{Prob}\left(\bigcap_{i=1}^{n} \overline{A}_i\right) = 1 - (1-p)^n \geqslant 1 - e^{-pn}.$$

The mutual independence is a very strong requirement: we require that $\text{Prob}\left(\bigcap_{i\in I} A_i\right) = p^{|I|}$ for *all* subsets $I \subseteq \{1, \ldots, n\}$. It appears, however, that a reasonable estimate may be obtained also in the case when $\text{Prob}\left(\bigcap_{i\in I} A_i\right)$ is only "near" to $p^{|I|}$ for the sets I of size up to some number k; in this case the events $A_1, \ldots, A_n$ are also called *almost k-wise independent*.

Lemma 19.3 (Razborov 1988). *Let $n > 2k$ be any natural numbers, let $0 < p, \delta < 1$, and let $A_1, \ldots, A_n$ be events such that, for every subset $I \subseteq \{1, \ldots, n\}$ of size at most k,*

$$\left|\text{Prob}\left(\bigcap_{i\in I} A_i\right) - p^{|I|}\right| \leqslant \delta.$$

Then

$$\text{Prob}\left(\bigcup_{i=1}^{n} A_i\right) \geqslant 1 - e^{-pn} - \binom{n}{k+1}(\delta k + p^k).$$

Note that if the events are k-wise independent, then $\delta = 0$ and the obtained estimate is only by an additive factor $\binom{n}{k+1}p^k$ worse than that for mutual independence.

Proof. Let us first consider the case where k is even. Let $B_1, \ldots, B_n$ be independent events, each having the success probability p. Applying the Bonferroni inequalities to $\text{Prob}\left(\bigcup_{i=1}^{n} A_i\right)$ and $\text{Prob}\left(\bigcup_{i=1}^{n} B_i\right)$ (see Exercise 3.7), we obtain that

$$\text{Prob}\left(\bigcup_{i=1}^{n} A_i\right) \geqslant \sum_{\nu=1}^{k}(-1)^{\nu+1} \sum_{|I|=\nu} \text{Prob}\left(\bigcap_{i\in I} A_i\right) \qquad (19.3)$$

and

$$\text{Prob}\left(\bigcup_{i=1}^{n} B_i\right) \leqslant \sum_{\nu=1}^{k}(-1)^{\nu+1}\sum_{|I|=\nu} p^{|I|} + \sum_{|I|=k+1} p^{k+1}. \tag{19.4}$$

The assumption of the lemma that $A_1, \ldots, A_n$ are almost k-wise independent implies that the right-hand side in (19.3) is at least

$$\sum_{\nu=1}^{k}(-1)^{\nu+1}\sum_{|I|=\nu} p^{|I|} - \delta k\binom{n}{k}. \tag{19.5}$$

On the other hand, the independence of $B_1, \ldots, B_n$ implies that

$$\text{Prob}\left(\bigcup_{i=1}^{n} B_i\right) = 1 - (1-p)^n \geqslant 1 - e^{-pn}. \tag{19.6}$$

Combining (19.3), (19.4), (19.5) and (19.6) yields

$$\text{Prob}\left(\bigcup_{i=1}^{n} A_i\right) \geqslant 1 - e^{-pn} - \delta k\binom{n}{k} - p^{k+1}\binom{n}{k+1}$$

$$\geqslant 1 - e^{-pn} - \binom{n}{k+1}(\delta k + p^{k+1}).$$

In the case where k is odd, we use the above argument with $k-1$ substituted for k. □

19.3 Applications

A striking feature of the Lovász sieve is the lack of conditions on the total number n of events – only the degree of their dependence is important. This is particularly useful when dealing with large families whose members share not too many points in common. Let us demonstrate this with several examples; more applications can be found in the book by Alon and Spencer (1992).

19.3.1 Colorings

First, let us consider 2-colorings of hypergraphs. Recall that a family of sets $\mathcal{F}$ is 2-*colorable* if it is possible to color the points of the underlying set in red and blue, so that no member of $\mathcal{F}$ is monochromatic. A family is k-*uniform* if all its members have size k.

In Chap. 6 (see Theorem 6.2) we proved that if the family $\mathcal{F}$ is relatively small then it is 2-colorable. We can re-phrase the proof in probabilistic terms.

Theorem 19.4. *Every k-uniform family of fewer than 2^{k-1} sets is 2-colorable.*

Proof. Suppose $\mathcal{F}$ is a k-uniform family with at most $2^{k-1} - 1$ sets. Consider a random coloring, each element independently colored red or blue with probability $1/2$. Any one member of $\mathcal{F}$ will then be monochromatic with probability $2 \cdot 2^{-k} = 2^{1-k}$, and so the probability that *some* member will be monochromatic, does not exceed $|\mathcal{F}| \cdot 2^{k-1}$, which is strictly smaller than 1. Therefore, at least one coloring must leave no member of $\mathcal{F}$ monochromatic.

$\square$

Now suppose that $\mathcal{F}$ has more than 2^k members. Then the above random coloring will be doomed since the chances of it to be a proper 2-coloring will tend to zero. Fortunately, we do not require a high probability of success, just a positive probability of success. For example, if $\mathcal{F}$ is a family of m mutually disjoint k-element subsets of some set, then the events $A_i=$"the ith member of $\mathcal{F}$ is monochromatic" are mutually independent, and so the probability that none of them holds is exactly $\left(1 - 2^{-(k-1)}\right)^m$, which is positive no matter how large m is. Therefore, $\mathcal{F}$ is 2-colorable.

Of course for general families $\mathcal{F}$, the events $A_1, \ldots, A_m$ are not independent as some pairs of members may intersect. In such situations the Lovász sieve shows its surprising power.

Theorem 19.5 (Erdős–Lovász 1975). *If every member of a k-uniform family intersects at most 2^{k-3} other members, then the family is 2-colorable.*

Proof. Suppose $\mathcal{F} = \{S_1, \ldots, S_m\}$ is a family of k-element subsets of some set X. Consider a random coloring of X, each point independently colored red or blue with probability $1/2$. Let A_i denote the event that S_i is monochromatic. Then $\text{Prob}\,(A_i) = p$ where $p = 2(1/2)^{|S_i|} = 2^{1-k}$. Our goal is to show that $\text{Prob}\,(\overline{A}_1 \cdots \overline{A}_m) > 0$. Define a dependency graph by joining A_i and A_j if and only if $S_i \cap S_j \neq \emptyset$. By the assumption, this graph has degree at most $d = 2^{k-3}$. Since $4dp = d2^{3-k} \leqslant 1$, Lemma 19.1 yields the result. $\square$

In the general (not necessarily uniform) case we have the following.

Theorem 19.6 (Beck 1980). *Let $\mathcal{F}$ be a family of sets, each of which has at least k ($k \geqslant 2$) points. Also suppose that for each point v,*

$$\sum_{S \in \mathcal{F}: v \in S} (1 - 1/k)^{-|S|} 2^{-|S|+1} \leqslant \frac{1}{k}.$$

Then $\mathcal{F}$ is 2-colorable.

Proof. Let $\mathcal{F} = \{S_1, \ldots, S_m\}$ and (again) color the points with red and blue at random, independently of each other and with probability $1/2$. Let A_i denote the event that S_i is monochromatic; hence $\text{Prob}\,(A_i) = 2^{-|S_i|+1}$. Consider the same dependency graph $G = (V, E)$ as above: $\{i, j\} \in E$ if and only if $S_i \cap S_j \neq \emptyset$. We shall prove that the condition of Lemma 19.2 is satisfied with

$$x_i \rightleftharpoons (1 - 1/k)^{-|S_i|} 2^{-|S_i|+1}.$$

Indeed, by the definition of the graph G, for every $i = 1, \ldots, m$ we have

$$x_i \prod_{\{i,j\} \in E} (1 - x_j) \geqslant x_i \prod_{v \in S_i} \prod_{j:v \in S_j} (1 - x_j)$$

$$\geqslant x_i \prod_{v \in S_i} \left[1 - \sum_{j:v \in S_j} x_j \right] \geqslant x_i (1 - 1/k)^{|S_i|},$$

since, by the condition of the theorem, $\sum_{j:v \in S_j} x_j \leqslant 1/k$. Thus,

$$x_i \prod_{\{i,j\} \in E} (1 - x_j) \geqslant x_i (1 - 1/k)^{|S_i|} = 2^{-|S_i|+1} = \mathrm{Prob}\,(A_i).$$

By the application of Lemma 19.2 we obtain $\mathrm{Prob}\left(\overline{A}_1 \overline{A}_2 \cdots \overline{A}_n\right) > 0$, i.e., there is a 2-coloring in which no set of $\mathcal{F}$ is monochromatic. □

Let $\mathcal{F}$ be a family of k-element sets and suppose that no point appears in more than l of its members. In Chap. 6 we have shown that then it is possible to color the points in $r = l(k-1) + 1$ colors so that no member of $\mathcal{F}$ contains two points of the same color (see Exercise 6.5). On the other hand, if we have only $r < k$ colors, then every member of $\mathcal{F}$ will always have at least k/r points of the same color. Is it possible, also in this case (when $r < k$) to find a coloring such that no member has much more than k/r points of one color? The following result says that, if $k = l$ and if we have about $k/\log k$ colors, then such a coloring exists.

Theorem 19.7 (Füredi–Kahn 1986). *Let k be sufficiently large. Let $\mathcal{F}$ be a k-uniform family of sets and suppose that no point belongs to more than k sets of $\mathcal{F}$. Then it is possible to color the points in $r = \lfloor k/\log k \rfloor$ colors so that every member of $\mathcal{F}$ has at most $v = \lceil 2e\log k \rceil$ points of the same color.*

In fact, Füredi and Kahn proved a stronger result, where $v = \lfloor 4.5 \log k \rfloor$ and the members of $\mathcal{F}$ have size at most k. The argument then is the same but requires more precise computations.

Proof. Color the points of X by r colors, each point getting a particular color randomly and independently with probability $1/r$. Let $A(S, i)$ denote the event that more than v points of S get color i. We are going to apply Lemma 19.1 to these events. Events $A(S, i)$ and $A(S', i')$ can be dependent only if $S \cap S' \neq \emptyset$. So, we consider the following dependency graph G for these events: the vertex set consists of the pairs (S, i) where $S \in \mathcal{F}$ and $1 \leqslant i \leqslant r$, and two vertices (S, i) and (S', i') are joined by the edge if and only if $S \cap S' \neq \emptyset$.

Let d be the maximum degree of G. By the condition on our family $\mathcal{F}$, every member can intersect at most $k(k-1)$ other members, implying that $d \leqslant (1 + k(k-1))r \leqslant k^3$. By Lemma 19.1, it remains to show that each of the events $A(S, i)$ can happen with probability at most $1/(4k^3)$.

Since $|S| = k$, the probability that only the points of a subset $I \subseteq S$ get color i, is $(1/r)^{|I|}(1 - 1/r)^{k-|I|}$. Summing over all subsets I of S, then the event $A(S, i)$ happens with probability at most

$$\sum_{t > v} \binom{k}{t} \left(\frac{1}{r}\right)^t \left(1 - \frac{1}{r}\right)^{k-t} \leqslant \binom{k}{v} \left(\frac{1}{r}\right)^v < \left(\frac{ek}{vr}\right)^v \leqslant 2^{-v} < k^{-4}.$$

By Lemma 19.1, with positive probability, none of the events $A(S, i)$ will happen, and the desired coloring exists. $\qquad\Box$

19.3.2 Hashing functions

Let N, n, k and t be positive integers and suppose (to avoid trivialities) that $N > n \geqslant k \geqslant 2$. Let X be a set of cardinality N and let Y be a set of cardinality n. We say that a function $f : X \to Y$ *separates* a subset S of X if f is an injection when restricted to S (i.e., $|f(S)| = |S|$); otherwise we say that f *reduces* S.

An (N, n, k)-*perfect hash family of size* t is a sequence $f_1, f_2, \ldots, f_t$ of functions from X to Y with the property that for all sets $S \subseteq X$ such that $|S| = k$, at least one of $f_1, f_2, \ldots, f_t$ separates S. Given N, n and k, what is the minimal t for which such a family exists? As shown in Mehlhorn (1984), a straightforward application of the counting sieve implies that such a family exists whenever

$$t > \frac{\log \binom{N}{k}}{\log n^k - \log \left(n^k - k!\binom{n}{k}\right)}.$$

For small values of t, this bound can be slightly improved using the Lovász sieve.

Theorem 19.8 (Blackburn 2000). *An (N, n, k)-perfect hash family of size t exists whenever*

$$t \geqslant \frac{2 + \log \left(\binom{N}{k} - \binom{N-k}{k}\right)}{\log n^k - \log \left(n^k - k!\binom{n}{k}\right)}.$$

Note that this bound is better whenever $\frac{3}{4}\binom{N}{k} \leqslant \binom{N-k}{k}$.

Proof. Let N, n, k and t be fixed. Let $f_1, \ldots, f_t$ be functions chosen uniformly and independently at random from the set of all functions from X to Y. For any subset $S \subseteq X$ such that $|S| = k$, let A_S be the event that S is reduced by *all* of $f_1, \ldots, f_t$. Note that these functions form an (N, n, k)-perfect hash family if and only if none of the events A_S occur.

It is not difficult to see that $\text{Prob}(A_S) = p$ where

$$p = \left(\frac{n^k - k!\binom{n}{k}}{n^k}\right)^t.$$

Let G be the graph whose vertices are identified with the k-element subsets S of X, and where vertices identified with S_1, S_2 are joined by an edge precisely when $S_1 \cap S_2 \neq \emptyset$. It is clear that G is a dependency graph of the events A_S, and the vertices of G have degree $d = \binom{N}{k} - \binom{N-k}{k}$. By Lemma 19.1, an (N, n, k)-perfect hash family exists provided that $4pd \leqslant 1$. Taking logarithms of both sides of this inequality and rearranging, we find that this condition is equivalent to the inequality in the statement of the theorem. $\Box$

Exercises

19.1.⁻ Show that the condition "$4pd \leqslant 1$" in Lemma 19.1 can be replaced by "$ep(d + 1) \leqslant 1$." *Hint*: Apply Lemma 19.2 with all $x_i = 1/(d+1)$ and use the estimate $1 - t < e^{-t}$.

19.2. Let $\mathcal{F}$ be a k-uniform k-regular family, i.e., each set has k points and each point belongs to k sets. Let $k \geqslant 10$. Show that then at least one 2-coloring of points leaves no set of $\mathcal{F}$ monochromatic.

19.3. Recall that a *clause* is an Or of literals, where each literal is a boolean variable x_i or its negation $\bar{x}_i$. Let $C_1, \dots, C_m$ be a collection of clauses, each consisting of k literals, and such that every variable appears (negated or not) in at most r clauses. Assume that $r \leqslant 2^{k-2}/k$. Prove that then there is an assignment which satisfies all the clauses $C_1, \dots, C_m$.

19.4. Recall that the van der Waerden number $W(2, k)$ is the least number n such that any coloring of $\{1, 2, \dots, n\}$ in two colors gives a monochromatic arithmetic progression with k terms. In the previous chapter we have proved that $W(2, k) > 2^{k/2}$ (see Theorem 18.2). Using Lemma 19.1 improve this bound to $W(2, k) > 2^k/(2ek)$.

Hint: Assume that $n \leqslant 2^k/(2ek)$ and observe that one progression with k terms intersects at most nk others.

19.5⁺ (Erdős–Lovász 1975). Consider the colorings of real numbers in r colors. Say that a set of numbers is *multicolored* if it contains elements of all r colors. Fix a finite set X of real numbers, and let m be such that

$$4rm(m-1)\left(1 - \frac{1}{r}\right)^m < 1.$$

Using Lemma 19.1 prove that then, for any set S of m numbers there is an r-coloring under which every translate $x + S \rightleftharpoons \{x + y : y \in S\}$, with $x \in X$, is multicolored.

Hint: Take a random r-coloring $Y = \bigcup_{x \in X}(x + S)$ which, for every point $y \in Y$, takes a particular color randomly and independently with probability $1/r$. Consider events A_x saying that $x + S$ is not multicolored, and show that $\text{Prob}(A_x) \leqslant r\left(1 - \frac{1}{r}\right)^m$. Also observe that for each point x there are at most $m(m-1)$ other points x' for which $(x + S) \cap (x' + S) \neq \emptyset$.

20. Linearity of Expectation

Let $X_1, \ldots, X_n$ be random variables, and $X = c_1 X_1 + \cdots + c_n X_n$. Linearity of expectation states that

$$\mathrm{E}[X] = c_1 \mathrm{E}[X_1] + \cdots + c_n \mathrm{E}[X_n].$$

The power of this principle comes from there being no restrictions on the dependence or independence of the X_i's. In applications we often use the fact that there must be a point in the probability space for which $X \geqslant \mathrm{E}[X]$ and a point for which $X \leqslant \mathrm{E}[X]$. This principle (known as the *pigeonhole property* of the expectation) is used in most arguments.

20.1 Hamilton paths in tournaments

Recall that a *tournament* is an oriented graph $T = (V, E)$ such that $(x, x) \notin E$ for all $x \in V$, and for any two vertices $x \neq y$ exactly one of (x, y) and (y, x) belongs to E. The vertices are players, each pair of which participates in a single match, and $(x, y) \in E$ if and only if x beats y. Given such a tournament, a *Hamiltonian path* in it is defined as a permutation $(x_1, x_2, \ldots, x_n)$ of players such that, for every i, x_i beats x_{i+1}.

It is easy to show (see Exercise 20.5) that every tournament contains a Hamiltonian path. On the other hand, there are tournaments with only one Hamiltonian path (the path itself). Are there tournaments with many Hamiltonian paths? The existence of such "rich" tournaments was proved by T. Szele in 1943. His proof is considered to be the first application of the probabilistic method in combinatorics.

Theorem 20.1 (Szele 1943). *There is a tournament T with n players and at least $n!/2^{n-1}$ Hamiltonian paths.*

Proof. Take a random tournament $\mathbf{T}$ (where the outcome of each game is determined by the flip of fair coin), and let X be the number of Hamiltonian paths in it. For each permutation $\pi = (x_1, x_2, \ldots, x_n)$ of players, let X_π denote the indicator random variable for the event "π is a Hamiltonian path in $\mathbf{T}$." Then $X = \sum X_\pi$, the summation over all $n!$ permutations π. For a given π, $\mathrm{E}[X_\pi] = 2^{-(n-1)}$, since that is the probability that the $n-1$ games x_i versus x_{i+1} all have the desired outcome. By the linearity of expectation,

$$E[X] = \sum_{\pi} E[X_{\pi}] = n!2^{-(n-1)}.$$

Since (by the pigeonhole property of the expectation) a random variable cannot always be smaller than its expectation, at least one tournament must have at least $E[X]$ Hamiltonian paths.   □

In the same paper, Szele also established an upper bound $O(n!/2^{3n/4})$ on the maximal possible number of Hamiltonian paths in any tournament with n players. Based on the solution of the well-known conjecture of H. Minc about the permanent of 0-1 matrices (found by Bregman in 1973), Alon (1990b) has essentially improved this upper bound to

$$cn^{3/2}n!/2^{n-1}, \tag{20.1}$$

where c is a positive constant independent of n.

20.2 Sum-free sets

Suppose we are given a finite set of nonzero integers, and are asked to mark a large as possible subset of them under the restriction that the sum of any two marked integers cannot be marked. It appears that (independent of what the given integers actually are!) we can always mark at least one-third of them. A subset B of an additive group is called *sum-free* if $x + y \notin B$ for all $x, y \in B$.

Theorem 20.2 (Erdős 1965). *Let $A \subseteq \mathbb{Z}$ be a set of N nonzero integers. Then there is a sum-free subset B of A with $|B| > N/3$.*

Proof. Let $p = 3k + 2$ be a prime, which satisfies $p > 2\max_{a \in A} |a|$. Write $S = \{k+1, k+2, \ldots, 2k+1\}$, and observe that S is a sum-free subset of the group $\mathbb{Z}_p$ (the integers modulo p), because, by the choice of p, the sum of any two numbers from S, taken modulo p, does not belong to S.

Let $\mathbf{t}$ be a number chosen from $\mathbb{Z}_p - \{0\} = \{1, 2, \ldots, p-1\}$ randomly and independently with probability $1/(p-1)$. Let $X_{a,j}$ be the indicator random variable for the event that $a \cdot \mathbf{t} = j \,(\text{mod}\,p)$ and consider the sum

$$X = \sum_{a \in A} \sum_{j \in S} X_{a,j}.$$

For every fixed $a \in A$, as t ranges over all numbers $1, 2, \ldots, p-1$, $a \cdot t$ ranges over all nonzero elements of $\mathbb{Z}_p$ and hence

$$E[X_{a,j}] = \text{Prob}\,(X_{a,j} = 1) = \frac{1}{p-1},$$

for all $a \in A$ and $j \in S$. By the linearity of expectation,

$$E[X] = \sum_{a \in A} \sum_{j \in S} E[X_{a,j}] = \frac{N \cdot |S|}{p-1} > \frac{N}{3},$$

because $|S| > (p-1)/3$. By the pigeonhole property of the expectation, there is some $t \in \{1, 2, \ldots, p-1\}$ for which $X > N/3$. Define

$$B = \{a \in A : a \cdot t = j \,(\mathrm{mod}\, p) \text{ for some } j \in S\}.$$

This set has $|B| > N/3$ numbers and is sum-free (even more, it is sum-free modulo p), because such is the set S.   □

It is not clear what is the largest constant that works in place of $1/3$ in the previous theorem. It is only known (see Alon and Kleitman 1990) that it must be smaller than $12/29$.

20.3 Dominating sets

A *dominating set* of vertices in a graph $G = (V, E)$ is a set $S \subseteq V$ such that every vertex of G belongs to S or has a neighbor in S.

Theorem 20.3 (Alon 1990c). *If $G = (V, E)$ is an n-vertex graph with minimum degree $d > 1$, then G has a dominating set with at most $n\frac{1+\ln(d+1)}{d+1}$ vertices.*

Proof. Form a random vertex subset $\mathbf{S} \subseteq V$ by including each vertex independently with probability $p \rightleftharpoons \ln(d+1)/(d+1)$. Given $\mathbf{S}$, let $\mathbf{T}$ be the set of vertices outside $\mathbf{S}$ having no neighbor in $\mathbf{S}$; adding $\mathbf{T}$ to $\mathbf{S}$ yields a dominating set. So, it remains to estimate the expected size of this union. Since each vertex appears in $\mathbf{S}$ with probability p, $\mathrm{E}\left[|\mathbf{S}|\right] = np$.

The random variable $|\mathbf{T}|$ is the sum $\sum_{v \in V} X_v$ of n indicator variables X_v for whether individual vertices v belong to $\mathbf{T}$. We have $X_v = 1$ if and only if v and its neighbors all fail to be in $\mathbf{S}$, the probability of which is bounded by $(1-p)^{d+1}$, since v has degree at least d. Hence, $\mathrm{E}\left[|\mathbf{T}|\right] = \sum_{v \in V} \mathrm{E}\left[X_v\right] \leqslant n(1-p)^{d+1}$. As $(1-p)^{d+1} \leqslant e^{-p(d+1)}$, we have

$$\mathrm{E}\left[|\mathbf{S} \cup \mathbf{T}|\right] \leqslant np + ne^{-p(d+1)} = n\frac{1+\ln(d+1)}{d+1}.$$

By the pigeonhole property of the expectation, there must be some S for which $S \cup T$ is a dominating set of size no larger than this.   □

20.4 The independence number

The independence number $\alpha(G)$ of a graph G is the maximum number of vertices with no edges between them. The following result is due to Caro (unpublished) and Wei (1981).

Theorem 20.4. *Let G be a graph on n vertices and let d_i denote the degree of the ith vertex. Then*

$$\alpha(G) \geq \sum_{i=1}^{n} \frac{1}{d_i + 1}. \tag{20.2}$$

Proof. (Alon–Spencer 1992). Let $V = \{1, \ldots, n\}$ and let $\pi : V \to V$ be a random permutation taking its values uniformly and independently with probability $1/n!$. This permutation corresponds to a random ordering of vertices in V. Let A_i be the event that all neighbors j of i in G are greater than i in the ordering, i.e., that $\pi(j) > \pi(i)$ for all d_i neighbors j of i. There are $\binom{n}{d_i+1}$ possibilities to choose a $(d_i + 1)$-element set $S \subseteq V$ of possible π-images of i and all its d_i neighbors. After that there are $(|S| - 1)! = d_i!$ possibilities to arrange the π-images of neighbors of i within S (the place of $\pi(i)$ is fixed – it must come first), and $(n - |S|)! = (n - d_i - 1)!$ possibilities to arrange the vertices outside S. Thus,

$$\text{Prob}\,(A_i) = \binom{n}{d_i + 1} \frac{d_i!(n - d_i - 1)!}{n!} = \frac{1}{d_i + 1}.$$

Let $\mathbf{U}$ be the set of those vertices i for which A_i holds. By linearity of expectation

$$\text{E}\,[|\mathbf{U}|] = \sum_{i=1}^{n} \text{Prob}\,(A_i) = \sum_{i=1}^{n} 1/(d_i + 1).$$

Thus, for some specific ordering, $|U| \geq \sum_{i=1}^{n} 1/(d_i + 1)$. Now let $\{i, j\}$ be an edge of G. Then either $\pi(i) < \pi(j)$ or $\pi(j) < \pi(i)$. In the first case $j \notin U$, and in the second case $i \notin U$. That is, U is an independent set. $\square$

The celebrated theorem due to P. Turán (1941) states: if a graph G has n vertices and has no k-clique then it has at most $(1 - (1/(k - 1))\,n^2/2$ edges (see Theorem 4.7). Its dual form states (see Exercise 4.7):

If G has n vertices and $nk/2$ edges, then $\alpha(G) \geq n/(k + 1)$.

This dual form of Turán's theorem also follows from Theorem 20.4: fixing the total number of edges, the sum $\sum_{i=1}^{n} 1/(d_i + 1)$ is minimized when the d_i's are as nearly equal as possible, and, by Theorem 1.7, $\frac{1}{2}\sum_{i=1}^{n} d_i$ is exactly the number of edges in G (see Exercise 20.7 for more precise proof).

20.5 Low degree polynomials

In this section we consider polynomials $f(x_1, \ldots, x_n)$ on n variables over the field $\mathbb{F}_2$. Such a polynomial has degree at most d if it can be written in the form

$$f(x_1, \ldots, x_n) = a_0 + \sum_{i=1}^{m} \prod_{j \in S_i} x_j,$$

where $a_0 \in \{0, 1\}$ and $S_1, \ldots, S_m$ are subsets of $\{1, \ldots, n\}$ of size at most d; here and throughout the section the sum is modulo 2.

If $f_1, \ldots, f_m$ are polynomials of degree at most d, then their product can have degree up to dm. The following result says that the product can still be approximated quite well by a polynomial of relatively small degree.

Lemma 20.5 (Razborov 1987). *Let $f = \prod_{i=1}^{m} f_i$, where $f_1, \ldots, f_m$ are polynomials of degree at most d over $\mathbb{F}_2$. Then, for any $r \geqslant 1$, there exists a polynomial g of degree at most dr such that g differs from f on at most 2^{n-r} inputs.*

Proof. Let $\mathbf{S}$ be a random subset of $\{1, \ldots, m\}$, that is, we choose $\mathbf{S}$ randomly from the family of all 2^m subsets with probability 2^{-m}. Let $\mathbf{S}_1, \ldots, \mathbf{S}_r$ be independent copies of $\mathbf{S}$. Consider a (random) function of the form

$$\mathbf{g} = \prod_{j=1}^{r} \mathbf{h}_j, \quad \text{where} \quad \mathbf{h}_j = 1 - \sum_{i \in \mathbf{S}_j} (1 - f_i). \tag{20.3}$$

We claim that, for every (fixed) input $a \in \{0, 1\}^n$,

$$\text{Prob}\,(\mathbf{g}(a) \neq f(a)) \leqslant 2^{-r}. \tag{20.4}$$

Indeed, if $f(a) = 1$ then all $f_i(a) = 1$, and hence, $\mathbf{g}(a) = 1$ with probability 1. Suppose now that $f(a) = 0$. Then $f_{i_0}(a) = 0$ for at least one i_0. Since each of the sets $\mathbf{S}_1, \ldots, \mathbf{S}_r$ contains i_0 with probability $1/2$, we have that $\text{Prob}\,(\mathbf{h}_j(a) = 1) \leqslant 1/2$ for all $j = 1, \ldots, r$ (consult Exercise 17.2 for this conclusion). Hence,

$$\text{Prob}\,(\mathbf{g}(a) = 0) = 1 - \text{Prob}\,(\mathbf{h}_1(a) = \ldots = \mathbf{h}_r(a) = 1) \geqslant 1 - 2^{-r},$$

as claimed.

For an input vector $a \in \{0, 1\}^n$, let X_a denote the indicator random variable for the event that $\mathbf{g}(a) \neq f(a)$, and let X be the sum of X_a over all a. By (20.4) and the linearity of expectation, the expected number of inputs on which $\mathbf{g}$ differs from f is

$$\text{E}\,[X] = \sum_a \text{E}\,[X_a] = \sum_a \text{Prob}\,(X_a = 1) \leqslant 2^{n-r}.$$

By the pigeonhole principle of expectation, there must be a point in the probability space for which this holds. This point is a polynomial of the form (20.3); it has degree at most dr and differs from f on at most 2^{n-r} inputs. $\qquad\square$

Razborov (1987) used this lemma to prove that the majority function cannot be computed by constant depth polynomial size circuits with And, Or and Parity gates. The idea is as follows.

If f can be computed by a depth-c circuit of size ℓ then, by Lemma 20.5, there exists a polynomial g of degree at most r^c such that g differs from f on at most $\ell \cdot 2^{n-r}$ inputs. The desired lower bound is then obtained by showing that the majority function cannot be approximated sufficiently well by such polynomials (see Lemma 14.8). Taking r to be about $n^{1/(2c)}$ and making necessary computations this leads to a lower bound $\ell \geqslant 2^{\Omega(n^{1/(2c)})}$. This final step requires some routine calculations, and we omit it.

20.6 Maximum satisfiability

In most of the above applications it was enough to take a uniform distribution, that is, every object had one and the same probability of appearing. In this section we will consider the situation where the distribution essentially depends on the specific properties of a given family of objects.

An *And-Or formula* or a *CNF* (or simply, a *formula*) over a set of variables $x_1, \ldots, x_n$ is an And of an arbitrary number of *clauses*, where a clause is an Or of an arbitrary number of *literals*, each literal being either a variable x_i or a negated variable $\overline{x}_i$. For example:

$$F = (x_1 \vee \overline{x}_3)(\overline{x}_1 \vee x_2 \vee \overline{x}_3)(\overline{x}_2)(\overline{x}_1 \vee \overline{x}_2).$$

An *assignment* is a mapping which assigns each variable one of the values 0 or 1. We can look at such assignments as binary vectors $v = (v_1, \ldots, v_n) \in \{0, 1\}^n$, where v_i is the value assigned to x_i. If y is a literal, then we say that v *satisfies* y if either $y = x_i$ and $v_i = 1$, or $y = \overline{x}_i$ and $v_i = 0$. An assignment satisfies a clause if it satisfies at least one of its literals. An assignment satisfies a formula if it satisfies each of its clauses. For the formula above, the assignment $v = (1, 0, 0)$ is satisfying. A formula is *satisfiable* if at least one assignment satisfies it. A formula F is *k-satisfiable* if any subset of k clauses of F is satisfiable.

It is an interesting "Helly-type" phenomenon, first established by Lieberher and Specker (1981), which says that if a formula is 3-satisfiable then at least 2/3 of its clauses are simultaneously satisfiable. For 2-satisfiable formulas this fraction is $2/(1 + \sqrt{5}) > 0.618$ (the inverse of the golden ratio). The original proof of these facts was rather involved. Yannakakis (1994) has found a very simple proof of these bounds using the probabilistic method.

Theorem 20.6 (Yannakakis 1994). *If F is a 3-satisfiable formula then at least 2/3 fraction of its clauses are simultaneously satisfiable.*

Proof. Given a 3-satisfiable formula F, define a random assignment $\mathbf{v} = (v_1, \ldots, v_n)$, where each bit v_i takes its value independently from other bits and with probability

$$\text{Prob}\,(v_i = 1) = \begin{cases} 2/3 & \text{if } F \text{ contains a unary clause } (x_i); \\ 1/3 & \text{if } F \text{ contains a unary clause } (\overline{x}_i); \\ 1/2 & \text{otherwise.} \end{cases}$$

Note that this definition is consistent since it is impossible to have the unary clauses (x_i) and $(\overline{x}_i)$ in the same 3-satisfiable formula. Simple (but crucial) observation is that each singular literal $y \in \{x_i, \overline{x}_i\}$, which appears in the formula F, is falsified with probability $\leqslant 2/3$ (independent of whether this literal forms a unary clause or not). To see this, let $y = x_i$ and $p = \mathrm{Prob}\,(v_i = 0)$. We have three possibilities:

- either (x_i) is a unary clause of F, and in this case $p = 1 - 2/3 = 1/3$;
- or F contains a unary clause $(\overline{x}_i)$, and in this case $p = 1 - 1/3 = 2/3$;
- or neither x_i nor $\overline{x}_i$ appears in a unary clause, in which case $p = 1/2$.

Using this observation, we can prove the following fact.

Claim 20.7. *Every clause is satisfied by* **v** *with probability at least* $2/3$.

For unary clauses the claim is trivial. On the other hand, if C contains three or more literals, then, by the above observation, each of these literals can be falsified with probability at most $2/3$, and hence, the clause is satisfied with probability at least $1 - (2/3)^3 = 0.7037... > 2/3$; for longer clauses probabilities are even better.

It remains to consider binary clauses. Assume w.l.o.g. that $C = (x_1 \vee x_2)$. If at least one of x_1 and x_2 is satisfied with probability $1/2$ then the clause C is satisfied with probability $1 - \mathrm{Prob}\,(v_1 = 0) \cdot \mathrm{Prob}\,(v_2 = 0) \geqslant 1 - \frac{1}{2} \cdot \frac{2}{3} = \frac{2}{3}$. Thus, the only bad case would be when both literals x_1 and x_2 are satisfied only with probability $1/3$. But this is impossible because it would mean that the formula F contains the clauses $(x_1 \vee x_2), (\overline{x}_1), (\overline{x}_2)$, which contradicts the fact that F is 3-satisfiable.

We now conclude the proof of the theorem in a standard manner. Suppose that F consists of the clauses $C_1, \ldots, C_m$. Let X_i denote the indicator random variable for the event "the ith clause C_i is satisfied by **v**". Then $X = \sum_{i=1}^m X_i$ is the total number of satisfied clauses of F. By Claim 20.7, $\mathrm{Prob}\,(X_i = 1) \geqslant 2/3$ for each i, and by the linearity of expectation, $\mathrm{E}\,[X] = \sum_{i=1}^m \mathrm{E}\,[X_i] \geqslant \frac{2m}{3}$. By the pigeonhole property of the expectation, at least one assignment **v** must satisfy so many clauses of F, as desired. $\qquad\square$

It is worth mentioning that, for large values of k, the right fraction for all k-satisfiable formulas is $3/4$. Namely, Trevisan (1997) has proved that, if r_k stands for the largest real such that in any k-satisfiable formula at least r_k-th fraction of its clauses are satisfied simultaneously, then $\lim_{k \to \infty} r_k = 3/4$.

20.7 Hashing functions

A set V of vectors of length t over an alphabet $A = \{1, \ldots, n\}$ is called k-*separated* if for every k distinct vectors there is a coordinate in which they are all distinct. How many vectors can such a set have?

This question is equivalent to the question about the maximum size $N = N(n, k, t)$ of a domain for which there exists a family of (n, k) *hashing functions* with t members, that is, a family of t partial functions $f_1, \ldots, f_t$ mapping a domain of size N into a set of size n so that every subset of k elements of the domain is mapped in a one-to-one fashion by at least one of the functions. To see this equivalence, it is enough to consider the set of vectors $(f_1(x), \ldots, f_t(x))$ for each point x of the domain.

The problem of estimating $N(n, k, t)$, which is motivated by the numerous applications of perfect hashing in theoretical computer science, has received a considerable amount of attention. The interesting case is when the number t of hash functions is much bigger than the size n of the target set (and, of course, $n \geqslant k$). The following are the best known estimates for $N(n, k, t)$:

$$\frac{1}{k-1} \log \frac{1}{1 - g(n, k)} \lesssim \frac{1}{t} \log N(n, k, t) \qquad (20.5)$$

and

$$\frac{1}{t} \log N(n, k, t) \lesssim \min_{1 \leqslant r \leqslant k-1} g(n, r) \log \frac{n - r + 1}{k - r}, \qquad (20.6)$$

where

$$g(n, k) \rightleftharpoons \frac{(n)_k}{n^k} = \frac{n(n-1) \cdots (n - k + 1)}{n^k}.$$

The lower bound (20.5), proved by Fredman and Komlós (1984), can be derived using a probabilistic argument (the *deletion method*) discussed in Chap. 21: one chooses an appropriate number of vectors randomly, shows that the expected number of non-separated k-tuples is small, and omits a vector from each such "bad" k-tuple. The proof of the upper bound (20.6) is much more difficult. For $r = k - 1$, a slightly weaker version of this bound was proved in Fredman and Komlós (1984), and then extended to (20.6) by Körner and Marton (1988). All these proofs rely on certain techniques from information theory.

A short and simple probabilistic proof of (20.6), which requires no information theoretic tools, was found by Nilli (1994) (c/o Noga Alon), and we describe it below.

The proof is based on the following lemma.

Lemma 20.8. *Let U be a set of m vectors of length t over the alphabet $B \cup \{*\}$, where $B = \{1, \ldots, b\}$, and let x_v denote the number of non-$*$ coordinates of $v \in U$. Let $\bar{x} = \sum x_v / m$ be the average value of x_v. If for every d distinct vectors in U there is a coordinate in which they all are different from $*$ and are all distinct, then*

$$m \leqslant (d - 1) \left(\frac{b}{d - 1} \right)^{\bar{x}}.$$

Proof. For every coordinate i, choose randomly and independently a subset $\mathbf{D}_i$ of cardinality $d - 1$ of $B - \{*\}$. Call a vector $v \in U$ *consistent* if for every i, $v_i \in \mathbf{D}_i \cup \{*\}$. Since each set $\mathbf{D}_i$ has size $d - 1$, the assumption clearly implies that for any choice of the sets D_i there are no more than $d - 1$ consistent vectors. On the other hand, for a fixed vector v and its coordinate i, $\operatorname{Prob}(v_i \in \mathbf{D}_i) = (d - 1)/b$. So, each vector v is consistent with probability $((d - 1)/b)^{x_v}$ and, by the linearity of expectation, the expected number of consistent vectors in U is

$$\sum_{v \in V} \left(\frac{d - 1}{b}\right)^{x_v} \geq m \left(\frac{d - 1}{b}\right)^{\overline{x}},$$

where the inequality follows from Jensen's inequality (see Proposition 1.10), since the function $g(z) = ((d - 1)/b)^z$ is convex.   □

Proof of (20.6). Let V be a k-separated set of $N = N(n, k, t)$ vectors of length t over the alphabet $A = \{1, \ldots, n\}$ and suppose $1 \leq r \leq k - 1$. We say that a set of r vectors is *separated at coordinate i* if their values in this coordinate are all distinct. Let $\alpha(n, r)$ be the fraction of r-subsets of vectors that is separated by a given coordinate. The Muirhead inequality (Hardy et al. 1952, p. 44) implies that $\alpha(n, r) \lesssim g(n, r)$ (see Exercise 20.16 for a slightly worse estimate).

Let $\mathbf{R} \subset V$ be a randomly chosen set of $r - 1$ vectors, and let $\mathbf{v}$ be a random vector in $V - \mathbf{R}$. Let $x_{\mathbf{v}}$ denote the number of coordinates of $\mathbf{v}$ in which all the vectors in $\mathbf{R} \cup \{\mathbf{v}\}$ have distinct values. Hence, the expectation of $x_{\mathbf{v}}$ is at most $t \cdot \alpha(n, r) \lesssim t \cdot g(n, r)$. By the pigeonhole property of the expectation, there is a fixed set R for which the average value $\overline{x}$ of x_v for $v \in V - R$ is at most $(1 + o(1))g(n, r) \cdot t$.

We now define a set U of $N - r + 1$ vectors over $\{1, 2, \ldots, n - r + 1, *\}$ as follows. Associate with each $v \in V - R$ a vector $v' \in U$ whose ith coordinate is $*$ unless $R \cup \{v\}$ is separated at this coordinate. In this later case, we define $v'_i = p$ if the value of v_i is the pth largest element in the set $A - \{u_i : u \in R\}$. Note that for every set S of $d = k - r + 1$ vectors in U there must be a coordinate in which the values of all these vectors are distinct and differ from $*$, since otherwise the k-element set $R \cup S$ would be not separated. Therefore, by Lemma 20.8 with $b = n - r + 1$, $d = k - r + 1$ and $m = |U| = N - r + 1$,

$$N - r + 1 \leq (k - r) \cdot \left(\frac{n - r + 1}{k - r}\right)^{(1 + o(1))g(n, r) \cdot t},$$

implying (20.6).   □

20.8 Submodular complexity measures

So far we have used the probabilistic argument to prove that some combinatorial objects exist. In this section we present one application which (although negative in the conclusion) is very interesting *methodologically*: it demonstrates how probabilistic methods can be used to prove that some kind of attempts to solve a problem cannot lead to a success!

In order to prove that some boolean function f is hard to compute (requires a large circuit or formula size), one tries to find some clever "combinatorial" measure μ on the set of all boolean functions satisfying two conditions: $\mu(f)$ is a lower bound on the size of any circuit computing f, and $\mu(f)$ can be non-trivially bounded from below at some *explicit* boolean functions f.

We have already seen that in the case of monotone models (circuits, formulas, span programs) such an approach works quite well. Namely, we can take $\mu(f)$ to be:

- the minimal number t for which the function f is t-simple, in the case of monotone circuits (Sect. 10.6);
- the minimal number of monotone rectangles covering $f^{-1}(1) \times f^{-1}(0)$, in the case of monotone formulas (Sect. 15.2.2);
- the maximal number of minterms of f forming a self-avoiding set system, in the case of monotone span programs (Sect. 16.4).

But (at least directly) none of these measures can be applied in the non-monotone case. It is therefore important to understand if some type of measures μ are inherently bad, i.e., cannot lead to non-trivial lower bounds in the general (non-monotone) case, or we are just not very lucky in applying them. And probabilistic arguments can help us to find an answer.

To demonstrate this, let us consider the case of boolean formulas. Recall that such formulas are defined inductively as follows: (i) every boolean variable x_i and its negation $\overline{x}_i$ is a formula of size 1, and (ii) if F_1 and F_2 are formulas of size l_1 and l_2, then both $F_1 \wedge F_2$ and $F_1 \vee F_2$ are formulas of size $l_1 + l_2$. Let $L(f)$ denote the minimum size of a formula computing f. Then, by the definition, $L(f) \geqslant \mu(f)$ for any measure μ such that

$$\mu(x_i) \leqslant 1, \ \mu(\overline{x}_i) \leqslant 1 \ \ (1 \leqslant i \leqslant n); \tag{20.7}$$

$$\mu(f \vee g) \leqslant \mu(f) + \mu(g) \ \text{ for each } f, g; \tag{20.8}$$

$$\mu(f \wedge g) \leqslant \mu(f) + \mu(g) \ \text{ for each } f, g; \tag{20.9}$$

Example 20.9. (Due to M. Paterson). Associate with a boolean function f a bipartite graph $G = (U, V, E)$, where $U = f^{-1}(0)$, $V = f^{-1}(1)$ and $(a, b) \in E$ if and only if vectors a and b differ in exactly one coordinate. Define

$$\mu(f) = \frac{|E|^2}{|U| \cdot |V|}.$$

This is the measure used by Khrapchenko (1971) to prove that the parity of n variables requires formulas of size n^2 (see Theorem 15.13). It can be shown (do this!) that this measure satisfies all three conditions (20.7), (20.8) and (20.9). However, we already know (see Exercise 15.6) that $\mu(f_n) \leqslant n^2$ for every function f_n on n variables.

There were many attempts to modify Khrapchenko's measure to get larger lower bounds. In Sect. 15.2.2 we defined one such measure in terms of matrix rank. Razborov (1992) has proved that this measure μ satisfies the following *submodularity condition*:

$$\mu(f \wedge g) + \mu(f \vee g) \leqslant \mu(f) + \mu(g) \quad \text{for each } f, g. \tag{20.10}$$

Note that this condition is stronger than both (20.8) and (20.9). In Sect. 15.2.2 we have shown that a version of this measure, corresponding to *monotone formulas, can* lead to super-polynomial lower bounds. But attempts to apply this measure for non-monotone formulas have failed. The following theorem explains the reason for this failure. Let F_n be the family of all boolean functions on n variables.

Theorem 20.10 (Razborov 1992). *If μ is a submodular measure on F_n, then $\mu(f_n) \leqslant O(n)$ for each $f_n \in F_n$.*

Proof. Let $\mathbf{g}_d$ be a random boolean function in d variables $x_1, \ldots, x_d$. That is, we choose $\mathbf{g}_d$ randomly and uniformly from F_d. We are going to prove by induction on d that

$$E\left[\mu(\mathbf{g}_d)\right] \leqslant d + 1. \tag{20.11}$$

Given a variable x_i, set $x_i^1 \rightleftharpoons x_i$ and $x_i^0 \rightleftharpoons \overline{x}_i$.
Base. $d = 1$. Here we have $\mu(g(x_1)) \leqslant 2$ for any $g(x_1)$. This follows from (20.7) if g is a variable x_1 or its negation $\overline{x}_1$. By (20.10) and (20.7) we have

$$\mu(0) + \mu(1) = \mu(x_1 \wedge \overline{x}_1) + \mu(x_1 \vee \overline{x}_1) \leqslant \mu(x_1) + \mu(\overline{x}_1) \leqslant 2$$

which proves $\mu(g(x_1)) \leqslant 2$ in the remaining case when g is a constant.

Inductive step. Assume that (20.11) is already proved for d. Let the symbol $\approx$ mean that two random functions have the same distribution. Note that

$$\mathbf{g}_{d+1} \approx \left(\mathbf{g}_d^0 \wedge x_{d+1}^0\right) \vee \left(\mathbf{g}_d^1 \wedge x_{d+1}^1\right), \tag{20.12}$$

where $\mathbf{g}_d^0$ and $\mathbf{g}_d^1$ are two independent copies of $\mathbf{g}_d$. By duality,

$$\mathbf{g}_{d+1} \approx \left(\mathbf{g}_d^0 \vee x_{d+1}^0\right) \wedge \left(\mathbf{g}_d^1 \vee x_{d+1}^1\right). \tag{20.13}$$

By the linearity of expectation, we obtain from (20.12) and (20.8) (remember that the latter is a consequence of (20.10)) that

$$E\left[\mu(\mathbf{g}_{d+1})\right] \leqslant E\left[\mu\left(\mathbf{g}_d^0 \wedge x_{d+1}^0\right)\right] + E\left[\mu\left(\mathbf{g}_d^1 \wedge x_{d+1}^1\right)\right] \tag{20.14}$$

and similarly from (20.13) and (20.9),

$$E\left[\mu(\mathbf{g}_{d+1})\right] \leqslant E\left[\mu\left(\mathbf{g}_d^0 \vee x_{d+1}^0\right)\right] + E\left[\mu\left(\mathbf{g}_d^1 \vee x_{d+1}^1\right)\right]. \tag{20.15}$$

Summing (20.14), (20.15) and applying consecutively (20.10), (20.7) and the inductive assumption (20.11), we obtain

$$2 \cdot \mathrm{E}\left[\mu(\mathbf{g}_{d+1})\right] \leqslant \mathrm{E}\left[\mu\left(\mathbf{g}_d^0 \wedge x_{d+1}^0\right)\right] + \mathrm{E}\left[\mu\left(\mathbf{g}_d^0 \vee x_{d+1}^0\right)\right] +$$
$$\mathrm{E}\left[\mu\left(\mathbf{g}_d^1 \wedge x_{d+1}^1\right)\right] + \mathrm{E}\left[\mu\left(\mathbf{g}_d^1 \vee x_{d+1}^1\right)\right]$$
$$\leqslant \mathrm{E}\left[\mu(\mathbf{g}_d^0)\right] + \mu(x_{d+1}^0) + \mathrm{E}\left[\mu(\mathbf{g}_d^1)\right] + \mu(x_{d+1}^1)$$
$$\leqslant 2 \cdot \mathrm{E}\left[\mu(\mathbf{g}_d)\right] + 2$$
$$\leqslant 2d + 4.$$

This completes the proof of (20.11). But this inequality only says that the expected value of $\mu(\mathbf{g}_n)$ does not exceed $n + 1$ for a *random* function $\mathbf{g}_n$, whereas our goal is to give an upper bound on $\mu(f_n)$ for *each* function f_n. So, we must somehow "derandomize" this result. To achieve this goal, observe that every function $f_n \in F_n$ can be expressed in the form

$$f_n = (\mathbf{g}_n \wedge (\mathbf{g}_n \oplus f_n \oplus 1)) \vee ((\mathbf{g}_n \oplus 1) \wedge (\mathbf{g}_n \oplus f_n)). \tag{20.16}$$

But $\mathbf{g}_n \approx \mathbf{g}_n \oplus f_n \oplus 1 \approx \mathbf{g}_n \oplus 1 \approx \mathbf{g}_n \oplus f_n$. So, applying to (20.16) the inequalities (20.8) and (20.9), averaging the result over $\mathbf{g}_n$ and applying (20.11) with $d = n$, we obtain $\mu(f_n) = \mathrm{E}\left[\mu(f_n)\right] \leqslant 4 \cdot \mathrm{E}\left[\mu(\mathbf{g}_n)\right] \leqslant 4n + 4$, as desired.   □

20.9 Discrepancy

Let $X_1, \ldots, X_k$ be n-element sets, and $X = X_1 \times \cdots \times X_k$. A subset T of X is called a *cylinder* in the ith dimension if membership in T_i does not depend on the ith coordinate. That is, $(x_1, \ldots, x_i, \ldots, x_k) \in T_i$ implies that $(x_1, \ldots, x_i', \ldots, x_k) \in T_i$ for all $x_i' \in X_i$. A subset $T \subseteq X$ is a *cylinder intersection* if it is an intersection $T = T_1 \cap T_2 \cap \cdots \cap T_k$, where T_i is a cylinder in the ith dimension. The (normalized) *discrepancy* of a function $f : X \to \{-1, 1\}$ on a set T is defined by

$$\mathrm{disc}_T(f) = \frac{1}{|X|} \sum_{x \in T} f(x).$$

The *discrepancy* $\mathrm{disc}(f)$ of f is the maximum, over all cylinder intersections T, of the absolute value $|\mathrm{disc}_T(f)|$.

The importance of this measure stems from the fact that functions with small discrepancy have large *multi-party communication complexity*. (We will discuss this in Sect. 29.3.2 devoted to multi-party games.) However, this fact alone does not give immediate lower bounds for the multi-party communication complexity, because $\mathrm{disc}(f)$ is very hard to estimate. Fortunately, the discrepancy can be bounded from above using the following more tractable measure.

A *cube* is defined to be a multi-set $D = \{a_1, b_1\} \times \{a_2, b_2\} \times \cdots \times \{a_k, b_k\}$, where $a_i, b_i \in X_i$ (not necessarily distinct) for all i. Being a multi-set means that one element can occur several times. Thus, for example, the cube $D = \{a_1, a_1\} \times \cdots \times \{a_k, a_k\}$ has 2^k elements. For a cube D, define its *sign* $f(D)$ to be

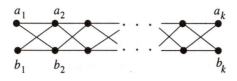

Fig. 20.1. A cube

$$f(D) = \prod_{x \in D} f(x).$$

Hence, $f(D) = 1$ if and only if $f(x) = 1$ for an even number of vectors $x \in D$. We choose a cube D at random according to the uniform distribution. This can be done by choosing $\mathbf{a}_i, \mathbf{b}_i \in X_i$ for each i according to the uniform distribution. Let

$$\mathcal{E}(f) = \mathrm{E}\left[f(\mathbf{D})\right]$$

be the expected value of the sign of a random cube $\mathbf{D}$. To stress the fact that the expectation is taken over a particular random object (this time, over $\mathbf{D}$) we will also write $\mathrm{E}_{\mathbf{D}}\left[f(\mathbf{D})\right]$ instead of $\mathrm{E}\left[f(\mathbf{D})\right]$.

The following result was proved in Chung (1990) and generalizes a similar result from Babai et al. (1992).

Theorem 20.11. *For every* $f : X \to \{-1, 1\}$,

$$\mathrm{disc}(f) \leqslant \mathcal{E}(f)^{-2^k}.$$

The theorem is very useful because $\mathcal{E}(f)$ is a much simpler object than $\mathrm{disc}(f)$. For many functions f, it is very easy to compute $\mathcal{E}(f)$ exactly. In Chung and Tetali (1993), $\mathcal{E}(f)$ was computed for some explicit functions, resulting in highest known lower bounds for the multi-party communication complexity of these functions. A new example of such a function is given in Sect. 20.9.1. The example is due to Raz (2000) who also has found a new and easier proof of the theorem itself.

Proof (due to Raz 2000). We first give a lower bound for $\mathcal{E}(f)$ in terms of the "absolute discrepancy"

$$\Delta(f) = \mathrm{E}\left[f(\mathbf{x})\right],$$

where $\mathbf{x}$ is a random vector uniformly distributed over X.

Claim 20.12. *For all* $f : X \to \{-1, 1\}$, $\mathcal{E}(f) \geqslant |\Delta(f)|^{2^k}$.

Proof. Let $\mathbf{D} = \{\mathbf{a}_1, \mathbf{b}_1\} \times \cdots \times \{\mathbf{a}_k, \mathbf{b}_k\}$ be a random cube. That is, for each i, $\mathbf{a}_i \in X_i$ and $\mathbf{b}_i \in X_i$ are chosen according to the uniform distribution. Let also $\mathbf{D}' = \{\mathbf{a}_1, \mathbf{b}_1\} \times \cdots \times \{\mathbf{a}_{k-1}, \mathbf{b}_{k-1}\}$. Then

$$\mathcal{E}(f) = \mathrm{E}\left[f(\mathbf{D})\right] = \mathrm{E}\left[f(\mathbf{D}' \times \{\mathbf{a}_k, \mathbf{b}_k\})\right]$$
$$= \mathrm{E}\left[f(\mathbf{D}' \times \{\mathbf{a}_k\}) \cdot f(\mathbf{D}' \times \{\mathbf{b}_k\})\right].$$

For any function g defined on X_k we have

$$\mathrm{E}\left[g(\mathbf{a}_k) \cdot g(\mathbf{b}_k)\right] = \mathrm{E}\left[g(\mathbf{a}_k)\right] \cdot \mathrm{E}\left[g(\mathbf{b}_k)\right] = (\mathrm{E}\left[g(\mathbf{a}_k)\right])^2.$$

For fixed D', take $g(\mathbf{a}_k) = f(D' \times \{\mathbf{a}_k\})$ to get

$$\mathcal{E}(f) = \mathrm{E}_{\mathbf{D}}\left[f(\mathbf{D})\right] = \mathrm{E}_{\mathbf{D}'}\mathrm{E}_{\mathbf{a}_k,\mathbf{b}_k}\left[f(\mathbf{D}' \times \{\mathbf{a}_k, \mathbf{b}_k\})\right]$$
$$= \mathrm{E}_{\mathbf{D}'}\left(\mathrm{E}_{\mathbf{a}_k}\left[f(\mathbf{D}' \times \{\mathbf{a}_k\})\right]\right)^2.$$

By the Cauchy–Schwarz inequality, for any random variable ξ, $\mathrm{E}\left[\xi^2\right] \geqslant \mathrm{E}\left[\xi\right]^2$. Therefore,

$$\mathcal{E}(f) \geqslant \left(\mathrm{E}_{\mathbf{D}'}\mathrm{E}_{\mathbf{a}_k}\left[f(\mathbf{D}' \times \{\mathbf{a}_k\})\right]\right)^2 = \left(\mathrm{E}\left[f(\mathbf{D}' \times \{\mathbf{a}_k\})\right]\right)^2.$$

Repeat the same argument k times to get

$$\mathcal{E}(f) \geqslant \left(\mathrm{E}\left[f(\{\mathbf{a}_1\} \times \{\mathbf{a}_2\} \times \cdots \times \{\mathbf{a}_k\})\right]\right)^{2^k}$$
$$= \left(\mathrm{E}\left[f(\mathbf{a}_1, \mathbf{a}_2, \ldots, \mathbf{a}_k)\right]\right)^{2^k} = |\Delta(f)|^{2^k}.$$

$\square$

Given two functions f and g, we denote by $f \cdot g$ their pointwise product, i.e., $f \cdot g(x) = f(x) \cdot g(x)$. A function g is *cylindrical* if it does not depend on at least one of its input variables, i.e., if there is an i such that $g(x_1, \ldots, x_i, \ldots, x_k) = g(x_1, \ldots, x_i', \ldots, x_k)$ for all $(x_1, \ldots, x_i, \ldots, x_k) \in X$ and $x_i' \in X_i$.

Claim 20.13. *For all $f, g : X \to \{-1, 1\}$, if g is cylindrical then*

$$\mathcal{E}(f \cdot g) = \mathcal{E}(f).$$

Proof. By the definition of the sign function, for every cube D we have $f \cdot g(D) = f(D) \cdot g(D)$. We can assume that g does not depend on x_k. Then for all $x_1, \ldots, x_{k-1}, a_k, b_k$ we have $g(x_1, \ldots, x_{k-1}, a_k) = g(x_1, \ldots, x_{k-1}, b_k)$. Therefore, g takes the value 1 on an even number of elements of D (remember that D is a multi-set). Hence $g(D) = 1$ and so $f \cdot g(D) = f(D)$. Since this holds for every cube D, we are done. $\square$

Claim 20.14. *For every $f : X \to \{-1, 1\}$ there exists $h : X \to \{-1, 1\}$ such that $\mathcal{E}(h) = \mathcal{E}(f)$ and $|\Delta(h)| \geqslant \mathrm{disc}(f)$.*

Proof. Take a cylinder intersection $T = T_1 \cap T_2 \cap \cdots \cap T_k$ with $|\mathrm{disc}_T(f)| = \mathrm{disc}(f)$. The idea is to define a random function $\mathbf{g} : X \to \{-1, 1\}$ such that $\mathrm{E}\left[\mathbf{g}(x)\right] = \mathrm{E}_{\mathbf{g}}\left[\mathbf{g}(x)\right]$ is the characteristic function of T.

For every $i = 1, \ldots, k$, define $\mathbf{g}_i : X \to \{-1, 1\}$, as a random function, in the following way: with probability $1/2$, $\mathbf{g}_i$ is the constant function 1, and with probability $1/2$, $\mathbf{g}_i(x)$ takes the value 1 on all elements $x \in T_i$, and -1 otherwise. Then for $x \in T_i$, $\mathbf{g}_i(x) = 1$ with probability 1, while for $x \notin T_i$, $\mathbf{g}_i(x) = 1$ with probability $1/2$ and $\mathbf{g}_i(x) = -1$ with probability $1/2$.

Define $\mathbf{g} = \mathbf{g}_1 \cdots \mathbf{g}_k$. Then for $x \in T$, $\mathbf{g}(x) = 1$ with probability 1, while for $x \notin T$, $\mathbf{g}(x) = 1$ with probability $1/2$ and $\mathbf{g}(x) = -1$ with probability

1/2 (this is so because the functions $\mathbf{g}_i$ are independent of each other, and $x \notin T$ iff $x \notin T_i$ for at least one i). Thus, the expectation $\mathrm{E}\,[\mathbf{g}(x)]$ takes the value 1 on all $x \in T$, and takes the value 0 on all $x \notin T$, i.e., $\mathrm{E}\,[\mathbf{g}(x)]$ is the characteristic function of the set T.

Let now $\mathbf{x}$ be a random vector uniformly distributed in X. Then

$$\mathrm{E}_{\mathbf{g}}\,[\Delta(f \cdot \mathbf{g})] = \mathrm{E}_{\mathbf{g}}\mathrm{E}_{\mathbf{x}}\,[f \cdot \mathbf{g}(\mathbf{x})] = \mathrm{E}_{\mathbf{g}}\mathrm{E}_{\mathbf{x}}\,[f(\mathbf{x}) \cdot \mathbf{g}(\mathbf{x})]$$
$$= \mathrm{E}_{\mathbf{x}}\mathrm{E}_{\mathbf{g}}\,[f(\mathbf{x}) \cdot \mathbf{g}(\mathbf{x})] = \mathrm{E}_{\mathbf{x}}\,[f(\mathbf{x}) \cdot \mathrm{E}_{\mathbf{g}}\,[\mathbf{g}(\mathbf{x})]] = \mathrm{disc}_T(f),$$

and by convexity,

$$\mathrm{E}\,[|\Delta(f \cdot \mathbf{g})|] \geqslant |\mathrm{E}\,[\Delta(f \cdot \mathbf{g})]| = |\mathrm{disc}_T(f)| = \mathrm{disc}(f).$$

By the pigeonhole property of the expectation, there exists a function $g = g_1 \cdots g_k$ for which $|\Delta(f \cdot g)| \geqslant \mathrm{disc}(f)$. On the other hand, since $g_1, \ldots, g_k$ are cylindrical, by the previous claim we have $\mathcal{E}(f \cdot g) = \mathcal{E}(f)$. Claim 20.14 follows by taking $h = f \cdot g$. $\qquad\square$

Now, Theorem 20.11 is an immediate consequence of Claims 20.12 and 20.14:

$$\mathcal{E}(f) = \mathcal{E}(h) \geqslant |\Delta(h)|^{2^k} \geqslant \mathrm{disc}(f)^{2^k}.$$

$\qquad\square$

20.9.1 Example: matrix multiplication

Let $X = X_1 \times \cdots X_k$, where each X_i is the set of all $m \times m$ matrices over the field $\mathbb{F}_2$; hence, $|X_i| = n = 2^{m^2}$. For $x_1 \in X_1, \ldots, x_k \in X_k$, denote by $x_1 \cdots x_k$ the product of $x_1, \ldots, x_k$ as matrices over $\mathbb{F}_2$. Let $F(x_1, \ldots, x_k)$ be a boolean function whose value is the element in the first row and the first column of the product $x_1 \cdots x_k$. Define the function $f : X \to \{-1, 1\}$ by

$$f(x_1, \ldots, x_k) = (-1)^{F(x_1, \ldots, x_k)} = 1 - 2F(x_1, \ldots, x_k).$$

Theorem 20.15 (Raz 2000). $\mathrm{disc}(f) \leqslant \left(\dfrac{k-1}{\sqrt{\log_2 n}}\right)^{1/2^k}.$

Proof. For every cube $D = \{a_1, b_1\} \times \cdots \times \{a_k, b_k\}$,

$$f(D) = \prod_{x \in D} f(x) = \prod_{x \in D} (-1)^{F(x)} = (-1)^{\bigoplus_{x \in D} F(x)},$$

where $\oplus$ denotes the addition over $\mathbb{F}_2$. Since F is linear in each variable,

$$f(D) = (-1)^{F(a_1 \oplus b_1, \ldots, a_k \oplus b_k)} = 1 - 2F(a_1 \oplus b_1, \ldots, a_k \oplus b_k),$$

where $a_i \oplus b_i$ denotes the sum of matrices a_i and b_i over $\mathbb{F}_2$. If we choose $\mathbf{D}$ at random according to the uniform distribution, then $(\mathbf{a}_1 \oplus \mathbf{b}_1, \ldots, \mathbf{a}_k \oplus \mathbf{b}_k)$ is a random vector $(\mathbf{x}_1, \ldots, \mathbf{x}_k)$ uniformly distributed over X. Therefore,

$$\mathcal{E}(f) = \mathrm{E}\left[f(\mathbf{D})\right] = \mathrm{E}\left[1 - 2F(\mathbf{a}_1 \oplus \mathbf{b}_1, \ldots, \mathbf{a}_k \oplus \mathbf{b}_k)\right]$$
$$= \mathrm{E}\left[1 - 2F(\mathbf{x}_1, \ldots, \mathbf{x}_k)\right] = \mathrm{E}\left[f(\mathbf{x}_1, \ldots, \mathbf{x}_k))\right] = \Delta(f).$$

To estimate $\Delta(f)$, let E_d denote the event that the first row of the matrix $\mathbf{x}_1 \cdots \mathbf{x}_d$ contains only 0's. Define $p_d = \mathrm{Prob}\,(E_d)$. Since p_1 is determined by $\mathbf{x}_1$ and since $\mathbf{x}_1$ is uniformly distributed, we have $p_1 = \mathrm{Prob}\,(E_1) = 2^{-m}$. Clearly we also have $\mathrm{Prob}\,(E_{d+1} \mid E_d) = 1$. On the other hand, since $\mathbf{x}_{d+1}$ is uniformly distributed, $\mathrm{Prob}\,(E_{d+1} \mid \neg E_d) = 2^{-m}$ (see Exercise 17.2). Therefore, for all $1 \leqslant d < k$,

$$p_{d+1} = \mathrm{Prob}\,(E_{d+1} \mid E_d) \cdot \mathrm{Prob}\,(E_d) + \mathrm{Prob}\,(E_{d+1} \mid \neg E_d) \cdot \mathrm{Prob}\,(\neg E_d)$$
$$= p_d + (1 - p_d) \cdot 2^{-m} \leqslant p_d + 2^{-m},$$

implying that $p_d \leqslant d \cdot 2^{-m}$ for all $d = 1, \ldots, k$.

If E_{k-1} occurs then $F(\mathbf{x}_1, \ldots, \mathbf{x}_k)$ is always 0, and hence, $f(\mathbf{x}_1, \ldots, \mathbf{x}_k)$ is always 1. If E_{k-1} does not occur then, since the first column of $\mathbf{x}_k$ is uniformly distributed, $F(\mathbf{x}_1, \ldots, \mathbf{x}_k)$ is uniformly distributed over $\{0, 1\}$, and hence, $f(\mathbf{x}_1, \ldots, \mathbf{x}_k)$ is uniformly distributed over $\{-1, 1\}$. Therefore,

$$\mathcal{E}(f) = \Delta(f) = p_{k-1} \leqslant (k-1) \cdot 2^{-m},$$

and Theorem 20.11 yields the desired upper bound on $\mathrm{disc}(f)$.   □

Exercises

20.1.⁻ We have n letters going to n different persons and n envelopes with their addresses. We insert each letter into an envelope independently from each other at random (several letters may go in the same envelope). What is the expected number of correct matches? (Answer: $E = 1$.)

20.2.⁻ There are k people in a lift at the ground floor. Each wants to get off at a random floor of one of the n upper floors. What is the expected number of lift stops?

Hint: Consider the indicator random variables X_i for the events that at least one person is off at the ith floor, and apply the linearity of expectation. Answer: $E = n(1 - (1 - 1/n)^k)$.

20.3. Let Ω be a uniform sample space, and let $X : \Omega \to \{0, 1, \ldots, M\}$ be a random variable with the expectation $\mu = M - a$ for some a. Prove that then, for any $1 \leqslant b \leqslant M$, $\mathrm{Prob}\,(X \geqslant M - b) \geqslant (b - a)/b$.

Sketch: Let B be the set of those points $\omega \in \Omega$ for which $X(\omega) < M - b$. Then $\mathrm{Prob}\,(B) \cdot (M - b) + \mathrm{Prob}\,(\overline{B}) \cdot M \geqslant M - a$, or $\mathrm{Prob}\,(B) \leqslant a/b$.

20.4. Let $\mathbf{T}$ be a random tournament chosen uniformly among all tournaments on n players. Then, by Szele's theorem, the expected number μ of Hamiltonian paths in it is $n!2^{-(n-1)}$. Use the argument of the previous exercise and Alon's upper bound (20.1) to prove that, with probability at least $\Omega(n^{-3/2})$, $\mathbf{T}$ has at least $n!2^{-n}$ Hamiltonian paths.

Hint: Let $M = \Delta \cdot \mu$ with $\Delta = cn^{3/2}$, and take $a = M(1-1/\Delta)$, $b = M(1-1/2\Delta)$.

20.5$^-$ (Redéi 1934). Prove that every tournament contains a Hamiltonian path. *Hint:* Every path missing at least one player can be prolonged by adding him to that path.

20.6$^{(!)}$ (Alon 1990c). Design an algorithm which, given a graph $G = (V, E)$ of minimal degree d, constructs a dominating set $S \subseteq V$ of size $|S| \leqslant n\frac{1+\ln(d+1)}{d+1}$, whose existence is guaranteed by Theorem 20.3.

Sketch: For $S \subseteq V$, let $D(S)$ be the set of vertices dominated by S (i.e., $v \in D(S)$ if either $v \in S$ or v is joined by an edge with some vertex in S). Let $N(S) = V - D(S)$. First show that, given $S \subseteq V$, there exists a vertex in $V - S$ which dominates at least $|N(S)|(d+1)/n$ vertices in $N(S)$. Now construct the desired set S by iteratively adding a vertex with the maximum number of neighbors undominated by the vertices already chosen. Prove that at most $n/(d+1)$ vertices remain undominated after $n\ln(d+1)/(d+1)$ steps, such that adding them yields a dominating set of size at most $n\frac{1+\ln(d+1)}{d+1}$.

20.7. Show that Theorem 20.4 implies Turán's theorem: if a graph G has n vertices and $nk/2$ edges, then $\alpha(G) \geqslant n/(k+1)$ (see Exercise 4.7).

Hint: Use the Cauchy–Schwarz inequality $\left(\sum_{i=1}^{n} a_i b_i\right)^2 \leqslant \left(\sum_{i=1}^{n} a_i^2\right)\left(\sum_{i=1}^{n} b_i^2\right)$ with $a_i = (d_i + 1)^{1/2}$ and $b_i = 1/a_i$.

20.8. Prove the Lieberher-Specker result for 2-satisfiable formulas: if F is a 2-satisfiable formula then at least γ-fraction of its clauses are simultaneously satisfiable, where $\gamma = (\sqrt{5} - 1)/2$.

Sketch (Yannakakis 1994): Define the probability of a literal y to be satisfied to be: a ($a > 1/2$) if y occurs in a unary clause, and $1/2$ otherwise. Observe that then the probability that a clause C is satisfied is a if C is a unary clause, and at least $1 - a^2$ otherwise (at worst, a clause will be a disjunction of two literals whose negations appear as unary clauses); verify that $a = 1 - a^2$ for $a = \gamma$.

20.9.$^-$ Prove that for any And-Or formula there is an input which satisfies at least half of its clauses. Is this bound optimal?

20.10.$^-$ Given a graph $G = (V, E)$, define the And-Or formula

$$F_G = \bigwedge_{\{i,j\} \notin E} (\overline{x}_i \vee \overline{x}_j).$$

Each assignment $v = (v_1, \ldots, v_n) \in \{0, 1\}^n$ can be interpreted as an incidence vector of the set of vertices $S_v = \{i : v_i = 1\}$. Show that S_v is a clique in G if and only if v satisfies the formula F_G.

20.11. Recall that the *length* (or the *norm*) $\|v\|$ of a vector $v \in \mathbb{R}^n$ is the square root of the scalar product $\langle v, v \rangle$. Prove that for any vectors $v_1, \ldots, v_n$ in $\{+1, -1\}^n$ there are scalars $\epsilon_1, \ldots, \epsilon_n \in \{+1, -1\}$ such that

$$\|\epsilon_1 v_1 + \cdots + \epsilon_n v_n\| \leqslant n.$$

Sketch: Choose the ϵ_i's independently at random to be $+1$ or -1 with probability $1/2$, and use the linearity of expectation to evaluate the expected length of the vector $\sum \epsilon_i v_i$ by computing the square of that quantity. When doing this, use the fact that ϵ_i and ϵ_j are independent and therefore $\mathrm{E}[\epsilon_i \cdot \epsilon_j] = \mathrm{E}[\epsilon_i] \cdot \mathrm{E}[\epsilon_j] = 0$.

20.12. Prove the following generalization of the previous result. Let $v_1, \ldots, v_n$ be vectors in $\{+1, -1\}^n$; $p_1 \ldots, p_n$ be real numbers in $[0, 1]$, and set $w = p_1 v_1 + \cdots + p_n v_n$. Then there exist $\epsilon_1, \ldots, \epsilon_n \in \{0, 1\}$ such that, setting $v = \epsilon_1 v_1 + \cdots + \epsilon_n v_n$, we have $\|w - v\| \leqslant n/2$.

Hint: Pick ϵ_i's independently with $\mathrm{Prob}(\epsilon_i = 1) = p_i$ and $\mathrm{Prob}(\epsilon_i = 0) = 1 - p_i$. Consider a random variable $X = \|w - v\|^2$, and prove that $\mathrm{E}[X] \leqslant n^2/4$.

20.13 (Chandra et al. 1983). Theorem 18.5 says that there exist (n, k)-universal sets of 0-1 vectors of size at most $r = k2^k \log n$. Give an alternative proof of this result using the linearity of expectation.

Hint: Choose a random set $\mathbf{A}$ uniformly and independently from the family of all r-element subsets of $\{0, 1\}^n$; the probability of one particular subset to be chosen is hence $\binom{2^n}{r}^{-1}$. For a set S of k coordinates and a vector $u \in \{0, 1\}^S$, let $X_{S,u}$ denote the indicator random variable for the event $u \notin \mathbf{A}\!\upharpoonright_S$, and let X be the sum of these random variables over all S and u. Show that for $r = k2^k \ln n$, $\mathrm{E}[X] < 1$.

20.14. Compare the bound $k2^k \log n$ in the previous exercise (and in Theorem 18.5) with those in Exercises 18.4 and 18.5. Are these bounds similar by accident?

20.15. Let K_n be a complete graph on n vertices, and E a subset of its edges. Say that an edge *joins* a pair V_1, V_2 of (not necessarily disjoint) subsets of vertices if one its endpoint belongs to V_1 and the other to V_2; in particular, both endpoints may belong to $V_1 \cap V_2$. Show that for any constant $c > 0$ there exists a set E of $|E| = \Theta(n)$ edges in K_n such that: (i) E has no triangles and only $O(n)$ paths of constant length, and (ii) every two subsets of vertices of size at least cn are joined by at least $\Omega(n)$ edges from E.

Hint: Pick edges at random with probability $\Theta(1/n)$.

20.16. Let $r \leqslant n \leqslant N^{1/2}$, and let w be a vector of length N over an alphabet of size n. Let $\mu(w, r)$ be the fraction of r-element subsets $i_1, \ldots, i_r$ of (distinct) coordinates such that all the letters $w_{i_1}, \ldots, w_{i_r}$ are distinct. Show that $\mu(w, r) \leqslant r \cdot (n)_r \cdot n^{-r}$.

Hint: Consider the case when each letter occurs in w the same number of times to show that $\mu(w, r) \leqslant \binom{n}{r} \left(\frac{N}{n}\right)^r \binom{N}{r}^{-1}$, and use (1.13).

21. The Deletion Method

As described in previous sections, the basic probabilistic method works as follows: trying to prove that an object with certain properties exists, one defines an appropriate probability space of objects and then shows that the desired properties hold in this space with positive probability. In this section, we consider situations where the "random" object does not have all the desired properties but may have a few "blemishes." With a small alteration, we remove the blemishes, giving the desired structure.

21.1 Ramsey numbers

Recall that the Ramsey number $R(k)$ is the smallest number n such that, for any red/blue coloring of the edges of a complete graph K_n on n vertices, all the edges of some induced subgraph on k vertices will have the same color. We already know (see Theorem 18.1 from Chap. 18) that $R(k) > n$, as long as $\binom{n}{k} 2^{1-\binom{k}{2}} < 1$. Small alterations allow us to slightly improve this result.

Theorem 21.1. *For any integer n, $R(k) > n - \binom{n}{k} 2^{1-\binom{k}{2}}$.*

Proof (Spencer 1987). Consider a random red/blue coloring of the edges of K_n. For any set S of k vertices, let X_S be the indicator random variable for the event that the induced subgraph of K_n on S is monochromatic. Let $X = \sum X_S$, the sum over all such S. From linearity of expectation,

$$E[X] = \sum_S E[X_S] = m \quad \text{with} \quad m = \binom{n}{k} 2^{1-\binom{k}{2}}.$$

Thus, there exists a coloring for which $X \leqslant m$, i.e., which leaves at most m k-cliques monochromatic. Fix such a coloring and remove from K_n one vertex from each monochromatic k-clique. At most m vertices have been removed, so s vertices remain with $s \geqslant n - m$. The clique K_s on these remaining vertices has no monochromatic subclique on k vertices, meaning that $R(k) > s \geqslant n - m$. □

21.2 Independent sets

Here is a short argument that gives roughly half of Turán's celebrated theorem. A set of vertices in a graph is *independent* if no two vertices from this set are joined by an edge; the *independence number* $\alpha(G)$ is the maximum number of vertices in G with no edges between them. Turán's theorem states: if G has n vertices and $nk/2$ edges, then $\alpha(G) \geqslant n/(k+1)$. We can get "halfway" to this result with the deletion method.

Theorem 21.2. *If a graph $G = (V, E)$ has n vertices and $nk/2$ edges, then $\alpha(G) \geqslant n/2k$.*

Proof (Spencer 1987). Form a random subset $\mathbf{S} \subseteq V$ of vertices by including each vertex independently with probability p. That is, we take a coin which comes up heads with probability p and flip it for each $x \in V$ to see if x is "chosen" to be in $\mathbf{S}$. Let X denote the number of vertices in $\mathbf{S}$, and let Y denote the number of edges of G, both ends of which lie in $\mathbf{S}$. For each edge $e \in E$, let Y_e be the indicator random variable for the event $e \subseteq \mathbf{S}$. Then for each edge $e \in E$, $\mathrm{E}[Y_e] = p^2$ as two vertices must be chosen for e to be inside $\mathbf{S}$. So, by linearity of expectation,

$$\mathrm{E}[Y] = \sum_{e \in E} \mathrm{E}[Y_e] = \frac{nk}{2} p^2.$$

Clearly, $\mathrm{E}[X] = np$, so, again by linearity of expectation,

$$\mathrm{E}[X - Y] = np - \frac{nk}{2} p^2.$$

We choose $p = 1/k$ to maximize this quantity, giving

$$\mathrm{E}[X - Y] = \frac{n}{k} - \frac{n}{2k} = \frac{n}{2k}.$$

Thus, there exists at least one point in the probability space for which the difference $X - Y$ is at least $n/2k$. That is, there is a set S which has at least $n/2k$ more vertices than edges. Delete one vertex from each edge from S leaving a set S'. This set S' is independent and has at least $n/2k$ vertices. $\qquad\square$

A similar argument can be used, for example, to show that if a graph has relatively few edges, then "almost all" sufficiently large subsets of vertices have a large independent part. Say that a graph on n vertices is *sparse* if the number of edges in it does not exceed kn for some constant k.

Proposition 21.3. *Let $G = (V, E)$ be a sparse graph on n vertices, and $V_1, \ldots, V_N$ subsets of V, each of size at least cn for some constant $c > 0$. Then there exist constants $c_1, c_2 > 0$ such that if $N \leqslant 2^{c_1 n}$, then there exists an independent set $S \subseteq V$ with $|S \cap V_i| \geqslant c_2 n$ for all $i = 1, \ldots, N$.*

Proof. We know that $|E| \leqslant kn$ for some constant k. If $k < c/2$, there is nothing to prove. So, assume that $k \geqslant c/2$ and form a random subset $\mathbf{S} \subseteq V$ of vertices by including each vertex independently with probability $p = c/(8k)$. Let Y denote the number of edges of G, both ends of which lie in $\mathbf{S}$. For each edge $e \in E$, let Y_e be the indicator random variable for the event $e \subseteq \mathbf{S}$. By linearity of expectation,

$$\mathrm{E}[Y] = \sum_{e \in E} \mathrm{E}[Y_e] = p^2 \cdot |E| \leqslant p^2 kn,$$

and by Markov's inequality,

$$\mathrm{Prob}\left(Y \leqslant 2p^2 kn\right) \geqslant \mathrm{Prob}\left(Y \leqslant 2 \cdot \mathrm{E}[Y]\right) > 1/2. \tag{21.1}$$

For each i, the expected number of vertices in $\mathbf{S} \cap V_i$ is $p \cdot |V_i| \geqslant cpn$. By Chernoff's inequality (17.7),

$$\mathrm{Prob}\left(|\mathbf{S} \cap V_i| \leqslant cpn/2\right) \leqslant e^{-cpn/8}.$$

Hence, if $N \leqslant 2^{cpn/8}$, then

$$\mathrm{Prob}\left(|\mathbf{S} \cap V_i| \geqslant cpn/2 \text{ for all } i\right) \geqslant 1 - N \cdot e^{-cpn/8} > 1/2. \tag{21.2}$$

Fix a set S satisfying both (21.1) and (21.2), and remove one vertex from each edge lying within this set. The obtained set S' is independent and still contains at least $cpn/2 - 2p^2 kn \geqslant cpn/4$ vertices in each of the sets $V_1, \ldots, V_N$, as desired. $\qquad\square$

21.3 Coloring large-girth graphs

A *cycle* of length k in a graph $G = (V, E)$ is a sequence of vertices $v_1, v_2, \ldots, v_k$ such that: $v_1 = v_k$, $v_i \neq v_j$ for all $1 < i < j \leqslant k$, and all the edges $\{v_i, v_{i+1}\}$ belong to E. The *girth* $g(G)$ of a graph is the length of the shortest cycle in G. Recall also that the *chromatic number* $\chi(G)$ is the minimal number of colors which we need to color the vertices of G so that no two vertices of the same color are joined by the edge.

A striking example of the deletion method is the proof that for fixed k and l, there are graphs with girth at least l and chromatic number at least k. This result was proved by Erdős, and is highly unintuitive. If the girth is large there is no simple reason why the graph could not be colored with a few colors: locally it is easy to color such a graph with three colors. Thus, we can force the chromatic number only by some global considerations and the deletion method helps in doing this. The proof we present here is a simplification of the Erdős proof, due to Alon and Spencer (1992).

Theorem 21.4 (Erdős 1959). *For all k, l there exists a finite graph G with $\chi(G) \geqslant k$ and $g(G) \geqslant l$.*

Since every color class must be an independent set, we have an obvious relation: $\chi(G) \geqslant n/\alpha(G)$. So, instead of showing that $\chi(G) \geqslant k$ it is sufficient to show that $\alpha(G) \leqslant n/k$. We will use this simple trick in the proof below.

Proof (Alon–Spencer 1992). Fix $\theta < 1/l$. Let n be large enough and let $\mathbf{G}$ be a random graph on n vertices, where each pair of nodes is joined by an edge with independent probability $p = n^{\theta-1}$. Let X be the number of cycles in $\mathbf{G}$ of length at most l. How many cycles $v_1, v_2, \ldots, v_i, v_1$ of length i can our graph have?

There are $(n)_i = n(n-1)\cdots(n-i+1)$ sequences $v_1, v_2, \ldots, v_i$ of distinct vertices, and each cycle is identified by $2i$ of those sequences: there are two possibilities to choose the "direction" and i possibilities to choose the first vertex of the cycle. Thus, for $3 \leqslant i \leqslant t$ there are $(n)_i/2i \leqslant n^i/2i$ potential cycles of length i, each of which is in $\mathbf{G}$ with probability p^i. By the linearity of the expectation

$$\mathrm{E}\left[X\right] = \sum_{i=3}^{l} \frac{(n)_i}{2i} p^i \leqslant \sum_{i=3}^{l} \frac{n^{\theta i}}{2i} = o(n)$$

as $\theta l < 1$. By Markov's inequality,

$$\mathrm{Prob}\left(X \geqslant n/2\right) \leqslant \frac{2\mathrm{E}\left[X\right]}{n} = o(1).$$

Set $x = \lceil \frac{3}{p} \ln n \rceil$, so that

$$\mathrm{Prob}\left(\alpha(\mathbf{G}) \geqslant x\right) \leqslant \binom{n}{x}(1-p)^{\binom{x}{2}} < \left[ne^{-p(x-1)/2}\right]^x = o(1).$$

Let n be sufficiently large so that both these events have probability less than $1/2$. Then there is a specific G with less than $n/2$ "short" cycles, i.e., cycles of length less than l, and with $\alpha(G) < x \leqslant 3n^{1-\theta} \ln n$. Remove from G a vertex from each short cycle. This gives a graph G' which has no short cycles and still has at least $n/2$ vertices. Hence, G' has girth greater than l and $\alpha(G') \leqslant \alpha(G)$ (since $\alpha(G)$ cannot grow when we delete vertices). Thus

$$\chi(G') \geqslant \frac{|G'|}{\alpha(G')} \geqslant \frac{n/2}{3n^{1-\theta}\ln n} = \frac{n^\theta}{6\ln n}.$$

To complete the proof, it remains to take n sufficiently large so that this is greater than k. $\square$

21.4 Point sets without obtuse triangles

Around 1950, Erdős conjectured that every set of more than 2^n points in $\mathbb{R}^n$ determines at least one *obtuse angle*, that is, an angle that is strictly greater than $\pi/2$. In other words, any set of points in $\mathbb{R}^n$ which only has acute angles (including right angles) has size at most 2^n.

In 1962, Danzer and Grünbaum proved this conjecture. They also constructed configurations of $2n - 1$ points in $\mathbb{R}^n$ with only acute angles, and conjectured that this may be best possible. But 21 years later, Erdős and Füredi – using a probabilistic argument – disproved this conjecture. It appears that, if the dimension n is high, the bound $2n - 1$ is not even near to the truth.

Theorem 21.5 (Erdős–Füredi 1983). *For every $n \geqslant 1$ there is a set of at least $m = \lfloor \frac{1}{2}(\frac{2}{\sqrt{3}})^n \rfloor$ points in the n-dimensional Euclidean space $\mathbb{R}^n$, such that all angles determined by three points from the set are strictly less than $\pi/2$.*

The theorem is an easy consequence of the following lemma.

Lemma 21.6. *For every $n \geqslant 1$ there is a family $\mathcal{F}$ of $m = \lfloor \frac{1}{2}(\frac{2}{\sqrt{3}})^n \rfloor$ subsets of $\{1, \ldots, n\}$, such that there are no three distinct members A, B, C of $\mathcal{F}$ satisfying*

$$A \cap B \subseteq C \subseteq A \cup B. \tag{21.3}$$

Proof of Lemma 21.6. Let $\mathbf{A}$ be a random subset of $\{1, \ldots, n\}$, where each element appears randomly and independently with probability $1/2$. Let $\mathcal{A}$ be a family of $2m$ independent copies of $\mathbf{A}$. For a triple $\mathbf{A}, \mathbf{B}, \mathbf{C}$ of sets in $\mathcal{A}$, what is the probability that they satisfy the condition (21.3)? This condition just means that for each $i = 1, \ldots, n$, neither $i \in \mathbf{A} \cap \mathbf{B}, i \notin \mathbf{C}$ nor $i \notin \mathbf{A} \cup \mathbf{B}, i \in \mathbf{C}$ hold. For each i, these two events are independent, and each happens with probability $(1/2)^3 = 1/8$. Therefore, the probability that the sets $\mathbf{A}, \mathbf{B}, \mathbf{C}$ satisfy the condition (21.3) is precisely $(1 - 2/8)^n = (3/4)^n$. Since there are $3\binom{2m}{3}$ possible triples $\mathbf{A}, \mathbf{B}, \mathbf{C}$ (there are 3 possibilities to choose $\mathbf{C}$ in a triple), the expected number of triples that satisfy (21.3) is

$$3\binom{2m}{3}(3/4)^n = m(2m - 1)(2m - 2)(3/4)^n < m(2m)^2(3/4)^n \leqslant m,$$

where the last inequality follows from the choice of m.

Thus, there is a choice of a family $\mathcal{A}$ of $2m$ subsets of $\{1, \ldots, n\}$ in which the number of triples A, B, C satisfying (21.3) is at most m. By deleting one set from each such triple we obtain a family $\mathcal{F}$ of at least $2m - m = m$ subsets satisfying the assertion of the lemma. Notice that the members of $\mathcal{F}$ are all distinct since (21.3) is trivially satisfied if $A = C$. This completes the proof of the lemma. $\qquad\square$

Proof of Theorem 21.5. We select the points of a set X in $\mathbb{R}^n$ from the points of the n-dimensional cube $\{0, 1\}^n$. We view the vertices of the cube, which are 0-1 vectors of length n, as the incidence vectors of subsets of an n-element set.

It is easy to verify that the three points a, b and c of the n-cube, corresponding to the sets A, B and C, respectively, determine a right angle at c if and only if (21.3) holds.

Indeed, the angle θ at c is the angle between the vectors $u = a - c$ and $v = b - c$. This angle can be computed from

$$\cos \theta = \frac{\langle u, v \rangle}{\|u\| \cdot \|v\|},$$

where $\langle u, v \rangle = \sum_{i=1}^{n} u_i v_i$ is the scalar product and $\|u\| = \left(\sum_{i=1}^{n} u_i^2 \right)^{1/2}$ is the norm of u. Since a, b and c are 0-1 vectors, the angle θ can be right if and only if $\langle u, v \rangle = 0$. This can happen if and only if $(a_i - c_i)(b_i - c_i) = 0$ for all $i = 1, \ldots, n$, which in its turn can happen if and only if for each i neither $a_i = b_i = 0, c_i = 1$ nor $a_i = b_i = 1, c_i = 0$ hold. This is precisely the condition (21.3), and the result follows immediately from Lemma 21.6.   □

21.5 Covering designs

An (n, k, l) *covering design*, (or a *covering*), is a family $\mathcal{F}$ of k-subsets of an n-element set such that every l-subset is contained in at least one of $A \in \mathcal{F}$. Let $M(n, k, l)$ denote the minimal cardinality of such a design. A simple counting argument (Exercise 1.22) shows that

$$M(n, k, l) \geqslant \binom{n}{l} \bigg/ \binom{k}{l}.$$

In 1985, Rödl proved a long-standing conjecture of Erdős and Hanani that for fixed k and l, coverings of size $\binom{n}{l} / \binom{k}{l}(1 + o(1))$ exist. We will use the deletion method to come near to this bound.

Theorem 21.7. $M(n, k, l) \leqslant 2 \binom{n}{l} \bigg/ \binom{k}{l} \left[1 + \ln \binom{k}{l} \right].$

Proof. Let $\mathcal{F}$ be a random k-uniform family, where each k-set is placed in $\mathcal{F}$ independently with probability p. Set

$$x \rightleftharpoons \ln \binom{k}{l} \quad \text{and} \quad p \rightleftharpoons x \cdot \binom{n-l}{k-l}.$$

For each l-set S, let A_S denote the event that $\mathcal{F}$ does *not* cover S. As S is contained in $\binom{n-l}{k-l}$ k-sets,

$$\text{Prob}\,(A_S) = (1 - p)^{\binom{n-l}{k-l}} \sim e^{-p\binom{n-l}{k-l}} = e^{-x}.$$

Let Z denote the number of l-sets S not covered by $\mathcal{F}$. As $\text{Prob}\,(A_S) \sim e^{-x}$, linearity of expectation gives $\text{E}\,[Z] \sim \binom{n}{l} e^{-x}$. By Markov's inequality, $\text{Prob}\,(Z > 2\text{E}\,[Z]) < 1/2$. On the other hand, the size of $\mathcal{F}$ has the binomial distribution $B(N, p)$ with $N = \binom{n}{k}$. So, $\text{E}\,[|\mathcal{F}|] = p\binom{n}{k}$ and, by Markov's inequality, $\mathcal{F}$ has more than $2 \cdot \text{E}\,[|\mathcal{F}|]$ members with probability $< 1/2$.

Thus, with positive probability $|\mathcal{F}| \leqslant 2p\binom{n}{k}$ *and* fewer than $2\binom{n}{l}e^{-x}$ l-sets are not covered by $\mathcal{F}$. Fix such an $\mathcal{F}$. For each l-set S not covered by $\mathcal{F}$ add to $\mathcal{F}$ an arbitrary k-set containing S. The new family $\mathcal{F}^*$ now covers all the l-sets (we have "corrected errors") and has size at most

$$2\binom{n}{l}e^{-x} + 2p\binom{n}{k} = 2\cdot\binom{n}{l}\bigg/\binom{k}{l}\left[1 + \ln\binom{k}{l}\right].$$

(Here we have used the identity $\binom{n}{k}\binom{k}{l} = \binom{n}{l}\binom{n-l}{k-l}$ proved in Exercise 1.10.)

$\square$

The proof above is a simplified version of the argument due to Spencer (1987). Actually, the factor 2 in Theorem 21.7 is not essential – Spencer has shown that it can be removed by applying the *second moment method* (which we will consider in the next chapter); in our situation this method says that $|\mathcal{F}| \sim E\left[|\mathcal{F}|\right]$ *almost always*.

21.6 Affine cubes of integers

A collection C of integers is called an *affine d-cube* if there exist $d+1$ positive integers $x_0, x_1, \ldots, x_d$ so that

$$C = \left\{x_0 + \sum_{i\in I} x_i : I \subseteq \{1, 2, \ldots, d\}\right\}.$$

Such a cube is *replete* if all the sums are distinct, i.e., if $|C| = 2^d$. If an affine cube is generated by $x_0, x_1, \ldots, x_d$ then we write $C = C(x_0, x_1, \ldots, x_d)$. For example, $C(1, 1, 1) = \{1, 2, 3\}$ is not replete, while $C(1, 3, 9) = \{1, 4, 10, 13\}$ is a replete affine 2-cube. Note also that $C(x_0, x_1, \ldots, x_d)$ may be different from, say, $C(x_1, x_0, \ldots, x_d)$.

Typical extremal problems related to affine cubes are the following:

1. *Partition problem*: Given r and d, what is the smallest integer $n = H(r, d)$ such that, for any partition of $\{1, \ldots, n\}$ into r classes, at least one class contains an affine d-cube?
2. *Density problem*: Given a set of integers $A \subseteq \{1, 2, \ldots\}$, how large must A be to contain an affine d-cube?

Concerning the first (partition) problem, the existence of $H(r, d)$ for all r and d was first proved by Hilbert (1892). This result was then strengthened by van der Waerden (1927), whose celebrated theorem says that, for every r and d there is an integer $n = W(r, d)$ such that, for any partition of $\{1, \ldots, n\}$ into r classes, at least one class contains an arithmetic progression with d terms. We will prove this theorem in Chap. 29. Note that it implies Hilbert's result because any such progression $a, a+b, a+2b, \ldots, a+(d-1)b$ is also an affine $(d-1)$-cube $C(a, b, b, \ldots, b)$.

At present not much is known about the rate of growth of van der Waerden's function $W(r, d)$ except that it is primitive recursive (Shelah 1988). For small values of r and d we know better bounds. Using the counting sieve we have proved that $W(2,d) > 2^{d/2}$ (see Theorem 18.2). Behrend (1949) proved a lower bound for the density of sets without arithmetic progressions of length three.

Theorem 21.8 (Behrend 1949). *There exists a constant c such that for n sufficiently large, there exists a subset $A \subseteq \{1,\ldots,n\}$ such that $|A| \geqslant ne^{-c\sqrt{\ln n}}$ and A contains no arithmetic progression with 3 terms.*

If we consider affine cubes instead of arithmetic progressions then the situation is better. In particular, the following bounds for Hilbert's function $H(r,d)$ are known. Brown et al. (1985) have shown that there exist constants $\epsilon > 0$ and $c > 0$ such that

$$r^{\epsilon d} \leqslant H(r,d) \leqslant r^{c^d},$$

where $c \sim 2.6$ follows from Hilbert's original proof. Quite recently, these bounds were improved by Gunderson and Rödl (1998):

Theorem 21.9. *For any integers $d \geqslant 3$ and $r \geqslant 2$,*

$$r^{(1-\epsilon)(2^d-1)/d} \leqslant H(r,d) \leqslant (2r)^{2^{d-1}}, \tag{21.4}$$

where $\epsilon \to 0$ as $r \to \infty$.

The proof of the upper bound in (21.4) is based on the following density result, known as *Szemerédi's cube lemma* whose strengthened version, found by Graham (1981) and Graham, Rothschild and Spencer (1990), is as follows (see Sect. 28.3 for the proof).

Lemma 21.10. *Let $d \geqslant 2$ be given. Then, for every sufficiently large n, every subset A of $\{1,\ldots,n\}$ of size $|A| \geqslant (4n)^{1-1/2^{d-1}}$ contains an affine d-cube.*

The proof of the lower bound in (21.4) is based on the following lemma, whose proof gives one more illustration of the deletion method at work.

Lemma 21.11 (Gunderson–Rödl 1998). *For each $d \geqslant 2$ and every set X of positive integers, there exists an $A \subseteq X$ with*

$$|A| \geqslant \frac{1}{8}|X|^{1-d/(2^d-1)},$$

which does not contain any replete affine d-cubes.

Proof. Fix $d \geqslant 3$, a set X, and let

$$p = |X|^{-d/(2^d-1)}.$$

Without loss of generality, we can assume that X is large enough so that $\frac{1}{8}pN > 2^d - 1$, because if not, then any set A of at most $2^d - 1$ elements would satisfy the lemma.

Let $\mathbf{Y}$ be a random subset of X whose elements are chosen independently with probability p.

Since any replete affine d-cube $C = C(x_0, x_1, \ldots, x_d)$ is uniquely determined by $d + 1$ distinct integers $x_0, x_0 + x_1, \ldots, x_0 + x_d$ in X, the expected number of replete affine d-cubes in $\mathbf{Y}$ is bounded above by

$$\binom{|X|}{d+1} \cdot \text{Prob}\,(C \subseteq \mathbf{Y}) = \binom{|X|}{d+1} \cdot p^{2^d}.$$

Therefore (by Markov's inequality), with probability at least $1/2$, the number of replete affine d-cubes in $\mathbf{Y}$ does not exceed

$$2\binom{N}{d+1}p^{2^d} < \frac{1}{3}|X|\,p, \tag{21.5}$$

because

$$2 \cdot \binom{|X|}{d+1} \cdot p^{2^d} < 2\left(\frac{e\,|X|}{d+1}\right)^{d+1} \cdot p^{2^d}$$

$$= 2\left(\frac{e}{d+1}\right)^{d+1}|X|^{d+1} \cdot |X|^{-d2^d/(2^d-1)}$$

$$\leqslant \frac{1}{3}|X|^{1-d/(2^d-1)} = \frac{1}{3}|X|\,p.$$

On the other hand, the number $|\mathbf{Y}|$ of elements in a random subset $\mathbf{Y}$ of X is a binomially distributed random variable with expectation $|X|\,p$. It can be shown (see Exercise 21.1) that then

$$\text{Prob}\left(|\mathbf{Y}| \leqslant \frac{1}{2}|X|\,p\right) < 2\left(\frac{2}{e}\right)^{|X|p/2}. \tag{21.6}$$

Since $p\,|X| > 8(2^d - 1)$ and $d \geqslant 2$, this implies that

$$\text{Prob}\left(|\mathbf{Y}| \geqslant \lfloor \tfrac{1}{2}|X|p\rfloor\right) > \frac{1}{2}. \tag{21.7}$$

Hence, there must exist an instance $Y \subseteq X$ of $\mathbf{Y}$ satisfying both above events; fix such Y. Due to (21.5) and (21.7), this set has at least $\lfloor \frac{1}{2}|X|\,p \rfloor$ elements and contains fewer that $\frac{1}{3}|X|\,p$ replete affine d-cubes. Deleting an element from each of these cubes, we get a set $A \subseteq Y$ with no replete affine d-cubes such that

$$|A| > |Y| - \frac{1}{3}|X|\,p \geqslant \lfloor \tfrac{1}{2}|X|\,p\rfloor - \frac{1}{3}|X|\,p > \frac{1}{3}|X|\,p,$$

where the last inequality follows from our assumption that $|X|\,p$ is larger than $8(2^d - 1) \geqslant 24$. $\square$

Exercises

21.1. Let $S_n = X_1 + \cdots + X_n$ where X_i are independent random variables with $\mathrm{Prob}\,(X_i = 1) = p$ and $\mathrm{Prob}\,(X_i = 0) = 1 - p$. Show that

$$\mathrm{Prob}\,(S_n \leqslant pn/2) < 2(2/e)^{pn/2}.$$

Hint: Recall that $\mathrm{Prob}\,(S_n \leqslant pn/2) = \sum_{k \leqslant pn/2} \binom{n}{k} p^k (1 - p)^{n-k}$; show that each term in this sum is more than two times larger than the previous one, and hence, $\mathrm{Prob}\,(S_n = pn/2 - k) < 2^{-k}\mathrm{Prob}\,(S_n = pn/2)$, for $k < pn/2$. Sum up and use the estimate for $\binom{n}{pn}$ obtained in Exercise 1.13.

21.2.⁻ Show that every graph on n vertices has an independent set of size at least $n/(2d)$, where d is the average degree of its vertices.

Hint: Combine Theorem 21.2 with Euler's theorem from Chap. 1.

21.3 (Gunderson–Rödl 1998). Prove that if a finite collection X of distinct positive integers contains an affine d-cube then either X contains a replete affine d-cube or an arithmetic progression of length 3 (or both).

Hint: First, show that if $C = C(x_0, x_1, \ldots, x_d) \subseteq X$ is an affine d-cube, but is not replete (i.e., $|C| < 2^d$) then we can find two subsets $I, J \subset [n]$ such that $I \neq J$ but $\sum_{i \in I - J} x_i = \sum_{j \in J - I} x_j \neq 0$, and consider the triple of integers

$$x_0 + \sum_{i \in I \cap J} x_i, \quad x_0 + \sum_{i \in I} x_i, \quad x_0 + \sum_{i \in I \cup J} x_i.$$

21.4$^{(!)}$ (*Zarankiewicz's problem*; Erdős–Spencer 1974). Let $k_a(n)$ be the minimal k such that all $n \times n$ 0-1 matrices containing more than k ones contain an $a \times a$ submatrix consisting entirely of ones (the "all-ones" submatrix). Prove that for every constant $a \geqslant 2$ there is an $\epsilon > 0$ such that $k_a(n) \geqslant \epsilon n^{2-2/a}$.

Sketch: Argue as in the proof of Theorem 21.2. Take a random $n \times n$ 0-1 matrix $\mathbf{A}$, each entry of which takes value 1 independently and with probability $p = n^{-2/a}$. Associate with each $a \times a$ submatrix e of $\mathbf{A}$ the indicator random variable Y_e for the event "e is an all-ones submatrix." Switch one entry of each such submatrix to 0 and argue that in the resulting matrix we still can expect at least $n^2 p - \binom{n}{a}^2 p^{a^2}$ ones. (A more accurate choice of p leads to a somewhat better bound $k_a(n) \geqslant \epsilon n^{2-2/(a+1)}$.)

21.5 (Reiman 1958). Improve the above bound on $k_a(n)$ when $a = 2$: show that $k_2(n) \geqslant (1 - o(1))n^{3/2}$.

Hint: Let $n = p^2 + p + 1$ for a prime p, and consider the incidence matrix of lines in a projective plane of order p (see Sect. 13.4).

21.6 (Spencer 1990). Let $\mathcal{F}$ be an r-uniform family of subsets on an n-element set. Suppose that, in average, each element belongs to d members of $\mathcal{F}$. Prove that there exists a set S of elements such that S is independent (i.e., contains no member of $\mathcal{F}$) and has size $|S| \geqslant (1 - 1/r) \cdot n \cdot d^{-1/(r-1)}$.

Hint: Argue as in the proof of Theorem 21.2 with $p = d^{-1/(r-1)}$.

22. The Second Moment Method

The pigeonhole property of expectation says that a random variable X cannot always be smaller (or always greater) than its expectation $E[X]$. The second moment property tells us more: if the variance of X is much smaller than $E[X]^2$ then X "almost always" equals $E[X]$, i.e., the values of X are concentrated around its expectation.

22.1 The method

Let X be a random variable and $\mathrm{Var}[X]$ be its variance,

$$\mathrm{Var}[X] = E\left[(X - E[X])^2\right] = E[X^2] - E[X]^2.$$

The *second moment method* uses the fact that, if $\mathrm{Var}[X] = o(E[X]^2)$, then $X \sim E[X]$, i.e., X is almost always almost equal to its expectation $E[X]$. This follows from Chebyshev's inequality

$$\mathrm{Prob}\left(|X - E[X]| \geqslant \lambda\right) \leqslant \mathrm{Var}[X]/\lambda^2. \tag{22.1}$$

In particular, by setting $\lambda = E[X]$,

$$\mathrm{Prob}\left(X = 0\right) \leqslant \frac{\mathrm{Var}[X]}{E[X]^2}.$$

If $X = X_1 + \cdots + X_n$ is a sum of random variables, then the variance can be computed by the formula

$$\mathrm{Var}[X] = \sum_{i,j=1}^{n} \mathrm{Cov}(X_i, X_j) = \sum_{i=1}^{n} \mathrm{Var}[X_i] + \sum_{i \neq j} \mathrm{Cov}(X_i, X_j),$$

where $\mathrm{Cov}(X_i, X_j)$ is the *covariance* and is defined as

$$\mathrm{Cov}(X_i, X_j) = E[X_i X_j] - E[X_i] E[X_j].$$

In general, if X_i and X_j are independent, then $E[X_i X_j] = E[X_i] \cdot E[X_j]$, and hence, $\mathrm{Cov}(X_i, X_j) = 0$. This often considerably simplifies the variance calculations.

Let us mention that there are several forms of Chebyshev's inequality – the usual form (stated above), and the following, less standard form (see, for example, Hardy–Littlewood–Polya (1952), Theorem 43):

Proposition 22.1. *Let* $a_1, \ldots, a_n$ *be a non-decreasing sequence and* $b_1, \ldots, b_n$ *be a non-increasing sequence of non-negative numbers. Then,*

$$\sum_{i=1}^{n} a_i b_i \leqslant \frac{1}{n} \left(\sum_{i=1}^{n} a_i \right) \left(\sum_{i=1}^{n} b_i \right).$$

22.2 Separators

In the complexity theory we often face the following problem. We have a set of players, each of whom can see some small portion of input bits, and we want to split the players into two groups so that for each group there is a large set of "forbidden" bits which are seen by *no* member of that group. To avoid trivial situations, we also assume that every bit is seen by at least one player. This question can be formalized as follows.

Let $\mathcal{F} = \{F_1, \ldots, F_m\}$ be a family of subsets of some set X. By a *separator* for $\mathcal{F}$ we will mean a pair (S, T) of disjoint subsets of X such that each member of $\mathcal{F}$ is disjoint from either S or from T; the *size* of such a separator is the minimum of $|S|$ and $|T|$.

To approach the question, raised at the beginning, interpret X as the set of bits and let F_i be the set of bits seen by the ith player. The problem is, given $\mathcal{F}$, to make the sets of "forbidden bits" S and T as large as possible. Intuitively, if no bit is seen by too many players then these sets should be large. Using the averaging principle for partitions (a prototype of Markov's inequality), we have shown in Exercise 2.12 that this is true if no bit is seen by too many players. Using the second moment method, Beame, Saks, and Thathachar (1998) have shown that we may relax this condition, and only require that on *average*, no bit is seen by too many players.

The *degree* d_x of a point x in $\mathcal{F}$ is the number of members of $\mathcal{F}$ that contain x. The *average degree* of $\mathcal{F}$ is

$$d = \frac{1}{|X|} \sum_{x \in X} d_x.$$

Theorem 22.2 (Beame–Saks–Thathachar 1998). *Let* $\mathcal{F}$ *be a family of non-empty sets of an n-element set, each containing at most r points. Let d be the average degree of* $\mathcal{F}$ *. Then,* $\mathcal{F}$ *has a separator of size at least* $(1 - \delta)2^{-d}n$, *where*

$$\delta = \sqrt{\frac{dr2^{d+1}}{n}}.$$

In particular, if $4rd2^{d+1} \leqslant n$, then $\mathcal{F}$ contains a separator of size at least $n/2^{d+1}$.

Proof. Let $X = \bigcup_{F \in \mathcal{F}} F$ and $n = |X|$. Color each set $F \in \mathcal{F}$ red or blue uniformly and independently with probability $1/2$. Define $\mathbf{S}$ (respectively, $\mathbf{T}$)

to be the set of points x such that *every* set that contains x is colored red (respectively, blue). Since every element of X occurs in at least one set, it follows that $\mathbf{S}$ and $\mathbf{T}$ are disjoint. Moreover, for each $F \in \mathcal{F}$, either $F \cap \mathbf{S}$ or $F \cap \mathbf{T}$ is empty. To complete the proof, we show that with positive probability both $\mathbf{S}$ and $\mathbf{T}$ have at least $(1 - \delta)2^{-d}n$ elements.

Let Z_x be the indicator random variable for the event "$x \in \mathbf{S}$." By the definition of $\mathbf{S}$, this event occurs with probability 2^{-d_x}, implying that

$$E[Z_x] = \text{Prob}(Z_x = 1) = 2^{-d_x}.$$

Let $Z = \sum_x Z_x$ and observe that $Z = |\mathbf{S}|$. Using the arithmetic-geometric mean inequality (1.10), we obtain

$$E[Z] = \sum_x E[Z_x] = \sum_x 2^{-d_x} \geqslant n \cdot 2^{-\sum_x d_x/n} = n2^{-d}. \tag{22.2}$$

Using the second moment argument, we show below that Z is close to its expected value with high probability. By Chebyshev's inequality (22.1) we need only to upper-bound the variance

$$\text{Var}[Z] = \sum_x \text{Var}[Z_x] + \sum_{x \neq y} \text{Cov}(Z_x, Z_y), \tag{22.3}$$

where $\text{Cov}(Z_x, Z_y) = E[Z_x Z_y] - E[Z_x]E[Z_y]$ is the covariance of Z_x and Z_y. Consider the first term in the right-hand side of (22.3). For any x, Z_x is a Bernoulli random variable, so $\text{Var}[Z_x] = E[Z_x] - E[Z_x]^2 \leqslant E[Z_x]$, implying that

$$\sum_x \text{Var}[Z_x] \leqslant E[Z]. \tag{22.4}$$

To bound the second term in the right-hand side of (22.3), observe that if no member of $\mathcal{F}$ contains both x and y, then Z_x and Z_y are independent, implying that $\text{Cov}(Z_x, Z_y) = 0$. Thus, we are only interested in those pairs (x, y) such that *some* member of $\mathcal{F}$ contains both x and y. For any fixed x, the number of such pairs (x, y) is at most $(r - 1)d_x$. For each such pair,

$$\text{Cov}(Z_x, Z_y) \leqslant E[Z_x Z_y] \leqslant E[Z_x] \leqslant 2^{-d_x}.$$

Therefore,

$$\sum_{x \neq y} \text{Cov}(Z_x, Z_y) \leqslant (r - 1) \sum_x d_x 2^{-d_x}.$$

The last term above can be bounded as follows. Order the x's so that the sequence $\{d_x\}$ is non-decreasing. Now the second Chebyshev inequality (Proposition 22.1) can be applied to the sequences $\{d_x\}$ and $\{2^{-d_x}\}$. We obtain

$$\sum_{x \neq y} \text{Cov}(Z_x, Z_y) \leqslant \frac{r-1}{n} \left(\sum_x 2^{-d_x} \right) \left(\sum_x d_x \right) = d(r - 1)E[Z] \tag{22.5}$$

because $(\sum_x d_x)/n = d$, and $\sum_x 2^{-d_x} = \mathrm{E}\,[Z]$. Substitute the bounds (22.4) and (22.5) into (22.3). We obtain

$$\mathrm{Var}\,[Z] \leqslant (d(r-1)+1)\mathrm{E}\,[Z] \leqslant dr\mathrm{E}\,[Z]\,,$$

where the last inequality holds because each $x \in X$ occurs in at least one set, implying that $d = \sum_x d_x/n \geqslant 1$. Using Chebyshev's inequality (22.1), we have

$$\mathrm{Prob}\,(Z < (1-\delta)\cdot\mathrm{E}\,[Z]) < \frac{\mathrm{Var}\,[Z]}{\delta^2\cdot\mathrm{E}\,[Z]^2} \leqslant \frac{dr}{\delta^2\mathrm{E}\,[Z]}.$$

Substituting for δ its value as given by the statement of the theorem, and using the inequality (22.2), we obtain

$$\mathrm{Prob}\,(Z < (1-\delta)\cdot\mathrm{E}\,[Z]) < \frac{drn}{dr2^{d+1}\mathrm{E}\,[Z]} = \frac{n}{2^{d+1}\mathrm{E}\,[Z]} \leqslant \frac{n}{2^{d+1}2^{-d}n} = \frac{1}{2}.$$

In a similar fashion, we obtain $\mathrm{Prob}\,(|\mathbf{T}| < (1-\delta)\cdot\mathrm{E}\,[Z]) < 1/2$. Thus, with positive probability, both $\mathbf{S}$ and $\mathbf{T}$ have size at least

$$(1-\delta)\cdot\mathrm{E}\,[Z] \geqslant (1-\delta)2^{-d}n.$$

We conclude that there is a coloring of the sets in $\mathcal{F}$ such that the induced S and T satisfy the theorem.   □

22.3 Threshold for cliques

The second moment method is a useful tool for determining the *threshold function* of an event, i.e., a threshold such that below it, the probability of the event tends to 0, and above it, the probability tends to 1.

A *k-clique* is a complete graph on k vertices. We write $\omega(G) \geqslant k$ if the graph G contains a k-clique. A *random graph* $\mathbf{G}(n,p)$ is a graph on n vertices where each edge appears independently with probability p.

How large does p have to be before a random graph G is very likely to contain a 4-clique? The answer is remarkable: there is a sharply defined *threshold value* of p such that, if p is above this value then G is almost certain to contain a 4-clique, and if p is below it then G is almost certain *not* to contain such a clique.

Theorem 22.3. *The threshold for a random graph* $\mathbf{G}(n,p)$ *to contain a 4-clique is* $p = n^{-2/3}$.

Proof. For a subset of four vertices S, let A_S be the event that S induces a clique in $\mathbf{G}(n,p)$, and let X_S be the indicator variable of A_S. Clearly, $\mathrm{Prob}\,(X_S = 1) = p^6$ for every S (every 4-clique has six edges). We define $X = \sum X_S$. Our goal is to show that $\mathrm{Prob}\,(X \geqslant 1)$ tends to 0 if $p \ll n^{-2/3}$, and to 1 if $p \gg n^{-2/3}$.

The first claim follows from Markov's inequality:

$$\text{Prob}\,(X \geqslant 1) \leqslant \text{E}\,[X] = \binom{n}{4} \cdot p^6 \sim \frac{n^4 p^6}{24} \to 0, \quad \text{if } p \ll n^{-2/3}.$$

To prove the second claim, suppose $p \gg n^{-2/3}$. We must show that $\text{Prob}\,(X = 0)$ tends to 0. By the second moment method,

$$\text{Prob}\,(X = 0) \leqslant \frac{\text{Var}\,[X]}{\text{E}\,[X]^2},$$

so all what we need is to estimate the variance. As we have already mentioned above, the variance can be written in the form

$$\text{Var}\,[X] = \sum_S \text{Var}\,[X_S] + \sum_{S \neq T} \text{Cov}\,(X_S, X_Y). \tag{22.6}$$

Since X_S is an indicator random variable, its variance $\text{Var}\,[X_S] = \text{E}\,[X_S] - \text{E}\,[X_S]^2 \leqslant \text{E}\,[X_S] = p^6$ as six different edges must lie in $\mathbf{G}(n, p)$. Since there are $\binom{n}{4} = O(n^4)$ sets S, the total contribution of the first sum in (22.6) is $O(n^4 p^6)$. Let us now estimate the contribution of pairs $S \neq T$.

If the events A_S and A_T are independent, then $\text{Cov}\,(X_S, X_T) = 0$, and these pairs contribute nothing. Now, since $S \neq T$, the events A_S and A_T can be dependent if and only if the cliques S and T have common edges – that is, if and only if $|S \cap T| = 2$ or $|S \cap T| = 3$.

There are $O(n^6)$ pairs S, T with $|S \cap T| = 2$ and for each of these $\text{Cov}\,(X_S, X_T) \leqslant \text{E}\,[X_S X_T] = p^{11}$ as $S \cup T$ has 11 different edges. So, the total contribution of these pairs is $O(n^6 p^{11})$.

Similarly, there are $O(n^5)$ pairs S, T with $|S \cap T| = 3$ and for each of these $\text{E}\,[X_S X_T] = p^9$, since in this case $S \cup T$ has 9 different edges; and the total contribution of these pairs is $O(n^5 p^9)$.

Putting all this together in (22.6) gives

$$\text{Var}\,[X] = O\left(n^4 p^6 + n^6 p^{11} + n^5 p^9\right) = o(n^8 p^{12}) = o(\text{E}\,[X]^2),$$

since $p \gg n^{-2/3}$. Therefore, $\text{Prob}\,(X = 0) = o(1)$ and hence $\mathbf{G}(n, p)$ is almost certain to contain a 4-clique. $\square$

Let us consider a more general question: given a graph G, what is the threshold function for the property that a random graph $\mathbf{G}(n, p)$ contains a copy of G as an induced subgraph? (Recall that induced subgraphs are obtained by deleting vertices together with all the edges incident to them.) In the previous section we have solved this question for the case when G is a complete graph on 4 vertices. What about other graphs? Using the second moment method, a surprisingly general answer to this question was found.

The *density* of a graph $G = (V, E)$ is the fraction $d(G) \rightleftharpoons |E|/|V|$. The *subgraph density* $m(G)$ of a graph is the maximum density $d(H)$ of its subgraph $H = (V', E')$. A graph G is *balanced* if $d(H) \leqslant d(G)$ for all subgraphs H of G.

It appears that $n^{-1/m(G)}$ is the right threshold for an arbitrary (!) balanced graph G. This fundamental result was proved by Erdős and Rényi (1960). Bollobás (1981) and Ruciński and Vince (1986) extended it by removing this restriction on G (of being balanced).

Theorem 22.4 (Erdős–Rényi 1960). *Let G be a balanced graph on k vertices. The threshold for the property that a random graph $\mathbf{G}(n,p)$ contains a subgraph isomorphic to G is $n^{-1/m(G)}$.*

This theorem can be proved in a similar way to Theorem 22.3. But the computations are more involved, and we omit the proof – the interested reader can find it in the book of Alon and Spencer (1992) or in the survey paper of Karoński (1995).

Exercises

22.1.⁻ Let $k > 0$ be an integer, and let $p = p(n)$ be a function of n such that $p \geqslant (6k \ln n)/n$ for large n. Prove that "almost surely" the random graph $\mathbf{G} = \mathbf{G}(n,p)$ has no large independent set of vertices. Namely, show that $\mathrm{Prob}\left(\alpha(\mathbf{G}) \geqslant \frac{n}{2k}\right) \to 0$ as $n \to \infty$.

22.2.⁻ Let $\epsilon > 0$ and $p = p(n) > 0$, and let $r \geqslant (1 + \epsilon)(2 \ln n)/p$ be an integer-valued function of n. Show that "almost surely" the random graph $\mathbf{G}(n,p)$ does *not* contain r independent vertices.

22.3. A *forest* is a graph without non-trivial cycles (i.e., cycles of length $\geqslant 3$). Prove that, if $np \to 0$ as $n \to \infty$, then $\mathrm{Prob}\left(\mathbf{G}(n,p) \text{ is a forest}\right) \to 1$.

Hint: Count the expected number of cycles and apply Markov's inequality.

22.4. Let p be constant. Show that with a probability approaching 1 the graph $\mathbf{G}(n,p)$ has the property that every pair of its vertices has a common neighbor, i.e., a vertex, adjacent to both of them.

Hint: Consider indicator random variables X_{ij} for the event that i and j do not have a common neighbor. Argue that $\mathrm{E}\left[X_{ij}\right] = (1 - p^2)^{n-2}$ and apply Markov's inequality.

22.5. Prove that $p = \ln n/n$ is a threshold probability for the disappearance of isolated vertices.

Sketch: Consider the random variable $X = X_1 + \ldots + X_n$ where X_i indicates whether vertex i is isolated in $\mathbf{G}(n,p)$. When estimating the variance of X, observe that $X_i X_j = 1$ iff both vertices are isolated. That requires forbidding $2(n - 2) + 1$ edges, so $\mathrm{E}\left[X_i X_j\right] = (1 - p)^{2n-3}$.

23. The Entropy Function

Entropy is a basic concept of information theory. In this chapter we will consider some applications of this concept in combinatorics. Most of these applications were collected in the survey by Alon (1994a).

Let X be a random variable taking values in some range B, and let p_b denote the probability that the value of X is b. The *binary entropy* of X, denoted by $H[X]$ is defined by

$$H[X] = \sum_{b \in B} -p_b \log_2 p_b$$

where $0 \log_2 0$ is interpreted as 0.

23.1 Basic properties

Entropy has the following basic properties:

(a) If $|B| = 2^t$ then $H[X] \leqslant t$. Convexity arguments show that if the p_b's are smaller then the entropy must be larger. The extreme case being $p_b = 2^{-t}$ for all $b \in B$, then $H[X] = t$. Moreover,

$$\sum_{b \in S} -p_b \log_2 p_b \leqslant \log_2 |S|$$

for any subset $S \subseteq B$.

(b) Entropy has the *concentration property*. If $H[X] \leqslant t$ then there must be some b for which

$$\text{Prob} \, (X = b) \geqslant 2^{-t}.$$

(c) Entropy is *subadditive*: if $X = (X_1, \ldots, X_n)$ then

$$H[X] \leqslant \sum_{i=1}^{n} H[X_i].$$

The first two properties follow directly from the definition. The last needs a proof, and we give it in Sect. 23.2.

If E is some event, it is natural to define the *conditional entropy* of X given E by

$$H[X|E] \rightleftharpoons \sum_{b \in B} -\text{Prob}(X = b \mid E) \cdot \log_2 \text{Prob}(X = b \mid E).$$

In the same way, if Y is any other random variable taking values in some range A, we define the conditional entropy of X given Y by

$$H[X|Y] \rightleftharpoons \sum_{a \in A} H[X|Y = a] \cdot \text{Prob}(Y = a).$$

We think of $H[X|Y]$ as the uncertainty of X given a particular value of Y, averaged over the range of values that Y can take. Fairly direct consequences of definitions are:

(d) $H[X|X] = 0$;
(e) $H[X|Y] = 0$ if and only if $X = f(Y)$ for some function f;
(f) $H[X|Y] = H[X]$ if X and Y are independent;
(g) $H[X|Y, Z] \leqslant H[X|Y]$.

The main property of conditional entropy is the following:

$$H[X, Y] = H[Y] + H[X|Y]. \tag{23.1}$$

This equation also follows (though not so immediately) from definitions, and we leave the proof as an exercise. Using this equality we can derive the following analogue of the inequality $\text{Prob}(A \cap B) \leqslant \text{Prob}(A) + \text{Prob}(B) - \text{Prob}(A \cup B)$:

$$H[X, Y, Z] \leqslant H[X, Y] + H[Y, Z] - H[Y]. \tag{23.2}$$

Indeed

$$\begin{aligned} H[X, Y, Z] &= H[X, Y] + H[Z|X, Y] \leqslant H[X, Y] + H[Z|Y] \\ &= H[X, Y] + H[Y, Z] - H[Y]. \end{aligned}$$

23.2 Subadditivity

The following inequality is very useful when dealing with the entropy function.

Lemma 23.1. Let $x_1, \ldots, x_n$ and $y_1, \ldots, y_n$ be real numbers in the interval $[0, 1]$ such that $\sum x_i = 1$ and $\sum y_i \leqslant 1$. Then $\sum x_i \log_2 x_i \geqslant \sum x_i \log_2 y_i$.

Proof. Multiplying both sides by $\ln 2$ we may assume that all logarithms are natural. Since the logarithm is a convex function we have $\ln x \leqslant x - 1$ with equality iff $x = 1$. Hence

$$\sum_{i=1}^{n} x_i \ln y_i - \sum_{i=1}^{n} x_i \ln x_i = \sum x_i \ln \left(\frac{y_i}{x_i} \right) \leqslant \sum x_i \left(\frac{y_i}{x_i} - 1 \right)$$

$$= \sum y_i - \sum x_i \leqslant 0. \qquad \square$$

Just like expectation has the additivity property for the *sums* $X = X_1 + \cdots + X_n$, entropy has similar (though weaker) property for *strings* $X = (X_1, \ldots, X_n)$ of random variables.

Theorem 23.2. *If X and Y are two random variables taking only finitely many values, then $H[X,Y] \leq H[X] + H[Y]$, with equality only holding when X and Y are independent.*

Proof. Suppose X and Y take their values in A and B, respectively. Let $p_{a,b}$ denote the probability that $(X,Y) = (a,b)$, p_a denote the probability that $X = a$ and p_b denote the probability that $Y = b$. Since $\sum_{b \in B} p_{a,b} = p_a$ and $\sum_{a \in A} p_{a,b} = p_b$, we have

$$H[X] + H[Y] = \sum_{a \in A} -p_a \log p_a + \sum_{b \in B} -p_b \log p_b$$

$$= \sum_{a \in A} \sum_{b \in B} -p_{a,b} \log p_a + \sum_{b \in B} \sum_{a \in A} -p_{a,b} \log p_b$$

$$= \sum_{(a,b) \in A \times B} -p_{a,b} \log(p_a \cdot p_b).$$

Since $\sum_{a,b} p_a \cdot p_b = \sum_a p_a \left(\sum_b p_b \right) = 1$, we can apply Lemma 23.1 to get

$$H[X] + H[Y] \geq \sum_{(a,b) \in A \times B} -p_{a,b} \log p_{a,b} = H[X,Y].$$

Equality holds only when $p_{a,b} = p_a p_b$ for all $a \in A$ and $b \in B$. But this is exactly the condition that X and Y are independent. $\qquad \square$

Theorem 23.2 generalizes readily to more than two variables (we leave the proof as an exercise).

Theorem 23.3. *Let $X = (X_1, \ldots, X_n)$ be a random variable taking values in the set $B = B_1 \times B_2 \times \cdots \times B_n$, where each of the coordinates X_i of X is a random variable taking values in B_i. Then $H[X] \leq \sum_{i=1}^n H[X_i]$ with equality only holding when $X_1, \ldots, X_n$ are mutually independent.*

An interesting extension was proved by Chung, Frankl, Graham, and Shearer (1986). As in that theorem, let $X = (X_1, \ldots, X_n)$ be a random variable taking values in the set $B = B_1 \times \cdots \times B_n$, where each X_i is a random variable taking values in B_i. Also assume that all $B_i = \{0, 1\}$. For a subset of coordinates S, let X_S denote the random variable $(X_i)_{i \in S}$.

Theorem 23.4 (Generalized Subadditivity). *Let $X = (X_1, \ldots, X_n)$ and B be as above and let $S_1, \ldots, S_m$ be subsets of $[n] = \{1, \ldots, n\}$ such that every $i \in [n]$ belongs to at least k of $S_1, \ldots, S_m$. Then*

$$H[X] \leq \frac{1}{k} \sum_{i=1}^m H[X_{S_i}]. \tag{23.3}$$

Proof. For $k = 1$ the assertion follows from Theorem 23.3. Now assume $k > 1$. Let ν denote the minimum number of S_i's whose union is $[n]$. We will prove (23.3) by induction on k and ν. If $\nu = 1$ then, say $S_1 = [n]$, and every point of $[n]$ belongs to at least $k - 1$ of the sets $S_2, \ldots, S_m$. By induction (on k) we have in this case that

$$(k - 1)H[X] \leqslant \sum_{i=2}^{m} H[X_{S_i}]$$

and consequently, $k \cdot H[X] \leqslant \sum_{i=1}^{m} H[X_{S_i}]$ since $X = X_{S_1}$.

Suppose $\nu > 1$. We may assume w.l.o.g. that $S_1 \cup S_2 \cup \cdots \cup S_\nu = [n]$. Let $S_1' \rightleftharpoons S_1 \cup S_2$ and $S_2' \rightleftharpoons S_1 \cap S_2$. Clearly, every element of $[n]$ is in at least k of the sets $S_1', S_2', S_3, \ldots, S_m$. Moreover, already $\nu - 1$ of these sets cover $[n]$ because $[n] = S_1' \cup S_3 \cup \cdots \cup S_\nu$. By induction (on ν),

$$k \cdot H[X] \leqslant \sum_{i=3}^{m} H[X_{S_i}] + H[X_{S_1'}] + H[X_{S_2'}].$$

Since by (23.2) we have (we address this conclusion in the exercises) that

$$H[X_{S_1 \cup S_2}] \leqslant H[X_{S_1}] + H[X_{S_2}] - H[X_{S_1 \cap S_2}], \tag{23.4}$$

the desired inequality (23.3) follows.   □

23.3 Combinatorial applications

Theorem 23.3 was used by Kleitman, Shearer, and Sturtevant (1981) to derive several interesting applications in extremal set theory. Their basic idea can be illustrated by the following simple corollary of Theorem 23.3.

Corollary 23.5. *Let $\mathcal{F}$ be a family of subsets of $\{1, 2, \ldots, n\}$ and let p_i denote the fraction of sets in $\mathcal{F}$ that contain i. Then*

$$|\mathcal{F}| \leqslant 2^{\sum_{i=1}^{n} H(p_i)},$$

where $H(y) \rightleftharpoons -y \log_2 y - (1 - y) \log_2(1 - y)$.

Proof. Associate each set $F \in \mathcal{F}$ with its incidence vector v_F, which is a binary vector of length n. Let $X = (X_1, \ldots, X_n)$ be the random variable taking values in $\{0, 1\}^n$, where $\text{Prob}(X = v_F) = 1/|\mathcal{F}|$ for all $F \in \mathcal{F}$. Clearly,

$$H[X] = |\mathcal{F}| \left(-\frac{1}{|\mathcal{F}|} \log_2 \frac{1}{|\mathcal{F}|} \right) = \log_2 |\mathcal{F}|,$$

and since here $H[X_i] = H(p_i)$ for all $1 \leqslant i \leqslant n$, the result follows from Theorem 23.3.   □

This corollary supplies a quick proof for the following well-known estimate (cf. Exercise 1.14).

Corollary 23.6. *For every integer n and for every real $0 < p \leqslant 1/2$,*

$$\sum_{i \leqslant np} \binom{n}{i} \leqslant 2^{nH(p)}.$$

Proof (due to P. Frankl). Let $\mathcal{F}$ be the family of all subsets of cardinality at most pn of $\{1, 2, \ldots, n\}$. If p_i is the fraction of subsets of $\mathcal{F}$ that contain i then $p_1 = \ldots = p_n$. Counting in two ways we obtain that $\sum_{i=1}^{n} p_i \leqslant pn$, and hence $p_i \leqslant p$ for all i. Since the function $H(p)$ is increasing for $0 \leqslant p \leqslant 1/2$ this, together with Corollary 23.5, implies that

$$\sum_{i \leqslant np} \binom{n}{i} = |\mathcal{F}| \leqslant 2^{\sum_{i=1}^{n} H(p_i)} \leqslant 2^{nH(p)},$$

as needed.   □

The following easy consequence of Theorem 23.4 tells us that a family cannot have many members if its "projections" are small.

Theorem 23.7. *Let Ω be a finite set and let $\mathcal{S} = \{S_1, \ldots S_m\}$ be subsets of Ω such that each element of Ω is contained in at least k members of $\mathcal{S}$. Let $\mathcal{F}$ be a family of subsets of Ω. For each $1 \leqslant i \leqslant m$ define the projection of $\mathcal{F}$ onto the set S_i by $\mathcal{F}_i \rightleftharpoons \{E \cap S_i : E \in \mathcal{F}\}$. Then*

$$|\mathcal{F}| \leqslant \left(\prod_{i=1}^{m} |\mathcal{F}_i| \right)^{1/k}.$$

Proof. Suppose $\Omega = \{1, \ldots, n\}$, and define $B_i = \{0, 1\}$ for $1 \leqslant i \leqslant n$. Let $X = (X_1, \ldots, X_n)$ be the random variable taking values in $B = B_1 \times B_2 \times \cdots \times B_n$, where for each $E \in \mathcal{F}$, X is equal to the incidence vector of E with probability $1/|\mathcal{F}|$. By Theorem 23.4, $kH[X] \leqslant \sum_{i=1}^{m} H[X_{S_i}]$. But $H[X] = \log_2 |\mathcal{F}|$, whereas $H[X_{S_i}] \leqslant \log_2 |\mathcal{F}_i|$, implying the desired result.   □

The generalized subadditivity of the entropy function can be used to prove some non-trivial "intersection theorems" (cf. Chap. 8). The following three results were obtained by Chung, Frankl, Graham, and Shearer (1986).

Recall that a family $\mathcal{F}$ of subsets of some set Ω is *intersecting* if $F \cap F' \neq \emptyset$ for all $F, F' \in \mathcal{F}$. If $\mathcal{F}$ is intersecting then $|\mathcal{F}| \leqslant 2^{|\Omega|-1}$, since $\mathcal{F}$ cannot contain both a set and its complement. Moreover, this is optimal (just take the family of all subsets of Ω containing one fixed point). To make the question more interesting we can require that the members of $\mathcal{F}$ not just intersect, but that these intersections contain at least one of the given configurations.

We first consider one easy example. Let as before $[n] \rightleftharpoons \{1, 2, \ldots, n\}$.

Theorem 23.8. *Suppose that $\mathcal{F}$ is a family of subsets of $[n]$ such that the intersection of any two of its members contains a pair of consecutive numbers, i.e., for all $F, F' \in \mathcal{F}$ there is some $1 \leqslant i < n$ such that $F \cap F' \supseteq \{i, i+1\}$. Then $|\mathcal{F}| \leqslant 2^{n-2}$.*

Note that this upper bound is optimal: just let $\mathcal{F}$ be the family of all subsets containing the set $\{1,2\}$

Proof. Let S_0 and S_1 be the set of all even and odd numbers in $[n]$, respectively. Consider the projections $\mathcal{F}_\epsilon \rightleftharpoons \{F \cap S_\epsilon : F \in \mathcal{F}\}$ of our family $\mathcal{F}$ onto these two sets, $\epsilon = 0, 1$. Note that if $G, G' \in \mathcal{F}_\epsilon$ then their intersection has the form $G \cap G' = (F \cap S_\epsilon) \cap (F' \cap S_\epsilon) = (F \cap F') \cap S_\epsilon$ for some $F, F' \in \mathcal{F}$, and hence, is non-empty because $F \cap F'$ must contain a pair of consecutive numbers. Thus, both of the families $\mathcal{F}_\epsilon$ are intersecting, and hence, $|\mathcal{F}_\epsilon| \leqslant 2^{|S_\epsilon|-1}$ for both $\epsilon = 0, 1$. Using Theorem 23.7 (with $k = 1$) we conclude that

$$|\mathcal{F}| \leqslant |\mathcal{F}_0| \cdot |\mathcal{F}_1| \leqslant 2^{|S_0|-1} \cdot 2^{|S_1|-1} = 2^{n-2},$$

as desired. $\qquad\square$

Theorem 23.7 also has not so trivial consequences. For example, one may take Ω to be the set of all $\binom{n}{2}$ edges of a complete graph K_n. We can look at the subsets F of Ω as (labeled) subgraphs $G = ([n], E)$ of K_n. (Dealing with "labeled" subgraphs means that we do not identify isomorphic ones.)

Theorem 23.9. *Suppose that $\mathcal{F}$ is a family of (labeled) subgraphs of K_n such that for all $F, F' \in \mathcal{F}$, the graph $F \cap F'$ does not contain any isolated vertices. Then $|\mathcal{F}| \leqslant 2^{\binom{n}{2}-\frac{n}{2}}$.*

Note that this requirement is rather severe: each of the n vertices must be incident with at least one edge in $F \cap F'$. Is this upper bound optimal? (See Exercise 23.9).

Proof. Choose S_i to be the star at the ith vertex, i.e., S_i consists of all $n - 1$ edges $\{i, j\}$, $j \neq i$. Clearly, every edge is in exactly two of $S_1, \ldots, S_n$. Consider the projections $\mathcal{F}_i \rightleftharpoons \{F \cap S_i : F \in \mathcal{F}\}$ of $\mathcal{F}$ onto these stars, $i = 1, 2, \ldots, n$. Note that if $G, G' \in \mathcal{F}_i$ then their intersection has the form $G \cap G' = (F \cap S_i) \cap (F' \cap S_i) = (F \cap F') \cap S_i$ for some $F, F' \in \mathcal{F}$, and hence, is non-empty because the vertex i cannot be isolated in the subgraph $F \cap F'$. Thus, each of the families $\mathcal{F}_i$ is intersecting, and hence, $|\mathcal{F}_i| \leqslant 2^{|S_i|-1} = 2^{n-2}$ for all $i = 1, 2, \ldots, n$. Applying Theorem 23.7 (with $k = 2$) we conclude that

$$|\mathcal{F}| \leqslant \left(\prod_{i=1}^n |\mathcal{F}_i| \right)^{1/2} \leqslant 2^{n(n-2)/2} = 2^{\binom{n}{2}-\frac{n}{2}}.$$

$\qquad\square$

Let us say that a family $\mathcal{F}$ of subgraphs of K_n is *triangle-intersecting* if $F \cap F'$ contains a triangle for all $F, F' \in \mathcal{F}$.

Theorem 23.10. *Let $n \geqslant 4$ be even and let $\mathcal{F}$ be a family of (labeled) subgraphs of K_n. If $\mathcal{F}$ is triangle-intersecting then $|\mathcal{F}| \leqslant 2^{\binom{n}{2}-2}$.*

It is not known if this bound is optimal, i.e., if $2^{\binom{n}{2}-2}$ can be replaced by $2^{\binom{n}{2}-3}$, the number of subgraphs of K_n containing a fixed triangle.

Proof. We choose S_i, $1 \leqslant i \leqslant m = \frac{1}{2}\binom{n}{n/2}$, to be all possible disjoint unions of two complete (labeled) subgraphs on $n/2$ vertices each. (That is, each S_i has the form $K_U \cup K_{\overline{U}}$ for some subset of vertices $U \subseteq \{1, \ldots, n\}$ with $|U| = n/2$.) Each of the families $\mathcal{F}_i = \{F \cap S_i : F \in \mathcal{F}\}$ is intersecting because no triangle can lie entirely in a bipartite graph. Therefore $|\mathcal{F}_i| \leqslant 2^{|S_i|-1}$. Each of the graphs S_i has $s = 2\binom{n/2}{2}$ edges. Each edge of K_n is in $k = \binom{n-2}{n/2}$ of S_i's. By Theorem 23.7,

$$|\mathcal{F}| \leqslant \left(\prod_{i=1}^{m} 2^{|S_i|-1} \right)^{1/k} = 2^{(s-1)m/k}.$$

Substituting the values of s, m and k, we conclude that

$$\frac{(s-1)m}{k} = \frac{1}{2}\left(2\binom{n/2}{2} - 1 \right)\binom{n}{n/2} \Big/ \binom{n-2}{n/2}$$

$$\leqslant \binom{n}{2} - \frac{n(n-1)}{n(n/2-1)} \leqslant \binom{n}{2} - 2,$$

as desired.   □

Exercises

23.1.⁻ Prove the properties (a), (b) and (d)—(g) of the entropy.

23.2.⁻ Prove the equation (23.1).

23.3. Let X be a random variable taking its values in some set B, and let $Y = f(X)$. Prove that $H[Y] \leqslant H[X]$. Show that equality holds if and only if f is one-to-one on the set of all $b \in B$ such that $\text{Prob}\,(X = b) > 0$.

23.4.⁻ Show that for any random variables $H[X, Y] \leqslant H[X]$.

23.5. Show that, for any random variable X, $H[X^2|X] = 0$, but give an example to show that $H[X|X^2]$ is not always zero.

Hint: Let X take the values $+1$ and -1 with equal probability.

23.6. The random variable Y takes the integer values $1, 2, \ldots, 2n$ with equal probability. The random variable X is defined by $X = 0$ if Y is even, and $X = 1$ if Y is odd. Show that $H[Y|X] = H[Y] - 1$, but that $H[X|Y] = 0$.

23.7. Prove the inequality (23.4).

Hint: Take $X = X_{S_1-S_2}$, $Y = X_{S_1 \cap S_2}$, $Z = X_{S_2-S_1}$ and apply (23.2).

23.8. Use Lemma 23.1 to prove that the entropy function is *convex down*, i.e., that for any real number $\alpha \in [0, 1]$, $H[\alpha X + (1-\alpha)Y] \geqslant \alpha H[X] + (1-\alpha)H[Y]$.

23.9. Show that for even n the bound of Theorem 23.9 is best possible.

Hint: Consider the family of all subgraphs of K_n containing a fixed matching.

24. Random Walks

There are n rocks forming a circle on a river, and there is a frog on one of those rocks. The frog jumps up himself with probability $1/3$, jumps up to the right rock with probability $1/3$ and jumps to the left one with probability $1/3$, too. The frog problem consists of knowing: where will the frog be after t times?

This is, perhaps, the most popular illustration of what is known as a *random walk*, a concept which arises in many models of mathematics and physics. The reader can find such applications in any standard probability book containing a chapter about random walks or Markov chains. Besides these, random walks have found interesting applications in the theory of computing, as well. In this chapter we present some of them.

24.1 Satisfying assignments for 2-CNF

A k-*CNF* (conjunctive normal form) is an And of an arbitrary number of *clauses*, each being an Or of k literals; a *literal* is a variable x_i or its negation $\overline{x}_i$. Given such a CNF F, we seek an assignment such that all the clauses are satisfied. For example, if $F = (x_1 \vee \overline{x}_2)(x_2 \vee x_3)(\overline{x}_1 \vee \overline{x}_3)$, then $(1, 1, 0)$ is a satisfying assignment for this CNF.

Let F be a 2-CNF, and suppose that we know that it is satisfiable. How quickly can we find a satisfying assignment? Papadimitriou (1991) proposed the following simple randomized procedure.

Suppose we start with an arbitrary assignment of values to the literals. As long as there is a clause that is unsatisfied, we modify the current assignment as follows: we choose an arbitrary unsatisfied clause and pick one of the (two) literals in it uniformly at random; the new assignment is obtained by complementing the value of the chosen literal. After each step we check if there is an unsatisfied clause; if not, the algorithm terminates successfully with a satisfying assignment.

Theorem 24.1 (Papadimitriou 1991). *Suppose that F is a satisfiable 2-CNF in n variables. Then, with probability at least $1/2$, the above algorithm will find a satisfying assignment in $2n^2$ steps.*

Proof. Fix an arbitrary satisfying assignment $a \in \{0,1\}^n$ for F, and refer to the values assigned by a to the literals as the "correct values."

The progress of the above algorithm can be represented by a particle moving between the integers $\{0, 1, \ldots, n\}$ on the real line. The position of the particle indicates how many variables in the current solution have "incorrect values," i.e., values different from those in a. At each iteration, we complement the current value of one of the literals of some unsatisfied clause, so that the particle's position changes by 1 at each step. In particular, a particle currently in position i, for $0 < i < n$, can only move to positions $i-1$ or $i+1$:

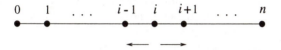

Let $t(i)$ denote the expected number of steps which a particle, started in position i, makes until it reaches position 0. Our goal is to show that $t(i) \leqslant n^2$ for all i.

A particle at location n can only move to $n-1$, and the process terminates when the particle reaches position 0 (although it may terminate earlier at some other position with a satisfying assignment other than a). Hence, $t(n) \leqslant t(n-1) + 1$ and $t(0) = 0$. In general, we have that

$$t(i) = p_{i,i-1} \cdot (1 + t(i-1)) + p_{i,i+1} \cdot (1 + t(i+1)),$$

where $p_{i,j}$ is the probability with which the particle moves from position i to position $j \in \{i-1, i+1\}$.

The crucial observation is the following: in an unsatisfied clause at least one of the literals has an incorrect value. Thus, with probability at least $1/2$ we decrease the number of variables having false values. The motion of the particle thus resembles a random walk on the line where the particle moves from the ith position $(0 < i < n)$ to position $i-1$ with probability $p_{i,i-1} \geqslant 1/2$. This implies that

$$t(i) \leqslant \frac{t(i-1) + t(i+1)}{2} + 1.$$

Replace the obtained inequalities by equations

$$x(0) = 0,$$
$$x(i) = \frac{x(i-1) + x(i+1)}{2} + 1,$$
$$x(n) = x(n-1) + 1.$$

This resolves to $x(1) = 2n - 1$, $x(2) = 4n - 4$ and in general $x(i) = 2in - i^2$. Therefore, $t(i) \leqslant x(i) \leqslant x(n) = n^2$, as desired.

By Markov's inequality, a random variable can take a value 2 times larger than its expectation only with probability $< 1/2$. Thus, the probability that the particle will make more than $2 \cdot t(i)$ steps to reach position 0 from position

i, is smaller than $1/2$. Hence, with probability at least $1/2$ the process will terminate in at most $2n^2$ steps, as claimed.   □

24.2 The best bet for simpletons

Let us consider a random walk in an infinite binary tree. Starting at the root, the walk follows the 0-edge or 1-edge independently with probability $1/2$. Given a 0-1 word a of length k, what is the probability that the walk will follow the sub-path a a given number of times? An intriguing aspect of this question is that the answer depends not only on the length of the word a, but also on the structure of its overlaps. This is so-called *overlapping words paradox*. The paradox can be explained by the following game investigated by John Conway.

Suppose we have two players, Alice and Bob. Alice selects a 0-1 word a of length $|a| = k$. Bob, knowing what a is, selects another word b of length k. The players then flip a (fair) coin until either a or b appears as a block of consecutive outcomes. That player wins whose word appears first.

At first glance it might seem that both players have the same chance of winning (the coin flips are independent). To see that this may be not the case, assume that Alice selects 00. Then Bob can select 10. After two tosses either Alice wins (00) or Bob wins (10), or the game will continue (01 or 11). However, in this last case it makes no sense for Alice to continue because Bob will win anyway. Hence, for $a = 00$ the odds of Bob over Alice are 3 : 1. On the other hand, would Alice have selected, say 01, the players would have an even break.

Even if somebody realizes that some words are "stronger" than others in this game, it maight seem that Alice should win after choosing the "strongest" word. The intriguing aspect is that for $k \geqslant 3$, no matter what a is, Bob can select b that beats a. Thus, the best bet for simpletons is a non-transitive game: a beats b and b beats c does not imply that a beats c. This can be shown (see Exercise 24.3) using the following surprisingly simple formula to compute the odds that Bob will win over Alice.

Given a set X of words, $P(X) = \sum_{x \in X} 2^{-|x|}$ is the probability that the coin flipping will produce a word in X. For two words a, b define $P_{ab} = P(H_{ab})$, where H_{ab} denotes the set of all words x such that $a = x \cdot y$ for some (non-empty) prefix y of b; hence, the empty word belongs to H_{ab} but $a \notin H_{ab}$. Let $W(a, b)$ be the odds that b will win over a. The example above shows that $W(00, 10) = 3$.

Theorem 24.2 (J. Conway). *For every $k \geqslant 1$ and any two 0-1 words $a \neq b$ of length k,*

$$W(a, b) = \frac{P_{aa} - P_{ab}}{P_{bb} - P_{ba}}.$$

Conway's proof of this formula was newer published. The proofs were given independently by Li (1980) and by Guibas and Odlyzko (1981). Later, Pevzner (1993) found a simple and elegant proof, and we present it below.

Proof (due to Pevzner 1993). A word s is an *A-victory* if it does not contain the word b, and the word a appears only at the end (as a suffix of s). A word s is an *A-previctory* if $s \cdot a$ is an *A*-victory. We define S_a to be the set of all *A*-previctories. Note that $2^{-k} \cdot P(S_a)$ is the probability that Alice will win. *B-victories*, *B-previctories* and the set S_b of all *B*-previctories are defined similarly. Given two sets of words X and Y, let $X \cdot Y = \{x \cdot y : x \in X, y \in Y\}$ denote the set of all concatenations of words from X with words from Y.

The idea is to show that the set $S = \{s : s \text{ contains neither } a \text{ nor } b\}$ of all "no-victory" words can be represented as

$$S = (S_a \cdot H_{aa}) \cup (S_b \cdot H_{ba}) \tag{24.1}$$

and as

$$S = (S_b \cdot H_{bb}) \cup (S_a \cdot H_{ab}). \tag{24.2}$$

To show (24.1), observe that every word $w = s \cdot a$ with $s \in S$ corresponds to either Bob's victory (if b appears before a in w) or Alice's victory (if a appears before b or b does not appear in w at all). In the first case s belongs to $S_b \cdot H_{ba}$ (and does not belong to $S_a \cdot H_{aa}$), whereas in the second s belongs to $S_a \cdot H_{aa}$ (and does not belong to $S_b \cdot H_{ba}$).

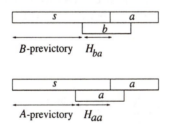

Fig. 24.1. Representations of words from $S \cdot \{a\}$

The proof of (24.2) is similar. Every word $w = s \cdot b$ with $s \in S$ corresponds to either Alice's victory (if a appears before b in w) or Bob's victory (if b appears before a or a does not appear in w at all). In the first case $s \in S_a \cdot H_{ab}$, whereas in the second $s \in S_b \cdot H_{bb}$.

By (24.1), the overall probability of words in S is

$$P(S) = P(S_a \cdot H_{aa}) + P(S_b \cdot H_{ba}) = P(S_a) \cdot P(H_{aa}) + P(S_b) \cdot P(H_{ba}),$$

and by (24.2),

$$P(S) = P(S_b \cdot H_{bb}) + P(S_a \cdot H_{ab}) = P(S_b) \cdot P(H_{bb}) + P(S_a) \cdot P(H_{ab}).$$

Hence, $W(a, b) = P(S_b)/P(S_a) = (P_{aa} - P_{ab}) / (P_{bb} - P_{ba})$. $\square$

24.3 Small formulas for complicated functions

Let V be a linear space over $\mathbb{F}_2$ of dimension d, and let $\mathbf{v}$ be a random vector in V. Starting with $\mathbf{v}$, let us "walk" over V by adding independent copies of $\mathbf{v}$. (Being an independent copy of $\mathbf{v}$ does not mean being identical to $\mathbf{v}$, but rather having the same distribution.) What is the probability that we will reach a particular vector $v \in V$? More formally, define

$$\mathbf{v}^{(r)} = \mathbf{v}_1 \oplus \mathbf{v}_2 \oplus \cdots \oplus \mathbf{v}_r,$$

where $\mathbf{v}_1, \mathbf{v}_2, \ldots, \mathbf{v}_r$ are independent copies of $\mathbf{v}$. What can be said about the distribution of $\mathbf{v}^{(r)}$ as $r \to \infty$? It turns out that, if $\mathrm{Prob}\,(\mathbf{v} = 0) > 0$ and $\mathbf{v}$ is not concentrated in some proper subspace of V, then the distribution of $\mathbf{v}^{(r)}$ converges to a uniform distribution, as $r \to \infty$. That is, we will reach each vector of V with almost the same probability!

Lemma 24.3 (Razborov 1988). *Let $\{b_1, \ldots, b_d\}$ be a basis of V and*

$$p = \min\{\mathrm{Prob}\,(\mathbf{v} = 0), \mathrm{Prob}\,(\mathbf{v} = b_1), \ldots, \mathrm{Prob}\,(\mathbf{v} = b_d)\}.$$

Then, for every vector $u \in V$ and for all $r \geqslant 1$,

$$\left| \mathrm{Prob}\left(\mathbf{v}^{(r)} = u\right) - 2^{-d} \right| \leqslant e^{-2pr}.$$

Proof. Let $\langle x, y \rangle = x_1 y_1 \oplus \cdots \oplus x_n y_n$ be the scalar product of vectors x, y over $\mathbb{F}_2$; hence $\langle x, y \rangle = 1$ if and only if the vectors x and y have an odd number of 1's in common. For a vector $w \in V$, let $p_w = \mathrm{Prob}\,(\mathbf{v} = w)$ and set

$$\Delta_v \rightleftharpoons \sum_{w \in V} p_w (-1)^{\langle w, v \rangle}. \tag{24.3}$$

Then, for every $u \in V$,

$$\sum_{v \in V} \Delta_v (-1)^{\langle u, v \rangle} = \sum_{v \in V} \sum_{w \in V} p_w (-1)^{\langle u \oplus w, v \rangle}$$
$$= \sum_{w \in V} p_w \sum_{v \in V} (-1)^{\langle u \oplus w, v \rangle} = 2^d p_u,$$

since

$$\sum_{v \in V} (-1)^{\langle x, v \rangle} = \begin{cases} 0 & \text{if } x \neq 0; \\ 2^d & \text{if } x = 0. \end{cases} \tag{24.4}$$

Therefore,

$$p_u = 2^{-d} \sum_{v \in V} \Delta_v (-1)^{\langle u, v \rangle}. \tag{24.5}$$

Fix an arbitrary vector $u \in V$. We claim that

$$\mathrm{Prob}\left(\mathbf{v}^{(r)} = u\right) = 2^{-d} \sum_{v \in V} \Delta_v^r (-1)^{\langle u, v \rangle}. \tag{24.6}$$

We show this by induction on r. If $r = 1$ then equation (24.6) is just equation(24.5). Suppose now that equation (24.6) holds for $\mathbf{v}^{(r-1)}$, and prove it for $\mathbf{v}^{(r)}$. We have

$$\text{Prob}\left(\mathbf{v}^{(r)} = u\right) = \text{Prob}\left(\mathbf{v}^{(r-1)} \oplus \mathbf{v}_r = u\right)$$

$$= \sum_{w \in V} \text{Prob}\left(\mathbf{v}^{(r-1)} = w\right) \cdot \text{Prob}\left(\mathbf{v}_r = u \oplus w\right)$$

$$= \sum_w \left(2^{-d} \sum_v \Delta_v^{r-1}(-1)^{\langle w,v \rangle}\right)\left(2^{-d} \sum_{v'} \Delta_{v'}(-1)^{\langle u \oplus w, v' \rangle}\right)$$

$$= 2^{-2d} \sum_w \sum_v \sum_{v'} \Delta_v^{r-1} \Delta_{v'} (-1)^{\langle w,v \rangle \oplus \langle u,v' \rangle \oplus \langle w,v' \rangle}$$

$$= 2^{-2d} \sum_v \sum_{v'} \Delta_v^{r-1} \Delta_{v'} (-1)^{\langle u,v' \rangle} \sum_w (-1)^{\langle w, v \oplus v' \rangle}$$

$$= 2^{-d} \sum_v \Delta_v^r (-1)^{\langle u,v \rangle}.$$

Here, the last equality follows from (24.4), because the last sum is 2^d only if $v' = v$, and is 0 otherwise.

By (24.3), $\Delta_0 = 1$. For each other vector $v \neq 0$, there exist vectors w_1, w_2 in $\{0, b_1, \ldots, b_d\}$ such that $\langle w_1, v \rangle \neq \langle w_2, v \rangle$, implying that

$$|\Delta_v| \leqslant 1 - 2p$$

(if $A + x = 1$ then $A - x = 1 - 2x$). This together with (24.6) yields the desired estimate:

$$\left|\text{Prob}\left(\mathbf{v}^{(r)} = u\right) - 2^{-d}\right| \leqslant \max_{\substack{v \in V \\ v \neq 0}} |\Delta_v^r| \leqslant (1 - 2p)^r \leqslant e^{-2pr}.$$

$\square$

Razborov (1988) used Lemma 24.3 to establish the following (counterintuitive) phenomenon.

Fix an arbitrary linear order $\preceq$ on $\{0, 1\}^n$; for example, we can take $\preceq$ to be the lexicographic order: given two vectors $a \neq b$, look for the smallest index i on which these vectors differ, and set $a \prec b$ if and only if $a_i < b_i$. After we fix such an order on the n-cube, every boolean function $f(x_1, \ldots, x_n, \ldots, x_{2n})$ on $2n$ variables defines the following undirected graph $G(f)$:

- the vertices of $G(f)$ are vectors in $\{0, 1\}^n$, and
- two vertices $a \neq b$ are joined by an edge if and only if $a \prec b$ and $f(a, b) = 1$.

Intuitively, if the graph $G(f)$ has a "complicated" combinatorial structure then the function f should be "hard" to compute (e.g., requires large circuits or formulas). It turns out that this intuition may be false!

Consider, for example, Ramsey graphs. Recall that a *clique* of size t in a graph is a set of t vertices, each pair of which is joined by an edge. Similarly, an *independent set* of size t is a set of t vertices with no edge between them.

Let $\rho(G)$ be the maximum number t such that G has either a clique or an independent set (or both) of size t. For every large enough n, we have the following information about this number:

(a) for every graph G on 2^n vertices, $\rho(G) > n/2$; this follows from Ramsey's theorem (see (27.3));
(b) there exists a graph G on 2^n vertices such that $\rho(G) \leqslant 2n$; we have proved this using the probabilistic argument (see Theorem 18.1); let us call these graphs *strongly Ramsey*;
(c) it is possible, for infinitely many values of n, to *construct* a graph G on 2^n vertices such that $\rho(G) \leqslant 2^{n^{1/2+\epsilon}}$.

The graphs, mentioned in (c), were constructed by Frankl and Wilson (1981) using a powerful linear algebra method (see Theorem 14.15 for the construction). This – the difficulty to explicitly construct graphs with small $\rho(G)$ – serves as a serious indication that strongly Ramsey graphs have a rather "complicated" combinatorial structure.

It is therefore surprising that boolean functions corresponding to these graphs may be computed by very small boolean formulas of depth-3 with And and Parity gates. These formulas have the form:

$$F = F_1 \oplus F_2 \oplus \cdots \oplus F_r,$$

where each F_i has the form

$$F_i = \bigwedge_{j=1}^{m} \bigoplus_{k=1}^{n} \lambda_{ijk} x_k \oplus \lambda_{ij}, \quad \text{with } \lambda_{ijk}, \lambda_{ij} \in \{0,1\}.$$

The *size* of such a formula F is the number rmn of literals in it. Let $L(f)$ denote the minimum size of such a formula computing f.

Theorem 24.4 (Razborov 1988). *There exists a sequence f_n of boolean functions in $2n$ variables such that: the graph $G(f_n)$ is strongly Ramsey, but $L(f_n) = O(n^5 \log n)$.*

This result is even more surprising because Razborov (1987) has proved earlier that the model of constant depth formulae is rather weak: some seemingly "simple" boolean functions, like the majority function (which outputs 1 if and only if the input vector contains more 1's than 0's) require constant depth formulas of size exponential in n (cf. Sect. 20.5).

The proof of Theorem 24.4 is based on the following lemma about the distribution of a random depth-3 formula, which may be derived from Lemma 24.3.

Let **h** be a random boolean formula in n variables given by:

$$\mathbf{h} = \lambda_0 \oplus \lambda_1 x_1 \oplus \lambda_2 x_2 \oplus \cdots \oplus \lambda_n x_n,$$

where all λ_i's are independent random variables taking their values from $\{0,1\}$ with probability $1/2$. Let

$$g = \mathbf{h}_1 \wedge \mathbf{h}_2 \wedge \cdots \wedge \mathbf{h}_m,$$

where $\mathbf{h}_1, \mathbf{h}_2, \ldots, \mathbf{h}_m$ are independent copies of $\mathbf{h}$. Finally, let $\mathbf{f} = \mathbf{f}_{n,m,r}$ be a random boolean function given by

$$\mathbf{f} = \mathbf{g}_1 \oplus \mathbf{g}_2 \oplus \cdots \oplus \mathbf{g}_r,$$

where $\mathbf{g}_1, \mathbf{g}_2, \ldots, \mathbf{g}_r$ are independent copies of $\mathbf{g}$.

If $f : \{0,1\}^n \to \{0,1\}$ is a boolean function and $E \subseteq \{0,1\}^n$, then f_E denotes the function f restricted to E.

Lemma 24.5. *Let m, r, d be natural numbers; $d \leqslant 2^{m-1}$. Let $E \subseteq \{0,1\}^n$, $|E| = d$ and $\phi : E \to \{0,1\}$ be a boolean function defined on E. Then*

$$\left| \mathrm{Prob}\,(\mathbf{f}_E = \phi) - 2^{-d} \right| \leqslant e^{-r/2^m}.$$

Proof. Recall that $\mathbf{f}$ is a sum modulo 2 of r independent copies of $\mathbf{g}$. We are going to apply Lemma 24.3 in the situation when $V = \{0,1\}^E$ is the linear space of all boolean functions defined on E, and $\mathbf{v} = \mathbf{g}_E$.

As the basis of our space V we take all $d = |E|$ boolean functions χ_a, $a \in E$, such that $\chi_a(b) = 1$ if and only if $a = b$.

It is clear that for every $a \in \{0,1\}^n$, $\mathrm{Prob}\,(\mathbf{h}(a) = 1) = 1/2$. Moreover, for $a \neq b$, $\mathbf{h}(a)$ and $\mathbf{h}(b)$ are two different linear forms $\neq 0$ of independent parameters $\lambda_0, \lambda_1, \ldots, \lambda_n$ that are uniformly distributed on $\{0,1\}$. Therefore, $\mathbf{h}(a)$ and $\mathbf{h}(b)$ are independent, implying that $\mathrm{Prob}\,(\mathbf{h}(a) = 1 \mid \mathbf{h}(b) = 1) = 1/2$, for any $a \neq b$. Since $\mathbf{g}$ is an And of m independent copies of $\mathbf{h}$, we obtain

$$\mathrm{Prob}\,(\mathbf{g}(a) = 1 \mid \mathbf{g}(b) = 1) = \mathrm{Prob}\,(\mathbf{g}(a) = 1) = 2^{-m}.$$

Using this and the condition $d \leqslant 2^{m-1}$ we obtain that

$$\mathrm{Prob}\,(\mathbf{g}_E \equiv 0) = 1 - \mathrm{Prob}\,(\exists a \in E : \mathbf{g}(a) = 1) \geqslant 1 - |E| \cdot 2^{-m} \geqslant 1/2$$

and

$$
\begin{aligned}
\mathrm{Prob}\,(\mathbf{g}_E = \chi_a) &= \mathrm{Prob}\,(\mathbf{g}(a) = 1 \text{ and } \mathbf{g}(b) = 0 \text{ for all } b \in E, b \neq a) \\
&= \mathrm{Prob}\,(\mathbf{g}(a) = 1) \cdot \mathrm{Prob}\,(\forall b \neq a : \mathbf{g}(b) = 0 \mid \mathbf{g}(a) = 1) \\
&\geqslant 2^{-m} \left[1 - \sum_{b \in E, b \neq a} \mathrm{Prob}\,(\mathbf{g}(b) = 1 \mid \mathbf{g}(a) = 1) \right] \\
&\geqslant 2^{-m} \left(1 - d \cdot 2^{-m} \right) \geqslant 2^{-m-1}.
\end{aligned}
$$

Thus, the minimum of $\mathrm{Prob}\,(\mathbf{g}_E \equiv 0)$ and $\mathrm{Prob}\,(\mathbf{g}_E = \chi_a)$ $(a \in E)$ is at least 2^{-m-1}, and we can apply Lemma 24.3 with $p = 2^{-m-1}$. This completes the proof of Lemma 24.5. $\qquad\square$

Proof of Theorem 24.4. Set $m \rightleftharpoons \lfloor 2 \log_2 n + 3 \rfloor$, $r \rightleftharpoons \lfloor 40 n^4 \rfloor$ and consider the random boolean function $\mathbf{f} = \mathbf{f}_{2n,m,r}$. Then $\rho(G(\mathbf{f})) \geqslant 2n + 1$ if and only if there is a subset of vertices $V \subseteq \{0,1\}^n$ of size $|V| = 2n + 1$ which either forms a clique or is an independent set in $G(\mathbf{f})$. This event happens only if

$f_E \equiv 1$ or $f_E \equiv 0$, where E is the set of all pairs (a, b) such that $a, b \in V$ and $a \prec b$. As $|V| = 2n + 1$, we have

$$|E| \leqslant 1 + 2 + \cdots + 2n = n(2n + 1),$$

and hence, the condition $d \leqslant 2^{m-1}$ of Lemma 24.5 is satisfied (with $d = n(2n + 1)$). Applying this lemma, we obtain that

$$\mathrm{Prob}\,(f_E \equiv 1 \vee f_E \equiv 0) = O(e^{-r/2^m}) = O(2^{-n(2n+1)}),$$

and hence,

$$\mathrm{Prob}\,(\rho(G(\mathbf{f})) \geqslant 2n + 1) \leqslant \binom{2^n}{2n+1} \cdot \mathrm{Prob}\,(f_E \equiv 1 \vee f_E \equiv 0)$$

$$= O\left(\binom{2^n}{2n+1} \cdot 2^{-n(2n+1)}\right) = o(1).$$

This means that there exists a depth-3 formula $f = f_{2n,m,r}$ of size $2nmr = O(n^5 \log n)$ for which $\rho(G(f)) \leqslant 2n$, as desired.   □

24.4 Random walks and search problems

Let $G = (V, E)$ be a directed graph without cycles (a *digraph*). The *degree* $d(v)$ of a vertex v is the number of outgoing edges. A *random walk* of length k in G starting at vertex v_0 is a randomly chosen sequence of vertices $\mathbf{p} = (v_0, v_1, \ldots, v_k)$ where each v_{i+1} is chosen, randomly and independently, among the $d(v_i)$ neighbors of v_i with probability $1/d(v_i)$. We say that the walk $\mathbf{p}$ *visits* a vertex $v \in V$ if this vertex belongs to the walk, i.e., if $v = v_i$ for some i. Below we describe one interesting application of random walks in the analysis of *search problems* found by Razborov, Wigderson, and Yao (1997).

Suppose we have a digraph $G = (V, E)$ with a fixed vertex (called the root), and have some "local" information about it (e.g. about the degrees of its vertices). We would like to find a lower bound on the total number $|V|$ of its vertices. To do this we choose a particular subset $U \subseteq V$ of vertices such that every path, starting in the root:

(i) visits at least one vertex in U, and
(ii) goes through k vertices of degree at least d, before it visits U.

By (i), a random walk will visit at least one vertex in U with probability 1, and by (ii), the walk visits a particular vertex in U with probability at most d^{-k}. This implies that $|U| \geqslant d^k$. Let us now look at how this idea works in concrete situations.

24.4.1 Long words over a small alphabet

Consider the following *Word Search* problem $W(n, m)$. Let A be a set of n elements, called *letters*. A *word* is a string $x = (x_1, \ldots, x_m)$ of m (not necessarily distinct) letters. We assume that $m \geqslant n + 1$. By the pigeonhole principle each such word contains at least two equal letters, i.e., there must be a pair $i \neq j$ for which $x_i = x_j$. The problem is, given a word to find such a pair of equal letters. Suppose that during this search procedure we can only make tests of the form "$x_i = a$" with $1 \leqslant i \leqslant n$ and $a \in A$.

If we have no further restrictions, the task is easy: simply make all $n\binom{m}{2}$ tests "$x_i = a$" and "$x_j = a$" over all $i \neq j$ and $a \in A$. The situation changes drastically if our letters are *sensitive* in that each of them disappears from the word (e.g. is replaced by some dummy letter $* \notin A$) immediately after it is tested for the first time.

More formally, a *search graph* for $W(n, m)$ is a digraph $G = (V, E)$ with a fixed vertex s (called the root), each vertex of which has outdegree n and is labeled by an integer $i \in \{1, \ldots, n\}$; the n outgoing edges are labeled by n distinct letters from A. Sinks (i.e., degree 0 vertices) are labeled by possible *solutions*, i.e., by expressions of the form (i, j, a) where $1 \leqslant i \neq j \leqslant m$ and $a \in A$. Such a solution is *admissible* for a word $x = (x_1, \ldots, x_m)$ if $x_i = x_j = a$.

Every word determines a (unique) path from the root s to a sink in a natural manner: suppose we have reached a vertex v, and i is its label; then we proceed along the edge labeled by that letter $a \in A$ for which $x_i = a$. In this case we say that the path makes the assignment $x_i = a$. This way we always will reach some sink. The graph *solves* the word search problem if every word leads to a sink labeled by a solution admissible for this word. A search graph is *read-once* if in every path from the root s no two vertices have the same label. (Note that the graph itself may have many vertices with the same label; the restriction is on their appearance along the paths).

Theorem 24.6. *Any read-once search graph solving the word search problem $W(n, m)$ must have at least $n^{\Omega(n)}$ vertices.*

Proof. We can assume $n \geqslant 3$. Let $G = (V, E)$ be a read-once search graph solving $W(n, m)$. For any vertex v, denote by $J(v)$ the set of all letters $a \in A$ such that for some *fixed* $i \in [m]$, *every* path from the root s to v makes the assignment $x_i = a$. Note that if $e = (v, u)$ is an edge (directed from v to u), then $|J(u)| \leqslant |J(v)| + 1$. We call an edge e labeled by a letter a and outgoing of v, *legal* if $a \notin J(v)$ and *illegal* otherwise.

Claim 24.7. *There is no path from the root to a sink consisting entirely of legal edges.*

Proof of Claim 24.7. Consider some path p from the root s to a sink t labeled by (i, j, a). Then p must contain at least two edges labeled by a. Let $e = (v, u)$

be the last edge along p with this property. We are going to show that e is illegal.

Replacing, if necessary, i by j, we may assume that i is *not* the label of v. Every path from s to v must make the assignment $x_i = a$: otherwise we could combine it with the segment of p beginning at v, and (keeping in mind that G is a read-once graph) get a path from s to t that does not make the assignment $x_i = a$, contrary to the assumption that G solves $W(n,m)$. Hence, $a \in J(v)$, and e is illegal.   □

Now we define a random walk $\mathbf{p}$ in G which at any non-sink vertex v traverses all outgoing *legal* edges with equal probabilities. Claim 24.7 implies that the walk will reach a sink vertex with zero probability, and hence, with probability 1 the walk $\mathbf{p}$ actually arrives at a vertex v such that $J(v) = A$. Also, the value $|J(v)|$ can increase, decrease, or stay the same along each edge in $\mathbf{p}$. But every time it increases, it increases by at most one in a step. Thus, with probability 1, $\mathbf{p}$ visits some node v such that $|J(v)| = k$ where $k \rightleftharpoons \lceil |A|/2 \rceil = \lceil n/2 \rceil$. Let $\mathbf{v}$ be the *first* such node along $\mathbf{p}$. It is only left to show that for every specific vertex v_0 with $|J(v_0)| = k$, $\mathrm{Prob}\,(\mathbf{v} = v) \leqslant n^{-\Omega(n)}$ (and hence, there must be at least $n^{\Omega(n)}$ such v_0).

Consider any v_0 with $J(v_0) = \{a_1, \ldots, a_k\}$, and let $i_1, \ldots, i_k$ be integers such that every path from s to v_0 has made all the assignments $x_{i_1} = a_1, \ldots, x_{i_k} = a_k$. Clearly, $i_1, \ldots, i_k$ are also distinct (since G is read-once). Then $\mathbf{v} = v_0$ implies, in particular, that before arriving at the vertex v_0, the walk $\mathbf{p}$ must have tested all the letters $x_{i_1}, \ldots, x_{i_k}$ (possibly in a different order) and every time make the "right" decision $x_{i_\nu} = a_\nu$. Moreover, since $\mathbf{v}$ was chosen to be the *first* vertex along $\mathbf{p}$ with $|J(v)| = k$, $\mathbf{p}$ must make these decisions at vertices v with at least k outgoing legal edges. This implies that, for each ν, the probability of making the decision $x_{i_\nu} = a_\nu$ is at most $1/k$. It follows that $\mathrm{Prob}\,(\mathbf{p} = v_0) \leqslant (1/k)^k \leqslant n^{-\Omega(n)}$, as desired.   □

24.4.2 Short words over a large alphabet

So far we have considered the situation when $m > n$, i.e., when words are longer than the total number of letters. What can be said about the dual situation in which $m < n$? In this case every word misses at least one letter, and the corresponding search problem is to find such a letter. So, in this case the search problem $W^*(n,m)$ is, given a word $x = (x_1, \ldots, x_m)$ to find a letter $a \notin x$. The corresponding model of search graphs is the same with one difference: this time sinks are labeled by letters from A (hence possible answers have the form "$a \notin x$").

Recall that such a graph is *read-once* if for every path p no two of its vertices have the same label. Let $L(p) \subseteq \{1, \ldots, m\}$ be the set of those labels which appear along p. A read-once graph is *uniform* if: (i) for a path p beginning at the root, $L(p)$ only depends on the terminal vertex v of p (accordingly, we denote it by $L(v)$), and (ii) for every sink t, $L(t) = \{1, \ldots, m\}$.

It can be shown that every read-once graph can be converted into an equivalent uniform graph by increasing the size (i.e., the number of vertices) by at most a factor of m.

Theorem 24.8. *Any read-once search graph solving the word search problem* $W^*(n,m)$ *must have at least* $2^{\Omega(m/\log n)}$ *vertices.*

Proof. Let $G = (V, E)$ be a read-once search graph solving $W^*(n,m)$. As noted, above we can assume that G is uniform (the resulting bound will only be by a factor of m worse). For a vertex v of G denote by $I(v)$ the set of all letters $a \in A$ such that no of the assignments $x_i = a$, with $i \in L(v)$, is made along any path from the root s to v. Note that $I(s) = A$, $I(v)$ can only decrease along edges, and $a \in I(t)$ for every sink t labeled by a. We need the following dual version of Claim 24.7 (its proof is left as Exercise 24.4):

Claim 24.9. $I(v) \neq \emptyset$ *for every node* v.

We call an edge outgoing of v and labeled by a, *legal* if $a \in I(v)$ and *illegal* otherwise. Define **p** to be the same random walk as in the proof of Theorem 24.6 (with the new notion of legal edge, of course). The remark above implies that **p** arrives with probability 1, at a sink **t**. Since G is uniform, **p** has length m (with probability 1). Let $k = \lceil \log n \rceil$, and $s = v_0, \mathbf{v}_1, \ldots, \mathbf{v}_k = \mathbf{t}$ be the nodes along **p** that divide this random path into segments of length at least $\lfloor m/k \rfloor$ each. Since $|I(v_0)| = |A| = n$, $|I(\mathbf{v}_k)| \geqslant 1$ and $I(\mathbf{v}_\nu)$ is strictly decreasing (with probability 1), we have for some $0 \leqslant \nu \leqslant k - 1$,

$$|I(\mathbf{v}_{\nu+1})| \geqslant \tfrac{1}{2}|I(\mathbf{v}_\nu)|.$$

Say that a pair (u_0, u_1) of vertices is *good* if

$$|L(u_1) - L(u_0)| \geqslant \lfloor m/k \rfloor, \quad I(u_1) \subseteq I(u_0) \quad \text{and} \quad |I(u_1)| \geqslant \tfrac{1}{2}|I(u_0)|.$$

Similarly to the proof of Theorem 24.6, we are left to show that for any specific good pair (u_0, u_1) we have

$$\text{Prob}\,(u_0 \text{ and } u_1 \text{ belong to } \mathbf{p} \text{ in this order}) \leqslant 2^{-\lfloor m/k \rfloor}. \tag{24.7}$$

To show this, observe that any successful **p** (i.e., a walk for which the above probability does not vanish) between u_0 and u_1 can visit only those vertices v for which

$$I(u_0) \supseteq I(v) \supseteq I(u_1).$$

At any such vertex v, there are $|I(v)| \geqslant |I(u_1)|$ outgoing legal edges, but starting from this vertex, at most $|I(v) - I(u_1)|$ of them can lead the walk **p** to the vertex u_1 (recall that $a \in I(u_1)$ requires that the letter a be tested along *no* path from the source to u_1). Thus, the probability for the walk to make the "right" decision at every individual vertex v (between u_0 and u_1) is at most

$$\frac{|I(v) - I(u_1)|}{|I(v)|} \leqslant 1 - \frac{|I(u_1)|}{|I(u_0)|} \leqslant \frac{1}{2},$$

and on its way from u_0 to u_1 the walk must make at least $\lfloor m/k \rfloor$ of them. This completes the proof of the bound (24.7), and thus, the proof of the theorem. □

Exercises

24.1.⁻ Show that the lower bound, given by Theorem 24.6 is not far from the optimum: n^{n+1} vertices are enough. *Hint*: Ignore all but the first $n+1$ letters in input words.

24.2. Let $P = (p_{i,j})$ be a *transition matrix* of a random walk on an undirected graph $G = (V, E)$ with n vertices $V = \{1, \dots, n\}$. That is, $p_{i,j} = 1/d(i)$ if $\{i, j\} \in E$, and $p_{i,j} = 0$ otherwise. Let H_{ij} be the expected number of steps needed to visit state j for the first time, when starting from state i. Define the vector $\pi = (\pi_1, \dots, \pi_n)$ by

$$\pi_i \rightleftharpoons \frac{d(i)}{2|E|} \qquad \text{for } i = 1, \dots, n.$$

Show that $\sum_{i=1}^{n} \pi_i = 1$; $\pi \cdot P = \pi$, and $H_{ii} = 1/\pi_i$ for all $i = 1, \dots, n$.

24.3.$^{(!)}$ Describe a winning strategy for Bob in the best bet for simpletons.

> *Hint*: Use Conway's formula (Theorem 24.2). If $a = x_1 \cdot x_2 \cdots x_k$, then try $a = ? \cdot x_1 \cdots x_{k-1}$.

24.4. Prove Claim 24.9.

> *Hint*: Due to the read-once property, $a \in I(v)$ means that, after reaching the vertex v, the letter a is still among the possible solutions, independent of the path by which we reached v.

24.5⁺ (Razborov 1988). Let $V = \{1, \dots, N\}$ be a set of players, $N = 2^n$. A *tournament* is an oriented graph $T = (V, E)$ such that $(i, i) \notin E$ for all $i \in V$, and for any two players $i \neq j$ exactly one of (i, j) and (j, i) belongs to E. A tournament is *transitive* if there exists a permutation σ of the players so that $(i, j) \in E$ if and only if $\sigma(i) < \sigma(j)$. Let $v(T)$ be the largest number of players in a transitive subtournament of T. It is known that: (a) $v(T) \geqslant n+1$ for every tournament T, and (b) tournaments T with $v(T) \leqslant 2n + 1$ exist.

Every boolean function $f(x, y)$ on $2n$ variables defines a tournament $T(f)$ with $N = 2^n$ players in a natural way: players are vectors in $\{0,1\}^n$, and player a beats player b iff either $a \prec b$ and $f(a, b) = 1$, or $b \prec a$ and $f(b, a) = 0$. Prove that there exists a sequence of boolean functions $f_n(x, y)$ such that $L(f_n) = O(n^5 \log n)$ and $v(T(f)) \leqslant 2n + 1$.

> *Hint*: Argue as in the proof of Theorem 24.4. Instead of sets V of size $2n + 1$ take sets of size $2n + 2$, and instead of the event "$\mathbf{f}_E = 1$ or $\mathbf{f}_E = 0$" consider the event "$\mathbf{f}_E$ induces a transitive subtournament of $T(\mathbf{f})$."

25. Randomized Algorithms

We have seen how coin flipping can help to design *proofs*. Randomness can also help to design quite efficient *algorithms*. In this chapter we will demonstrate the idea with several examples. Comprehensive guides to the state of the art of randomized algorithms are the books by Alon and Spencer (1992), and by Motwani and Raghavan (1995).

25.1 Zeroes of multivariate polynomials

Consider the following problem. We are given a polynomial $f(x_1, \ldots, x_n)$ with n variables, and would like to decide if $f = 0$ (i.e., if f is identically 0 as a function in n variables). This situation may occur, for example, when we have a matrix with entries that contain polynomials, and we would like to know whether or not the determinant of this matrix equals zero. Trying to compute the determinant explicitly may not be possible, since it might take exponential time even to write down the output.

The basic idea of what is known as a *randomized algorithm* is that we write *random* numbers in place of the variables and compute the value of the polynomial. Now if the value computed is not zero then we know the answer: $f \neq 0$. But what happens if we get zero? Well, we just hit a root of f and try again. Another root? Try once more. After a number of runs we are tired and would like to stop with the answer $f = 0$. How big will the error of such a decision be?

If we could give real values to the variables, chosen according to the uniform distribution, for example, in the interval $[0, 1]$, then the probability of error would be zero (since the number of possible roots is finite). In reality however, we must compute with discrete values; therefore we assume that the values of the variables are chosen from among the integers of an interval $\{0, 1, \ldots, N - 1\}$, independently and with uniform distribution. In this case, the probability of error will not be zero but it will be small if N is large enough. This is the meaning of the following result about the zeroes of multivariate polynomials.

Let $\mathbb{F}$ be a field and $f(x_1, \ldots, x_n)$ be a multivariate polynomial over $\mathbb{F}$. The *degree* of f is the maximum exponent of any variable when f is written as a sum of monomials (i.e., products of the variables). Polynomials of degree 1 are called *multilinear*. Note the difference from the *total degree* of a polynomial, where one first adds the exponents of the variables in each monomial and then takes the maximum over these sums. For example, the polynomial $f = x_1^3 x_2 + x_1 x_2^2 x_3^2$ has degree 3 but its total degree is 5.

Given some polynomial f written as an arithmetic expression, we want to find out whether f is in fact the zero polynomial. Efficient *probabilistic* zero-tests were developed by Schwartz (1980) and Zippel (1979). Their test is based on the following result (also known as the *Schwartz Lemma*):

Lemma 25.1 (Schwartz 1980). *Let $f(x_1, \ldots, x_n)$ be a polynomial of total degree d over field $\mathbb{F}$ that is not the zero polynomial. Let $S \subseteq \mathbb{F}$ with $|S| \geqslant d$. Then there are at least $|S|^n - d \cdot |S|^{n-1}$ points $(a_1, \ldots, a_n) \in S^n$ such that $f(a_1, \ldots, a_n) \neq 0$.*

Proof. We proceed by induction on n, the number of variables. The statement is true for $n = 1$ since the number of roots of f does not exceed its degree. Let now $n \geqslant 2$ and arrange f according to the powers of x_n:

$$f = f_0 + f_1 x_n + f_2 x_n^2 + \cdots + f_t x_n^t$$

where $f_0, \ldots, f_t$ are polynomials of the $n - 1$ variables $x_1, \ldots, x_{n-1}$, the term f_t is not identically 0, and $t \leqslant d$. Our goal is to estimate for how many of the points $(a, b) \in S^{n-1} \times S$, $f(a, b) = 0$. We distinguish between two cases:

1. $f_t(a) = 0$. Since f_t is non-zero and has total degree $\leqslant d - t$, we have by the induction hypothesis, that it can vanish on at most $(d - t)|S|^{n-2}$ points in S^{n-1}. Therefore, in this case, there are at most $(d - t)|S|^{n-1}$ points $(a, b) \in S^{n-1} \times S$ for which $f(a, b) = 0$ and $f_t(a) = 0$.
2. $f_t(a) \neq 0$. For every (fixed) point $a \in S^{n-1}$ for which $f_t(a) \neq 0$, the polynomial $f(a, x_n)$ is a polynomial in one variable of degree t, and it is not identically zero. Therefore it has at most t roots. Since there are at most $|S|^{n-1}$ such points a, the number of points $(a, b) \in S^{n-1} \times S$ for which $f(a, b) = 0$ and $f_t(a) \neq 0$, does not exceed $t \cdot |S|^{n-1}$.

Thus, there are at most $(d-t)|S|^{n-1} + t \cdot |S|^{n-1} = d \cdot |S|^{n-1}$ points $(a, b) \in S^n$ for which $f(a, b) = 0$. $\qquad\square$

If f has degree d then the total degree of f can be as large as dn. Thus, even for multilinear polynomials, the bound given by Lemma 25.1 is trivial if $|S| \leqslant n$. This case is covered by the following result.

Lemma 25.2 (Zippel 1979). *Let $f(x_1, \ldots, x_n)$ be a polynomial of degree d over field $\mathbb{F}$ that is not the zero polynomial. Let $S \subseteq \mathbb{F}$ with $|S| \geqslant d$. Then there are at least $(|S| - d)^n$ points $(a_1, \ldots, a_n) \in S^n$ such that $f(a_1, \ldots, a_n) \neq 0$.*

Proof. The proof is by induction on n. For $n = 1$ the theorem is true because a degree d polynomial can have at most d roots.

Let $n > 1$ and let $f(x_1, \ldots, x_n)$ be a non-zero polynomial of degree d. Let furthermore $(a_1, \ldots, a_n) \in \mathbb{F}^n$ be a point on which $f(a_1, \ldots, a_n) \neq 0$. We consider two polynomials, both are subfunctions of f:

$$f_0(x_1, \ldots, x_{n-1}) \rightleftharpoons f(x_1, \ldots, x_{n-1}, a_n)$$
$$f_1(x_n) \rightleftharpoons f(a_1, \ldots, a_{n-1}, x_n).$$

By construction, both polynomials are non-zero and have degree bounded by d. The first polynomial f_0 has $n-1$ variables and therefore differs from 0 on at least $(|S|-d)^{n-1}$ points in S^{n-1}, by the induction hypothesis. Similarly, f_1 has one variable and therefore at least $|S|-d$ non-zero points. For each of the $|S|-d$ choices for a_n where f_1 is non-zero, the corresponding polynomial f_0 has at least $(|S|-d)^{n-1}$ non-zero points. Therefore, the number of non-zero points of the original polynomial on S^n is at least $(|S|-d)^{n-1}(|S|-d) = (|S|-d)^n$. □

An important consequence from Lemma 25.2 is the following.

Corollary 25.3. *Let S be any subset of $\mathbb{F}$ that has $d+1$ elements. Then any nonzero polynomial of degree d has a nonzero point in S^n.*

By enlarging S even further, we can achieve that any polynomial f does not vanish on *most* points of S^n. This provides a tool for the probabilistic test.

Corollary 25.4. *Let $f(x_1, \ldots, x_n)$ be a nonzero polynomial in n variables of degree d, and $S \subseteq \mathbb{F}$ a finite set. Let $r_1, \ldots, r_n$ be chosen at random in S independently of each other according to the uniform distribution. Then*

$$\mathrm{Prob}\,(f(r_1, \ldots, r_n) \neq 0) \geqslant \left(1 - \frac{d}{|S|}\right)^n.$$

This fact suggest the following *randomized* algorithm (i.e., one that uses randomness) to decide whether a polynomial f is identically zero:

Fix a set $S \subseteq \mathbb{F}$ with $|S| = 2dn$. Compute $f(r_1, \ldots, r_n)$ with values r_i chosen randomly and independently according to the uniform distribution in S. If we do *not* get the value 0 we stop with answer $f \neq 0$. If we get value 0 we repeat the computation. If we get 0 value 100 times we stop and declare that $f = 0$.

If $f = 0$, then the algorithm will determine this. Assuming $f \neq 0$, what is the probability that one iteration will not discover this? At most $1 - \left(1 - \frac{1}{2n}\right)^n$, according to Corollary 25.4. By inequality (1.3), this probability does not exceed $1 - e^{-1/2 - O(1/n)} < 1/2$. With 100 experiments repeated independently of each other, the probability that this occurs *every* time is less than 2^{-100}. So, if the algorithm does not prove that $f \neq 0$, we can be pretty certain that actually $f = 0$. Not 100% certain, but if we lose the bet, we would know that an experiment that had only two possible outcomes ended with the one that had probability 2^{-100}. This should compensate for our trouble: we found a needle in a haystack!

25.2 Verifying the equality of long strings

Suppose that Alice maintains a large database of information. Bob maintains a second copy of the database. Periodically, they must compare their databases for consistency. Because the transmission between Alice and Bob is expensive, they would like to discover the presence of inconsistency without transmitting the entire database between them. Denote Alice's data by the sequence $a = a_0 \cdots a_{n-1}$ and Bob's data by the sequence $b = b_0 \cdots b_{n-1}$ where $a_i, b_i \in \{0, 1\}$. It is clear that any deterministic consistency check that transmits fewer than n bits will fail (just because an adversary can modify the unsent bits). Using randomness it is possible to design a strategy that detects an inconsistency with high probability (at least $1 - n^{-1}$) while transmitting much fewer than n bits, namely - only $O(\log n)$ bits.

Think of the strings a and b as polynomials over the field $\mathbb{F}_p$ where p is a prime such that $n^2 < p < 2n^2$ (theorems regarding the density of primes guarantee the existence of such p). That is, consider polynomials

$$A(x) = a_0 + a_1 x + \ldots + a_{n-1} x^{n-1} \pmod{p},$$
$$B(x) = b_0 + b_1 x + \ldots + b_{n-1} x^{n-1} \pmod{p}.$$

In order to detect if $a = b$, Alice and Bob use the following strategy:

Alice picks uniformly at random a number $\mathbf{r}$ in $\mathbb{F}$ and sends to Bob the numbers $\mathbf{r}$ and $A(\mathbf{r})$. Bob responds with 1 if $A(\mathbf{r}) = B(\mathbf{r})$ and with 0 otherwise. The number of bits transmitted is $1 + 2 \log p = O(\log n)$.

If $a = b$ then $A(\mathbf{r}) = B(\mathbf{r})$ for all $\mathbf{r}$, so the output is always 1. If $a \neq b$ we have two distinct polynomials $A(x)$ and $B(x)$ of degree at most $n - 1$. By Lemma 25.2, the probability of error is

$$\text{Prob}\left(A(\mathbf{r}) = B(\mathbf{r})\right) \leqslant \frac{n-1}{|\mathbb{F}|} = \frac{n-1}{p} \leqslant \frac{1}{n}.$$

25.3 The equivalence of branching programs

A labeled directed acyclic graph $P = (V, E)$ may be used to represent a boolean function in n variables $x_1, \ldots, x_n$ as follows. There are two fixed vertices in V; one is called the *start* vertex, and the other the *finish* vertex. Every vertex of V has out-degree 0, 1 or 2. Each edge is labelled by a variable x_i or its negation $\bar{x}_i$; moreover, if two edges leave a vertex, one must be labeled by a variable x_i and the other by its negation $\bar{x}_i$. Such a graph P is called a *branching program*. A *product term* is the And of all labels on a path leading from the start vertex to the finish vertex. (Note that some product terms may be trivial, i.e., evaluate to 0; this happens when the term contains some variable and its negation.) The function f_P computed by P is the Or of all its

Fig. 25.1. Read-once branching programs computing the parity function $x \oplus y \oplus z$ and the majority function $xy \vee xz \vee yz$

product terms. A program P is a *read-once program* if along any (directed) path, every variable occurs – negated or not – at most once.

Given two branching programs P and Q on n variables, how can we verify if they are equivalent, i.e., if they compute the same boolean function? The simplest way would be to test if $P(a) = Q(a)$ for all inputs $a \in \{0,1\}^n$. But this is a bad strategy because even for a moderate number of variables, say $n = 80$, we would be forced to test 2^{80} inputs; even assuming that every input can be tested in 10^{-10} seconds, we would need more than 100 years to make all the tests.

Fortunately, as shown by Blum, Chandra, and Wegman (1980), at least in the read-once case, randomness makes things more realistic. We can associate with a branching program P a polynomial f_P (over any field $\mathbb{F}$ with at least two elements). To get such a polynomial, associate with each edge e the polynomial $f_e \rightleftharpoons x_i$ if e is labeled by x_i, and $f_e \rightleftharpoons 1 - x_i$ if e is labeled by $\overline{x}_i$. The polynomial associated with a path is the product of polynomials of its edges; the polynomial f_P associated with the whole program P is the sum of polynomials associated with its accepting paths. The resulting polynomial f_P computes some function from $\mathbb{F}^n$ to $\mathbb{F}$. Of importance here is that on $\{0,1\}^n$, the polynomial f_P and the program P have the same values. Thus, two programs P and Q are equivalent if and only if $f_P(a) - f_Q(a) = 0$ for all $a \in \{0,1\}^n$.

Now assume that both programs P and Q are read-once. Then the polynomial $f_P - f_Q$ is multilinear and, by Corollary 25.3, for every $S \subseteq \mathbb{F}$ with $|S| \geqslant 2$, this polynomial vanishes on S^n if and only if it does so on $\{0,1\}^n$. This way, the equivalence of two read-once programs P and Q reduces to verifying a polynomial identity $f_P - f_Q = 0$, where we can apply the randomized algorithm described above. Assuming that P and Q compute different boolean functions, what is the error probability of this algorithm? Taking $|S| = 2n$, Corollary 25.4 ensures that this error probability is

$$\text{Prob}\left(f_P(\mathbf{r}) = f_Q(\mathbf{r})\right) \leqslant 1 - \left(1 - \frac{1}{|S|}\right)^n \sim 1 - e^{-1/2} < 1/2.$$

One thing is needed for us to be able to actually carry out this algorithm: we must be able to evaluate the polynomial f_P on any input in polynomial time. The size of the input is the size (e.g. the number of vertices) of the graph $P = (V, E)$, which can be exponentially smaller than the number of monomials of f_P. To get rid of this problem, we can evaluate the polynomial f_P on a particular input $(r_1, \ldots, r_n) \in \mathbb{F}^n$ by traversing the underlying graph P as follows.

1. Assign the start-vertex the value 1.
2. Assign each edge the value of its corresponding polynomial; hence, if the edge is labeled by a variable x_i (or its negation $\overline{x}_i$) then its value is r_i (or $1 - r_i$, respectively).
3. Assign each vertex the sum over all edges entering it of the product of the value of each such edge, times the value of the vertex at that edge's tail.

It is easy to see that this way the finish-vertex will get the value $f_P(r_1, \ldots, r_n)$. During this procedure we make $|E| - |V| + 1$ additions since all vertices except the start have an entering edge; the number of multiplications equals $|E|$ since the value of each edge is multiplied by the value of the vertex at its tail.

25.4 A min-cut algorithm

Let $G = (V, E)$ be a connected, undirected multigraph with n vertices. The graph being a *multigraph* means that it may contain several edges between any two vertices. A *cut* is a set of edges whose removal results in a graph broken into two or more components. A *min-cut* is a cut of minimal cardinality. We will now study the following (surprisingly simple) randomized algorithm for finding such a cut, proposed by Karger (1993).

We repeat the following step: pick an edge uniformly at random, delete this edge (together with all other edges joining the same pair of vertices, if there are such) and merge the two vertices at its end points. If as a result there are several edges between some pairs of (newly formed) vertices, retain all of them. We refer to this process as the *contraction* of that edge (see Fig. 25.2).

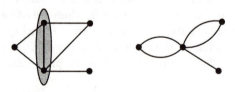

Fig. 25.2. Contraction of the edge

The crucial observation is that an edge contraction does not reduce the min-cut size in G (see Exercise 25.3). The algorithm continues the contraction process until only two vertices remain; at this point, the set of edges between these two vertices is a cut in G and the algorithm outputs it as a candidate for a min-cut.

With what probability does this algorithm find a min-cut? For the analysis we make the following observation:

If a graph $G = (V, E)$ has n vertices and has no cut of size smaller than k, then it has at least $kn/2$ edges.

Indeed, if G has no cut of size smaller than k then, in particular, each of its vertices $v \in V$ has degree $d(v) \geqslant k$. Since, by Euler's theorem (see Theorem 1.7), $\sum_{v \in V} d(v) = 2|E|$, this implies that $2|E| \geqslant k \cdot |V| = kn$, and hence, $|E| \geqslant kn/2$.

Now, let C be a min-cut in $G = (V, E)$, and let k be its size. Let A_i be the event of *not* picking an edge of C at the ith step ($i = 1, \ldots, n-2$). Thus,

$$\text{Prob (the algorithm outputs } C) = \text{Prob}\left(A_1 \cap A_2 \cap \cdots \cap A_{n-2}\right).$$

Since G has n vertices, the above observation tells us that it has at least $kn/2$ edges. Thus, the probability that the edge randomly chosen in the first step is in C is at most $|C|/|E| \leqslant k/(kn/2) = 2/n$, so that

$$\text{Prob}\left(A_1\right) \geqslant 1 - \frac{2}{n}.$$

Assuming that A_1 occurs, during the second step there are at least $k(n-1)/2$ edges (because there are $n-1$ vertices), so the probability of picking an edge of C is at most $2/(n-1)$; hence

$$\text{Prob}\left(A_2 \mid A_1\right) \geqslant 1 - \frac{2}{n-1}.$$

At the ith step, the number of remaining vertices is $n - (i-1)$. The size of a min-cut is still k, so the graph has at least $k(n-i+1)/2$ edges remaining at this step. Thus

$$\text{Prob}\left(A_i \; \middle| \; \bigcap_{j=1}^{i-1} A_j\right) \geqslant 1 - \frac{2}{n-i+1}.$$

By the definition of conditional probability,

$$\text{Prob}\left(\bigcap_{j=1}^{n-2} A_j\right) = \text{Prob}\left(A_1\right) \cdot \text{Prob}\left(A_2 \mid A_1\right) \cdots \text{Prob}\left(A_i \; \middle| \; \bigcap_{j=1}^{n-2} A_j\right),$$

and hence, the algorithm outputs the min-cut C with probability at least

$$\prod_{i=1}^{n-2} \left(1 - \frac{2}{n-i+1}\right) = \prod_{j=n}^{3} \frac{j-2}{j} = \frac{2}{n(n-1)}.$$

Thus, any specific min-cut of G is output by the algorithm with probability larger than $2/n^2$. To reduce the error probability we can, for example, repeat the algorithm $n^2/2$ steps. The error probability (i.e., the probability that the algorithm rejects the specific min-cut C each time) is at most $\left(1 - 2/n^2\right)^{n^2/2} \leqslant e^{-1} < 0.37$. Further repetitions will make the error probability arbitrarily small.

Exercises

25.1$^+$ (Schwartz 1980). Prove the following version of Lemma 25.1. Let $f(x_1, \ldots, x_n)$ be a multivariate polynomial over a field $\mathbb{F}$ with the *degree sequence* $(d_1, \ldots, d_n)$, which is defined as follows: let d_1 be the maximum exponent of x_1 in f, and let $f_1(x_2, \ldots, x_n)$ be the coefficient of $x_1^{d_1}$ in f; then, let d_2 be the maximum exponent of x_2 in f_1, and $f_2(x_3, \ldots, x_n)$ be the coefficient of $x_2^{d_2}$ in f_1; and so on. Suppose that f is not the zero polynomial, and let $S_1, \ldots, S_n \subseteq \mathbb{F}$ be arbitrary subsets. For $\mathbf{r}_i \in S_i$ chosen independently and uniformly at random, show that

$$\mathrm{Prob}\left(f(\mathbf{r}_1, \ldots, \mathbf{r}_n) = 0\right) \leqslant \frac{d_1}{|S_1|} + \frac{d_2}{|S_2|} + \cdots + \frac{d_n}{|S_n|}.$$

25.2. Show that the bound of Lemma 25.2 is best possible.

Hint: Consider the polynomial $f(x_1, \ldots, x_n) = \prod_{i=1}^{d}(x_1 - i) \cdots \prod_{i=1}^{d}(x_n - i)$.

25.3. Verify that the contraction of edges in Karger's algorithm does not reduce the min-cut size.

Hint: Show that every cut in the graph at any intermediate stage is also a cut in the original graph: contracting an edge $e = \{u, v\}$ simply restricts attention to cuts of the original graph in which u and v are on the same side.

25.4.$^-$ Let G be a graph which consists of a cycle on n vertices. Show that Karger's algorithm produces each of the min-cuts in G with probability $\binom{n}{2}^{-1}$. How many min-cuts does G have?

25.5 (R. Freivalds 1977). Suppose that somebody gives us three $n \times n$ matrices A, B, C with real entries and claims that $C = A \cdot B$. We are too busy to verify this claim exactly and do the following. We take a random vector $\mathbf{r}$ of length n whose entries are integers chosen uniformly from the interval $\{0, 1, \ldots, N - 1\}$, and check if $A \cdot (B \cdot \mathbf{r}) = C \cdot \mathbf{r}$. If this is true we accept the claim, otherwise we reject it. How large must N be set to make the probability of false acceptance smaller than $1/100$?

Hint: Consider the matrix $X = A \cdot B - C$. If $C \neq A \cdot B$ then X has a row $x \neq 0$. Take a scalar product of this row with the random vector $\mathbf{r}$, observe that $\mathrm{Prob}\,(X \cdot \mathbf{r} = 0) \leqslant \mathrm{Prob}\,(x \cdot \mathbf{r} = 0)$, and apply Lemma 25.2.

26. Derandomization

Given a randomized algorithm $\mathcal{A}$, a natural approach towards *derandomizing* it is to find a method for searching the associated sample space Ω for a good point ω with respect to a given input instance x; a point is good for x if $\mathcal{A}(x, \omega) = f(x)$. Given such a point ω, the algorithm $\mathcal{A}(x, \omega)$ becomes a deterministic algorithm and it is guaranteed to find the correct solution. The problem faced in searching the sample space is that it is usually exponential in size.

Two general methods for searching the sample space have emerged in recent years. One of them – the *method of conditional probabilities* – starts with a "trivial" (in most situations, uniform) sample space Ω whose size is usually exponential, and the idea is to make some non-trivial search procedure within it. The other method – the *method of small sample spaces* – first tries to design some "small" sample space Ω' (say, of polynomial size), and then performs an exhaustive search of it. In this chapter we will shortly discuss both these approaches.

26.1 The method of conditional probabilities

The aim of this method is to convert probabilistic proofs of *existence* of combinatorial structures into efficient deterministic algorithms for their actual *construction*. The idea is to perform a binary search of the sample space Ω for a good point. At each step, the current sample space is split into two equal halves and the *conditional probability* of obtaining a good point is computed for each half. The search is then restricted to the half where the conditional probability is higher. The search terminates when only one sample point (which must be good) remains. This method is applicable to large sample spaces Ω since it requires only $\log_2 |\Omega|$ steps. In situations, where the corresponding conditional probabilities can be effectively computed (or at least approximated) this approach works pretty well. To explain the idea, let us consider the following problem.

Given a 3-CNF formula $F(x_1, \ldots, x_n)$, we want to find an assignment of values 0 or 1 to $x_1, \ldots, x_n$ satisfying as many clauses as possible. If we assign to each variable the value 0 or 1 at random independently and with equal probability, then we may expect that at least 7/8 fraction of clauses will be satisfied, just because each clause is satisfied with probability $1 - 2^{-3} = 7/8$ (see Proposition 26.2).

But where is the assignment? The argument above guarantees only the *existence* of such an assignment and gives no idea about how to *find* it. An exhaustive search will always lead us to the desired assignment. But this dummy strategy will require exponential (in n) number of steps. Can we do better? It appears that we can "derandomize" the probabilistic *proof* of existence so that it leads to a *deterministic* algorithm which is only polynomial in the length of the input formula.

Before we turn to a formal description of the method, let us first try to solve our special problem with the help of a chimpanzee (this beautiful explanation is due to Maurice Cochand).

We build a binary tree whose 2^n leaves correspond to the 2^n possible assignments. Leaves are close to the sky, as they should be. Going up to the left branch at level i corresponds to choose the value 0 for x_i, going up to the right gives x_i the value 1.

In oder to motivate the chimpanzee for this fascinating problem, we attach at every leaf of the tree a black box containing a number of bananas equal to the number of clauses satisfied by the assignment corresponding to that leaf. We do then invite the chimpanzee to go up in oder to bring down one of the black boxes, making him most clear the potential benefit of the operation.

We repeat this experiment many times, with many different trees corresponding to as many formulas F, having different number of variables and clauses. The chimpanzee *newer* looked at the list of clauses (although he was allowed to do it), did not even care about the number of variables. He moved up quickly along the tree, and *always* brought back a box having a number of bananas at least equal to $7/8$ times the number of clauses!

We asked him for his secret (he definitely had one, this was more than luck!). For a number of bananas we do not dare to mentioned here, he gave the following answer:

"Embracingly simple," he said. "At every junction I do the same: because of the weight, the branch supporting the subtree having the biggest number of bananas is not as steep as the other one, there I go!"

26.1.1 A general frame

Suppose we have a sample space Ω, and assume, for simplicity, that it is symmetric (i.e., each point has probability $2^{-|\Omega|}$) and that $\Omega = \{0,1\}^n$. Let $A_1, \ldots, A_m$ be a collection of events, and consider the random variable $X = X_1 + \cdots + X_m$ where X_i is the indicator random variable for A_i. Hence, $\mathrm{E}[X] = \sum_{i=1}^{m} \mathrm{Prob}(A_i)$. Also suppose that we have a *proof* that $\mathrm{E}[X] \geqslant k$. So, there is a point $(\epsilon_1, \ldots, \epsilon_n)$ in the sample space in which at least k of the events hold. Our objective is to find such a point *deterministically*.

Introduce n random variables $Y_1, \ldots, Y_n$ where each Y_i takes value 0 or 1 independently with equal probability. We find the bits $\epsilon_1, \epsilon_2, \ldots$ sequentially as follows. Assume $\epsilon_1, \ldots, \epsilon_{j-1}$ have already been fixed. Our goal is to choose ϵ_j. We make this choice based on the value of "conditional" expectation

$$\mathrm{E}\left[X \,|\epsilon_1, \ldots, \epsilon_j\right] = \sum_{i=1}^{m} \mathrm{Prob}\left(A_i \mid \epsilon_1, \ldots, \epsilon_j\right),$$

where here and in what follows "$\epsilon_1, \ldots, \epsilon_j$" stands for the event that $Y_1 = \epsilon_1, \ldots, Y_j = \epsilon_j$. By Adam's Theorem (Exercise 17.1), for each choice of $\epsilon_1, \ldots, \epsilon_j$ and for each event A_i, the conditional probability

$$\mathrm{Prob}\left(A_i \mid \epsilon_1, \ldots, \epsilon_j\right)$$

of the event A_i given the values $Y_1 = \epsilon_1, \ldots, Y_j = \epsilon_j$ is obviously the average

$$\frac{\mathrm{Prob}\left(A_i \mid \epsilon_1, \ldots, \epsilon_j, 0\right) + \mathrm{Prob}\left(A_i \mid \epsilon_1, \ldots, \epsilon_j, 1\right)}{2}.$$

of the two conditional probabilities corresponding to the two possible choices for Y_{j+1}. Consequently,

$$\mathrm{E}\left[X \,|\epsilon_1, \ldots, \epsilon_j\right] = \frac{\mathrm{E}\left[X \,|\epsilon_1, \ldots, \epsilon_j, 0\right] + \mathrm{E}\left[X \,|\epsilon_1, \ldots, \epsilon_j, 1\right]}{2}$$
$$\leqslant \max\left\{\mathrm{E}\left[X \,|\epsilon_1, \ldots, \epsilon_j, 0\right], \mathrm{E}\left[X \,|\epsilon_1, \ldots, \epsilon_j, 1\right]\right\}.$$

Therefore, if the values ϵ_{j+1} are chosen, each one in its turn, so as to maximize the value of $\mathrm{E}\left[X \,|\epsilon_1, \ldots, \epsilon_{j+1}\right]$, then this value cannot decrease. Since this value is k at the beginning, it follows that it is at least k at the end. But at the end, each ϵ_i is fixed, and hence the value of $\mathrm{E}\left[X \,|\epsilon_1, \ldots, \epsilon_n\right]$ is precisely the number of events that hold at the point $(\epsilon_1, \ldots, \epsilon_n)$, showing that our procedure works.

Note that the procedure above is efficient provided n is not too large (as is usually the case in combinatorial examples) and, more importantly, provided the conditional probabilities $\mathrm{Prob}\left(A_i \mid \epsilon_1, \ldots, \epsilon_j\right)$ can be computed efficiently.

To see how the "chimpanzee algorithm" from the previous section fits in this general frame, just observe that the total weight of the bananas in the subtree reached by the chimpanzee after j moves $\epsilon_1, \ldots, \epsilon_j$ is the number of clauses X in F, times the conditional expectation $\mathrm{E}\left[X \,|\epsilon_1, \ldots, \epsilon_j\right]$.

26.1.2 Splitting graphs

A widely applied remark of Paul Erdős is that a graph with m edges always contains a bipartite subgraph of at least $m/2$ edges. This fact has a quick probabilistic proof.

Theorem 26.1 (Erdős 1965c). *Every graph with m edges always contains a bipartite subgraph of at least $m/2$ edges.*

Proof. Let $G = (V, E)$ with the vertex set $V = \{1, \ldots, n\}$. Take a random subset $\mathbf{U} \subseteq V$ given by $\mathrm{Prob}\,(i \in \mathbf{U}) = 1/2$, these probabilities being mutually independent. Call an edge $e = \{i, j\}$ *crossing* if exactly one of i, j is in $\mathbf{U}$. Let X be the number of crossing edges. Then $X = \sum_{e \in E} X_e$, where X_e is the indicator random variable for the edge e being crossing. For a given edge e, $\mathrm{E}\,[X_e] = 1/2$ as two fair coin flips have probability $1/2$ of being different. By linearity of expectation,

$$\mathrm{E}\,[X] = \sum_{e \in E} \mathrm{E}\,[X_e] = \frac{|E|}{2}.$$

Thus, $X \geqslant |E|/2$ for some choice of $\mathbf{U}$, and the set of those (corresponding to this particular $\mathbf{U}$) edges forms the desired bipartite subgraph. $\square$

The proof of this theorem gives us a *randomized* algorithm to find a bipartite subgraph whose expected number of edges is at least $|E|/2$. Moreover, Luby (1986) has shown how it can be converted to a linear time *deterministic* algorithm.

We use the conditional expectations to derandomize the algorithm. Introduce n random variables $Y_1, \ldots, Y_n$ where $Y_i = 1$ if $i \in \mathbf{U}$, and $Y_i = 0$, otherwise. Select an $\epsilon_1 \in \{0, 1\}$ such that $\mathrm{E}\,[X \,|\, Y_1 = \epsilon_1] \geqslant \mathrm{E}\,[X \,|\, Y_1 = \epsilon_1 \oplus 1]$, and set $Y_1 = \epsilon_1$. Repeat this process for all Y_i's. At the end we have an assignment for all Y_i's such that $\mathrm{E}\,[X \,|\, Y_1 = \epsilon_1, \ldots, Y_n = \epsilon_n] \geqslant |E|/2$. But X is no longer a random variable at this point (since $\mathbf{U}$ is completely defined), so $X \geqslant |E|/2$.

What is the running time? To determine each ϵ_i, we need to count the number of edges between vertices in current U and vertex i, the number of edges between vertices in current $V - U$ and vertex i. If the former is smaller than the latter, we set $\epsilon_i = 1$; otherwise, we set $\epsilon_i = 0$. This means that we only need to check the edges incident to vertex i to determine ϵ_i. So the running time of this algorithm is $O(n + |E|)$.

26.1.3 Maximum satisfiability: the algorithmic aspect

Recall that an *And-Or formula* (or simply, a *formula*) over a set of variables $x_1, \ldots, x_n$ is an And $F = \bigwedge_{j=1}^m C_j$ of an arbitrary number of *clauses*, where a clause C_j is an Or of an arbitrary number of *literals*, each atom being either a variable x_i or a negated variable $\overline{x}_i$. An *assignment* is a mapping which assigns each variable one of the values 0 or 1. An assignment satisfies a clause if it satisfies at least one of its atoms.

Given such a formula $F = \bigwedge_{j=1}^m C_j$, our goal is to find an assignment which satisfies as many of the clauses as possible. We already know (see Exercise 20.9) that we can always satisfy at least half of clauses. In Sect. 20.6 we have proved that we can even satisfy a 2/3 fraction of the clauses, if the formula F is 3-satisfiable, i.e., if any three clauses of F are satisfiable. We can also prove a similar result in terms of the length of clauses.

Proposition 26.2. *For any formula $F = C_1 \wedge C_2 \wedge \cdots \wedge C_m$ there is an assignment that satisfies at least $(1 - 2^{-l})$ fraction of its clauses, where l is the minimum length of a clause in F.*

Proof. Assign each variable x_i the value 0 or 1 independently at random with probability $1/2$. Hence, our sample space is the set of all assignments $w \in \{0,1\}^n$ of boolean variables $x_1, \ldots, x_n$ with a uniform distribution $\text{Prob}(\mathbf{w} = w) = 2^{-n}$.

Let ℓ_j denote the length of (i.e., the number of literals in) the clause C_j, $j = 1, \ldots, m$, and let Z_j be the indicator random variable for the event that the clause C_j is satisfied. Then

$$\text{E}[Z_j] = \text{Prob}(C_j(\mathbf{w}) = 1) = 1 - \text{Prob}(C_j(\mathbf{w}) = 0) = 1 - 2^{-\ell_j}.$$

Let Z be the total number of satisfied clauses. Then $Z = \sum_{j=1}^{m} Z_j$ and, by the linearity of the expectation,

$$\text{E}[Z] = \sum_{j=1}^{m} \text{E}[Z_j] = \sum_{j=1}^{m} (1 - 2^{-\ell_j}) \geqslant m(1 - 2^{-l}),$$

as desired. $\qquad\square$

Let us now show how to construct the desired assignment $x_1 = w_1, \ldots,$ $x_n = w_n$ via an appropriate derandomization of that proof. The resulting algorithm is essentially a probabilistic version of the algorithm due to Johnson (1974).

Suppose that $w_1, \ldots, w_i$ are already defined, and denote:

$$e_i \rightleftharpoons \text{E}[Z \,|\, x_1 = w_1, \ldots, x_i = w_i] \,;$$
$$e_{i0} \rightleftharpoons \text{E}[Z \,|\, x_1 = w_1, \ldots, x_i = w_i, x_{i+1} = 0] \,;$$
$$e_{i1} \rightleftharpoons \text{E}[Z \,|\, x_1 = w_1, \ldots, x_i = w_i, x_{i+1} = 1] \,.$$

Hence $e_0 = \text{E}[Z]$ and, as above, $e_i = (e_{i0} + e_{i1})/2$, which implies that $e_i \leqslant \max\{e_{i0}, e_{i1}\}$. Therefore, we take in the $(i+1)$-st step such a value w_{i+1} for x_{i+1}, which does not decrease the conditional expectation: $e_{i+1} = e_{iw_{i+1}} \geqslant e_i$. After n steps we will find an assignment w which satisfies at least

$$\text{E}[Z \,|\, x_1 = w_1, \ldots, x_n = w_n] \geqslant e_0 = \text{E}[Z] \geqslant m(1 - 2^{-l})$$

clauses.

Again, the question remains: what is the running time of this algorithm? To estimate it, let us observe that, in fact, there is no need to compute conditional expectations e_{i0} and e_{i1} explicitly. We set $x_{i+1} = 1$ precisely when $e_{i1} - e_i \geqslant 0$. The assignment to x_{i+1} affects only the clauses containing either this variable or its negation. Each of the clauses C_j containing x_{i+1} will be satisfied, which increases e_i by $1 - (1 - 2^{-\ell_j}) = 2^{-\ell_j}$ (the probability that C_j could be falsified later). All the clauses C_j containing $\overline{x}_{i+1}$ will be shortened by one literal, which decreases their probability to be satisfied from $1 - 2^{-\ell_j}$ to $1 - 2^{-(\ell_j - 1)}$ – hence, again by $2^{-\ell_j}$. It is therefore enough to compare these effects:

$$e_{i1} - e_i = \sum_{C_j \in \mathcal{C}: C_j \ni x_{i+1}} 2^{-\ell_j} - \sum_{C_j \in \mathcal{C}: C_j \ni \overline{x}_{i+1}} 2^{-\ell_j},$$

where $\mathcal{C}$ is the set of not yet satisfied clauses and ℓ_j is the actual length of the jth clause C_j (from which some literals could be already eliminated via setting them to 0). The algorithm now is easy: set x_{i+1} to 0 or 1 according to whether the difference above is negative or not. If the queries "$C_j \ni x_i$" may be realized in constant time, then the overall running time of this algorithm is $O(nm)$.

26.2 The method of small sample spaces

Let f be some function; for simplicity assume that its domain is the n-cube $\{0,1\}^n$. A randomized algorithm $\mathcal{A}$ for f has a probability space (Ω, P) associated with it, where Ω is the sample space and P is some probability measure. Each point $\omega \in \Omega$ corresponds to a sequence of coin flips performed by the algorithm; hence we assume that each sample point appears with uniform probability $2^{-|\Omega|}$. A point ω *good* for some input instance x, if $\mathcal{A}(x, \omega)$ computes the correct solution, i.e., if $\mathcal{A}(x, \omega) = f(x)$. The algorithm computes f if

$$\text{Prob}\left(\mathcal{A}(x, \omega) = f(x)\right) \geq 1/2$$

for each input $x \in \{0,1\}^n$.

As we already mentioned above, the problem with randomized algorithms is that usually their sample spaces are of exponential size, so that an exhaustive search is impossible. It appears however, that a (uniform) sample space Ω associated with a randomized algorithm *always* contains a polynomial-sized subspace $S \subseteq \Omega$ which still has a good point for each possible input, i.e., for every input x there is a point $\omega \in S$ such that $\mathcal{A}(x, \omega) = f(x)$.

Theorem 26.3 (Adleman 1978). *There exists a set $S \subseteq \Omega$ of size $|S| \leq n$ such that for every input $x \in \{0,1\}^n$ there is at least one good point in S.*

Proof. Let $M = (m_{x,\omega})$ be a $2^n \times |\Omega|$ 0-1 matrix whose rows are labeled by inputs and columns by points in the sample space, such that $m_{x,\omega} = 1$ if ω is a good point for x, and $m_{x,\omega} = 0$, otherwise.

The desired set $S \subseteq \Omega$ is constructed iteratively. Initially, S is empty. Since $\text{Prob}\left(\mathcal{A}(x, \omega) = f(x)\right) \geq 1/2$, each row of M has at least $|\Omega|/2$ ones. Thus, for at least one column ω in M, at least half of its entries must be 1's, and hence, there exists a point $\omega \in \Omega$ that is good for at least half of the inputs x. We add this point ω to S, delete all the rows in M corresponding to inputs for which ω is good, and repeat the argument for the resulting submatrix. After at most $\log_2 2^n = n$ iterations there will be no rows, and the obtained set S of points has the desired properties. □

Unfortunately, the above result is highly non-constructive and it cannot be used to actually derandomize algorithms. This difficulty was overcome in certain cases by constructing a (different) polynomial-sized sample spaces.

One way of constructing the new sample space is by showing that the probabilistic choices of the randomized algorithm are only required to be k-wise independent; then a sample space of size $O(n^k)$ suffices. Another way is to consider "small bias" probability spaces, i.e., to construct small probability spaces that "behave similarly" to larger probability spaces in certain senses.

Let $X_1, \ldots, X_n$ be random variables taking their values in a finite set S. These variables are k-*wise independent* if any k of them are mutually independent, i.e., if for every sequence $(s_{i_1}, \ldots, s_{i_k})$ of k values $s_{i_j} \in S$,

$$\text{Prob}\,(X_{i_1} = s_{i_1}, \ldots, X_{i_k} = s_{i_k}) = \prod_{j=1}^{k} \text{Prob}\,(X_{i_j} = s_{i_j}).$$

To illustrate how k-wise independence can help us to derandomize probabilistic proofs, let us look more carefully at the proof of Theorem 26.1. This theorem states that every graph $G = (V, E)$ contains a bipartite subgraph of at least $|E|/2$ edges. That is, there is a subset $U \subseteq V$ of vertices such that at least one half of the edges in E join vertices from U with those from $V - U$. We used random variables to produce the desired (random) subset $\mathbf{U}$. Namely, for each vertex $i \in V$, we flip a coin to decide whether to include this vertex into the set $\mathbf{U}$ or not. This requires $n = |V|$ coin flips, and hence, the whole sample space is huge – it has 2^n points.

A closer look at this proof shows that the independence is used only to conclude that, for any two vertices $i \neq j$, the events $i \in \mathbf{U}$ and $j \in \mathbf{U}$ are independent. We need this independence to show that

$$\text{Prob}\,(i \in \mathbf{U}, j \notin \mathbf{U}) = \text{Prob}\,(i \in \mathbf{U}) \cdot \text{Prob}\,(j \notin \mathbf{U}) = 1/4.$$

So, in this proof 2-wise independence of the indicator random variables X_i for the events "$i \in \mathbf{U}$" suffices. This observation allows us to substantially reduce the size of a sample space as follows (see Exercise 26.3 for more direct construction).

Look at our random set of vertices $\mathbf{U} \subseteq V$ as a sequence of random colorings $X_1, \ldots, X_n : V \to \{0, 1\}$, where $X_i = 1$ iff $i \in \mathbf{U}$. Our goal is to construct as small as possible sample space for these colorings, in which they are pairwise independent.

Suppose for simplicity that $n = 2^d$ for some d, and identify the vertices with the elements of the field $\mathbb{F}_n$. Choose two elements a and b of this field randomly and independently, and define for each element i the random variable $Z_i = a \cdot i + b$.

Claim 26.4. $Z_1, \ldots, Z_n$ are 2-*wise independent.*

Proof.

$$\text{Prob}\,(Z_i = x, Z_j = y) = \text{Prob}\,(ai + b = x, aj + b = y)$$
$$= \text{Prob}\left(a = \frac{x-y}{i-j}, b = \frac{yi - xj}{i-j}\right) = \frac{1}{n^2}$$
$$= \text{Prob}\,(Z_i = x) \cdot \text{Prob}\,(Z_j = y).$$

□

Encode the elements of $\mathbb{F}_n$ by binary strings of length d, and let X_i be the first bit of the code of ith element. By the claim, random variables X_i are also 2-wise independent and uniform on $\{0, 1\}$. Now color the vertex i by the color X_i. Each coloring so obtained is defined by the pair (a, b) of elements from our field $\mathbb{F}_n$. Thus, the whole sample space has size only n^2, and we can perform an exhaustive search of it to find the desired coloring.

Another example is Theorem 18.1 from Chap. 18 saying that the edges of K_n can be colored in two colors so that we get no monochromatic $K_{2\log n}$. In this proof, a variable X_i gives the color of the ith edge. Their independence is used only to estimate the probability that all the edges of some fixed clique on $2\log n$ vertices, receive the same color. Hence, once again, the k-wise independence with $k = \binom{2\log n}{2} = O((\log n)^2)$ is sufficient. The reader is encouraged to convince himself/herself that in most of the previous proofs k-wise independence with $k \ll n$ works.

Now suppose that for some probabilistic proof, k-wise independence is enough. One may expect that then a sample space Ω of a size much smaller than 2^n would suffice. How much?

In combinatorial terms, we are looking for a set $\Omega \subseteq \{0,1\}^n$ with the following property: for every set of k coordinates, each vector from $\{0,1\}^k$ is a projection (onto these k coordinates) *of one and the same* number of vectors in Ω. Thus, if we let $X = (X_1, \ldots, X_n)$ be a string chosen uniformly from Ω then, for any k indices $i_1 < i_2 < \cdots < i_k$ and any k-bit string $\alpha \in \{0,1\}^k$,

$$\text{Prob}\,((X_{i_1}, X_{i_2}, \ldots, X_{i_k}) = \alpha) = 2^{-k},$$

i.e., the coordinates X_i are k-wise independent.

Taking the duals of binary BCH codes, it is possible, for every fixed k, to construct a k-wise independent sample space Ω of size $|\Omega| = O(n^{\lfloor k/2 \rfloor})$. The construction can be found, for example, in the book of Alon and Spencer (1992); see also Exercise 14.3. The idea is to show that the dual code is not only (n, k)-universal (as we have proved in Sect. 14.4.1) but is such in a very strong sense: for every set of k coordinates, every 0-1 vector of length k is a projection of one and the same number of code words onto these coordinates.

It is natural to ask if this construction is optimal. It appears that, indeed, the bound $n^{\lfloor k/2 \rfloor}$ cannot be improved, up to a constant factor (depending on k). Say that a random variable is *almost constant* if it takes a single value with probability 1. Let $m(n, k)$ denote the following sum of binomial coefficients:

$$m(n,k) = \sum_{i=0}^{k/2} \binom{n}{i} \qquad \text{if } k \text{ is even,}$$

and

$$m(n,k) = \sum_{i=0}^{(k-1)/2} \binom{n}{i} + \binom{n-1}{(k-1)/2} \qquad \text{if } k \text{ is odd.}$$

Observe that for every fixed k, $m(n,k) = \Omega(n^{\lfloor k/2 \rfloor})$.

Theorem 26.5 (Alon–Babai–Itai 1986). *Assume that the random variables $X_1, \ldots, X_n$ over a sample space Ω are k-wise independent and none of them is almost constant. Then $|\Omega| \geqslant m(n,k)$.*

Note that we assume neither that the variables X_i are $(0,1)$-variables nor that Ω is a symmetric space.

Proof. We can assume that the expected value of each X_i is 0 (since otherwise we can replace X_i by $X_i - \mathrm{E}[X_i]$). For a subset $S \subseteq \{1, \ldots, n\}$, define

$$\alpha_S = \prod_{i \in S} X_i.$$

Since no X_i is almost constant and since the variables are k-wise independent,

$$\mathrm{E}[\alpha_S \alpha_S] = \prod_{i \in S} \mathrm{E}[X_i^2] > 0 \qquad (26.1)$$

for all S satisfying $|S| \leqslant k$. Similarly (and since $\mathrm{E}[X_i] = 0$), for all $S \neq T$ satisfying $|S \cup T| \leqslant k$, we have

$$\mathrm{E}[\alpha_S \cdot \alpha_T] = \prod_{i \in S \cap T} \mathrm{E}[X_i^2] \cdot \prod_{i \in (S \cup T) - (S \cap T)} \mathrm{E}[X_i] = 0. \qquad (26.2)$$

Now let $S_1, \ldots, S_m$ be all the subsets of $\{1, \ldots, n\}$ such that the union of each two is of size at most k. Then $m = m(n,k)$. (Take all sets of size at most $k/2$, and if k is odd add all the subsets of size $(k+1)/2$ containing 1.)

To complete the proof, we show that the functions $\alpha_{S_1}, \ldots, \alpha_{S_m}$ (considered as real vectors of length $|\Omega|$) are linearly independent. Since their number m cannot then exceed the dimension $|\Omega|$, this will imply the result.

To prove the linear independence, take a linear combination $\sum_{i=1}^m \lambda_i \alpha_{S_i}$. Then for every j, multiplying by α_{S_j} and computing expected values we obtain, by (26.2),

$$0 = \sum_{i=1}^m \lambda_i \mathrm{E}[\alpha_{S_i} \cdot \alpha_{S_j}] = \lambda_j \mathrm{E}[\alpha_{S_j} \cdot \alpha_{S_j}].$$

By (26.1), this implies that $\lambda_j = 0$ for all j, and the required linear independence follows. □

26.3 Sum-free sets: the algorithmic aspect

In previous sections we considered two general approaches toward derandomizing of probabilistic proofs. In this section we will give one example to demonstrate that sometimes the desired polynomial-time algorithm is hidden in the existence proof *itself*.

A subset B of an additive group is called *sum-free* if $x + y \notin B$ for all $x, y \in B$. Erdős (1965) and Alon and Kleitman (1990) have proved that every set $A = \{a_1, \ldots, a_N\}$ of integers has a sum-free subset B, with $|B| > N/3$. The proof is probabilistic (see Theorem 20.2) and the question was whether there exists a deterministic algorithm for the selection of such a subset B, which runs in time polynomial in the (binary) size of the problem, that is in $\ell = \sum_{i=1}^{N} \log_2 |a_i|$.

Kolountzakis (1994) has shown that, with a slight modification, the proof of Theorem 20.2 can be transformed to such an algorithm.

For a prime p let (as before) $\mathbb{Z}_p = \{0, 1, \ldots, p-1\}$ be the field of the integers mod p, and let $\mathbb{Z}_p^\times = \{1, \ldots, p-1\}$ be the corresponding multiplicative group in $\mathbb{Z}_p$.

Theorem 26.6 (Kolountzakis 1994). *Let $p = 3k+2$ be a prime number and $w(x)$ a nonnegative function defined on $\mathbb{Z}_p^\times$. Define $W = \sum_{x \in \mathbb{Z}_p^\times} w(x)$ and assume $W > 0$. Then there is a sum-free subset E of $\mathbb{Z}_p^\times$ for which*

$$\sum_{x \in E} w(x) > \frac{1}{3} W. \tag{26.3}$$

Proof. Write $S = \{k+1, k+2, \ldots, 2k+1\}$, and observe that S is a sum-free subset in $\mathbb{Z}_p$ and $|S| > (p-1)/3$. Let the random variable $\mathbf{t}$ be uniformly distributed in $\mathbb{Z}_p^\times$, and write $f(\mathbf{t}) = \sum w(x)$, where the sum is over all x for which $x \cdot \mathbf{t} \in S$, and the product $x \cdot \mathbf{t}$ is computed in $\mathbb{Z}_p$. Since $\mathbb{Z}_p^\times$ is a multiplicative group, we have

$$\mathrm{E}\left[f(\mathbf{t})\right] = W \cdot (|S|/(p-1) > W/3.$$

By the pigeonhole property of the expectation, there is some $t \in \mathbb{Z}_p^\times$ for which $f(t) > W/3$. Define $E = t^{-1}S$. This set is sum-free and (26.3) true for it. $\square$

We now turn this proof into an algorithm. Given a set $A = \{a_1, \ldots, a_N\}$ of integers of (binary) size $\ell = \sum_{i=1}^{N} \log_2 |a_i|$, our goal is to find a sum-free subset B, with $|B| > N/3$, in time polynomial in ℓ. We assume that ℓ is large.

First, observe that the number of prime factors of an integer x is at most $\log_2 x$. This means that the number of prime factors which appear in the factorization of *any* element of A is at most ℓ. The Prime Number Theorem says that for every pair b, c of relatively prime positive integers, the number of primes $p \leqslant x$ such that p is of the form $p = bk + c$, asymptotically equals to $x/(\varphi(b) \cdot \ln x)$, where $\varphi(b) = |\{y \in \mathbb{Z}_b : \gcd(y, b) = 1\}|$ is the Euler totient function. In our case $b = 3$ and $c = 2$; hence, $\varphi(b) = 2$. Thus, there is a prime p of the form $p = 3k + 2$, not greater than $3\ell \log_2 \ell$, which does not divide any member of A.

Define now

$$w(x) = |\{t \in A : t = x \bmod p\}| .$$

Since p does not divide any member of A, we have $W = N$ and, using Theorem 26.6, we can find a sum-free subset $E \subseteq \mathbb{Z}_p^\times$ for which the set

$$B = \{t \in A : t \bmod p \in E\}$$

has more than $N/3$ elements. This set B is sum-free since $x + y = z$ for some $x, y, z \in B$ would imply $x + y = z \bmod p$ and E would not be sum-free.

In summary, the steps of our algorithm are the following.

1. Compute all primes up to $3\ell \log_2 \ell$.
2. Find a prime $p = 3k + 2$ which divides no element of A.
3. Compute the values $w(x)$ for all $x \in \mathbb{Z}_p^\times$.
4. Find by exhaustive search a $t \in \mathbb{Z}_p^\times$ for which $f(t) > N/3$ (Theorem 26.6 guarantees that such t exists) and compute the set $E = t^{-1}S$.
5. Construct the set $B = \{t \in A : t \bmod p \in E\}$.

It is easy to verify (do this!) that all these steps can be carried out in time polynomial in ℓ.

Exercises

26.1.[1] Use the method of conditional probabilities to derandomize the proof of Theorem 18.1 and Theorem 18.2.

26.2.[-] Let $G = (V, E)$ be a graph with $n = 2m$ vertices. Improve the lower bound $|E|/2$ on the size of a cut in G (proved in Theorem 26.1) to $|E| \geqslant m/(2m - 1)$.

Hint: Follow the argument of Theorem 26.1 with another probability space: choose $\mathbf{U} \subseteq V$ uniformly from among all m-element subsets of V. Observe that then any edge has probability $m/(2m - 1)$ of being crossing.

26.3.[-] Let $\mathbf{r}$ be a random vector uniformly distributed in $\mathbb{F}_2^d$. With each vector $a \in \mathbb{F}_2^d$ associate a random variable $X_a = \langle a, \mathbf{r} \rangle$ whose value is the scalar product over $\mathbb{F}_2$ of this vector with $\mathbf{r}$. Show that these random variables are 2-wise independent. Hint: Exercise 17.2.

26.4. Let $\log m = o(\sqrt{n})$, $m > 4$, and let H be an $m \times n$ 0-1 matrix, the average density of (i.e., the average number of 1's in) each row of which does not exceed p, $0 \leqslant p < 1$. Show that then, for every constant $\delta > 0$, there is an $m \times t$ submatrix H' of H such that $t = O(\log m/\delta^2)$ and each row of H' has average density at most $p + \delta$.

Sketch: Let ξ be a random variable uniformly distributed in $\{1,\ldots,n\}$ and let $\xi_1,\ldots,\xi_t$ be its independent copies, $t = \lceil 4p\log m/\delta^2 \rceil$. First, observe that with probability strictly larger than $1/2$ all the selected columns $\xi_1,\ldots,\xi_t$ are distinct. Next, fix a row $x_1,\ldots,x_n$ of H, and consider the 0-1 random variables $X_i = x_{\xi_i}$, for $i = 1,\ldots,t$. Observe that $\mathrm{Prob}\,(X_i = 1) \leqslant p$ and apply the Chernoff inequality (17.6) to show that the average density $(\sum X_i)/t$ of 1's in the selected columns can exceed $p + \delta$ with probability at most $1/m^2$. Since we have only m rows, with probability at least $1 - 1/m > 1/2$, all the rows of the selected submatrix will have average density at most $p + \delta$.

26.5. Let $f(x)$ be a boolean function on n variables $x = (x_1,\ldots,x_n)$. Let $F(x,y)$ be a formula with an additional set of boolean variables $y = (y_1,\ldots,y_m)$. The *size* $|F|$ of the formula F is the number of literals in it. Let $\mathbf{y}$ be a random vector taking its values in $\{0,1\}^m$ independently and with equal probability 2^{-m}. Suppose that $F(x,y)$ computes f with (one-sided) *failure probability* p. That is, for every input $a \in \{0,1\}^n$, $\mathrm{Prob}\,(F(a,\mathbf{y}) \neq f(a))$ is zero if $f(a) = 0$, and is at most p if $f(a) = 1$.

(a) Use Adleman's theorem to show that, if $p \leqslant 1/2$, then f can be computed by a usual (deterministic) formula of size $O(n \cdot |F|)$.

(b) The formula $F(a,\mathbf{y})$ can be written in the form

$$F(a,\mathbf{y}) = \sum F(x,b) \cdot X_b,$$

where the sum is over all $b \in \{0,1\}^m$ and X_b is the indicator random variable for the event "$\mathbf{y} = b$." This formula uses m random bits (to chose a particular formula $F(x,b)$). Use Exercise 26.4 to essentially reduce this number of random bits until $O\left(\log(m/\delta^2)\right)$ at the cost of a slight increase of failure probability by δ. Namely, prove that there is a subset $B = \{b_1,\ldots,b_t\}$ of $t = O(m/\delta^2)$ vectors such that the formula

$$F'(x,\mathbf{z}) = \sum_{i=1}^{t} F(x,b_i) \cdot Y_i$$

computes the same boolean function f with failure probability at most $p + 1/4$; here $\mathbf{z}$ is a random variable taking its values in $\{1,\ldots,t\}$ independently and with equal probability, and Y_i is the indicator random variable for the event "$\mathbf{z} = i$."

Part V

Fragments of Ramsey Theory

27. Ramsey's Theorem

In 1930 Frank Plumpton Ramsey had written a paper *On a problem in formal logic* which initiated a part of discrete mathematics nowadays known as Ramsey Theory. At about the same time B.L. van der Waerden (1927) proved his famous Ramsey-type result on arithmetical progressions. A few years later Ramsey's theorem was rediscovered by P. Erdős and G. Szekeres (1935) while working on a problem in geometry. In 1963 A.W. Hales and R.I. Jewett revealed the combinatorial core of van der Waerden's theorem and proved a general result which turned this collection of separate ingenious results into Ramsey Theory. In the next three chapters we present some of the most basic facts of this theory.

27.1 Colorings and Ramsey numbers

The subject uniformly named *Ramsey Theory* deals with colorings of objects. The pigeonhole principle is the simplest "result" in this theory: in terms of colorings it states that, if $r + 1$ objects are colored with r different colors, then there must be two objects with the same color. In order to move towards Ramsey's Theorem, we quantify the result further as follows.

Proposition 27.1. *Suppose $n \geqslant rs - r + 1$. Let n objects be colored with r different colors. Then there exist s objects all with the same color. Moreover, the inequality is best possible.*

Proof. If the conclusion is false, then there are at most $s - 1$ objects of each color, hence at most $r(s - 1)$ altogether, contrary to assumption. To see that the result is best possible, observe that a set with at most $r(s - 1)$ points can be divided into r groups with at most $s - 1$ points in each group; hence none of the groups contains s points.   □

This is a "1-dimensional" Ramsey-type result. Things become more complicated if we move to larger dimensions, i.e., if we color *subsets* instead of points. Roughly put, the famous theorem of Ramsey (1930) says the following: for every coloring of the k-subsets of a sufficiently large set X, there is an s-element subset of X whose k-subsets all have the same color. This statement is quantified using so-called *Ramsey numbers*.

Definition 27.2. Let $r, k, s_1, \ldots, s_r$ be given positive integers, $s_1, \ldots, s_r \geqslant k$. Then $R_r(k; s_1, \ldots, s_r)$ denotes the smallest number n with the property that, if the k-subsets of an n-set are colored with r colors $1, \ldots, r$, then for some $i \in \{1, \ldots, r\}$, there is an s_i-set, all of whose k-subsets have color i. If $s_1 = s_2 = \ldots = s_r = s$ then this number is denoted by $R_r(k; s)$.

Thus, the pigeonhole principle states that $R_r(1; 2) = 1 + r$ and Proposition 27.1 that $R_r(1; s) = 1 + r(s - 1)$. These are 1-dimensional results since $k = 1$.

27.2 Ramsey's theorem for graphs

The 2-dimensional case ($k = 2$) corresponds to coloring the edges of a graph. Moreover, if we consider 2-colorings ($r = 2$) then the corresponding Ramsey number is denoted by $R(s, t)$, i.e., $R(s, t) \rightleftharpoons R_2(2; s, t)$.

To warm-up let us consider the following simple game. Mark six points on the paper, no three in line. There are two players; one has a Red pencil the other Blue. Each player's turn consists in drawing a line with his/her pencil between two of the points which haven't already been joined. (The crossing of lines is allowed). The player's goal is to create a triangle in his/her color. If you try to play it with a friend, you will notice that it always end in a win for one player: a draw is not possible. Is this really so? In terms of Ramsey numbers, we ask if $R(3, 3) \leqslant 6$: we have $r = 2$ colors, edges are k-sets with $k = 2$ and we are looking for a monochromatic 3-set. Prove that indeed, $R(3, 3) = 6$. (Hint: see Fig. 27.1.)

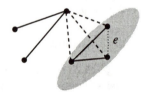

Fig. 27.1. What is the color of e?

You have just shown that the number $R(s, t)$ exists if $s = t = 3$. This is a very special case of the well-known version of Ramsey's theorem for graphs, which says that $R(s, t)$ exists for any natural numbers s and t.

Let $G = (V, E)$ be an undirected graph. A subset $S \subseteq V$ is a *clique* of G if any two vertices of S are adjacent. Similarly, a subset $T \subseteq V$ is an *independent set* of G if no two vertices of T are adjacent in G.

Theorem 27.3. *For any natural numbers s and t there exists a natural number $n = R(s, t)$ such that in any graph on n or more vertices, there exists either a clique of s vertices or an independent set of t vertices.*

Proof. To prove the existence of the desired number $n = R(s,t)$, it is sufficient to show, by induction on $s+t$, that $R(s,t)$ is bounded. For the base case, it is easy to verify that $R(1,t) = R(s,1) = 1$. For $s > 1$ and $t > 1$, let us prove that

$$R(s,t) \leqslant R(s,t-1) + R(s-1,t).$$ (27.1)

Let $G = (V,E)$ be a graph on $n = R(s,t-1) + R(s-1,t)$ vertices. Take an arbitrary vertex $x \in V$, and split $V - \{x\}$ into two subsets S and T, where each vertex of S is nonadjacent to x and each vertex of T is adjacent to x (see Fig. 27.2). Since

$$R(s,t-1) + R(s-1,t) = |S| + |T| + 1,$$

we have either $|S| \geqslant R(s,t-1)$ or $|T| \geqslant R(s-1,t)$.

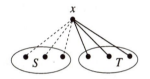

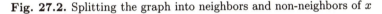

Fig. 27.2. Splitting the graph into neighbors and non-neighbors of x

Let $|S| \geqslant R(s,t-1)$, and consider the induced subgraph $G[S]$ of G: this is a graph on vertices S, in which two vertices are adjacent if and only if they are such in G. Since the graph $G[S]$ has at least $R(s,t-1)$ vertices, by the induction hypothesis, it contains either a clique on s vertices or an independent set of $t-1$ vertices. Moreover, we know that x is not adjacent to any vertex of S in G. By adding this vertex to S, we conclude that the subgraph $G[S \cup \{x\}]$ (and hence, the graph G itself) contains either a clique of s vertices or an independent set of t vertices.

The case when $|T| \geqslant R(s-1,t)$ is analogous. $\square$

The recurrence (27.1) implies (see Exercise 27.6)

$$R(s,t) \leqslant \binom{s+t-2}{s-1}.$$ (27.2)

The lower bound on $R(t,t)$ of order $t2^{t/2}$ was proved in Chap. 18 (Theorem 18.1) using the probabilistic method. Thus,

$$c_1 t2^{t/2} \leqslant R(t,t) \leqslant \binom{2t-2}{t-1} \sim c_2 4^t / \sqrt{t}.$$ (27.3)

The gap is still large, and in recent years, relatively little progress has been made. Tight bounds are known only for $s = 3$:

$$c_1 \frac{t^2}{\log t} \leqslant R(3,t) \leqslant c_2 \frac{t^2}{\log t}.$$

The upper bound is due to Ajtai, Komlós, and Szemerédi (1980) and the lower bound was proved by Kim (1995) using a probabilistic argument.

We have proved Theorem 27.3 by induction on $s+t$. The same result can also be proved using so-called *induced coloring argument*. This argument is encountered frequently in Ramsey theory; we also will use it in next sections. To explain the idea, let us prove the following weaker version of (27.3):

$$R(t,t) \leqslant 2^{2t-1}.$$

Proof. Take $n = 2^{2t-1}$ and consider a clique K_n on a vertex set V, $|V| = n$. Fix an arbitrary coloring χ of the edges of K_n in red and blue. Our goal is to show that K_n must contain a monochromatic clique on t vertices.

Set $S_1 = V$, and construct sets S_i and elements $x_i \in S_i$ as follows.

1. Having chosen S_i, select $x_i \in S_i$ arbitrarily.
2. Having selected $x_i \in S_i$, split the set $S_i - \{x_i\}$ into two sets S_i^0 and S_i^1, where S_i^0 consists of all vertices $y \in S_i$ joined with x_i by a red edge.
3. Set S_{i+1} equal to the larger of S_i^0, S_i^1.

Since $|S_i^0| + |S_i^1| = |S_i| - 1$, we always have $|S_{i+1}| \geqslant (|S_i| - 1)/2$. Since $|S_1| = 2^{2t-1}$ was sufficiently large, we may select $X = \{x_1, x_2, \ldots, x_{2t-1}\}$ before this procedure terminates. Define a coloring $\chi' : X \to \{0,1\}$ by:

$$\chi'(x_i) = 0 \quad \text{if and only if} \quad S_{i+1} = S_i^0.$$

Since we have $2t - 1$ vertices in X and only two colors, at least t of the vertices must receive the same color, say, 0. Take $T = \{x_i : \chi'(x_i) = 0\}$; hence, $|T| \geqslant t$. It remains to show that every pair of vertices in T is joined by a red edge. To see this, take $x_i, x_j \in T$, $i < j$. Since $\chi'(x_i) = 0$, all edges joining x_i with the vertices in S_{i+1} are red. Since $x_j \in S_j \subseteq S_{i+1}$, we are done. $\qquad\square$

27.3 Ramsey's theorem for sets

Recall that $R_r(k; s)$ denotes the smallest number n with the property that, if the k-subsets of an n-set are colored with r colors, then there is an s-set all of whose k-subsets have the same color. In its unabridged form, the celebrated result of Ramsey (1930) states that the function $R_r(k; s)$ is well-defined for all values of r, k and $s \geqslant k$.

Theorem 27.4 (Ramsey 1930). *Let r, k, s be given positive integers, $s \geqslant k$. Then there is a number $n = R_r(k; s)$ with the following property. If k-subsets of an n-set are colored with r colors, then there is an s-set all of whose k-subsets have the same color.*

First, we show that it is enough to prove this theorem for two colors, because $R_{r+1}(k; s) \leqslant R_r(k; R_2(k; s))$.

To see this, let $N = R_r(k; R_2(k; s))$ and let an arbitrary coloring of k-subsets of an N-element set X with $r + 1$ colors $0, 1, \ldots, r$ be given. Then consider this as an r-coloring simply by identifying the colors 0 and 1. (This is known as the "mixing colors" trick.) By the choice of N, either there exists an $R_2(k; s)$-element subset, all whose k-subsets receive one of the colors $2, \ldots, r$ (and we are done), or there exists an $R_2(k; s)$-element subset Y with each its k-subsets in color 0 or 1. According to the size of Y, all k-subsets of some its s-element subset must be monochromatic.

Thus, Theorem 27.4 follows from the following its "two-colors" version.

Theorem 27.5. *Let k, s, t be given positive integers; $s, t \geqslant k$. Then there is a number n with the following property. If k-subsets of an n-set are colored with two colors, then there is a subset of size $l \in \{s, t\}$, all whose k-subsets have the same color.*

Proof. In terms of Ramsey numbers, the theorem just claims that $R(k; s, t) \rightleftharpoons R_2(k; s, t)$ exists for all $s, t \geqslant k$. We will prove a stronger statement that $R(k; s, t) \leqslant n$, where

$$n \rightleftharpoons R(k-1; R(k; s-1, t), R(k; s, t-1)) + 1.$$

We prove this recurrence by induction on k and on s, t. Observe that, by the pigeonhole principle, $R(1; s, t) = s + t - 1$ for all s and t and, moreover, $R(k; x, k) = R(k; k, x) = x$ for all k and $x \geqslant k$. By induction, we may assume that the numbers $R(k; s-1, t)$ and $R(k; s, t-1)$ exist, and take an arbitrary n-element set X, where n is defined above.

Let χ be a coloring of k-subsets of X with two colors 0 and 1. Fix a point $x \in X$, and let $X' \rightleftharpoons X - \{x\}$. We define a new coloring χ' of the $(k-1)$-subsets A of X' by

$$\chi'(A) \rightleftharpoons \chi(A \cup \{x\}).$$

By the choice of n and by symmetry, we can assume to have found a subset $Y \subseteq X'$ such that $|Y| = R(k; s-1, t)$ and

$$\chi'(A) = 0 \text{ for all } (k-1)\text{-subsets } A \text{ of } Y.$$

Now consider how the original coloring χ acts on the k-subsets of Y. According to its size, the set Y must either contain a t-element subset, all whose k-subsets receive color 1 (and we are done), or it must contain an $(s-1)$-element subset Z, all whose k-subsets receive color 0. In this last case consider the s-element subset $Z \cup \{x\}$ and take an arbitrary its subset B of size k. If $x \notin B$ then B is a k-element subset of Z, and hence, $\chi(B) = 0$. If $x \in B$ then the set $A = B - \{x\}$ is a $(k-1)$-subset of Y, and hence again, $\chi(B) = \chi(A \cup \{x\}) = \chi'(A) = 0$. $\qquad\square$

27.4 Schur's theorem

The following result proved by I. Schur in 1916 may perhaps be considered as the earliest result in Ramsey theory.

Theorem 27.6 (Schur 1916). *For every r there exists a positive integer $n = S(r)$ such that for every partition of the set $\{1, \ldots, n\}$ into r classes one of the classes contains two numbers x and y together with their sum $x + y$.*

Schur established this lemma to prove that for each r the congruence $x^r + y^r = z^r \pmod{p}$ has solutions for all sufficiently large primes p.

We give two proofs of Schur's theorem. The first (original) one yields the upper bound $S(r) \leqslant er!$. The second one is an application of Ramsey's theorem.

First proof. Let $\chi : \{1, \ldots, n_0\} \to \{1, \ldots, r\}$ be an r-coloring of the first n_0 positive integers. Assume that there *do not* exists integers $x, y \leqslant n_0$ such that $\chi(x) = \chi(y) = \chi(x + y)$. Our goal is to show that then $n_0 < er!$.

Let c_0 be a color which appears most frequently among the n_0 elements, and let $x_0 < x_1 < \ldots < x_{n_1-1}$ be the elements of color c_0. Observe that $n_0 \leqslant rn_1$.

Consider the set $A_0 = \{x_i - x_0 : 1 \leqslant i < n_1\}$. By our assumption, $A_0 \cap \chi^{-1}(c_0) = \emptyset$, i.e., no of its elements receive the color c_0. Let c_1 be a color which appears most frequently among the elements of A_0, and let $y_0 < y_1 < \ldots < y_{n_2-1}$ be its elements of color c_1. Observe that $n_1 - 1 \leqslant (r-1)n_2$.

Consider the set $A_1 = \{y_i - y_0 : 1 \leqslant i < n_2\}$. By the assumption, $A_1 \cap \chi^{-1}(c_0) = \emptyset$ and $A_1 \cap \chi^{-1}(c_1) = \emptyset$. Let c_2 be a color which appears most frequently among the elements of A_1, and let $z_0 < z_1 < \ldots < z_{n_3-1}$ be its elements of color c_2. Observe that $n_2 - 1 \leqslant (r-2)n_3$.

Continue this procedure until some n_k becomes 1. Since we have only r colors, this happens at latest for $k = r$. Thus, we obtained the inequalities $n_0 \leqslant rn_1$ and $n_i \leqslant (r-i)n_{i+1} + 1$ for $i = 1, \ldots, k-1$, with $n_k = 1$. Putting them together we obtain that

$$n_0 \leqslant \sum_{i=0}^{r-1} r(r-1)(r-2) \cdots (r-i) = \sum_{i=0}^{r-1} \frac{r!}{i!} < er!.$$

$\square$

Second proof. Choose $n = R_r(2; 3)$, the number from Ramsey's Theorem 27.4. Let $\chi : \{1, \ldots, n\} \to \{1, \ldots, r\}$ be a fixed coloring. Define the coloring χ' of the pairs by $\chi'(\{x, y\}) = \chi(|x - y|)$. By the choice of n, there exists a χ'-monochromatic triangle with vertices $x < y < z$. However, then $\chi(y - x) = \chi(z - y) = \chi(z - x)$ and luckily $(y - x) + (z - y) = z - x$. $\square$

27.5 Geometric application: convex polygons

One of the earliest and most popular applications of Ramsey's theorem is due to Erdős and Szekeres (1935). In fact, this application was a first step in popularizing Ramsey's theorem.

Theorem 27.7 (Erdős–Szekeres 1935). *Let $m \geqslant 3$ be a positive integer. Then there exists a positive integer n such that any set of n points in the Euclidean plane, no three of which are collinear, contains m points which are the vertices of a convex m-gon.*

Proof (due to Johnson 1986). Choose $n = R_2(3; m)$, the number from the Ramsey's Theorem 27.4, and let A be any set of n points in the plane, no three of which are collinear (i.e., lie on a line). For $a, b, c \in A$, let $|abc|$ denote the number of points of A which lie in the interior of the triangle spanned by a, b and c. Define the 2-coloring χ of triples of points in A by $\chi(a, b, c) = 0$ if $|abc|$ is even and $\chi(a, b, c) = 1$ otherwise. By the choice of n, there exists an m-element subset $B \subseteq A$ such that all its 3-element subsets receive the same color. Then the points of B form a convex m-gon. Otherwise, there would be four points $a, b, c, d \in B$ such that d lies in the interior of the triangle abc (see Fig. 27.3). Since no three points of B are collinear, we have

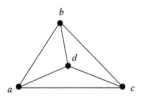

Fig. 27.3. Point d lies in no of the lines ab, bc and ac.

$|abc| = |abd| + |acd| + |bcd| + 1$, contradicting that the coloring χ is constant on all triples from B. □

Exercises

27.1. Prove the following general version of Schur's theorem. For every r and $l \geqslant 2$, there exists a positive integer n such that for every partition $A_1, \ldots, A_r$ of the set $\{1, \ldots, n\}$ into r classes one of the classes contains l (not necessarily distinct) numbers $x_1, \ldots, x_l$ such that $x_1 + \ldots + x_{l-1} = x_l$.

Hint: Take $n = R_r(2; l)$ and assign every pair $\{x, y\}$ the color i if $|x - y| \in A_i$.

27.2. Color all non-empty subsets (not the points!) of $[n] = \{1, \ldots, n\}$ with r colors. Prove that, if n is large enough, then there are two disjoint non-empty subsets A, B such that A, B and $A \cup B$ have the same color.

Sketch: Take $n = R_r(2; 3)$. Assume the non-empty subsets of $[n]$ are colored with r colors. Now color each pair $\{i, j\}$ $(1 \leqslant i < j < n)$ by the color of the interval $\{i, i + 1, \ldots, j - 1\}$. By Theorem 27.4, there exists a monochromatic triangle $x < y < z$. Take $A = \{x, x + 1, \ldots, y - 1\}$ and $B = \{y, y + 1, \ldots, z - 1\}$.

27.3$^+$ (Goodman 1959). Let G be a graph with n vertices and m edges. Let $t(G)$ denote the number of triangles contained in the graph G or in its complement. Prove that

$$t(G) \geqslant \binom{n}{3} + \frac{2m^2}{n} - m(n - 1).$$

Hint: Let t_i be the number of triples of vertices $\{i, j, k\}$ such that the vertex i is adjacent to precisely one of j or k. Observe that $t(G) \geqslant \binom{n}{3} - \frac{1}{2} \sum_i t_i$ and that $t_i = d_i(n - 1 - d_i)$, where d_i is the degree of the vertex i in G. Use the Cauchy–Schwarz inequality (14.3) and Euler's theorem (Theorem 1.7) to show that $\sum d_i^2 \geqslant \frac{1}{n} \left(\sum d_i\right)^2 = \frac{4m^2}{n}$.

27.4. Use the previous exercise to show that, for any coloring of the edges of a complete graph on n vertices with two colors, nearly a quarter of all triangles will always be monochromatic.

27.5.$^-$ Show that for every $r \geqslant 2$ there exists a constant $c = c(r)$ such that, if n is large enough, then for every r-coloring of the points $1, \ldots, n$, at least $c \cdot n^2$ of the pairs $\{i, j\}$ of points will receive the same color.

Hint: By the pigeonhole principle, every $(r + 1)$-subset of points contributes at least one monochromatic pair, and every pair is contained only in $\binom{n-2}{r-1}$ of such subsets.

27.6.$^-$ Prove the upper bound (27.2) on $R(s, t)$. *Hint*: $\binom{n}{k} = \binom{n-1}{k-1} + \binom{n-1}{k}$.

27.7. Prove that $R(3, 4) \leqslant 9$.

Sketch: Color the edges of K_9 in red and blue, and assume that there are no red triangles and no blue 4-cliques. Then each vertex is incident to precisely three red edges five blue edges. Thus, there are exactly $(9 \cdot 3)/2$ many red edges. But this should be an integer!

27.8.$^-$ Use the previous exercise to show that $R(4, 4) \leqslant 18$. *Hint*: (27.1).

28. Ramseyan Theorems for Numbers

In this chapter we will discuss several Ramsey-type problems in additive number theory.

28.1 Sum-free sets

In the wake of Shur's theorem, many people have studied so-called *sum-free sets*, i.e., subsets A of the positive integers such that for any $x, y \in A$, $x + y \notin A$, and their properties. In particular, people have examined the question of how large a sum-free set can be.

This question has a natural generalization to arbitrary abelian groups. An *abelian group* is a nonempty set G together with an operation $(x, y) \mapsto x + y$, called addition, which is associative $(x+y)+z = x+(y+z)$ and commutative $x + y = y + x$. Moreover, the operation must satisfy two conditions:

(i)   there must be an element $0 \in G$ (called a zero) such that $x + 0 = x$ for all $x \in G$;

(ii) every $x \in G$ must have an inverse $-x$ such that $(-x) + x = 0$.

Standard examples of abelian groups are: the set $\mathbb{Z}$ of integers and the set $\mathbb{Z}_n$ of residues modulo n.

For a subset S of an abelian group G, let $\alpha(S)$ denote the cardinality of the largest sum-free subset of S. The following upper bound is immediate (see Exercise 28.1): for any finite abelian group G,

$\alpha(G) \leqslant |G|/2.$

If we take $G = \mathbb{Z}_n$ for an even n, then the set

$A = \{1, 3, \ldots, n - 1\}$

is clearly sum-free, and hence, in this case $\alpha(G) = |G|/2$.

In the case of odd n we may derive a better upper bound using a beautiful theorem of Kneser about sums of finite subsets of an abelian group. Here we will prove a special case of this result which still has many applications in additive number theory and the proof of which is particularly simple.

If $A, B \subseteq G$ are subsets of an abelian group G, then by $A + B$ we denote the set of all its elements of the form $a + b$ with $a \in A$ and $b \in B$. A *subgroup* of G is a subset $H \subseteq G$ which itself forms a group; it is *proper* if $H \neq G$.

Theorem 28.1 (Kneser's Theorem). *Let G be an abelian group, $G \neq \{0\}$, and let A, B be nonempty finite subsets of G. If $|A| + |B| \leqslant |G|$, then there exists a proper subgroup H of G such that*

$$|A + B| \geqslant |A| + |B| - |H|.$$

Proof. We proceed by induction on $|B|$. If $|B| = 1$, then

$$|A + B| = |A| = |A| + |B| - 1 \geqslant |A| + |B| - |H|$$

for every subgroup H.

Let $|B| > 1$, and suppose that the theorem holds for all pairs A', B' of finite nonempty subsets of G such that $|B'| < |B|$. We distinguish two cases.
Case 1: $a + b - c \in A$ for *all* $a \in A$ and $b, c \in B$.

In this case $A + b - c = A$ for all $b, c \in B$. Let H be the subgroup of G generated by all elements of the form $b - c$, where $b, c \in B$. Then $|B| \leqslant |H|$ and $A + H = A \neq G$. Therefore, H is a proper subgroup of G, and

$$|A + B| \geqslant |A| \geqslant |A| + |B| - |H|.$$

Case 2: $a + b - c \notin A$ for *some* $a \in A$ and $b, c \in B$.

Let $e \rightleftharpoons a - c$ and define the subsets

$$A' = A \cup (B + e), \quad B' = B \cap (A - e).$$

Note that $b \notin B'$, and hence, B' is a proper subset of B, because otherwise b would have a form $x - a + c$ for some $x \in A$, and hence,

$$a + b - c = a + (x - a + c) - c = x \in A,$$

a contradiction. Also, $c \in B'$ (because $0 \in A - a$), and hence, B' is nonempty. Therefore, we can apply the induction hypothesis to A' and B', and deduce that there exists a proper subgroup H of G such that

$$|A' + B'| \geqslant |A'| + |B'| - |H|. \tag{28.1}$$

It remains to observe that

$$
\begin{aligned}
A' + B' &= [A \cup (B + e)] + [B \cap (A - e)] \\
&\subseteq (A + B) \cup [(B + e) + (A - e)] = A + B
\end{aligned}
$$

and

$$
\begin{aligned}
|A'| + |B'| &= |A \cup (B + e)| + |B \cap (A - e)| \\
&= |A \cup (B + e)| + |(B + e) \cap A| \\
&= |A| + |B + e| = |A| + |B|.
\end{aligned}
$$

$\square$

In the next section we will use the following fact which is a special version of Kneser's theorem (see Exercise 28.5):

Lemma 28.2 (Cauchy–Davenport). *If p is a prime, and A, B are two non-empty subsets of $\mathbb{Z}_p$, then*

$$|A + B| \geqslant \min\{p, |A| + |B| - 1\}.$$

Kneser's theorem immediately yields the following upper bound on the size of sum-free subset in abelian groups.

Corollary 28.3. *Let G be a finite abelian group and let p be the smallest prime divisor of $|G|$. Then $\alpha(G) \leqslant (p + 1)|G|/(3p)$.*

Proof. If $A \subseteq G$ is a sum-free set, then $|A + A| \leqslant |G| - |A|$ because $A + A$ and A are disjoint. Since $|A| \leqslant |G|/2$ (see Exercise 28.1), we can apply Theorem 28.1 and deduce

$$|G| - |A| \geqslant |A + A| \geqslant 2|A| - |H|$$

for some proper subgroup H of G. Since, by Lagrange's theorem, the order $|H|$ of any subgroup H of G divides the order $|G|$ of the group G, we have $|H| \leqslant |G|/p$. Therefore,

$$3|A| \leqslant |G| + |H| \leqslant (1 + 1/p)|G|,$$

and the desired result follows.   □

What about the lower bounds for $\alpha(G)$?

We have already seen that for $G = \mathbb{Z}_n$ with even n, we have an equality $\alpha(G) = |G|/2$. If $G = \mathbb{Z}$ is the group of integers, then

$$\alpha(S) > |S|/3$$

for any finite subset $S \subseteq \mathbb{Z} - \{0\}$ (we have proved this fact in Sect. 20 using the probabilistic argument). For other abelian groups the situation is not so clear.

The following naive argument shows that in this case also

$$\alpha(G) \geqslant \sqrt{|G|} - 1.$$

To show this, let A be a maximal sum-free subset. If $a \notin A$, then $A \cup \{a\}$ is not sum-free by the assumption, so we can write $a = s_1 + s_2$ for some $s_1, s_2 \in A$. Therefore $|G - A| \leqslant |A|^2$, from which the desired inequality $|A| \geqslant \sqrt{|G|} - 1$ follows.

Better lower bounds can be derived using an improvement of Theorem 28.1, also due to Kneser, which states that with the same hypotheses, either $|A + B| \geqslant |A| + |B|$ or $|A + B| \geqslant |A| + |B| - |H|$ for some proper subgroup H such that $H + A + B = A + H$ (see, for example, Street (1972) for the proof). Here we only mention that the best know lower bound for an arbitrary finite abelian group G is $\alpha(G) \geqslant 2|G|/7$. More information about the properties of sum-free sets can be found, for example, in Nathanson (1996).

28.2 Zero-sum sets

A sequence of (not necessarily distinct) numbers $b_1, \ldots, b_m$ is a *zero-sum sequence* (modulo n) if the sum $b_1 + \cdots + b_m$ is 0 (modulo n).

Proposition 28.4. *Suppose we are given a sequence of n integers $a_1, \ldots, a_n$, which need not be distinct. Then there is always a set of consecutive numbers $a_{r+1}, a_{r+2}, \ldots, a_s$ whose sum is divisible by n.*

Proof. Make n pigeonholes labeled from 0 up to $n-1$ and place the n sequences

$$(a_1), (a_1, a_2), \ldots, (a_1, a_2, \ldots, a_n)$$

into the pigeonholes corresponding to the remainder when the sum is divided by n. If any of these sequences is in the pigeonhole 0 then the sum of its numbers is divisible by n. If not, then the n sequences are in the $n - 1$ pigeonholes. By the pigeonhole principle some two of them, $(a_1, a_2, \ldots, a_r)$ and $(a_1, a_2, \ldots, a_s)$ with $r < s$, must lie in the same pigeonhole, meaning that the sum $a_{r+1} + a_{r+2} + \cdots + a_s$ is divisible by n. $\square$

By Proposition 28.4, every sequence of n numbers has a zero-sum subsequence modulo n. We know that this subsequence consists of consecutive numbers but we don't know how long it is. If we want to find a subsequence of n (not necessarily consecutive numbers) whose sum is divisible by n, how long must a given sequence be? The sequence $0^{n-1}1^{n-1}$ of $n - 1$ copies of 0 and $n - 1$ copies of 1 shows that the sequence must contain at least $2n - 1$ numbers. It turns out that *every* sequence of that many numbers already has the desired zero-sum subsequence!

Theorem 28.5 (Erdős–Ginzburg–Ziv 1961). *Any sequence of $2n - 1$ integers contains a subsequence of cardinality n the sum of whose elements is divisible by n.*

There are several different proofs of this theorem – the interested reader can find them, as well as some interesting extensions of this result to higher dimensions, in the paper of Alon and Dubiner (1993). The original proof was based on the Cauchy–Davenport lemma, which, as we have seen, is a special case of Kneser's theorem.

First proof of Theorem 28.5. We will first prove the theorem only for the case when $n = p$ is a prime number, and then show how the general case reduces to it.

Let $a_1 \leqslant a_2 \leqslant \ldots \leqslant a_{2p-1}$ be integers. If $a_i = a_{i+p-1}$ for some $i \leqslant p - 1$, then $a_i + a_{i+1} + \cdots + a_{i+p-1} = pa_i = 0$ (in $\mathbb{Z}_p$) and the desired result follows. Otherwise, define $A_i \rightleftharpoons \{a_i, a_{i+p-1}\}$ for $i = 1, \ldots, p-1$. By repeated application of the Cauchy-Davenport lemma, we conclude that

$$|A_1 + A_2 + \cdots + A_{p-1}| = p,$$

and hence, every number from $\mathbb{Z}_p$ is a sum of precisely $p-1$ of the first $2p-2$ elements of our sequence. In particular, the number $-a_{2p-1}$ is such a sum, supplying the required p-element subset whose sum is 0 in $\mathbb{Z}_p$.

The general case may be proved by induction on the number of primes in the prime factorization of n. Put $n = pm$ where p is a prime, and let $a_1, \ldots, a_{2n-1}$ be the given sequence. By the result for the prime case, each subset of $2p - 1$ members of the sequence contains a p-element subset whose sum is 0 modulo p. Therefore, we can find ℓ pairwise disjoint p-element subsets $I_1, \ldots, I_\ell$ of $\{1, \ldots, 2n-1\}$, where

$$\sum_{j \in I_i} a_j \equiv 0 \,(\mathrm{mod}\, p)$$

for each $i = 1, \ldots, \ell$. Moreover, $\ell \geqslant 2m - 1$ since otherwise the number of left elements would be still $2pm - 1 - (2m - 2)p = 2p - 1$, and we could choose the next subset $I_{\ell+1}$. Now define a sequence $b_1, \ldots, b_{2m-1}$ where

$$b_i = \sum_{j \in I_i} \frac{a_j}{p}$$

(recall that each of these sums is divisible by p). By the induction hypothesis this new sequence has a subset $\{b_i : i \in J\}$ of $|J| = m$ elements whose sum is divisible by m, and the union of the corresponding sets $\{a_j : j \in I_i\}$ with $i \in J$, supplies the desired n-element subset of our original sequence, whose sum is divisible by $n = pm$. $\qquad\square$

Theorem 28.5 can also be derived using another powerful tool – the classical result of Chevalley and Warning about the zeroes of multivariate polynomials (see Theorem 13.11).

Second proof of Theorem 28.5 (Alon 1995). We will prove the theorem only for a prime $n = p$; the general case reduces to it (see the first proof of Theorem 28.5).

Let $a_1, a_2, \ldots, a_{2p-1}$ be integers, and consider the following system of two polynomials in $2p - 1$ variables of degree $p - 1$ over $\mathbb{F}_p$:

$$\sum_{i=1}^{2p-1} a_i x_i^{p-1} = 0,$$

$$\sum_{i=1}^{2p-1} x_i^{p-1} = 0.$$

Since $2(p - 1) < 2p - 1$ and $x_1 = x_2 = \cdots = x_{2p-1} = 0$ is a common solution, Theorem 13.11 implies the existence of a nontrivial common solution $(y_1, \ldots, y_{2p-1})$. Since p is a prime, Fermat's little theorem (see Exercise 1.12) tells us that $x^{p-1} = 1$ in $\mathbb{F}_p$ for every $x \in \mathbb{F}_p$, $x \neq 0$. So, if we take $I = \{i : y_i \neq 0\}$ then the first equation ensures that $\sum_{i \in I} a_i = 0$, whereas the second ensures that $|I| \equiv 0 \,(\mathrm{mod}\, p)$, and hence, that $|I| = p$ because $|I| \leqslant 2p - 1$. This completes the proof of the theorem for the prime n. $\qquad\square$

28.3 Szemerédi's cube lemma

A collection C of integers is called an *affine d-cube* if there exist $d+1$ positive integers $x_0, x_1, \ldots, x_d$ so that

$$C = \left\{ x_0 + \sum_{i \in I} x_i : I \subseteq \{1, 2, \ldots, d\} \right\}.$$

If an affine cube is generated by $x_0, x_1, \ldots, x_d$ then we write

$$C = C(x_0, x_1, \ldots, x_d).$$

For example, every arithmetic progression $a, a + b, a + 2b, \ldots, a + db$ is an affine d-cube $C(a, b, b, \ldots, b)$.

We have already considered affine cubes in Sect. 21.6. Using a probabilistic argument we have proved (see Lemma 21.11) that in any set of n positive integers there is a subset A which has size

$$|A| \geqslant \tfrac{1}{8} n^{1-d/(2^d-1)}$$

and contains *no* replete affine d-cube (recall that a cube is replete if it contains 2^d elements). We have also mentioned a result, known as *Szemerédi's cube lemma*.

Lemma 28.6 (Szemerédi 1969). *For any $0 < \epsilon < 1$ and positive integer d, there exists $n_0 = n_0(\epsilon, d)$ such that, for all $n \geqslant n_0$, every subset A of $\{1, \ldots, n\}$ of size $|A| > \epsilon n$ contains an affine d-cube.*

As it often happens with important results, different proofs of this lemma were found. The arguments below are from Lovász (1979).

First we prove a Ramsey-type version of Szemerédi's lemma.

Lemma 28.7. *For every $d, r \geqslant 1$ there exists an $n = N(d, r)$ with the following property. If we color the set $\{1, \ldots, n\}$ in r colors then all the elements of at least one affine d-cube lying in this set will receive the same color.*

Proof. We argue by induction on d. The case $d = 1$ is obvious, so assume that $n = N(r, d - 1)$ exists and take $N(r, d) \rightleftharpoons r^n + n$.

Suppose $N \geqslant N(r, d)$ and $\{1, \ldots, N\}$ is colored in r colors. Consider the colors of the strings of n consecutive numbers

$$i, i+1, \ldots, i+n-1 \quad \text{for } 1 \leqslant i \leqslant r^n + 1.$$

We have $r^n + 1$ such strings of numbers but only r^n possible strings of their colors (we have only r colors in our disposal). By the pigeonhole principle, some two strings

$$i, i+1, \ldots, i+n-1;$$
$$j, j+1, \ldots, j+n-1,$$

with $i < j$, will receive the same sequence of colors. That is, for each x in $\{i, i+1, \ldots, i+n-1\}$, the numbers x and $x + (j - i)$ receive the same color.

By the choice of $n = N(r, d - 1)$, the set $\{i, i+1, \ldots, i+n-1\}$ contains a monochromatic affine $(d-1)$-cube $C(x_0, x_1, \ldots, x_{d-1})$. But then all the numbers of the affine d-cube $C(x_0, x_1, \ldots, x_{d-1}, j - i)$ have the same color. Since $j - i \leqslant r^n$, this cube lies in $\{1, \ldots, N\}$, and we are done. $\qquad\square$

Now we prove the "density-version" of Szemerédi's lemma.

Lemma 28.8. *Let $d \geqslant 2$ be given. Then, for every sufficiently large n, every subset A of $\{1, \ldots, n\}$ of size $|A| \geqslant (4n)^{1-1/2^{d-1}}$ contains an affine d-cube.*

Proof. We will iteratively use the following fact: if $B \subseteq \{1, \ldots, n\}$, $|B| \geqslant 2$, then there is a $i \geqslant 1$ such that the set

$$B_i \rightleftharpoons \{b \in B : b + i \in B\}$$

has size

$$|B_i| > \frac{|B|^2}{4n}. \qquad (28.2)$$

This immediately follows from the equality (see Exercise 28.7):

$$\sum_{i=1}^{n-1} |B_i| = \binom{|B|}{2}. \qquad (28.3)$$

Applying this fact to the set A, we will find $i_1 \geqslant 1$ such that

$$|A_{i_1}| > \frac{|A|^2}{4n} \geqslant \frac{(4n)^{2-2/2^{d-1}}}{4n} = (4n)^{1-1/2^{d-2}}.$$

Similarly, applying the fact to the set A_{i_1}, we will find $i_2 \geqslant 1$ such that

$$|A_{i_1,i_2}| = |(A_{i_1})_{i_2}| > \frac{|A_{i_1}|^2}{4n} \geqslant \frac{(4n)^{2-2/2^{d-2}}}{4n} = (4n)^{1-1/2^{d-3}}.$$

Continuing this process, we will find $i_1, i_2, \ldots, i_{d-1}$ such that

$$\left| A_{i_1,i_2,\ldots,i_{d-1}} \right| > (4n)^{1-1/2^{d-d}} = 1.$$

Since this set still has at least 2 elements, we can apply the fact once more and conclude that the set $A_{i_1,i_2,\ldots,i_d}$ contains at least one element b_0. Observe that

$$A_{i_1} = \{b : b \in A, \, b + i_1 \in A\},$$
$$A_{i_1,i_2} = \{b : b \in A, \, b + i_1 \in A, \, b + i_2 \in A, \, b + i_1 + i_2 \in A\}$$

determines an affine 2-cube $C(b, i_1, i_1) \subseteq A$, and so on. Hence, the last set $A_{i_1,i_2,\ldots,i_d}$ determines an affine d-cube $C(b_0, i_1, \ldots, i_d)$, and this cube lies entirely in A. $\qquad\square$

In the proof above no attemp was made to get the best constant. In particular, we were very generous when deriving (28.2) from (28.3). Using more precise estimate at this step, Gunderson and Rödl (1998) slightly improved the bound in Lemma 28.8 to

$$|A| \geqslant 2^{1-1/2^{d-1}} (\sqrt{n} + 1)^{2-1/2^{d-1}}.$$

Exercises

28.1.⁻ Show that for any finite abelian group G, $\alpha(G) \leqslant |G|/2$.

Hint: If S is a sum-free set then $S + S$ and S are disjoint.

28.2. Give a detailed proof of the last two statements in the proof of Kneser's theorem that $A' + B' \subseteq A + B$ and $|A'| + |B'| = |A| + |B|$.

28.3.⁻ Let G be a finite abelian group, and let A and B be subsets of G such that $|A| + |B| > |G|$. Show that then $A + B = G$.

Hint: For every $x \in G$, the set $A \cap (x - B)$ has at least $|A| + |B| - |G| \geqslant 1$ elements.

28.4. Let A and B be finite subsets of an abelian group G. For $x \in G$, let $r(x)$ be the number of representations of x as the sum of elements from A and B, that is, $r(x)$ is the number of ordered pairs $(a, b) \in A \times B$ such that $x = a + b$. Prove the following: if $|A| + |B| \geqslant |G| + t$ then $r(x) \geqslant t$.

Hint: Take an $x \in G$ and show that $|A \cap (x - B)| \geqslant |A| + |B| - |G| \geqslant t$.

28.5.⁻ Show that Knesers's theorem (Theorem 28.1) implies the Cauchy–Davenport lemma.

Hint: For a prime p, the only proper subgroup of $\mathbb{Z}_p$ is the trivial group $H = \{0\}$.

28.6. If A be a nonempty subset of an abelian group G, then its *stabilizer* is the set $H(A) = \{x \in G : x + A = A\}$. Show that A is a subgroup if and only if $H(A) = A$.

28.7. Prove (28.3).

Sketch: Consider a complete graph whose vertices are elements of B; label the edge, joining two elements $a < b$, by their difference $b - a$, and observe that $|B_i|$ is precisely the number of edges labeled by i.

28.8.⁻ Show that the number of sum-free subsets of $\{1, \ldots, n\}$ is at least $c2^{n/2}$ for some constant $c > 0$. *Hint*: We can take any set of odd numbers, or any set of numbers grater than $n/2$.

28.9 (Cameron–Erdős 1999). A sum-free subset S of $X = \{1, \ldots, n\}$ is *maximal* if none of the sets $S \cup \{x\}$ with $x \in X - S$ is sum-free. Show that the number of maximal sum-free subsets of X is at least $2^{\lfloor n/4 \rfloor}$.

Hint: Let m be either n or $n - 1$, whichever is even. Let S consist of m together with one of each pair of numbers $x, m - x$ for odd $x < m/2$. Show that every such set S is sum-free, and distinct sets S lie in distinct maximal sum-free sets.

29. The Hales–Jewett Theorem

In 1963 A. W. Hales and R. I. Jewett proved a very general result from which most of Ramsey-like theorems may be easily derived. Hales–Jewett theorem is presently one of the most useful techniques in Ramsey theory. Without this result, *Ramsey theory* would more properly be called *Ramseyan theorems*.

29.1 The theorem and its consequences

Let A be an alphabet of t symbols; we will usually take $A = \{0, 1, \ldots, t-1\}$. Let $* \notin A$ be a new symbol. Words are strings of symbols without $*$. Strings containing the symbol $*$ will be called *roots*. For a root $\tau \in (A \cup \{*\})^n$ and a symbol $a \in A$, we denote by $\tau(a)$ the word in A^n obtained from τ by replacing each occurrence of $*$ by a.

Let τ be a root. A *combinatorial line* rooted in τ is the set of t words

$$L_\tau = \{\tau(0), \tau(1), \ldots, \tau(t-1)\}.$$

For example, if $\tau = 0\,1\,*\,2\,*\,1$ is a root in the alphabet $A = \{0, 1, 2, 3, 4\}$ then the set of five words

$$L_\tau = \begin{cases} 0\,1\,0\,2\,0\,1 \\ 0\,1\,1\,2\,1\,1 \\ 0\,1\,2\,2\,2\,1 \\ 0\,1\,3\,2\,3\,1 \\ 0\,1\,4\,2\,4\,1 \end{cases}$$

is a combinatorial line rooted in τ. Clearly, there are $(|A| + 1)^n - |A|^n$ lines in A^n (every root defines its own line).

To match the classical parametric representation $x = a + \lambda b$ of a line in $\mathbb{R}^n$, observe that it corresponds to a combinatorial line rooted in τ, where $a_i = 0, b_i = 1$ if $\tau_i = *$, and $a_i = \tau_i, b_i = 0$ if $\tau_i \neq *$. Figure 29.1 shows a line $x = a + \lambda b$ with $a = (0, 0, 0)$ and $b = (1, 0, 1)$. Thus, the symbol $*$ indicates the moving coordinate.

Theorem 29.1 (Hales–Jewett 1963). *Let A be a finite alphabet of t symbols and let r be a positive integer. Then there exists an integer $n = HJ(r, t)$ such that for every coloring of the cube A^n in r colors there exists a combinatorial line, which is monochromatic.*

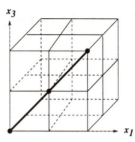

Fig. 29.1. A line $L = \{(0,0,0),(1,0,1),(2,0,2)\}$ rooted in $\tau = (*,0,*)$

In fact, Hales and Jewett have proved this result for more general configurations, known as *combinatorial spaces*. Each combinatorial line has only one set of moving coordinates, determined by the occurrences of the special symbol $*$. A natural generalization is to allow several (disjoint) sets of moving coordinates.

A *combinatorial m-space* $S_\tau \subseteq A^n$ is given by a word (a generalized root) $\tau \in (A \cup \{*_1, \ldots, *_m\})^n$, where $*_1, \ldots, *_m$ are distinct symbols not in the alphabet A. These symbols represent m mutually disjoint sets of moving coordinates. We require that each of these symbols occurs at least once in τ. Then S_τ is the set of all t^m words in A^n which can be obtained by simultaneously replacing each occurrence of these new symbols by symbols from A. Thus, combinatorial line is a just a combinatorial 1-space. In the case of two-letter alphabet $A = \{0,1\}$ there is a 1-1 correspondence between combinatorial spaces and subcubes of $\{0,1\}^n$.

Theorem 29.2. *Let A be a finite alphabet of t symbols and let m,r be positive integers. Then there exists an integer $n = HJ(m,r,t)$ such that for every coloring of the cube A^n in r colors there exists a combinatorial m-space, which is monochromatic.*

We will prove this result only for combinatorial lines (Theorem 29.1). On one hand, the proof in this case is easier and more transparent. On the other hand, the result can be extended to arbitrary combinatorial m-spaces by the induction on m (see Exercise 29.10).

But let us first look how this theorem implies some classical results of Ramsey Theory.

29.1.1 Van der Waerden's theorem

Recall that an arithmetic progression of length t is a sequence of t natural numbers $a, a + d, a + 2d, \ldots, a + (t-1)d$, each at the same distance $d \geqslant 1$ from the previous one. Thus, each arithmetic progression is a very regular configuration of numbers, just like are cliques in graphs.

In 1927 B.L. van der Waerden published a proof of the following Ramsey-type result for arithmetic progressions.

Theorem 29.3 (Van der Waerden 1927). *For every choice of positive integers r and t, there exists a positive integer $N = W(r,t)$ such that for every coloring of the set of integers $\{1, \ldots, N\}$ in r colors at least one arithmetic progression with t terms will be monochromatic.*

Proof. Take $N = n(t-1)$ where $n = HJ(r,t)$ is from the Hales–Jewett theorem. Take the alphabet $A = \{0, 1, \ldots, t-1\}$ and define a mapping $f : A^n \to \{1, \ldots, N\}$ which takes a word $x = (x_1, \ldots, x_n)$ to the sum

$$f(x) = x_1 + \ldots + x_n$$

of its letters. The mapping f induces a coloring of A^n in a natural manner: the color of a word $x \in A^n$ is the color of the number $f(x)$.

It is easy to see that every combinatorial line

$$L_\tau = \{\tau(0), \tau(1), \ldots, \tau(t-1)\}$$

is mapped to an arithmetic progression of length t, because the difference between the integers corresponding to strings $\tau(i+1)$ and $\tau(i)$ is the same (and is equal to the number of $*$'s in τ). By the Hales–Jewett theorem, there is a monochromatic line that, in turn, translates back to a monochromatic arithmetical progression of length t, as desired.   □

29.1.2 Gallai–Witt's Theorem

A multidimensional version of van der Waerden's theorem was proved independently by Gallai (= Grünwald), cf. Rado (1943), and Witt (1951).

A subset of vectors $U \subseteq \mathbb{Z}^m$ is a *homothetic copy* of a subset $V \subseteq \mathbb{Z}^m$ if there exists a vector $u \in \mathbb{Z}^m$ and a constant $\lambda \in \mathbb{Z}$, $\lambda > 0$ such that

$$U = u + \lambda V = \{u + \lambda v : v \in V\}. \tag{29.1}$$

Theorem 29.4 (Gallai–Witt). *Let the vectors of $\mathbb{Z}^m$ be finitely colored. Then every finite subset of $\mathbb{Z}^m$ has a homothetic copy which is monochromatic.*

Proof. Fix the number of colors r and a finite set of vectors $V = \{v_0, \ldots, v_{t-1}\}$ in $\mathbb{Z}^m$. We will consider these vectors as symbols of our alphabet $A = V$. Set $n = HJ(r,t)$ and consider the cube A^n; its elements are vectors $x = (x_1, \ldots, x_n)$, each of whose coordinates x_i is one of the vectors $v_0, \ldots, v_{t-1}$. As in the previous theorem, define a map $f : A^n \to \mathbb{Z}^m$ by

$$f(x) = x_1 + \ldots + x_n.$$

By the Hales–Jewett theorem, there is a monochromatic combinatorial line

$$L_\tau = \{\tau(v_0), \tau(v_1), \ldots, \tau(v_{t-1})\} \subseteq A^n.$$

Let $I = \{i : \tau_i \neq *\}$ be the set of coordinates, where the root τ has a value, different from $*$. Then $f(L_\tau)$ is the set of t vectors of the form $\lambda v_j + u$, $j = 0, 1, \ldots, t - 1$, where $u = \sum_{i \in I} v_i$ and $\lambda = n - |I| > 0$. Hence, $f(L_\tau)$ is a homothetic copy of $V = \{v_0, \ldots, v_{t-1}\}$, as desired. $\qquad\square$

29.2 Shelah's proof of HJT

Various proofs of Theorem 29.1 are known. The original proof of Hales and Jewett is quite short but provides an upper bound for the function $HJ(r, t)$ which grows extremely fast. Shelah (1988) found a fundamentally new proof which yields a much smaller (in particular, a primitive recursive) upper bound for $HJ(r, t)$. Here we will follow the compact version of Shelah's proof from A. Nilli (1990) (c/o Noga Alon).

For each fixed number of colors r, we apply an induction on the number of symbols t in our alphabet A. For $t = 1$ the theorem is trivial. Assuming it holds for $t - 1$ (and r) prove it for t. Set

$$n \rightleftharpoons HJ(r, t - 1),$$

and define the following increasing sequence of dimensions $N_1, \ldots, N_n$ by

$$N_1 \rightleftharpoons r^{t^n} \quad \text{and} \quad N_i \rightleftharpoons r^{t^{n + \sum_{j=1}^{i-1} N_j}}.$$

Put

$$N \rightleftharpoons N_1 + \cdots + N_n.$$

We will prove that the dimension N has the desired property, i.e., that $HJ(r, t) \leqslant N$. (This particular choice of the dimensions N_i will be important only in the proof of Claim 29.5 below).

To show this, let $A = \{0, 1, \ldots, t - 1\}$ be an alphabet of t symbols, and let $\chi : A^N \to \{1, \ldots, r\}$ be a coloring the N-cube A^N in r colors. Our goal is to prove that at least one combinatorial line will be monochromatic. The key to the whole proof is the following technical claim about the colors of neighboring words. We say that two words $a, b \in A^n$ are *neighbors* if they differ in exactly one coordinate, say, the ith in which $a_i = 0$ and $b_i = 1$:

$$a = a_1 \ldots a_{i-1}\, 0\, a_{i+1} \ldots a_n$$
$$b = a_1 \ldots a_{i-1}\, 1\, a_{i+1} \ldots a_n$$

For a word $a = a_1 a_2 \ldots a_n$ over A of length n and a sequence of n roots

$$\tau = \tau_1 \tau_2 \ldots \tau_n,$$

the ith of which has length N_i, let $\tau(a)$ denote the corresponding word of length N:

$$\tau(a) = \tau_1(a_1)\tau_2(a_2) \ldots \tau_n(a_n).$$

That is, we replace each occurrence of $*$ in τ_1 by a_1, each occurrence of $*$ in τ_2 by a_2, and so on.

Claim 29.5. *There exists a sequence of n roots $\tau = \tau_1\tau_2\ldots\tau_n$ as above, such that $\chi(\tau(a)) = \chi(\tau(b))$ for any two neighbors $a, b \in A^n$.*

Before proving the claim, let us look at how it implies the theorem. Using the coloring χ of the N-cube A^N, we define a coloring χ' of the n-cube $(A - \{0\})^n$ by:

$$\chi'(a) = \chi(\tau(a)),$$

where τ is from the claim. Since the alphabet $A - \{0\}$ has only $t - 1$ symbols and $n = HJ(r, t - 1)$, we can apply the induction to this coloring χ'. By the induction hypothesis there exists a root

$$\nu = \nu_1\nu_2\ldots\nu_n \in \big((A - \{0\}) \cup \{*\}\big)^n$$

such that the combinatorial line

$$L_\nu = \{\nu(1), \nu(2), \ldots, \nu(t - 1)\}$$

is monochromatic (with respect to χ'). Consider the string

$$\tau(\nu) = \tau_1(\nu_1)\tau_2(\nu_2)\ldots\tau_n(\nu_n).$$

This string has length N and is a root since ν is a root (and hence, has at least one $*$). We claim that the corresponding line

$$L_{\tau(\nu)} = \big\{\tau(\nu(0)), \tau(\nu(1)), \ldots, \tau(\nu(t - 1))\big\}$$

is monochromatic with respect to the original coloring χ. Indeed, the coloring χ' assigns the same color to all the words $\nu(1), \ldots, \nu(t - 1)$. Hence, by the definition of χ', the coloring χ assigns the same color to all the words $\tau(\nu(1)), \ldots, \tau(\nu(t - 1))$. If ν contains only one $*$, then $\tau(\nu(0))$ is a neighbor of $\tau(\nu(1))$ and, by the claim, receives the same color (under χ). If ν has more $*$'s, then we can still reach the word $\tau(\nu(0))$ from the word $\tau(\nu(1))$ by passing through a sequence of neighbors

$$\tau(\nu(1)) = \ldots 1 \ldots 1 \ldots 1 \ldots$$
$$\ldots 0 \ldots 1 \ldots 1 \ldots$$
$$\ldots 0 \ldots 0 \ldots 1 \ldots$$
$$\tau(\nu(0)) = \ldots 0 \ldots 0 \ldots 0 \ldots$$

and, by the claim, the word $\tau(\nu(0))$ will receive the color of $\tau(\nu(1))$. So, the whole line $L_{\tau(\nu)}$ is monochromatic, as desired.

It remains to prove the claim.

Proof of Claim 29.5. Recall that we want to find a sequence of n roots $\tau = \tau_1\tau_2\ldots\tau_n$, the ith of which has length N_i, and such that $\chi(\tau(a)) = \chi(\tau(b))$ for any two neighbors $a, b \in A^n$.

We prove the existence of required roots τ_i by backward induction on i. Suppose we have already defined the roots $\tau_{i+1}, \ldots, \tau_n$. Our goal is to define the root τ_i.

Let $L_{i-1} = \sum_{j=1}^{i-1} N_j$ be the length of the initial segment $\tau_1 \tau_2 \ldots \tau_{i-1}$ of the sequence of roots we are looking for. The length of the ith segment τ_i is N_i. For $k = 0, 1, \ldots, N_i$, let W_k denote the following word of length N_i:

$$W_k = \underbrace{0 \ldots 0}_{k} \underbrace{1 \ldots 1}_{N_i - k}.$$

For each $k = 0, 1, \ldots, N_i$, define the r-coloring χ_k of all words in $A^{L_{i-1}+n-i}$ as follows: let $\chi_k \left(x_1 x_2 \ldots x_{L_{i-1}} y_{i+1} \ldots y_n \right)$ be equal to

$$\chi \left(x_1 x_2 \ldots x_{L_{i-1}} W_k \tau_{i+1}(y_{i+1}) \ldots \tau_m(y_n) \right).$$

We have $N_i + 1$ colorings $\chi_0, \chi_1, \ldots, \chi_{N_i}$, each being chosen from the set of at most

$$r^{\#\{\text{of words}\}} = r^{t^{L_{i-1}+n-i}} \leqslant r^{t^{L_{i-1}+n}} = N_i$$

such colorings. By the pigeonhole principle, at least two of these colorings must coincide, i.e., $\chi_s = \chi_k$ for some $s < k$. Now define the desired root τ_i by

$$\tau_i = \underbrace{0 \ldots 0}_{s} \underbrace{* \ldots *}_{k-s} \underbrace{1 \ldots 1}_{N_i - k}.$$

One can easily check that the roots $\tau_1, \ldots, \tau_n$ defined by this procedure satisfy the assertion of the claim. Indeed, observe that

$$\tau_i(0) = W_k \quad \text{and} \quad \tau_i(1) = W_s.$$

Hence, if we take any two neighbors in the ith coordinate

$$a = a_1 \ldots a_{i-1} 0 \, a_{i+1} \ldots a_n$$
$$b = a_1 \ldots a_{i-1} 1 \, a_{i+1} \ldots a_n$$

then

$$\tau(a) = \tau_1(a_1) \ldots \tau_{i-1}(a_{i-1}) \, \tau_i(0) \, \tau_{i+1}(a_{i+1}) \ldots \tau_n(a_n),$$
$$\tau(b) = \tau_1(a_1) \ldots \tau_{i-1}(a_{i-1}) \, \tau_i(1) \, \tau_{i+1}(a_{i+1}) \ldots \tau_n(a_n),$$

and since $\chi_s = \chi_k$,

$$\chi(\tau(a)) = \chi\big(\tau_1(a_1) \ldots \tau_{i-1}(a_{i-1}) W_k \tau_{i+1}(a_{i+1}) \ldots \tau_m(a_n)\big)$$
$$= \chi_k\big(\tau_1(a_1) \ldots \tau_{i-1}(a_{i-1}) a_{i+1} \ldots a_n\big)$$
$$= \chi_s\big(\tau_1(a_1) \ldots \tau_{i-1}(a_{i-1}) a_{i+1} \ldots a_n\big)$$
$$= \chi\big(\tau_1(a_1) \ldots \tau_{i-1}(a_{i-1}) W_s \tau_{i+1}(a_{i+1}) \ldots \tau_m(a_n)\big) = \chi(\tau(b)).$$

This completes the proof of the claim, and thus, the proof of the Hales–Jewett theorem. $\square$

29.3 Application: multi-party games

In this section we present one of the earliest applications of Ramsey-type arguments to prove lower bounds on the multi-party communication complexity. The results are due to Chandra, Furst and Lipton (1983).

Let $X = X_1 \times X_2 \times \cdots \times X_k$, where $X_1, X_2, \ldots, X_k$ are n-element sets. There are k players $P_1, \ldots, P_k$ who wish to collaboratively evaluate a given function $f : X \to \{0, 1\}$ on every input $x \in X$. Each player has unlimited computational power and full knowledge of the function. However, each player has only partial information about the input $x = (x_1, \ldots, x_k)$: the ith player P_i has access to all the x_j's *except* x_i. We can imagine the situation as k poker players sitting around the table, and each one is holding a number to his/her forehead for the others to see. Thus, all players know the function f but their access to the input vector is restricted: the first player sees the string $(*, x_2, \ldots, x_k)$, the second sees $(x_1, *, x_3, \ldots, x_k)$, ..., the kth player sees $(x_1, \ldots, x_{k-1}, *)$.

Players can communicate by writing bits 0 and 1 on a blackboard. The blackboard is seen by all players. The game starts with the empty blackboard. For each string on the blackboard, the protocol either gives the value of the output (in that case the protocol is over), or specifies which player writes the next bit and what that bit should be as a function of the inputs this player knows (and the string on the board). During the computation on one input the blackboard is never erased, players simply append their messages. The objective is to compute the function with as small an amount of communication as possible.

The *communication complexity* of a k-party game for f is the minimal number $C_k(f)$ such that on every input $x \in X$ the players can decide whether $f(x) = 1$ or not, by writing at most $C_k(f)$ bits on the blackboard. Put otherwise, $C_k(f)$ is the minimal number of bits written on the blackboard on the worst-case input.

It is clear that $C_k(f) \leqslant \log_2 n + 1$ for any f: the first player writes the binary code of x_2, and the second player announces the result. But what about the lower bounds? The twist is that (for $k \geqslant 3$) the players share some inputs, and (at least potentially) can use this overlap to encode the information in some wicked and non-trivial way (see Theorem 29.12 and Exercises 29.6 and 29.7).

The lower bounds problem for $C_k(f)$ can be re-stated as a Ramsey type problem about the minimal number of colors in coloring of the hypercube which leaves no "forbidden sphere" monochromatic.

A *sphere* in X around a vector $x = (x_1, \ldots, x_k)$ is a set S of k vectors

$$x^i = (x_1, \ldots, x_{i-1}, y_i, x_{i+1}, \ldots, x_k) \text{ with } y_i \neq x_i \text{ for all } i = 1, \ldots, k.$$

The vector x is a *center* of this sphere. Such a sphere is *forbidden* (for f) if it lies entirely in one of the parts $f^{-1}(0)$ or $f^{-1}(1)$, whereas its center belongs to the other part. An r-coloring $c : X \to \{1, \ldots, r\}$ of X is *legal* (with respect to f) if it:

- leaves no forbidden sphere monochromatic;
- uses different colors for vectors in $f^{-1}(0)$ and in $f^{-1}(1)$.

Let $\chi_k(f)$ (or $\chi(f)$ if the number k of variables is clear from the context) denote the minimal number of colors in a legal coloring of X.

Proposition 29.6. *For every* $f : X \to \{0, 1\}$, $C_k(f) \geq \log_2 \chi_k(f)$.

Proof. Take an optimal protocol of the communication game for f. Color each vector $x \in X$ by the string, which is written on the blackboard at the end of communication between the players on the input x. We have $2^{C_k(f)}$ colors and it remains to verify that the coloring is legal for f.

To show this, assume that some forbidden sphere $S = \{x^1, \ldots, x^k\}$ around some vector x is monochromatic. Assume w.l.o.g. that $f(x^1) = \ldots = f(x^k) = 1$ and $f(x) = 0$. An important fact is that given the first l bits communicated by the players, the $(l+1)$-th bit of communication (transmitted, say, by the ith player) must be defined by a function which does not depend on the ith coordinate of the input: player P_i cannot see it. Therefore, for every l, there is an i $(1 \leq i \leq k)$ such that the $(l+1)$-th communicated bit is the same for both inputs x and x^i. Since on all inputs $x^1, \ldots, x^k$ the players behave in the same way (i.e., write the same string on the blackboard), it follows that they will also behave in the same way on the input x. But this means that the players will accept x, a contradiction. $\square$

29.3.1 Few players: the hyperplane problem

To illustrate how the connection between communication complexity and colorings works in concrete situations, we consider the case when

$$X = X_1 \times \cdots \times X_k \quad \text{with} \quad X_1 = \ldots = X_k = \{1, \ldots, n\}.$$

The *hyperplane problem* is the function $h : X \to \{0, 1\}$ such that $h(x) = 1$ if and only if x belongs to the hyperplane

$$H = \{x \in X : x_1 + \cdots + x_k = n\}.$$

For this special function h the lower bound, given by Proposition 29.6, is almost optimal (see Exercise 29.3 for a more general result).

Proposition 29.7. *For every* $k \geq 2$, $C_k(h) \leq \log_2 \chi_k(h) + k$.

Proof. By the definition of h, *every* sphere, lying entirely in the hyperplane H, is forbidden (because then its center must lie outside the hyperplane). Color the hyperplane H with $\chi_k(h)$ colors so that none of these spheres is monochromatic. Let $x = (x_1, \ldots, x_k)$ be an input vector. Each player P_i computes the vector

$$x^i = \left(x_1, \ldots, x_{i-1}, n - \sum_{j \neq i} x_j, x_{i+1}, \ldots, x_k \right).$$

If for some i, the ith coordinate $n - \sum_{j \neq i} x_j$ does not belong to $\{1, \ldots, n\}$, the ith player immediately announces the result: $h(x) = 0$. Otherwise they proceed as follows.

Using $\log_2 \chi_k(h)$ bits, the first player broadcasts the color of x^1. Then, each P_i in turn transmits a 1 if and only if the color of x^i matches the color of x^1. The process halts and accepts the input x if and only if every player broadcast a 1 in the last phase. It is easy to verify that this protocol computes h. To see this, let $S = \{x^1, \ldots, x^k\}$. If none of the players announced $h(x) = 0$ during the first stage, then either all the points of S coincide or S is a sphere around x lying entirely in the hyperplane H.

If $h(x) = 0$, then S is a forbidden sphere, and the players cannot see the same color on its points. Therefore, the players correctly reject that input.

If $h(x) = 1$, then $x \in H$, and hence, $x^1 = \ldots = x^k = x$. In this case the players agree that all points of S have the same color and correctly accept x.
□

Propositions 29.6 and 29.7 give us almost tight characterization $C_k(h) = \Theta\left(\log_2 \chi_k(h)\right)$ of the k-party communication complexity of the hyperplane problem for any fixed k. However, $\chi_k(h)$ is very hard to compute, or even to estimate. In particular, even the following question is not trivial: if the number k of players is fixed, then is $C_k(h)$ also bounded by a constant, as n grows?

Using Gallai–Witt's theorem this question can be answered negatively.

Theorem 29.8. *For every $k \geqslant 2$, the communication complexity of any k-party game for h is greater than any constant.*

Proof. By Proposition 29.6 it suffices to prove that for a fixed k, $\chi_k(h)$ is unbounded. To get a contradiction, assume that $\chi_k(h)$ is bounded by some constant r. We say that a sphere around a vector $x = (x_1, \ldots, x_k)$ is *nice* if it consists of vectors

$$x^i = (x_1, \ldots, x_{i-1}, x_i + \lambda, x_{i+1}, \ldots, x_k), \qquad i = 1, \ldots, k,$$

for some $\lambda \neq 0$. A *ball* around a vector x is a sphere plus the vector x itself; hence, a ball contains $k + 1$ vectors.

Define the projection p from H to $X_1 \times \cdots \times X_{k-1}$ by

$$p(x_1, \ldots, x_k) = (x_1, \ldots, x_{k-1}).$$

Since $\chi_k(h) \leqslant r$, there is an r-coloring c of the hyperplane H after which no sphere, lying entirely in H, is colored monochromatically. The mapping p is an injection. So c and p induce, in a natural way, an r-coloring of the points in $p(H)$. Consider a smaller set, $Y = Y_1 \times \cdots \times Y_{k-1}$ with all $Y_i = \{1, \ldots, m\}$ for $m = \lfloor (n-1)/k \rfloor$. It is a subset of $p(H)$ and is thus also r-colored via p.

Take the set of k vectors $V = \{0, e_1, \ldots, e_{k-1}\}$ in Y, where $e_i = (0, \ldots, 1, 0, \ldots, 0)$ with 1 in the ith coordinate. For large enough m, Gallai–Witt's theorem (Theorem 29.4) implies the existence of a homothetic copy

$$x + \lambda V = \{x, x + \lambda e_1, \ldots, x + \lambda e_{k-1}\}$$

with $\lambda > 0$, which is monochromatic. This copy is a nice ball around the vector x. Consider the vector

$$y = (x_1, \ldots, x_{k-1}, n - s - \lambda),$$

where $s = x_1 + x_2 + \cdots + x_{k-1}$. Since $0 < s \leqslant (k-1)m$ and $0 < \lambda \leqslant m$, the vector y belongs to X. We now have a contradiction since the vectors

$$p^{-1}(x), p^{-1}(x + \lambda e_1), \ldots, p^{-1}(x + \lambda e_{k-1})$$

form a monochromatic sphere around y in H, but $y \notin H$.   □

In this proof we have used Ramsey-type argument to obtain a *lower bound* on the k-party communication complexity of the hyperplane problem. It turns out that this type of arguments can also lead to surprising *upper bounds*: three players can solve the hyperplane problem by using only $O(\sqrt{\log_2 n})$ bits of communication! To prove this result (Theorem 29.12 below) Chandra et al. (1983) have related $\chi_k(h)$ to classic Ramsey numbers for arithmetic progressions.

Let $A_k(N)$ be the minimum number of colors needed to color $1, 2, \ldots, N$ such that no length-k arithmetic progression is colored monochromatically.

Proposition 29.9. *For $N = kn$, $\chi_k(h) \leqslant A_k(N)$.*

Proof. Define a map f from the hyperplane H to $\{1, \ldots, N\}$ by

$$f(x_1, x_2, \ldots, x_k) = x_1 + 2x_2 + \cdots + kx_k.$$

Color $1, \ldots, N$ with $r = A_k(N)$ colors, avoiding monochromatic length-k arithmetic progressions. Color each point $x \in H$ with the color of $f(x)$. We prove by contradiction that this coloring of H contains no monochromatic spheres.

Assume $x^1, \ldots, x^k$ is a monochromatic sphere in H around some vector y. By the definition, $x^i = y + \lambda_i e_i$ for some $\lambda_i \neq 0$ ($i = 1, \ldots, k$). Since each x^i is in the hyperplane H, it follows that $\lambda_1 = \lambda_2 = \ldots = \lambda_k = \lambda$. Consider the points $f(x^1), f(x^2), \ldots, f(x^k)$. The map f is linear, so $f(x^i) = f(y) + \lambda f(e_i)$. By the definition of f, $f(e_i) = i$; hence,

$$f(x^i) = f(y) + i \cdot \lambda \quad (i = 1, \ldots, k)$$

forms a monochromatic arithmetic progression of length k, which is a contradiction.   □

Determining the true rate of growth of $A_k(N)$ is a difficult problem. It is known that $A_k(N)$ is unbounded for fixed k, however, for $k \geqslant 4$ it can only be shown to grow extremely slowly. In the specific case of $k = 3$ it is possible to relate $A_k(N)$ to another Ramsey-like function.

Let $R_k(N)$ be the size of a largest subset $A \subset \{1, \dots, N\}$ that contains no arithmetic progression of length k.

Theorem 29.10. *For some constant $c > 0$,*

$$\frac{N}{R_k(N)} \leqslant A_k(N) \leqslant \frac{cN \ln N}{R_k(N)}.$$

Proof. Lower bound. Let $r = A_k(N)$. The set $\{1, \dots, N\}$ can be r-colored in such a way that there are no length-k monochromatic arithmetic progressions. Some color must be used at least N/r times, and hence, some color class has size at least N/r and has no arithmetic progression of length k. Therefore, $R_k(N) \geqslant N/r$, implying that $A_k(N) = r \geqslant N/R_k(N)$.

Upper bound. Assume that S is a subset of $\{1, \dots, N\}$ that contains no arithmetic progressions of length k. We demonstrate that the upper bound on $A_k(N)$ holds by proving that it is possible to r-color the set $\{1, \dots, N\}$ so that it has no length-k monochromatic progressions, provided $r \leqslant cN \ln N/|S|$. Since the set S has no arithmetic progression of length k, it is enough to show that we can cover the set $\{1, \dots, N\}$ by at most $\ell = cN \ln N/|S|$ translates $S + t_1, \dots, S + t_\ell$ of S; here, as before, $S + t = \{a + t : a \in S\}$.

Claim 29.11. *Let $S \subseteq \{1, \dots, N\}$. No more than $O(N \log N/|S|)$ translates of S are needed to cover $\{1, \dots, N\}$.*

Proof of Claim 29.11. We use a probabilistic argument. Pick the numbers $t_1, \dots, t_\ell$ in the interval $-N$ to N at random independently and uniformly; hence, $\mathrm{Prob}\,(t_i = a) = 1/(2N)$ for each number a. The probability that a value $x \in \{1, \dots, N\}$ is not covered by a particular translation is at most $1 - |S|/(2N)$ (cf. the proof of Theorem 20.2). Therefore, the expected number of missed values is bounded above by

$$N \cdot \left(1 - \frac{|S|}{2N}\right)^\ell.$$

For large enough c, choosing $\ell = cN \ln N/|S|$ makes this less than 1 and the claim is proved. $\qquad\square$

We have already mentioned a result due to Behrend (1949) which says that $R_3(N) \geqslant N\mathrm{e}^{-O(\sqrt{\ln N})}$ (see Theorem 21.8). Combining this result with Theorem 29.10, Proposition 29.9, and Proposition 29.6 we obtain:

Theorem 29.12. *Three players can solve the hyperplane problem by communicating only $O(\sqrt{\log n})$ bits.*

Proof. Taking $N = 3n$, from the above we obtain

$$\chi_3(h) \leqslant A_3(N) \leqslant cN \ln N/R_3(N) \leqslant (\ln n) \cdot \mathrm{e}^{O(\sqrt{\log n})}.$$

Therefore, $3 + \log_2 \chi_k(h)$, the number of bits that have to be communicated, is $O(\sqrt{\log n})$. $\qquad\square$

29.3.2 Many players: the matrix product problem

One of the highest known lower bounds for the communication complexity of k-party games has the form $\Omega\left(4^{-k}\log_2 n\right)$. This bound was proved by Babai, Nisan, and Szegedy (1992) for so-called *generalized inner product* function GIP_n. In this case each X_i consists of all 0-1 (column) vectors of length $m = \lfloor\log_2 n\rfloor$; hence, the input space X consists of all binary $m \times k$ matrices. Given such a matrix $x \in X$, $GIP_n(x) = 1$ if and only if the number of all-1 rows in x is odd. Grolmusz (1994) has shown that this lower bound for GIP_n is almost optimal (see Exercise 29.6 and 29.7).

Similar lower bounds for other explicit functions were obtained by Chung (1990), Chung and Tetali (1993), Babai, Hayes and Kimmel (1998), Raz (2000), and other authors (see Babai 1997 for a survey). However, so far, no non-trivial lower bound is known for the case when the number of players k is larger than $\log_2 \log_2 n$.

As we already mentioned in Sect. 20.9, lower bounds on the multi-party communication complexity of explicit functions $F : X \rightarrow \{0,1\}$ can be obtained by bounding the discrepancy of the corresponding functions $f : X \rightarrow \{-1,1\}$ defined by $f(x) = 1 - 2 \cdot F(x)$. Now we explain this.

Recall that a subset $T \subseteq X$ is a *cylinder* in the ith dimension if membership in T_i does not depend on the ith coordinate. A subset $T \subseteq X$ is a *cylinder intersection* if it is an intersection $T = T_1 \cap T_2 \cap \cdots \cap T_k$, where T_i is a cylinder in the ith dimension. The *discrepancy* of a function $f : X \rightarrow \{-1,1\}$ on a set T is $\mathrm{disc}_T(f) = \left(\sum_{x \in T} f(x)\right) / |X|$. The *discrepancy* $\mathrm{disc}(f)$ of f is the maximum, over all cylinder intersections T, of the absolute value $|\mathrm{disc}_T(f)|$.

It can be shown (see Exercise 29.9) that a set T is a cylinder intersection if and only if it does not separate a sphere from its center, i.e., if for every sphere S around a vector x, $S \subseteq T$ implies $x \in T$. Thus, a coloring $c : X \rightarrow \{1,\ldots,r\}$ is legal for a given function $F : X \rightarrow \{0,1\}$ if and only if each color class $T = c^{-1}(i)$ is a cylinder intersection and the function F is constant on T. Since this last event is equivalent to $|\mathrm{disc}_T(f)| = |T|/|X|$, no color class can have more than $|X| \cdot \mathrm{disc}(f)$ vectors. This implies that we need at least $1/\mathrm{disc}(f)$ colors, and by Proposition 29.6,

$$C_k(F) \geqslant \log_2 \chi_k(F) \geqslant \log_2 \left(1/\mathrm{disc}(f)\right).$$

An explicit function $f : X \rightarrow \{-1,1\}$ (the matrix product function), for which

$$\mathrm{disc}(f) \leqslant \left(\frac{k-1}{\sqrt{\log_2 n}}\right)^{1/2^k},$$

was described in Sect. 20.9.1 (see Theorem 20.15). Hence, for the corresponding $F(x) = (1 - f(x))/2$, we have

$$C_k(F) = \Omega\left(2^{-k}\log_2 n\right).$$

Exercises

29.1. $^-$ Show that $HJ(r,2) \leqslant r$. That is, for every coloring of the cube $\{0,1\}^r$ in r colors at least one combinatorial line is monochromatic.

Hint: Consider the words $0^i 1^{r-i}$ for $i = 0,1,\ldots,r$.

29.2. Let $N = W(2, t^2 +1)$, where $W(r,t)$ is the van der Waerden's function, and let χ be a coloring of $\{1,\ldots,N\}$ in two colors. Show that there exists a t-term arithmetic progression $\{a + i \cdot d : i = 0,1,\ldots,t-1\}$ which together with its difference d is monochromatic, i.e., $\chi(d) = \chi(a+i \cdot d)$ for every $i < t$.

Hint: Van der Waerden's theorem gives a monochromatic arithmetical progression $\{a + j \cdot d : j \leqslant t^2\}$ with t^2 terms. Then either some $j \cdot d$, with $1 \leqslant j \leqslant t$, gets the same color or all the numbers $d, 2d, \ldots, td$ get the opposite color.

29.3. Prove the following extension of Proposition 29.7 to an arbitrary function $f : X \rightarrow \{0,1\}$. For a vector $x \in X$, its *ith neighbor* is a vector $x^i = (x_1,\ldots,x'_i,\cdots,x_k)$, with $x'_i \in X_i$, such that $f(x^i) = 1$; if $f(x) = 1$, then x is a neighbor of itself. Let $c : X \rightarrow \{1,\ldots,r\}$ be a legal coloring (with respect to f). Let also $N(c) = \max\{N(c,x) : x \in X\}$, where $N(c,x)$ is the minimum, over all coordinates i, of the number of colors used by c to color the ith neighbors of x. Then

$$C_k(f) \leqslant \log_2 r + k + N(c) \log_2 N(c).$$

Hint: Given an input vector x, the ith player can privately compute the set R_i of colors used by c to color the ith neighbors of x. Show that $f(x) = 1$ if and only if $R_1 \cap R_2 \cap \cdots \cap R_k \neq \emptyset$.

29.4. $^-$ For a fixed vector $x \in X$, there are many (how many?) spheres around it. How many colors do we need to leave none of them monochromatic?

29.5. $^-$ Consider the function $f : X_1 \times X_2 \rightarrow \{0,1\}$ such that $f(x_1,x_2) = 1$ if and only if $x_1 = x_2$. Show that $\chi(f) = n$.

Hint: If $\chi(f) < n$ then some color class contains two distinct vectors (x_1,x_1) and (x_2,x_2). What about the color of (x_1,x_2)?

29.6. $^{(!)}$ (Grolmusz 1994). Consider the following k-party communication game. Input is an $m \times k$ 0-1 matrix A, and the ith player can see all A except its ith column. Suppose that the players a priori know that some string $v = (0,\ldots,0,1,\ldots,1)$ with the first 1 in position $t+1$, does not appear among the rows of A. Show that then the players can decide if the number of all-1 rows is even or odd by communicating only t bits.

Sketch: Let y_i denote the number of rows of A of the form $(0,\ldots,0,1,\ldots,1)$, where the first 1 occurs in position i. For every $i = 1,\ldots,t$, the ith player announces the parity of the number of rows of the form $(0,\ldots,0,*,1,\ldots,1)$, where the $*$ is at place i. Observe that this number is $y_i + y_{i+1}$. Subsequently, each player privately computes the mod 2 sum of all numbers announced. The result is $y_1 + y_{t+1} \pmod 2$, where $y_{t+1} = 0$.

29.7. Use the previous protocol to show that (without any assumption) the players can decide if the number of all-1 rows is even or odd by communicating only $O(km/2^k)$ bits.

Sketch: Divide the matrix A into blocks with at most $2^{k-1} - 1$ rows in each. For each block there will be a string v' of length $k - 1$ such that neither $(0, v')$ nor $(1, v')$ occurs among the rows in that block. Using k bits the first player can make the string $(0, v')$ known to all players, and we are in the situation of the previous exercise.

29.8 (due to Babai and Kimmel). Consider the following multiparty game with the *referee*. As before, we have an $m \times k$ 0-1 matrix A, and the ith player can see all A except its ith column. The restriction is that now the players do not communicate with each other but simultaneously write their messages on the blackboard. Using only this information (and without seeing the matrix A), an additional player (the referee) must compute the string $P(A) = (x_1, \ldots, x_m)$, where x_i is the sum modulo 2 of the number of 1's in the ith row of A. Let N be the maximal number of bits which any player is allowed to write on any input matrix. Prove that $N \geqslant m/k$.

Sketch: For a matrix A, let $f(A)$ be the string $(p_1, \ldots, p_k)$, where $p_i \in \{0, 1\}^N$ is the string written by the ith player on input A. For each possible answer $x = (x_1, \ldots, x_m)$ of the referee, fix a matrix A_x for which $P(A_x) = x$. The correctness of the communication protocol ensures that $f(A_x) \neq f(A_y)$ for all $x \neq y$; hence, $2^{Nk} \geqslant 2^m$.

29.9. Show that a set $T \subseteq X$ is a cylinder intersection if and only if, for every sphere S around a vector x, $S \subseteq T$ implies $x \in T$.

Hint: For the "if" part consider the sets T_i of all vectors $(x_1, \ldots, x_i, \ldots, x_k)$ such that $(x_1, \ldots, x'_i, \ldots, x_k) \in T$ for at least one $x'_i \in X_i$.

29.10.[+] Use Theorem 29.1 to derive the Hales–Jewett theorem for combinatorial m-spaces (Theorem 29.2).

Hint: Argue by induction on m. For the induction step $m \mapsto m + 1$ consider the words of length $M + N$, where $M = HJ(1, r, t)$ and $N = HJ(m, r^{t^M}, t)$. Split each such word a into two segments $a = (b, c)$ of length M and N, respectively, and color each tail c by the sequence of colors it gets by varying over all t^M possible initial pieces. By the choice of N, there exists a combinatorial m-space S, which is monochromatic with respect to this new coloring. Use this space (and its color) to define a coloring of initial segments b. By the choice of M, there exists a combinatorial line L, which is monochromatic with respect to this coloring. Combine the obtained monochromatic space S and line L into a monochromatic $(m + 1)$-space.

Epilog: What Next?

In this book we have touched some of the basic ideas and methods of extremal combinatorics. For further in-depth study of the subject I would recommend the following texts.

Extremal problems for graphs:

- B. Bollobás, *Extremal Graph Theory*. London Mathematical Society Monographs, No. 11, Academic Press, 1978.
- B. Bollobás, *Modern Graph Theory*. Graduate Texts in Mathematics, Vol. 184, Springer, 1998.

Sets systems and hypergraphs:

- I. Anderson, *Combinatorics of Finite Sets*. Oxford University Press, 1987.
- C. Berge, *Hypergraphs: Combinatorics of Finite Sets*. North-Holl. Math. Library, Vol. 45, Elsevier Sci. Publishers, 1989.
- B. Bollobás, *Combinatorics: Set Systems, Hypergraphs, Families of Vectors, and Combinatorial Probability*. Cambridge University Press, 1986.

The linear algebra method:

- L. Babai and P. Frankl, *Linear Algebra Methods in Combinatorics*. Preliminary Version 2 (September 1992), University of Chicago. This is the first (and, so far, the only) comprehensive text on this topic.

The probabilistic method:

- N. Alon and J. Spencer, *The Probabilistic Method*. Second Edition, Wiley, 2000.
- P. Erdös and J. Spencer, *Probabilistic Methods in Combinatorics*. Academic Press, 1974.
- J. Spencer, *Ten Lectures on the Probabilistic Method*. SIAM Regional Conference Series in Applied Mathematics, Vol. 52, Second Edition, 1993.

Randomized algorithms:

o R. Motwani and P. Raghavan, *Randomized Algorithms*. Cambridge University Press, 1995.

Ramsey theory:

o R.L. Graham, B.L. Rothschild, and J.H. Spencer, *Ramsey Theory*. Second Edition, Wiley, 1990.
o *Mathematics of Ramsey Theory*. J. Nesetril and V. Rödl (eds.), Algorithms and Combinatorics, Vol. 5, Springer, 1990.

Advanced sources:

o *Handbook of Combinatorics*. R. Graham, M. Grötschel, and L. Lovász (eds.), Elsevier Science, 1995.
o *Extremal Problems for Finite Sets*. P. Frankl, Z. Füredi, G. Katona, and D. Miklós (eds.), János Bolyai Mathematical Society, 1994.

Problem-solving

o L. Lovász, *Combinatorial Problems and Exercises*. Akadémiai Kiadó & North-Holland, 1979. For advanced problem-solvers this remains an invaluable source.
o Z. Michalewicz and D.B. Fogel, *How to Solve It: Modern Heuristics*. Corrected Second Printing, Springer, 2000. This successor of G. Polya's classic title *How to Solve It* (Princeton University Press, 1945) gives a compendium in heuristic problem solving.

References

Adleman, L. (1978): Two theorems on random polynomial time, in: *Proc. of 19th Ann. IEEE Symp. on Foundations of Comput. Sci.*, 75–83.

Ahlswede, R., El Gamal, A., and Pang, K.F. (1984): A two-family extremal problem in Hamming space, *Discrete Math.* **49**:1, 1–5.

Ahlswede, R., Cai, N., and Zhang, Z. (1994): A new direction in extremal theory, *J. Combin. Inform. System Sci.* **19**, no. 3–4, 269–280.

Ahlswede, R., Alon, N., Erdős, P., Ruszinkó, M., and Szekely, L.A. (1997): Intersecting systems, *Combinatorics, Probability & Computing* **6**:2, 127–138.

Ajtai, M. (1983), Σ_1^1-formulae on finite structures, *Ann. Pure and Appl. Logic* **24**, 1–48.

Ajtai, M., Komlós, J., and Szemerédi, E. (1980): A note on Ramsey numbers, *J. Combin. Theory (A)* **29**, 354–360.

Alon, N. (1983): On the density of sets of vectors, *Discrete Math.* **46**, 199–202.

Alon, N. (1985): An extremal problem for sets with applications to graph theory, *J. Combin. Theory (A)* **40**, 82–89.

Alon, N. (1986): Explicit constructions of exponential size families of k-independent sets, *Discrete Math.* **58**, 191–193.

Alon, N. (1990a): On the rigidity of Hadamard matrices. Unpublished manuscript.

Alon, N. (1990b): The maximum number of Hamiltonian paths in tournaments, *Combinatorica* **10**, 319–324.

Alon, N. (1990c): Transversal numbers of uniform hypergraphs, *Graphs and Combinatorics* **6**, 1–4.

Alon, N. (1994): Probabilistic methods in extremal finite set theory, in: *Extremal Problems for Finite Sets*, Bolyai Society Mathematical Studies, Vol. 3, 39–58, János Bolyai Math. Society.

Alon, N. (1995): Tools from higher algebra, in: *Handbook of Combinatorics*, R. Graham, M. Grötschel and L. Lovász (eds.), Elsevier Science, Vol. 2, 1749–1784.

Alon, N., Babai, L., and Itai, A. (1986): A fast and simple randomized parallel algorithm for the maximal independent set problem, *J. of Algorithms* **7**, 567–583.

Alon, N. and Boppana, R. (1987): The monotone circuit complexity of boolean functions, *Combinatorica* **7**:1, 1–22.

Alon, N. and Kleitman, D.J. (1990): Sum-free subsets, in: *A Tribute to Paul Erdős*, A. Baker, B. Bollobás, and A. Hajnál (eds.), Cambridge University Press, 13–26.

Alon, N. and Dubiner, M. (1993): Zero-sum sets of prescribed size, in: *Combinatorics, Paul Erdős is Eighty*, Vol. 1, 33–50.

Alon, N. and Spencer, J. (1992): *The Probabilistic Method*, Wiley, 1992. Second edition: Wiley, 2000.

Alon, N., Babai, L., and Suzuki, H. (1991): Multilinear polynomials and Frankl–Ray-Chaudhuri–Wilson type intersection theorems, *J. Combin. Theory (A)* **58**:2, 165–180.

Alon, N., Bergmann, E.E., Coppersmith, D., and Odlyzko, A.M. (1988): Balancing sets of vectors, *IEEE Trans. Information Theory IT-***34**, 128–130.

Anderson, I. (1987): *Combinatorics of Finite Sets*. Oxford University Press.

Anderson, I. and Honkala, I. (1997): *A short course in combinatorial designs*. Internet Edition, Spring 1997.

Andreev, A.E. (1985): On a method for obtaining lower bounds for the complexity of individual monotone functions, *Soviet Math. Dokl.* **31**:3, 530–534.

Andreev, A.E. (1987): On one method of obtaining effective lower bounds on monotone complexity, *Algebra i Logica* **26**:1, 3–26. (Russian)

Aspnes, J., Beigel, R., Furst, M., and Rudich, S. (1994): The expressive power of voting polynomials, *Combinatorica*, **14**:2, 1–14.

Babai, L. (1988): A short proof of the nonuniform Ray-Chaudhuri–Wilson inequality, *Combinatorica* **8**:1, 133–135.

Babai, L. (1997): Communication complexity, in: *Proc. of 22nd Internat. Symp. on Mathematical Foundations of Computer Science*, Lect. Notes in Comput. Sci., Vol. 1295, 5–18, Springer.

Babai, L., Frankl, P., and Simon, J. (1986): The complexity classes in communication complexity, in: *Proc. of 27th Ann. IEEE Symp. on Foundations of Comput. Sci.*, 337–347.

Babai, L. and Frankl, P. (1992): *Linear Algebra Methods in Combinatorics*. Preliminary Version 2, University of Chicago.

Babai, L., Fortnow, L., Levin, L.A., and Szegedy, M. (1991): Checking computations in polylogarithmic time, in: *Proc. of 23rd Ann. ACM Symp. on the Theory of Computing*, 21–31.

Babai, L., Nisan, N., and Szegedy, M. (1992): Multiparty protocols, pseudorandom generators for Logspace, and time-space trade-offs, *J. Comput. Syst. Sci.* **45**, 204–232.

Babai, L., Gál, A., Kollár, J., Rónyai, L., Szabó, T., and Wigderson, A. (1996): Extremal bipartite graphs and superpolynomial lower bounds for monotone span programs, in: *Proc. of 28th Ann. ACM Symp. on the Theory of Computing*, 603–611.

Babai, L., Hayes, T., and Kimmel, P. (1998): The cost of the missing bit: communication complexity with help, in: *Proc. of 30th Ann. ACM Symp. on the Theory of Computing*, 673–682.

Babai, L., Gál, A., and Wigderson, A. (1999): Superpolynomial lower bounds for monotone span programs, *Combinatorica* **19**:3, 301–319.

Barefoot, C., Casey, K., Fisher, D., Fraughnaugh, K., and Harary, F. (1995): Size in maximal triangle-free graphs and minimal graphs of diameter 2, *Discrete Math.* **138**, No. 1-3, 93–99.

Beame, P. and Pitassi, T. (1996): Simplified and improved resolution lower bounds, in: *Proc. of 37th Ann. IEEE Symp. on Foundations of Comput. Sci.*

Beame, P., Saks, M., and Thathachar, J. (1998): Time-space tradeoffs for branching programs, in: *Proc. of 39th Ann. IEEE Symp. on Foundations of Comput. Sci.*, 254–263.

Beck, J. (1978): On 3-chromatic hypergraphs, *Discrete Math.* **24**, 127–137.

Beck, J. (1980): A remark concerning arithmetic progressions, *J. Combin. Theory (A)* **29**, 376–379.

Behrend, F.A. (1946): On sets of integers which contain no three elements in arithmetic progression, *Proc. Nat. Acad. Sci.* **23**, 331–332.

Beimel, A., Gál, A., and Paterson, M. (1996): Lower bounds for monotone span programs, *Computational Complexity* **6** (1996/97), 29–45.

Berge, C. (1957): Two theorems in graph theory, *Proc. Nat. Acad. Sci.* **43**, 842–844.

Berge, C. (1989): *Hypergraphs.* Elsevier Science.

Birkhoff, G. (1946): Tres observaciones sobre el algebra lineal, *Rev. Fac. Ci. Exactas, Puras y Aplicadas Univ. Nac. Tucuman, Ser A* **5**, 147–151.

Blackburn, S.R. (2000): Perfect hash families: probabilistic methods and explicit constructions, *J. Combin. Theory (A)* **92**, 54–60.

Blake, A. (1937): *Canonical expressions in boolean algebra.* PhD thesis, University of Chicago.

Blokhuis, A. (1981): A new upper bound for the cardinality of 2-distance sets in Euclidean space. Eindhoven Univ. Technology, Mem. 1981–04.

Blokhuis, A. (1987): More on maximal intersecting families of finite sets, *J. Combin. Theory (A)* **44**, 299–303.

Blum, M., Chandra, A.K., and Wegman, M.N. (1980): Equivalence of free boolean graphs can be decided probabilistically in polynomial time, *Inform. Process. Letters* **10**, 80–82.

Blum, M. and Impagliazzo, R. (1987): Generic oracles and oracle classes, in: *Proc. of 28th Ann. IEEE Symp. on Foundations of Comput. Sci.*, 118–126.

Bollobás, B. (1965): On generalized graphs, *Acta Math. Acad. Sci. Hungar.* **16**, 447–452.

Bollobás, B. (1978): *Extremal Graph Theory.* LMS Monographs, No. 11, Academic Press.

Bollobás, B. (1981): Threshold functions for small subgraphs, *Math. Proc. Cambridge Philos. Soc.* **90**, 197–206.

Bollobás, B. (1986): *Combinatorics: Set Systems, Hypergraphs, Families of Vectors, and Combinatorial Probability.* Cambridge University Press.

Bollobás, B. and Thomason, A. (1981): Graphs which contain all small graphs, *European J. Combin.* **2**, 13–15.

Bondy, J.A. (1972): Induced subsets, *J. Combin. Theory (B)* **12**, 201–202.

Borodin, A., Dolev, D., Fich, F., and Paul, W. (1986): Bounds for width two branching programs, *SIAM J. Comput.* **15**:2, 549–560.

Borsuk, K. (1933): Drei Sätze über die n-dimensionale euklidische Sphäre, *Fund. Math.* **20**, 177–190.

Bose, R.C. (1949): A note on Fisher's inequality for balanced incomplete block designs, *Ann. Math. Stat.* **20**, 619–620.

Brower, A.E. and Schrijev, A. (1978): The blocking number of an affine space, *J. Combin. Theory (A)* **24**, 251–253.

Brown, T.C., Chung, F.R.K., Erdös, P., and R.L. Graham (1985): Quantitive forms of a theorem of Hilbert, *J. Combin. Theory (A)* **38**, 210–216.

Bruen, A. (1970): Baer subplanes and blocking sets, *Bull. Amer. Math. Soc.* **76**, 342–344.

Buss, S. (1987): Polynomial size proofs of the propositional pigeonhole principle, *J. Symbolic Logic* **52**, 916–927.

Buss, S. and Pitassi, T. (1998): Resolution and the weak pigeonhole principle, in: *Proc. of Workshop on Comput. Sci. Logic,* Annual Conf. of the EACSL, Lect. Notes in Comput. Sci., Vol. 1414, 149–158, Springer.

Cameron, P.J. (1994): *Combinatorics: Topics, Techniques, Algorithms.* Cambridge University Press.

Cameron, P.J. and Erdős, P. (1999): Notes on sum-free and related sets, *Combinatorics, Probability & Computing* **8**, 95–107.

Chandra, A.K., Furst, M., and Lipton, R.J. (1983): Multi-party protocols, in: *Proc. of 15th Ann. ACM Symp. on the Theory of Computing,* 94–99.

Chandra, A.K., Kou, L., Markowsky, G., and Zaks, S. (1983): On sets of boolean n-vectors with all k-projections surjective, *Acta Informatica* **20**:2, 103-111.

Chernoff, H. (1952): A measure of the asymptotic efficiency for tests of a hypothesis based on the sum of observations, *Ann. Math. Stat.* **23**, 493–509.

Chung F.R.K. (1990): Quasi-random classes of hypergraphs, *Random Structures and Algorithms* **1**, 363–382.

Chung, F.R.K. and Tetali, P. (1993): Communication complexity and quasi randomness, *SIAM J. Discrete Math.* **6**, 110–123.

Chung, F.R.K., Frankl, P., Graham, R.L., and Shearer, J.B. (1986): Some intersection theorems for ordered sets and graphs, *J. Combin. Theory (A)* **43**, 23-37.

Chvátal, V. and Szemerédi, E. (1988): Many hard examples for resolution, *J. of the ACM* **35**:4, 759–768.

Cohen, G., Honkala, I., Litsyn, S., and Lobstein, A. (1997): *Covering Codes.* Math. Library, Vol. 54, North-Holland.

Corrádi, K. (1969): Problem at Schweitzer competition, *Mat. Lapok* **20**, 159–162.

Danzer, L. and Grűnbaum, B. (1962): Über zwei Probleme bezüglich konvexer Körper von P. Erdös und von V. L. Klee, *Math. Zeitschrift* **79**, 95-99.

Davis, M. and Putnam, H. (1960): A computing procedure for quantification theory, *J. of the ACM* **7**:3, 210–215.

de Bruijn, N.G. and Erdős, P. (1948): On a combinatorial problem, *Indag. Math.* **10**, 421–423.

de Bruijn, N.G., Tengbergen, C., and Kruyswijk, D. (1952): On the set of divisors of a number, *Nieuw. Arch. Wiskunde* **23**, 191–193.

Delsarte, Ph. and Piret, Ph. (1985): An extension of an inequality by Ahlswede, El Gamal and Pang for pairs of binary codes, *Discrete Math.* **55**, 313–315.

Deuber, W.A., Erdős, P., Gunderson, D.S., Kostochka, A.V., and Meyer A.G. (1997): Intersection statements for systems of sets, *J. Combin. Theory (A)* **79**:1, 118–132.

Deza, M. (1973): Une propriété extrémale des plans projectifs finis dans une classe des codes équidistantes, *Discrete Math.* **6**, 343–352.

Deza, M., Frankl, P., and Singhi, N.M. (1983): On functions of strength t, *Combinatorica* **3**, 331–339.

Dirichlet, P.G.L. (1879): *Vorlesungen über Zahlentheorie.* Vieweg, Braunschweig.

Dilworth, R.P. (1950): A decomposition theorem for partially ordered sets, *Ann. of Math.* **51**, 161–165.

Dyachkov A.G. and Rykov V.V. (1982): Bounds on the length of disjunctive codes, *Problemy Peredachi Informatsii* **18**:3, 7–13. (Russian)

Egerváry, E. (1931): On combinatorial properties of matrices, *Mat. Lapok* **38**, 16–28. (Hungarian with German summary)

Erdős, P. (1945): On a lemma of Littlewood and Offord, *Bull. Amer. Math. Soc.* **51**, 898–902.

Erdős, P. (1947): Some remarks on the theory of graphs, *Bull. Amer. Math. Soc.* **53**, 292–294.

Erdős, P. (1959): Graph theory and probability, *Canadian J. Math.* **11**, 34–38.

Erdős, P. (1963): On a problem of graph theory, *Math. Gaz.* **47**, 220–223.

Erdős, P. (1963b): On a combinatorial problem I, *Nordisk Tidskr. Informationsbehandlung (BIT)* **11**, 5–10.

Erdős, P. (1964): On a combinatorial problem II, *Acta Math. Acad. Sci. Hungar.* **15**, 445–447.

Erdős, P. (1964b): On extremal problems of graphs and generalized graphs, *Israel J. Math.*, **2**, 183–190.

Erdős, P. (1965a): Extremal problems in number theory, *Proc. of the Symposium on Pure Mathematics*, **VIII**, AMS, 181–189.

Erdős, P. (1965c): On some extremal problems in graph theory, *Israel J. Math.*, **3** (1965), 113–116.

Erdős, P. (1967): On bipartite subgraphs of a graph, *Matematikai Lapok* **18**, 283–288. (Hungarian)

Erdős, P. and Szekeres, G. (1935): A combinatorial problem in geometry, *Composito Math.* **2**, 464–470.

Erdős, P. and Rado, R. (1960): Intersection theorems for systems of sets, *J. London Math. Soc.* **35**, 85–90.

Erdős, P. and Rényi, A. (1960): On the evolution of random graphs, *Publ. Math, Inst. Hung. Acad. Sci.* **5**, 17–61.

Erdős, P., Ginzburg, A., and Ziv, A. (1961): Theorem in the additive number theory, *Bull. Research Council Israel* **10F**, 41–43.

Erdős, P., Chao Ko and Rado, R. (1961): Intersection theorems for systems of finite sets, *Quart. J. Math. Oxford (2)* **12**, 313–320.

Erdős, P., Hajnal, A., and Moon, J.W. (1964): A problem in graph theory, *Acta Math. Acad. Sci. Hungar.* **71**, 1107–1110.

Erdős, P. and Kleitman, D.J. (1968): On coloring graphs to maximize the proportion of multicolored k-edges, *J. Combin. Theory (A)* **5**, 164–169.

Erdős, P. and Spencer, J. (1974): *Probabilistic Methods in Combinatorics.* Academic Press, New York and London, and Akadémiai Kiadó, Budapest.

Erdős, P. and Lovász, L. (1975): Problems and results on 3-chromatic hypergraphs and some related questions, in: *Infinite and Finite Sets*, Coll. Math. Soc. J. Bolyai, 609–627.

Erdős, P. and Füredi, Z. (1983): The greatest angle among n points in the d-dimensional Euclidean space, *Annals of Discrete Math.* **17**, 275–283.

Erdős, P., Frankl, P., and Füredi, Z. (1985): Families of finite sets in which no set is covered by the union of r others, *Israel J. Math.* **51**, 79–89.

Evans, T. (1960): Embedding incomplete Latin squares, *Amer. Math. Monthly* **67**, 958–961.

Frankl, P. (1977): A constructive lower bound for Ramsey numbers, *Ars Combin.* **3**, 297–302.

Frankl, P. (1982): An extremal problem for two families of sets, *European J. Combin.* **3**, 125–127.

Frankl, P. (1983): On the trace of finite sets, *J. Combin. Theory (A)* **34**, 41–45.

Frankl, P. (1988): Intersection and containment problems without size restrictions, in: *Algebraic, Extremal and Metric Combinatorics*, 1986, M. Deza and P. Frankl (eds.), Cambridge University Press.

Frankl, P. and Wilson, R.M. (1981): Intersection theorems with geometric consequences, *Combinatorica* **1**, 357–368.

Frankl, P. and Pach, J. (1983): On the number of sets in a null t-design, *European J. Combin.* **4**, 21–23.

Frankl, P. and Pach, J. (1984): On disjointly representable sets, *Combinatorica* **4**:1, 39–45.

Frankl, P. and Rödl, V. (1987): Forbidden intersections, *Trans. Amer. Math. Soc.* **300**, 259–286.

Frankl, P. and Tokushige, N. (1998): Some inequalities concerning cross-intersecting families, *Combinatorics, Probability & Computing* **7**:3, 247–260.

Fredman, M. and Komlós, J. (1984): On the size of separating systems and perfect hash functions, *SIAM J. Algebraic and Discrete Meth.* **5**, 61–68.

Freivalds, R. (1977): Probabilistic machines can use less running time, in: *Proc. of IFIP Congress 77*, 839–842. North-Holland, Amsterdam.

Friedman, J. (1984): Constructing $O(n \log n)$ size monotone formulae for the kth elementary symmetric polynomial of n boolean variables, in: *Proc. of 25th Ann. IEEE Symp. on Foundations of Comput. Sci.*, 506–515.

Frobenius, G. (1917): Über zerlegbare Determinanten, *Sitzungsber. Königl. Preuss. Acad. Wiss.* **XVIII**, 274–277.

Füredi, Z. (1980): On maximal intersecting families of finite sets, *J. Combin. Theory (A)* **28**, 282–289.

Füredi, Z. (1984): Geometrical solution of an intersection problem for two hypergraphs, *European J. Combin.* **5**, 133–136.

Füredi, Z. (1996): On r-cover-free families, *J. Combin. Theory (A)* **73**, 172–173.

Füredi, Z. and Kahn, J. (1989): On the dimension of ordered sets of bounded degree, *Order* **3**, 15–20.

Füredi, Z. and Tuza, Zs. (1985): Hypergraphs without large star, *Discrete Math.* **55**, 317–321.

Furst, M., Saxe, J., and Sipser, M. (1984): Parity, circuits and the polynomial time hierarchy, *Math. Syst. Theory* **17**, 13–27.

Gál, A. (1998): A characterization of span program size and improved lower bounds for monotone span programs, in: *Proc. of 30th Ann. ACM Symp. on the Theory of Computing*, 429–437.

Galvin, F. (1994): A proof of Dilworth's chain decomposition theorem, *Amer. Math. Monthly* **101**:4, 352–353.

Galvin, F. (1995): The list chromatic index of a bipartite multigraph, *J. Combin. Theory (B)* **63**, 153–158.

Galvin, F. (1997): On a theorem of J. Ossowski, *J. Combin. Theory (A)* **78**, 151–153.

Goodman, A.W. (1959): On sets of acquaintances and strangers at any party, *Amer. Math. Monthly* **66**, 778–783.

Graham, R.L. and Spencer, J. (1971): A constructive solution to a tournament problem, *Canad. Math. Bull.* **14**, 45–48.

Graham, R.L. and Kleitman, D.J. (1973): Intersecting paths in edge ordered graphs, *Period. Math. Hungar.* **3**, 141–148.

Graham, R., Rothschild, B., and Spencer, J. (1990): *Ramsey Theory*. Second Edition, Wiley.

Graham, R., Grötschel, M., and Lovász, L. (eds.) (1995): *Handbook of Combinatorics*. Elsevier Science, North-Holland.

Grigni, M. and M.Sipser, M. (1995): Monotone separation of logarithmic space from logarithmic depth, *J. Comput. Syst. Sci.* **50**, 433–437.

Grolmusz, V. (1994): The BNS lower bound for multi-party protocols is nearly optimal, *Information and Computation* **112**:1, 51–54.

Guibas, L.J. and Odlyzko, A.M. (1981): String overlaps, pattern matching and non-transitive games, *J. Combin. Theory (A)* **30**, 183–208.

Gunderson, D.S. and Rödl, V. (1998): Extremal problems for affine cubes of integers, *Combinatorics, Probability & Computing* **7**:1, 65–79.

Gyárfás, A. (1987): Partition covers and blocking sets in hypergraphs, *MTA SZ-TAKI Tanulmáyok* **71**, Budapest. (Hungarian)

Hagerup, T. and Rüb, Ch. (1989): A guided tour of Chernoff bounds, *Inform. Process. Letters* **33**, 305–308.

Haken, A. (1985): The intractability of resolution, *Theor. Comput. Sci.* **39**, 297–308.

Haken, A. (1995): Counting bottlenecks to show monotone P≠NP, in: *Proc. of 36th Ann. IEEE Symp. on Foundations of Comput. Sci.*, 36–40.

Haken, A. and Cook, S. (1999): An exponential lower bound for the size of monotone real circuits, *J. Comput. Syst. Sci.* **58**:2, 326–225.

Hales, A.W., and Jewett, R.I. (1963): Regularity and positional games, *Trans. Amer. Math. Soc.* **106**, 222–229.

Hall, P. (1935): On representatives of subsets, *J. London Math. Soc.* **10**, 26–30.

Hardy, G.H., Littlewood, J.E., and Polya, G. (1952): *Inequalities*. Cambridge University Press, 1952.

Hartmanis, J. and Hemachandra, L.A. (1991): One-way functions, robustness and non-isomorphism of NP-complete classes, *Theor. Comput. Sci.* **81**:1, 155–163.

Håstad, J. (1986): *Computational Limitations for Small Depth Circuits*. MIT Press.

Håstad, J. (1989): Almost optimal lower bounds for small depth circuits, in: *Advances in Computing Research*, S. Micali (ed.), Vol. 5, 143–170.

Håstad, J. (1993): The shrinkage exponent is 2, in: *Proc. of 34th Ann. IEEE Symp. on Foundations of Comput. Sci.*, 114–123.

Håstad, J., Jukna, S., and Pudlák, P. (1995): Top-down lower bounds for depth-three circuits, *Computational Complexity* **5**, 99–112.

Helly, E. (1923): Über Mengen konvexer Körper mit gemeinschaftlichen Punkten, *Jahresber. Deutsch. Math.-Verein* **32**, 175–176.

Hilbert, D. (1892): Über die Irreduzibilität ganzer rationaler Funktionen mit ganzzahligen Koeffizienten, *J. Reine Angew. Math.* **110**, 104–129.

Hopcroft, J.E. and Karp, R.M. (1973): An $n^{5/2}$ algorithm for maximum matching in bipartite graphs, *SIAM J. Comput.* **4**, 225–231.

Hwang, F.K. and Sós, V.T. (1987): Non-adaptive hypergeometric group testing, *Studia Sci. Math. Hungar.* **22**, 257–263.

Ibarra, O.H. and Moran, S. (1983): Probabilistic algorithms for deciding equivalence of stright-line programs, *J. of the ACM* **30**:1, 217–228.

Jamison, R. (1977): Covering finite fields with cosets of subspaces, *J. Combin. Theory (A)* **22**, 253–266.

Janssen, J.C.M. (1992): The Dinitz problem solved for rectangles, *Bull. Amer. Math. Soc.* **29**, 243–249.

Johnson, D.S. (1974): Approximation algorithms for combinatorial problems, *J. Comput. Syst. Sci.* **9**, 256–278.

Johnson, S. (1986): A new proof of the Erdős–Szekeres convex k-gon result, *J. Combin. Theory (A)* **42**, 318–319.

Jukna, S. (1997): Finite limits and monotone computations: the lower bounds criterion, in: *Proc. of 12th Ann. IEEE Conf. on Computational Complexity*, 302–313.

Jukna, S. (1999): Combinatorics of monotone computations, *Combinatorica* **19**:1, 65–85.

Kahn, J. and Kalai, G. (1993): A counterexample to Borsuk's conjecture, *Bull. Amer. Math. Soc.* **29**:1, 60–62.

Kalai, G. (1984): Intersection patterns of convex sets, *Israel J. Math.* **48**, 161–174.

Karchmer, M. and Wigderson, A. (1990): Monotone circuits for connectivity require super-logarithmic depth, *SIAM J. Discrete Math.* **3**, 255–265.

Karchmer, M. and Wigderson, A. (1993): On span programs, in: *Proc. of 8th Ann. IEEE Conf. on Structure in Complexity Theory*, 102–111.

Karger, D.R. (1993): Global min-cuts in RNC, and other ramifications of a simple min-cut algorithm, in: *Proc. 4th Ann. ACM-SIAM Symp. on Discrete Algorithms*, 21–30.

Karoński, M. (1995): Random graphs, in: *Handbook of Combinatorics*, R. Graham, M. Grötschel and L. Lovász (eds.), Elsevier-Science, Vol. 1, 351–380.

Katona, G.O. (1972): A simple proof of Erdős Ko Rado theorem, *J. Combin. Theory (B)* **13**, 183–184.

Kautz, W.H. and Singleton, R.C. (1964): Nonrandom binary superimposed codes, *IEEE Trans. Inform. Theory* **10**, 363–377.

Kelly, L.M. (1947): Elementary problems and solutions. Isosceles of n-points, *Amer. Math. Monthly* **54**, 227–229.

Khrapchenko, V.M. (1971): A method of determining lower bounds for the complexity of Π–schemes, *Math. Notes of the Acad. of Sci. of the USSR* **10**:1, 474–479.

Kim, J.H. (1995): The Ramsey number $R(3,t)$ has order of magnitude $t^2/\log t$, *Random Structures and Algorithms* **7**, 173–207.

Kleitman, D.J. (1966): Families of non-disjoint subsets, *J. Combin. Theory (A)* **1**, 153–155.

Kleitman, D.J. and Spencer, J. (1973): Families of k-independent sets, *Discrete Math.* **6**, 255–262.

Kleitman, D.J., Shearer, J. B., and Sturtevant, D. (1981): Intersection of k-element sets, *Combinatorica* **1**, 381–384.

Knuth, D.E. (1986): Efficient balanced codes, *IEEE Trans. Information Theory* *IT*-**32**, 51–53.

König, D. (1931): Graphen und Matrizen, *Mat. és Fiz. Lapok* **38**, 116–119. (Hungarian)

Körner, J. and Marton, K. (1988): New bounds for perfect hashing via information theory, *European J. Combin.* **9**, 523–530.

Kostochka, A.V. and Rödl, V. (1998): On large systems of sets with no large weak Δ-subsystem, *Combinatorica* **18**:2, 235–240.

Kostochka, A.V., Rödl, V., and Talysheva, L.A. (1999): On systems of small sets with no large Δ-system, *Combinatorics, Probability & Computing* **8**, 265–268.

Kővári, P., Sós, V.T. and Túran, P. (1954): On a problem of Zarankiewicz, *Colloq. Math.* **3**, 50–57.

Kolountzakis, M.N. (1994): Selection of a large sum-free subset in polynomial time, *Inform. Process. Letters* **49**, 255–256.

Kushilevitz, E., Linial, N., Rabinovitch, Y., and Saks, M. (1996): Witness sets for families of binary vectors, *J. Combin. Theory (A)* **73**:2, 376–380.

Larman, D.G. and Rogers, C.A. (1972): The realization of distances within sets in Euclidean space, *Mathematika* **12**, 1–24.

Larman, D.G., Rogers, C.A., and Seidel, J.J. (1977): On two-distance sets in Euclidean space, *Bull. London Math. Soc.* **9**, 261–267.

Li, S.Y.R. (1980): A martingale approach to the study of occurence of sequence patterns in repeated experiments, *Annals of Probability* **8**, 1171–1176.

Lieberher, K. and Specker, E. (1981): Complexity of partial satisfaction, *J. of the ACM* **28**:2, 411–422.

Lindstrom, B. (1993): Another theorem on families of sets, *Ars Combinatoria* **35**, 123–124.

Lipski, W. (1978): On strings containing all subsets as substrings, *Discrete Math.* **21**, 253–259.

Lovász, L. (1973): Coverings and colorings of hypergraphs, in: *Proc. 4th Southeastern Conf. on Comb.*, Utilitas Math., 3–12.

Lovász, L. (1977a): Certain duality principles in integer programming, *Ann. Discrete Math.* **1**, 363–374.

Lovász, L. (1977b): Flats in matroids and geometric graphs, in: *Combinatorial Surveys*, P.J. Cameron (ed.), Academic Press, 45–86.

Lovász, L. (1979): *Combinatorial Problems and Exercises.* Akadémiai Kiadó & North-Holland.

Lovász, L., Pelikán J., and Vesztergombi K. (1977): Kombinatorika. Tankönyvkiadó, Budapest, 1977. (German translation: Teubner, 1977)

Lovász, L., Shmoys, D.B., and Tardos, É. (1995): Combinatorics in computer science, in: *Handbook of Combinatorics*, R. Graham, M. Grötschel, and L. Lovász (eds.), Elsevier Science, Vol. 2, 2003–2038.

Lovász, L., Pyber, L., Welsh, D.J.A., and Ziegler, G.M. (1995): Combinatorics in pure mathematics, in: *Handbook of Combinatorics*, R. Graham, M. Grötschel, and L. Lovász (eds.), Elsevier Science, Vol. 2, 2039–2082.

Lubell, D. (1966): A short proof of Sperner's lemma, *J. Combin. Theory (A)* **1**:2, 402.

Luby, M. (1986): A simple parallel algorithm for the maximal independent set, *SIAM J. Comput.* **15**, 1036–1053.

MacWilliams, F.J. and Sloane, N.J.A. (1977): *The theory of error correcting codes.* North-Holland, Amsterdam.

Majumdar, K.N. (1953): On some theorems in combinatorics relating to incomplete block designs, *Ann. Math. Stat.* **24**, 377–389.

Mantel, W. (1907): Problem 28, *Wiskundige Opgaven* **10**, 60–61.

Mehlhorn, K. (1984): *Data Structures and Algorithms* 1: *Sorting and Searching.* EATCS Monographs on Theoretical Computer Science, Springer.

Menger, K. (1927): Zur allgemeinen Kurventheorie, *Fund. Math.* **10**, 95–115.

Meschalkin, L.D. (1963): A generalization of Sperner's theorem on the number of subsets of a finite set (in Russian), *Teor. Veroj. Primen.* **8**, 219–220. English translation in: *Theory of Probab. and its Appl.* **8** (1964), 204–205.

Moon, J.W. and Moser, L. (1962): On a problem of Turán, *Publ. Math. Inst. Hungar. Acd. Sci.* **7**, 283–286.

Motwani, R. and Raghavan, P. (1995): *Randomized Algorithms.* Cambridge University Press.

Motzkin, T.S. and Straus, E.G. (1965): Maxima for graphs and a new proof of a theorem of Turán, *Canadian J. Math.* **17**, 533–540.

Mulmuley, K., Vazirani, U., and Vazirani, V. (1987): Matching is as easy as matrix inversion, *Combinatorica* **7**, 105–114.

Nagy, Z. (1972): A certain constructive estimate of the Ramsey number, *Matematikai Lapok* **23**, 301–302. (Hungarian)

Nathanson, M.B. (1996): *Additive Number Theory: Inverse Problems and Geometry of Sumsets.* Graduate Texts in Mathematics, Vol. 165, Springer.

Nechiporuk, E.I. (1966): On a boolean function, *Doklady Akademii Nauk SSSR* **169**:4, 765–766 (in Russian). English translation in: *Soviet Math. Dokl.* **7**:4, 999–1000.

Nesetril, J. and Rödl, V. (eds.) (1990): *Mathematics of Ramsey Theory*, Algorithms and Combinatorics, Vol. 5, Springer.

Nilli, A. (1990): Shelah's proof of the Hales–Jewett theorem, in: *Mathematics of Ramsey Theory*, J. Nesetril and V. Rödl, (eds.), 150–151, Springer.

Nilli, A. (1994): Perfect hashing and probability, *Combinatorics, Probability & Computing* **3**:3, 407–409.

Nisan, N. (1989): CREW PRAM's and decision trees, in: *Proc. of 21st Ann. ACM Symp. on the Theory of Computing*, 327–335.

Ossowski, J. (1993): On a problem of F. Galvin, *Congr. Numer.* **96**, 65–74.

Papadimitriou, C.H. (1991): On selecting a satisfying truth assignment, in: *Proc. of 32nd Ann. IEEE Symp. on Foundations of Comput. Sci.*, 163–169.

Papadimitriou, C.H. and Sipser, M. (1984): Communication complexity, *J. Comput. Syst. Sci.* **28**:2, 260–269.

Paturi, R. and Zane, F. (1998): Dimension of projections in boolean functions, *SIAM J. Discrete Math.* **11**:4, 624–632.

Petrenjuk, A.Ya. (1968): On Fisher's inequality for tactical configurations, *Mat. Zametki* **4**, 417–425. (Russian)

Pevzner, P.A. (1993): DNA statistics, overlapping word paradox and Conway equation, in: *Proc. of 2nd Internat. Conf. on Bioinformatics, Supercomputing, and Complex Genome Analysis*, H.A. Lim, K.W. Fickett, C.R. Cantor, and R.J. Robbins (eds.), 61–68, World Scientific.

Plumstead, B.R., and Plumstead, J.B. (1985): Bounds for cube coloring, *SIAM J. Alg. Discr. Meth.* **6**:1, 73–78.

Pudlák, P. and Sgall, J. (1997): An upper bound for a communication game related to time-space tradeoffs, in: *The Mathematics of Paul Erdős*, R. Graham and J. Nesetril (eds.), Vol. I, Springer, 393–399.

Pudlák, P. and Sgall, J. (1998): Algebraic models of computation and interpolation for algebraic proof systems, in: *Proof Complexity and Feasible Arithmetics*, P.W. Beame and S.R. Buss (eds.), DIMACS Series in Discrete Math. and Theor. Comput. Sci., Vol. 39, 279–295.

Pudlák, P., Razborov, A., and Savický, P. (1988): Observations on rigidity of Hadamard matrices. Unpublished manuscript.

Quine, W.V. (1988): Fermat's last theorem in combinatorial form, *Amer. Math. Monthly* **95**:7, 636.

Rado, R. (1943): Note on combinatorial analysis, *Proc. London Math. Soc.* **48**, 122–160.

Ramsey, F.P. (1930): On a problem of formal logic, *Proc. of London Math. Soc.*, 2nd series, **30**, 264–286.

Ray-Chaudhuri, D.K., and Wilson, R.M. (1975): On *t*-designs, *Osaka J. Math.* **12**, 737–744.

Raz, R. (2000): The BNS–Chung criterion for multi-party communication complexity, *Computational Complexity* **9**, 113–122.

Raz, R. (2001): Resolution lower bounds for the weak pigeonhole principle, *Electronic Colloquium on Computational Complexity*, Report No. 21.

Razborov, A.A. (1985): Lower bounds for the monotone complexity of some boolean functions, *Soviet Math. Dokl.* **31**, 354-357.

Razborov, A.A. (1987): Lower bounds on the size of bounded-depth networks over a complete basis with logical addition, *Math. Notes of the Acad. of Sci. of the USSR* **41**:4, 333–338.

Razborov, A.A. (1988): Bounded-depth formulae over $\{\&, \oplus\}$ and some combinatorial problem, in: *Problems of Cybernetics, Complexity Theory and Applied Mathematical Logic*, S.I. Adian (ed.), VINITI, Moscow, 149–166. (Russian)

Razborov, A.A. (1990): Applications of matrix methods to the theory of lower bounds in computational complexity, *Combinatorica* **10**:1, 81–93.

Razborov, A.A. (1992): On submodular complexity measures, in: *Boolean Function Complexity*, M. Paterson (ed.), London Math. Society Lecture Note Series, Vol. 169, Cambridge University Press, 76–83.

Razborov, A.A. (1995): Bounded arithmetics and lower bounds in boolean complexity, in: *Feasible Mathematics II*, P. Clote and J. Remmel (eds.), Proc. of Workshop (Cornell University, Ithaca, NY, USA, May 28-30, 1992), Birkhäuser, Boston, MA.

Razborov, A., Wigderson, A., and Yao, A. (1997): Read-once branching programs, rectangular proofs of the pigeonhole principle and the transversal calculus, in: *Proc. of 29th Ann. ACM Symp. on the Theory of Computing*. To appear in *Combinatorica*.

Razborov, A. and Vereshchagin N.K. (1999): One property of cross-intersecting families, *Electronic Colloquium on Computational Complexity*, Report No. 14, 1999. Also appeared in: *Research Communications of the conference held in the memory of Paul Erdős* in Budapest, Hungary, July 4–11, 1999, 218–220.

Redéi, L. (1934): Ein kombinatorischer Satz, *Acta Litt. Szeged* **7**, 39–43.

Reiman, I. (1958): Über ein Problem von K. Zarankiewicz, *Acta Math. Acad. Sci. Hungar.* **9**, 269–279.

Robinson, J.A. (1965): A machine-oriented logic based on the resolution principle, *J. of the ACM* **12**:1, 23–41.

Rosenbloom, A. (1997): Monotone circuits are more powerful than monotone boolean circuits, *Inform. Process. Letters* **61**:3, 161–164.

Ruciński, A. and Vince, A. (1986): Strongly balanced graphs and random graphs, *J. Graph Theory* **10**, 251–264.

Ruszinkó, M. (1994): On the upper bound of the size of the r-cover-free families, *J. Combin. Theory (A)* **66**, 302–310.

Rychkov K.L. (1985): A modification of Khrapchenko's method and its application to lower bounds for π-schemes of code functions, in: *Metody Diskretnogo Analiza*, 42 (Novosibirsk), 91–98. (Russian)

Ryser, H.J. (1951): A combinatorial theorem with an application to Latin rectangles, *Proc. Amer. Math. Soc.* **2**, 550–552.

Sauer, N. (1972): On the density of families of sets, *J. Combin. Theory (A)* **13**, 145–147.

Schmidt, W.M. (1976): Equations over finite fields, an elementary approach, Lect. Notes in Math., Vol. 536, Springer.

Schur, I. (1916): Über die Kongruenz $x^m + y^m \equiv z^m \pmod{p}$, Jahresber. Deutsch. Match.-Verein 25, 114–116.

Schwartz, J.T. (1980): Fast probabilistic algorithms for verification of polynomial identities, J. of the ACM 27:4, 701–717.

Seidenberg, A. (1959): A simple proof of a theorem of Erdős and Szekeres, J. London Math. Soc. 34, 352.

Shelah, S. (1972): A combinatorial problem; stability and order for models and theories in infinitary languages, Pacific J. Math. 41, 247–261.

Shelah, S. (1988): Primitive recursive bounds for van der Waerden's numbers, J. Amer. Math. Soc. 1:3, 683–697.

Smetaniuk, B. (1981): A new construction on Latin squares I: A proof of Evans conjecture, Ars Combinatoria 11, 155–172.

Spencer, J. (1977): Asymptotic lower bounds for Ramsey functions, Discrete Math. 20, 69–76.

Spencer, J. (1987): Ten lectures on the probabilistic method, SIAM Regional Conference Series in Applied Mathematics, Vol. 52.

Spencer, J. (1990): Uncrowed graphs, in: Mathematics of Ramsey Theory, Nesetril, J., and Rödl, V. (eds.), 253–262, Springer.

Spencer, J. (1995): Probabilistic methods, in: Handbook of Combinatorics, R. Graham, M. Grötschel and L. Lovász (eds.), Elsevier-Science, Vol. 2, 1786–1817.

Sperner, E. (1928): Ein Satz über Untermengen einer endlichen Menge, Math. Z. 27, 544–548.

Pefold Street, A. (1972): Sum-free sets, in: Lect. Notes in Math., Vol. 292, 123–272, Springer.

Szele, T. (1943): Combinatorial investigations concerning complete directed graphs, Matematicko Fizicki Lapok 50, 223–256 (Hungarian). German translation in: Publ. Math. Debrecen 13 (1966), 145–168.

Szemerédi, E. (1969): On sets of integers containing no four elements in aritmethic progression, Acta. Math. Acad. Sci. Hungar. 20, 898–104.

Tardos, G. (1989): Query complexity, or why is it difficult to separate $NP^A \cap co - NP^A$ from P^A by a random oracle A? Combinatorica 8:4, 385–392.

Trevisan, L. (1997): On local versus global satisfiability, Electronic Colloquium on Computational Complexity, Report Nr. 12. To appear in: SIAM J. Discrete Math.

Tseitin, G.C. (1968): On the complexity of derivations in propositional calculus, in: Studies in mathematics and mathematical logic, Part II, Slisenko, A.O. (ed.), 115–125.

Turán, P. (1941): On an extremal problem in graph theory, Math. Fiz. Lapok 48, 436–452.

Tuza, Zs. (1984): Helly-type hypergraphs and Sperner families, European J. Combin. 5, 185–187.

Tuza, Zs. (1985): Critical hypergraphs and intersecting set-pair systems, J. Combin. Theory (B) 39, 134–145.

Tuza, Zs. (1989): The method of set-pairs for extremal problems in hypergraphs, in: *Finite and Infinite Sets*, A. Hajnal et al. (eds.), Colloq. Math. Soc. J. Bolyai, Vol. 37, 749–762, North-Holland, Amsterdam.

Tuza, Zs. (1989b): Intersection properties and extremal problems for set systems, in: *Irregularities of Partitions*, G. Halśz and V.T. Sós (eds.), *Algorithms and Combinatorics*, Vol. 8, 141–151, Springer.

Tuza, Zs. (1994): Applications of the set-pair method in extremal hypergraphs, in: *Extremal Problems for Finite Sets*, P. Frankl et al. (eds.), Bolyai Society Mathematical Studies, Vol. 3, 479–514, János Bolyai Math. Society.

Valiant, L.G. and Vazirani, V.V. (1986): NP is as easy as detecting unique solutions, *Theor. Comput. Sci.* **47**, 85–93.

Van der Waerden, B.L. (1927): Beweis einer Baudetschen Vermutung, *Nieuw Arch. Wisk.* **15**, 212–216.

Vapnik, V.N. and Chervonenkis, A.Ya. (1971): On the uniform convergence of relative frequencies of events to their probabilities, *Theory of Probability and Appl.* **XVI**, 264–280.

Von Neumann, J. (1953): A certain zero-sum two-person game equivalent to the optimal assignment problem, in: *Contributions to the Theory of Games*, Vol. II, H.W. Kuhn (ed.), *Ann. of Math. Stud.* **28**, 5–12.

Ward., C. and Szabó, S. (1994): On swell colored complete graphs, *Acta Math. Univ. Comenianae* **LXIII**(2), 303–308.

Wei, V.K. (1981): A lower bound on the stability number of a simple graph. Technical Memorandum No. 81-11217-9, Bell Laboratories.

Weil, A. (1948): Sur les courbes algébraiques et les variétés qui s'en déduisent, *Actualités Sci. Ind.* **1041** (Herman, Paris).

Welsh, D.J.A. and Powell, M.B. (1967): An upper bound for the chromatic number of a graph and its applications to timetabling problems, *Computer Journal* **10**, 85–87.

Wigderson, A. (1993): The fusion method for lower bounds in circuit complexity, in: *Combinatorics, Paul Erdős is Eighty*, Vol. 1, 453–468.

Wigderson, A. (1994): NL/poly $\subseteq \oplus$L/poly, in: *Proc. of 9th Ann. IEEE Conf. on Structure in Complexity Theory*, 59–62.

Witt, E. (1951): Ein kombinatorischer Satz der Elementargeometrie, *Math. Nachr.* **6**, 261–262.

Yamamoto, K. (1954): Logarithmic order of free distributive lattices, *J. Math. Soc. Japan* **6**, 343–353.

Yannakakis, M. (1994): On the approximation of maximum satisfiability, J. of Algorithms **17**, 475–502.

Yao, A.C. (1985): Separating the polynomial time hierarchy by oracles, in: *Proc. of 26th Ann. IEEE Symp. on Foundations of Comput. Sci.*, 1–10.

Zarankiewicz, K. (1951): Problem P 101, *Colloq. Math.* **2**, 116–131.

Zippel, R.E. (1979): Probabilistic algorithms for sparse polynomials, in: *Proc. of EUROSAM 79*, Lect. Notes in Comput. Sci., Vol. 72, 216–226, Springer.

Name Index

Subject Index

Monographs in Theoretical Computer Science · An EATCS Series

C. Calude
Information and Randomness
An Algorithmic Perspective

K. Jensen
Coloured Petri Nets
Basic Concepts, Analysis Methods
and Practical Use, Vol. 1
2nd ed.

K. Jensen
Coloured Petri Nets
Basic Concepts, *Analysis Methods*
and Practical Use, Vol. 2

K. Jensen
Coloured Petri Nets
Basic Concepts, Analysis Methods
and *Practical Use,* Vol. 3

A. Nait Abdallah
The Logic of Partial Information

Z. Fülöp, H. Vogler
Syntax-Directed Semantics
Formal Models
Based on Tree Transducers

A. de Luca, S. Varricchio
Finiteness and Regularity
in Semigroups
and Formal Languages

E. Best, R. Devillers, M. Koutny
Petri Net Algebra

Texts in Theoretical Computer Science · An EATCS Series

J. L. Balcázar, J. Díaz, J. Gabarró
Structural Complexity I
2nd ed. (see also overleaf, Vol. 22)

M. Garzon
Models of Massive Parallelism
Analysis of Cellular Automata
and Neural Networks

J. Hromkovič
Communication Complexity
and Parallel Computing

A. Leitsch
The Resolution Calculus

G. Păun, G. Rozenberg, A. Salomaa
DNA Computing
New Computing Paradigms

A. Salomaa
Public-Key Cryptography
2nd ed.

K. Sikkel
Parsing Schemata
A Framework for Specification
and Analysis of Parsing Algorithms

H. Vollmer
Introduction to Circuit Complexity
A Uniform Approach

W. Fokkink
Introduction to Process Algebra

K. Weihrauch
Computable Analysis
An Introduction

J. Hromkovič
Algorithmics for Hard Problems
Introduction to Combinatorial Optimization,
Randomization, Approximation, and Heuristics

S. Jukna
Extremal Combinatorics
With Applications in Computer Science

Former volumes appeared as
EATCS Monographs on Theoretical Computer Science

Printing: Weihert-Druck GmbH, Darmstadt
Binding: Buchbinderei Schäffer, Grünstadt